$34.00

D1401582

INTEGRATED PEST CONTROL IN CITRUS-GROVES

Proceedings of the Experts' Meeting / Acireale / 26-29 March 1985

INTEGRATED PEST CONTROL IN CITRUS-GROVES

Edited by
R.CAVALLORO
Commission of the European Communities, Joint Research Centre of Ispra

E.DI MARTINO
Citrus Experimental Institute of Acireale

Published for the Commission of the European Communities by

A.A.BALKEMA / ROTTERDAM / BOSTON / 1986

CEE: VI/52

The texts of the various papers in this volume were set individually
by typists under the supervision of each of the authors concerned.

CIP-DATA KONINKLIJKE BIBLIOTHEEK, DEN HAAG

Integrated

Integrated pest control in citrus-groves: proceedings of the experts' meeting, Acireale 26-29
March 1985 / ed. by R. Cavalloro, E. Di Martino. — Rotterdam [etc.]: Balkema. — Ill.
Publ. for the Commission of the European Communities.
ISBN 90-6191-678-X bound
SISO 635.2 UDC 634.3:632.93(063)
Subject heading: pest control; citrus-groves.

Publication arrangements: *P.P.Rotondó,* Commission of the European Communities,
Directorate-General Telecommunications, Information Industries & Innovation, Luxembourg

LEGAL NOTICE
Neither the Commission of the European Communities nor any person acting on behalf of
the Commission is responsible for the use which might be made of the following information.

EUR 9872 En

ISBN 90 6191 678 X

Published by A.A.Balkema, P.O.Box 1675, 3000 BR Rotterdam, Netherlands
Distributed in USA & Canada by A.A.Balkema Publishers, P.O.Box 230, Accord, MA 02018

Foreword

This volume collects together the contributions, the experiences, and the ideas which have been presented in an experts' meeting on the theme of the protection of citrus groves.

The occasion of the meeting permitted researchers in various disciplines, and from various Countries, to tackle just one subject of research, the citrus fruit, in the context of a single specific topic, the defence of the crop in the light of the most recent knowledge gained.

The meeting was divided into working sessions which considered basic research on damaging insects, mites, nematodes, weeds, fungal, bacterial and viral diseases, as well as the application of control techniques, subdivided into biological, biotechnical, chemical, cultural and genetic, all of which were proposed from an ecological point of view regarding the citrus agro-ecosystem.

Particular attention was paid to methods and strategies of integrated control, taking into account also the influence of some phytochemicals on the physiology of the citrus crop; in addition economic considerations were discussed.

The works presented are all reported "in extenso" and show an up-to-date picture of both the causes of damage and the most efficient methods of control.

About one hundred experts participated from nine Countries and three international organizations, with a total of eighty scientific contributions.

The meeting was organised by the Commission of the European Communities and was particularly wide-ranging both in the expertise of the participants and in their different places of origin.

The meeting was held at a highly suitable place, the Citrus Experimental Institute of Acireale, which can justly be considered a meeting point for experts in this field.

This work intends to make widely available the most recent knowledge on new phytosanitary techniques and on the situations in the different Countries, with regard to control measures. It certainly constitutes a point of reference for all the knowledge acquired in this specific sector of the defence of citrus groves against their noxious agents.

R. Cavalloro, E. Di Martino

ORGANIZING COMMITTEE

Cavalloro Raffaele, Principal Scientific Officer CEC Programme "Integrated Plant Protection", Ispra

La Malfa Giuseppe, President Citrus Experimental Institute, Acireale

Spina Paolo, Director Citrus Experimental Institute, Acireale

SCIENTIFIC COMMITTEE

Cavalloro Raffaele, Commission of the European Communities, Joint Research Centre of Ispra

Di Martino Enrico, Citrus Experimental Institute of Acireale

Prota Romolo, University of Sassari

SESSIONS' STRUCTURE

	Chairman	Secretary
Opening Session	P. Spina	E. Di Martino
Insect Pests	P. Jourdheuil	D. Benfatto
Other Pests	E. Di Martino	V. Lo Giudice
Diseases	C. Thanassoulopoulos	G. Lanza
Means of Control	R. Prota	A. Starrantino
Methodologies and Strategies of Control	L. Brader	G. Reforgiato Recupero
Economical Considerations of I.P.M. in Citrus groves	S.C. Misseri	F. Intrigliolo
Closing Session	R. Cavalloro	A. Caruso

LOCAL ORGANIZING COMMITTEE

Di Martino Enrico, Citrus Experimental Institute of Acireale

Di Martino Aleppo Emanuela, Citrus Experimental Institute of Acireale

Romeo Mario, Citrus Experimental Institute of Acireale

ORGANIZING SECRETARIAT

Citrus Experimental Institute - Corso Savoia, 190 - Acireale (CT)

PROCEEDING DESK

Rotondo Pier Paolo, CEC, Directorate-General Telecommunications, Information Industries & Innovation, Luxembourg

Table of contents

Session 2. *Other pests*

Session 3. *Diseases*

PART II. INTEGRATED PEST MANAGEMENT IN CITRUS-GROVES

Session 4. *Means of control*

A. *Biological*

B. *Biotechnical*

C. *Chemical*

Opening session and general reports

Chairman: P.Spina

Opening address

G. La Malfa

Citrus Experimental Institute, Acireale, Italy

Ladies and Gentlemen,

As President of the Board of this "Istituto Sperimentale per l'Agrumi-coltura", I am very pleased to open the first session of this Expert's Meeting of the Commission of the European Communities on the theme "Integrated Pest Control in Citrus Groves". I give you my kind regards, a warm welcome and the best wishes for your work. This also expresses the feelings of the Board members of this Citrus Research Institute. In addition, they explicitly gave me the responsibility of expressing their satisfaction, and their thanks to the Authorities of the C.E.C. and, particularly, to Prof. R. Cavalloro for the organization of this meeting in Acireale.

This Institute has always been eager to promote and to organize events that come within the scope of their main interests, that is to say research on citrus cultivation. Therefore, this event is all the more noteworthy in that it includes the participation of eminent scientists coming from many different countries of the world.

I am sure that the results of your work during this meeting will add a valuable contribution to the advance of our knowledge regarding pest control in citrus groves, using means which are efficacious, but not dangerous to mankind and to the environment.

Apart from those results which you will offer today to citrus growers, there is the added interest of the promotion of cooperation between the scientists of the various countries of the world, which will surely yield a rich harvest of interesting new ideas.

This is also an example of the acquired experience within the programmes supported by C.E.C in recent years and dealing with different topics which also involve many Italian institutions. Among these may be included the "Istituto di Agrumicoltura" and some of their researchers.

An important contribution to this programm was made by Prof. E. Di Martino, Director of the Biology and Pest Control Section of the Institute, who must be thanked for the meeting's organization.

Our thanks also go to Prof. P. Spina, Director, and to the scientific, technical and administrative staff of the Institute. In addition, let me take this opportunity to congratulate, together with you, the scientists of the Institute who were the co-writers of the very interesting "Trattato di Agrumicoltura" which was just published. I think that this meeting is a happy occasion for the first official public presentation of this paper, which is the result of much sustained and arduous work on the part of your colleagues at Acireale.

Finally, I wish also to thank the General Directorate responsible for Agricultural Production of the Italian Ministry of Agriculture, who sponsored this meeting, and all the people and institutions which have supported in different ways this event. Among them, and for everybody, I mention the Major of Acireale, who is among us to welcome you to the town.

Again, I wish personally to extend to you my best regards for a profitable meeting.

Welcome address

A.Quacquarelli
Experimental Institute for Plant Pathology, Roma, Italy

Ladies, Gentlemen, and dear Colleagues,

It is a great pleasure and satisfaction to see so many scientists of different countries attending this meeting.

As one of the Italian representatives of the CEE Programme "Energy in Agriculture", I am honoured to welcome all of you on the behalf of those Italian scientists who are involved either in "Plant Protection" in general or in "Integrated Plant Protection" in particular.

The contributions which we are going to deliver in this Meeting spring from our firm belief that "Integrated Control" is the "winning choice" and represents the "right key" to open the door to a new Agriculture.

In addition to all of the scientific consideration involved, let me say that I am proud to stress one other positive goal of this Meeting: we are all contributing to making a more united Europe!

Welcome address

P.Jourdheuil

INRA, Zoological & Biological Control Station, Antibes, France

Mr. President, Mr. Director, Ladies and Gentlemen,

On behalf of the West Palearctic Regional Section of the International Organization for Biological Control (IOBC/WPRS), I would like to welcome the numerous participants here in this hall.

As you know, the main aim of our organization has always been to promote biological and integrated control. With this in mind, the west palearctic section, which includes more than 20 western European and Mediterranean countries, makes a special effort:

a) to circulate scientific and technical information, notably by giving logistic aid to publish the review "Entomophaga" and by guaranteeing the regular appearance of bulletins and brochures, and

b) by organizing symposiums and, particularly, working parties specialized in a given type of cultivation or phytosanitary protection method. The IOBC/WPRS has been taking an interest in citrus farming for more than 20 years and has created an ad hoc working party, led until recently by C. Benassy, which has played an important role in the development of research on biosystematics, the ecology of fixed Homoptera and on the use of associated entomophaga. Now under the direction of R. Prota, this group is tending to increase the scope of its studies to include all citrus pests.

Towards the end of the 1970's, after the Commission of the European Communities had set up important research programs covering the same subject (olive growing and citrus groves, in particular), permanent consultation between the two organizations was instituted to assess how the available resources could best be used. So, joint meetings of the Commission and the IOBC were organized, like the International Symposium on Fruit Flies held in Athens in 1982 or the one in Pisa in 1984 on integrated pest control in olive groves, as were regular meetings in one particular geographical location like those held today on integrated pest control in citrus growing.

On behalf of our organization, I would like to thank the Istituto Sperimentale per l'Agrumicoltura, as well as the national, regional and municipal authorities in Italy for having contributed so generously and so efficiently to the success of this event and I wish all the participants an agreeable and fruitful stay in Acireale.

Welcome address from the mayor

M.Coco
Municipality of Acireale, Italy

On behalf of the local administration of Acireale, it is a great honour for me to welcome the illustrious researchers who are meeting here.

The subjects which will be discussed here have a particular relevance to the economy of citrus groves, where the intrinsic healthiness of the product and the maintenance of a hygienic level guaranteeing them against contamination by toxic chemical substances is sought, the aim being to defend the natural environment in such a way that nature itself, encouraged rather than hampered, can act with its own defence mechanisms, which in most cases are unobserved, but decisive.

These are all topics which involve me personally, both as a citizen of this municipality, which bases much of its economic resources on citrus production, and due to the effect which this has on the activities of many of our citizens who, through the possible solution of the problems mentioned above, could maintain and certainly improve the quality of their life in a world where the use of chemical agents to protect citrus fruits is wisely moderated, allowing natural forces to work in our favour.

These themes, among many others, have long been subjects of study for the Institute, which is hosting us and which this local administration considers one of its most prestigious and valued institutions.

This city is proud of being rich in culture and scholarship, of its school of high level and long tradition, of being the seat of the "Academy of Sciences, Letters and Fine Arts of the Zelanti and the Dafnici" which began its great tradition in the early 1700s, and is proud of having as its chief ornament, like a very pure pearl, the Experimental Institute for Citrus Groves, the only institution of university rank existing in this field. We are proud not only of hosting it, but of having been in various ways and in various periods, if not exactly its parent, at least its godmother or its cradle.

This institution was started long ago - in 1866 soon after the annexing to Italy of the Kingdom of the Two Sicilies - as the "Agrarian Assembly of the Acireale Area". For over half a century it experimented in forestry and vine-growing, and the municipality allowed it the use of the ex-Capuchin convent, including a hectare of land for its experiments.

As it became clear and was confirmed around the turn of the century that this land was markedly suitable for the cultivation of lemons, the "Agrarian Assembly" worked hard to push in this direction. On the initiative of a group of citizens of this city, in 1907 the Institute became the "Experimental Station for Fruit and Citrus Groves", under the control of the Ministry of Agriculture, Industry and Commerce and with the duties of research and experimentation, including studying the most suitable means for controlling the pests and diseases of fruit trees, particularly citrus fruit trees. It has been deeply involved in this field and its researchers have been at the forefront in contributions made in this area of research.

Thus, the Municipality of Acireale made available for this Institute in 1918 the 4-hectare field of Saint Salvatore, in which a suitable building was constructed in the 1920's for use by this station.

In 1967, when the experimental station was turned by decree of the Ministry of Agriculture into the Experimental Institute of Citrus Groves, the Municipality of Acireale sold both this building and the field next to it to the Institute at a token price.

One therefore has good reason to say that this Institute, which the Municipality considers the apple of its eye, is a creation of this city; we are proud of it and it has for over a century made the name of Acireale known with great honour and prestige in the world of scientists of agriculture.

I hope that I may be forgiven for recounting these memories, for they are mentioned only to reaffirm that we should be proud of this institution, by the good offices of which we are meeting here today. The administration which I represent would like to welcome the illustrious experts who have come here from many Mediterranean and non-Mediterranean countries. We wish you all a pleasant and productive stay, and hope that you will carry away warm memories of these regions. I should, in addition, like to express the Municipality's grateful recognition of the EEC, which has chosen our city as the site for this "Meeting of Experts on Integrated Control in the Defence of Citrus Groves Against Pests".

The Experimental Institute for Citrus-Groves in defending citrus fruits with particular reference to integrated control
General report

P.Spina
Citrus Experimental Institute, Acireale, Italy

It is now my privilege to begin the work of this "Experts meeting" which the Commission of the European Communities wished to take place in this old and glorious Institute founded in faraway 1907 by Prof. Luigi Savastano, which was until 1967 the Experimental Station for Citrus Fruit and Fruit Growing, and which I have the honour to direct.

We are flattered by the decision taken by the Commission of the European Communities through our friend, Prof. Raffaele Cavalloro, and we hope that these fruitful days of study will provide some interesting conclusions to the researchers, technicians and citrus fruit growers.

At this point I need only describe the merits of the themes entrusted to me, which I will try to do as briefly as possible, particularly to allow room for those who, better than me, will deal the problems which are the objects of these days of study.

Control of crop parasites is felt to be indispensabile to maintain and increase the level of agricultural production and to combat hunger in the World.

The number of phytophages which attacked citrus fruit up to immediately after the second World War was relatively small and the damage, which was very largely due to certain scales, was relatively unimportant, being efficiently controlled by inorganic means of control (phytochemicals) made up of polysulphides, which were soon replaced by synthetic chlorides, phosphorganics and mineral oils and oils of organic origin (nicotine), which were also efficient against aphids, while prussic acid was widely used to combat the scales infestations.

Up to this time, which may be considered almost idyllic because of the lack of wide spectrum acting insecticides with long residual toxicity, there were no serious disturbances of the biocenosis of citrus fruits. Such disturbances were almost unknown at this time, but were sufficiently investigated as far as cocchinella predators and parasites were concerned, which were found to survive after the use of these old means of control.

From 1951 with the introduction of synthetic chlorides and phosphorganics the mite Aceria sheldoni Ew. appeared in Sicily and Campania, as did the red spider (Tetranychus telarius L.), the coleopter Otiorrhynchus cribricollis Gyll., various other mites (Brevipalpus phoenicis Geijskes, B. californicus Banks and Hemitarsonemus latus Banks, the Aculus pelekassi K. and Panonychus citri (McGregor) and ever more severe attacks of various scales, (Saissetia oleae Bern., Ceroplastes sinensis Del Guercio, Ceroplastes rusci L., Aonidiella aurantii Mask., Parlatoria pergandei

Comst., P. zizyphus Lucas, Coccus hesperidum L., Dialeurodes citri Ashm. and Aleurothrixus floccosus Mask.), the hemipter Calocoris trivialis Costa (green bug) and various aphids (Aphis spiraecola Patch and Toxoptera aurantii Boyer d.F.).

All this occurred because the useful entomofauna were severally harmed by the indiscriminate use of powerful chemical treatments which destroyed Rodolia cardinalis (Muls.), Scutellista cyanea Motsh., Coccidiphaga scitula (Rbr), all the various Chilocorus, Exochomus, Stethorus, Scymnus, Diptera, Neuroptera and Thysanoptera and many other living organisms, which had more or less efficiently controlled the various phytophages listed above.

It was also found that the antiparasite treatments constituted a danger to man and to animals, in the form of acute and chronic poisoning, which occurred after the biological balance in nature between phytophages and their parasites and predators was upset. There was also a reappearance of phytophages and "stocks" of species were found which were more and more resistant to pesticides.

Many researchers have for some time sounded the alarm, asking for more care for public health and respect of the agroecosystem, i.e. for the rational use of chemical means, disturbing the natural control of phytophages as little as possible and thus using biological and integrated control which, by using useful means to encourage the diffusion and development of natural foes of phytophages, improve production both qualitatively and quantitatively.

One of the cardinal doctrines of integrated control is the use of selective phytochemicals, while there is no shortage of other control means (ionizing radiation to sterilise males, chemical sterilizers, antimeta-bolites, pheromones, selective attractors, repellents, hormones, etc.). These and other techniques and methods of partial and localized action could be found suitable in many cases containing phytophages within limits which were agriculturally not damaging and at the same time less dangerous for the environment and for foods.

This orientation is really attractive, of considerable interest for the future, and could be used with greater profit than chemical control in controlling the animal parasites of citrus fruits, although in our opinion it is only through biological control that the dangers of pollution which have an evergrowing effect on the society of the 1980s can be eliminated.

At this Institute studies have been carried out on the many phytophages of citrus fruits and also on biological control where possible. We should like to recall the studies of Costantino before the second World War on Cryptolaemus montrouzieri Muls. to combat Planococcus, the releasing of some Leptomastix dactylopii How. in long ago 1951 and the rearing and release of specimens of the same parasite in 1980. The work of the Institute continued until the 1960s with the collection and diffusion of Rodolia cardinalis, which began in the 1910s and which managed efficiently to control the spreading of Icerya purchasi in Sicilian citrus orchards.

Thus far we have considered as far as possible the useful parasitic and predator entomofauna and where possible as a priority the emphasis has been placed on biological control or the use of repellent mixtures

(Empoasca decedens P.) or adhesive substances to protect plants (ants), hydrocyanic fumigations (scales) and attracting substances (Ceratitis capitata Wied.). There is also considerable interest in research into the "safety times" of some phytochemicals and their migration in citrus fruits and the phytotoxicity of phytochemicals in citrus fruit growing and on the influence of some antiparasites on the colouring and quality of fruits and on the phytotoxicity.

Unfortunately although considerable successes have been obtained already, biological control at present, by itself, cannot fight all the phytophages which afflict citrus fruit growing, which are in fact quite numerous so that it is necessary to use integrated control in citrus groves.

If integrated control is to be successful one nevertheless needs a collection of complex knowledge of plants, phytophages, the zoophageous organisms present, the climate and their influence on all the agroecosystem and the phytochemicals and their influence on plants and on phytophages and zoophages. One also needs knowledge of crop growing techniques used in the business where research is being carried and of how much is connected with it and can interact with the plants and thus with the hosted biocenosis and above all of "damage thresholds".

The activity of the Institute which I have the honour to direct in the sector of integrated control has been important since the time of the old Experimental Station of Citrus Fruit and Fruit Growing.

One should remember here the sulphur-calcium paste used in the 1910s in the control of scales.

In this context the Institute has carried out studies and experiments on the following parasites:

- Various species of scales, by direct control of ants which encourage infestations and also by viscous rings or by the distribution of parasitoids. We have had interesting results with the use in winter of white mineral oil at doses between 2 and 3%, i.e. in a period when parasites and predators are less active, the plants are more open to receiving them and less sensitive to the physiological disorders.
- Various species of Aphids, by diffusion in the field of Coccinella septempuntata.
- Prays citri, combatted in its development by the anticipation of agricultural practices associated with "forcing" to produce "verdelli" lemons, while it is probable that Bacillus thuringiensis has some efficiency.
- Ceratitis capitata by the use of poisoned bait.
- Trips tabaci, which attacks small lemon fruits in May-June, which can be controlled efficiently by the anticipated weeding of the infested grass which harbours the parasite in the citrus orchard.
- Finally one should not forget the interesting control tests carried out in winter with 2% white mineral oil against Aceria sheldoni, Dialeurodes citri and against Aleurothrixus floccosus, not forgetting Cales noacki and various cocchinellae, aphelinids and other useful insects.

These are the most important items of research and experimentation which have been carried out by this Institute while many others are in progress.

Citrus protection activities in the European countries promoted by the Commission of the European Communities
General report

R.Cavalloro
CEC, Joint Research Centre, Ispra, Italy

Distinguished and dear Colleagues,

It is with great pleasure that I see here so many eminent experts on the defence of citrus cultivation against pests and diseases: this is a concrete sign of the interest which exists in cooperating in research into methods and protection techniques which are both efficient and at the same time have no undesirable side effects.

Our meeting is centred on integrated control, which can certainly be achieved in citrus groves when one knows the biocenoses which exist in the zone to be defended, the exact entity and the behaviour of the phytophagous species and diseases to be controlled, and the intensity of damage caused, and one can make judicious recourse to intervention strategies with cultural, biotechnical, biological and genetic methods, and with the use of selective chemical products.

The Commission of the European Communities has tackled the problem with zeal, by joint research programmes which were at first general in nature, with five-years association contracts (1971-1975) in which Germany, Hollan and Italy participated, and later (1979-1983) with more specific studies in which all the ten Member countries have taken part (Tab. I).

Tab. I - Joint research programmes carried out in citrus groves in the European Countries.

STUDY GROUP	1967	Experts' Meeting
PROGRAMME GROUP	1969	Experts' Meeting
"RADIOENTOMOLOGY" PROGRAMME (D-I-NL)	1971-1975	Association Contracts and Contact Group Meetings
INTEGRATED CONTROL GROUP	1975-1976	Experts' Meeting
"INTEGRATED AND BIOLOGICAL CONTROL" PROGRAMME (B-D-DK-F-GB-GR-I-IR-L-NL)	1979-1983	Common Contracts and Coordinated Actions
"INTEGRATED PLANT PROTECTION" PROGRAMME (B-D-DK-F-GB-GR-I-IR-L-NL)	1984-1988	Coordinated Actions

These common actions were developed, according to a formula which has been established in time, on a contractual basis with well-qualified national research Institutes, supported by financial contributions from the Commission of the European Communities which generally amounted to 50% of the total cost of the research, while the remaining part was the responsability of the contracting Institute.

The most directly involved Institute are obviously the French, Greek and Italian Institute, which have collaborated closely, with work mainly concentrated on carrying out pilot projects in Corsica, the Côte d'Azur, Crete, Sardinia and Sicily, in the field of integrated and biological control against the main pests of citrus groves (Tab. II).

Tab. II - "Integrated and biological control" programme: common activities on pest control in citrus-groves.

. INTEGRATED PEST CONTROL IN CITRUS GROVES
  Station de Zoologie et de Lutte Biologique - Antibes

. PILOT PROJECT OF BIOLOGICAL CONTROL IN LEMON GROVES
  Istituto di Entomologia agraria dell'Università - Catania

. INTEGRATED PEST CONTROL IN SARDINIA ORANGE GROVES
  Istituto di Entomologia agraria dell'Università - Sassari

. INTEGRATED CONTROL ON CITRUS MEALYBUG, PLANOCOCCUS CITRI RISSO
  Institute of Subtropical plants and Olive trees - Chania

The citrus agro-ecosystem is certainly one of the most complex with a very rich harmful arthropodofauna, and with a vast number of parasites and of predators. This situation has suggested, in the joint research, that there should be in particular an exploitation of the natural and biotechnical factors of control.

But great attention has also been paid to other important aspects, such as the definition of economic tolerance thresholds for the various phytophagous species, the use of non-polluting pesticides, and research into models for predicting attacks.

Of the most important studies carried out in the citrus protection in the European countries involved, one should remember:

- for Aphididae

the main harmful species are Toxoptera aurantii, Aphis citricola, A. gossypii, A. spiraecola; their noxious actions may be effectively impeded by the parasites Aphidinae, most important of which are Lysiphlebus testaceipes, L. fabarum, Aphidius matricariae, as well as the various predators <u>Crysopidae</u>, <u>Syrphidae</u>, and <u>Coccinellidae</u>;

- for <u>Aleyrodidae</u>

<u>Aleurothrixus</u> <u>floccosus</u> and <u>Dialeurodes</u> <u>citri</u> are the two species which dominate with attacks which are often impressive: the first species can be effectively controlled by <u>Cales</u> <u>noacki</u> and <u>Amitus</u> <u>spiniferus</u>, the second by the Aphelinid <u>Encarsia</u> <u>lahorensis</u> and by the Coccinellid <u>Clitostethus</u> <u>arcuatus</u>;

- for <u>Coccidae</u>

these are the predominant phytophages in citrus groves, with many species of Pseudococcine (<u>Planococcus</u> <u>citri</u>, Pseudococcus <u>calceolariae</u>, <u>P.</u> <u>maritimus</u>), of Lecanine (<u>Saessetia</u> <u>oleae</u> and some <u>Coccus</u>), of Diaspidine (especially <u>Aonidiella</u> <u>aurantii</u>, <u>Aspidiotus</u> <u>nerii</u>, <u>Chrysomphalus</u> <u>dictyospermi</u>, <u>C.</u> <u>ficus</u>, <u>Lepidosaphes</u> <u>beckii</u>, <u>L.</u> <u>gloverii</u>, <u>Parlatoria</u> <u>ziziphus</u>, <u>Unaspis</u> <u>yanonensis</u>). The richness of the biocenosis clearly shows the value of the antagonistic species: their characteristics make available an enormous potential for use in an effective biological control. The Pseudococcine are attacked by many entomophages among which predominate <u>Pseudaphycus</u> <u>maculipennis</u>, the very active Coccinellid <u>Nephus</u> <u>reunioni</u>, as well as <u>Neuroptera</u> and <u>Diptera</u>; the Lecanine are controlled by important entomophages such as <u>Metaphycus</u> <u>bartletti</u>, <u>M.</u> <u>helvolus</u>, <u>M.</u> <u>loundsburii</u>, <u>M.</u> <u>swirskii</u>, <u>Diversinervus</u> <u>elegans</u>, <u>Scutellista</u> <u>cyanea</u>; the Diaspidine are actively parasitized by various species of <u>Aphytis</u>.
The research into sexual pheromone of the Scales-insects has proved to be of great value for the practical applications;

- for <u>Lepidoptera</u>

the key species is <u>Prays</u> <u>citri</u>, but recent enquiries have reduced the damage imputed to them, which is found to be low on the product obtained even at attack levels of 70% of flowers; the use of pheromone traps, of entomopathogens like the <u>Bacillus</u> <u>thuringiensis</u>, and suitable cultural methods are highly effective;

- for <u>Diptera</u>

the key species is <u>Ceratitis</u> <u>capitata</u>, which has been widely studied in various environments. Chromotropic traps set with specific attractants prove to be of great selectivity and effectiveness, like the more sophisticated but efficient and positively demonstrated method of autocid control which uses the sterile insect technique;

- for the Mites

the swarming of Mites is clearly due to the careless use of non-selective chemical pesticides, which drastically eliminate the predator species. The main phytophagous Acarids known up to now are <u>Panonychus</u> <u>citri</u>, <u>Tetranychus</u> <u>urticae</u>, <u>Aceria</u> <u>sheldoni</u>, <u>Aculus</u> <u>pelekassi</u>. Their

17

effect is contained however by the activity of phytoseiids Mites (mainly Amblyseius stipulatus, Phytoseiulus permisilis, and various Typhlodromus), and by the use of white mineral oils.

Tab. III - Meetings dealing with citrus protection on integrated and biological pest control.

. INTEGRATED AND BIOLOGICAL CONTROL IN CITRUS-GROVES
  I - Firenze, 19 April 1979

. PROGRESS REPORT ON FIELD PILOT PROJECTS FOR INTEGRATED PEST CONTROL IN CITRUS CULTURE
  I - Catania, 12-13 June 1980

. STANDARDIZATION OF BIOTECHNICAL METHODS ON INTEGRATED PEST CONTROL IN CITRUS ORCHARDS
  F - San Giuliano / I - Siniscola, 4-6 November 1980

. PROGRESS REPORT AND EXCHANGE OF INFORMATION ON INTEGRATED PEST CONTROL IN CITRUS CULTURE
  I - Catania, 15-18 September 1981

. INTEGRATED CONTROL IN CITRUS: COMPARISON OF RESULTS ACHIEVED BY APPLYING A STANDARDIZED METHODOLOGY
  I - Siniscola-Muravera, 20-22 October 1982

. JOINT CEC/IOBC SYMPOSIUM ON FRUIT-FLIES OF ECONOMIC IMPORTANCE
  GR - Athens, 16-19 November 1982

. STATISTICAL AND MATHEMATICAL METHODS IN POPULATION DYNAMICS AND PEST CONTROL
  I- Parma, 26-28 October 1983

. GENERAL MEETING OF THE CONTRACTANTS IN "INTEGRATED AND BIOLOGICAL CONTROL" PROGRAMME
  B - Bruxelles, 16 November 1983

. JOINT CEC/IOBC AD-HOC MEETING ON FRUIT-FLIES OF ECONOMIC IMPORTANCE
  D - Hamburg, 23 August 1984

. JOINT CEC/EPPO/IOBC CONFERENCE ON INTEGRATED CROP PROTECTION: FROM PRINCIPLES TO PRACTICAL IMPLEMENTATION
  B - Bruxelles, 9-11 October 1984

There has also been intensive coordinated activity, carried out in parallel with these common actions especially dealt with meetings between contractors and meetings of experts in specific subjects, to discuss the state of progress of the research or on particular themes (Tab. III), besides the exhange of researchers involving not only the direct interested citrus-producing States, but also various instituts of other European countries (Fig. 1), symposia, study visits, etc.

Fig. 1 - Exchange of scientists on European countries dealing with citrus protection on integrated and biological pest control.

The definition of sampling methods and intervention thresholds for the main citrus pests is of considerable importance. They are today generally accepted and in full operation, linked with meetings of experts devoted to the standardization of biotechnical methods on the basis of experiments carried out in various environments, in order to direct the citrus protection rationally, towards a more incisive integrated pest control.

Recently, a new five-year "Integrated Plant Protection" programme (1984-1988) has been set up at the level of the Member countries of the European Communities. This, taking account of the positive results obtained from the work carried out up to now in citrus groves and from the clear orientations of control which have emerged, plans the continuation and enlargement of coordinated work, and our meeting can be seen precisely as part of this new line of action.

On the basis of the actions carried out by the Commission of the European Communities, integrated control with judicious recourse to

biological and biotechnical means and to selective products, offers
excellent possibilities for restraining the main citrus phytophages
within economically-tolerable threshold. I am sure that the work of our
highly-qualified meeting will confirm the orientations which have emerged so far.

I would like to welcome everyone cordially and wish everyone fruitful work in the name of the Commission of the European Communities,
Directorate General for Agriculture, which planned this meeting in this
prestigious institut of Acirelae, reference point for research in citrus groves not only in the countries of the European Communities, but
more generally for all the countries of the Mediterranean basin.

PART I. PRINCIPAL AGENTS NOXIOUS TO CITRUS CULTURE

Address of the session chairman

P.Jourdheuil
INRA, Zoological & Biological Control Station, Antibes, France

It is the entomologists who have the honour of opening this meeting with a session on "insect pests". In other words, phytophagous organisms are still of great economic importance, whether they are indigenous species, species which have been accidently introduced and whose distribution in the Mediterranean region is increasing, and even species not yet present but which have already been announced over our borders and cannot be ignored.

As well as general reports which will enable us to make a useful balance of our current knowledge of the taxonomy, biology, ecology and ethology of phytophagous insects which infest citrus trees, like Homoptera Aphides, Coccides Aleurodides or Psyllides, Thysanoptera, Lepidoptera, Fruit flies or Coleoptera, this session will also enable us to present several original papers relating to the geographical distribution, biological cycle, population dynamics and the influence of various factors on the biotic potential of a certain number of species.

Citrus scale insects

C.Benassy

INRA, Zoological & Biological Control Station, Laboratory 'E.Biliotti', Valbonne, France

Summary

Reference is made to the economic importance of citrus fruits in the Mediterranean region, after which the main species of destructive scale insects are mentioned and the characteristics of the damage which they cause are analysed. Such damage is reduced by effectively limiting population densities at the start of the season. The chemical control methods used in early intervention are based on the establishment of tolerance thresholds and the use of pheromones, while biological control involves the introduction of various exotic species. These different approaches are outlined on the basis of specific examples.

1. INTRODUCTION

Citrus fruits originally came from South East Asia, mainly from the hot foothills of the Himalayas (12) and were first planted in the Mediterranean region over a century ago.

However, the present-day citrus grove is a relatively recent phenomenon. One of the first major increases in areas under cultivation, mainly in the Middle East and North Africa, occurred in the years following the First World War, and from the 1950s more land was used for citrus fruit cultivation throughout the Mediterranean region.

About ten years ago the Mediterranean basin was already producing about 27% of all the citrus fruits harvested throughout the world. It has maintained this position since then, and the various producing countries - which in 1983 and 1984 exported 4,760,000 t of citrus fruits - account for 65% to 70%, depending on the season, of total world exports. This export figure is 9% up from the previous year (IRFA document).

Despite this favourable situation, however, most producer countries apply a strict policy of quality control and market only produce which is virtually free of damage or blemishes caused by pests.

Two groups of Arthropoda - i.e. mites and scale insects - are the most common citrus fruit pests. Although the latter tend to predominate in terms of the number of species encountered (Table I), the damage caused by both groups in some cases puts them more or less on a par economically.

TABLE I

Main species of scale harmful to citrus trees

Armored	Unarmored
Aspidiotini:	**Lecaninae:**
Aonidiella aurantii Mask. = California red scale	**Saissetia oleae** B. = Black scale **Coccus hesperidum** L. **Coccus pseudomagnoliarum** Kuw.
Aspidiotus nerii Bouché = White scale	**Ceroplastes floridensis** Comst. **Ceroplastes sinensis** Del Guercio
Chrysomphalus dictyospermi Morg. = Dictyospermum scale	
Chrysomphalus ficus Ashm. = Florida red scale	**Pseudococcinae:**
	Planococcus citri Risso **Pseudococcus maritimus** **P. calceolariae** Mask.
Parlatorini:	
Parlatoria pergandei Comst. = Grey scale	**Monophlebinae:**
Parlatoria ziziphi Lucas = Ebony scale	**Icerya purchasi** Muls. = Cottony-cushion scale
Diaspidini:	
Lepidosaphes beckii Newm. = Purple scale	
Lepidosaphes gloverii Pack. = Glover scale	
Unaspis yanonensis Kuw. = Japanese scale	

2. THE SPECIES AND THEIR DISTRIBUTION

Because of their particular ecological requirements, the various species, although non-specific for the most part, were never far behind the citrus trees as cultivation was extended world-wide (16).

For example, apart from a few species with strictly tropical or equatorial affinities, such as Aonidiella orientalis Newst., Selenaspidus articulatus Morg. or Unaspis citri Comst., the majority of the Diaspinae which alone account for half of the common scales, are now to be found mainly in the Mediterranean basin, where their area of dispersal appears to be well established.

There are others, though few in number, which have not yet attained their maximum potential dispersal: without effective control, these would cover a wider area every year. This is true in particular of Unaspis yanonensis Kuw., a species originating in the Far East and introduced from Japan to the Côte d'Azur in the 1960s (13). The centre of the infestation, initially restricted to a single locality on the coast of south-east France (9), spread very slowly west gradually reaching other coastal districts, though it has so far not reach Corsica.

Among the very long established species found around the Mediterranean, two belonging to the genus Chrysomphalus are spread over quite distinct areas. One, C.dictyospermi Morg. (Dictyospermum scale) is a major problem throughout the western Mediterranean, while the other, C. ficus (Florida red scale) is a very serious pest throughout the Middle East, including Egypt, where it is found together with the California red scale (Aonidiella aurantii Mask.) which for some decades has been spreading over the entire region. Its spread from Morocco since 1949 has been a recent demonstration of this, and a few scattered centres of A. aurantii were observed in Italy in 1960 (14). A few years later, the California red scale became a serious economic problem in Sicily, mainly in the region of Catania (23). It is also a fairly serious pest in Crete (3).

In the genus Parlatoria, the species P. ziziphi Lucas is restricted to the southern shores of the Mediterranean from southern Spain to Turkey, while P. pergandei, a polyphagous species, is found throughout the Mediterranean region, where long-established groves appear to favour frequent population explosions (19).

Owing to its specific ecological requirements, the purple scale (Lepidosaphes beckii Newm.) (17, 21) is restricted to coastal areas of high humidity: it is thus the most common species in coastal groves, though two other species, Lepidosaphes gloverii Pack. and Aspidiotus nerii Bouché, can multiply rapidly on occasion in such places.

As far as the unarmored scales are concerned, Mediterranean conditions are also favourable for various species of Pseudococcinae, mainly Planococcus citri Risso, and also Icerya purchasi Muls., which therefore tend to be found in coastal plantations throughout the area. The same applies, among the Lecaninae, to the two polyphagous species, i.e. Saissetia oleae Bern., which is common everywhere, and to a lesser extent Coccus hesperidum L., though the distribution of the other two species in the genus, i.e. C. aegaeus De Lotto, which is morphologically very similar to the previous one and is found in the Greek islands (6), and C. pseudo-magnolarium (Kuw), found in Turkey (35) and Sicily (27), has not yet been firmly established.

On the other hand the various species of Ceroplastes tend to occupy quite separate areas of citrus trees. C. sinensis Del Guercio is mainly restricted to the western Mediterranean, while its relative, C.floridensis Comst., is very common throughout the east from Egypt to Turkey, including Cyprus.

All citrus groves are thus host to large numbers of destructive scale insects, comprising a group of common species accompanied by varying numbers, depending on the country, of other species, one of which usually predominates.

Most of the research in the Mediterranean basin over the past ten years has therefore concentrated on L. beckii, A. aurantii and S. oleae, which are found throughout the region. The work has been coordinated by the working group on citrus scales and whiteflies of the OILB, the aim being to establish various control strategies (7).

All the methods used are designed to reduce population densities, and their main objective is to avoid damage to fruit, since this presents marketing problems.

3. DAMAGE

Most Diaspinae cause serious crop damage by attaching themselves to the skins of the fruit and producing permanent scales, such as in the case of L. beckii, A. aurantii, C. ficus, U. yanonensis, P. ziziphi, etc., or a fairly pronounced discolouration which is a constant reminder that an insect was attached earlier (mainly P. pergandei), even if the fruits are thoroughly cleaned at the packing station.

The large amounts of honey-dew secreted by the non-diaspine scales encourages the extensive development of fumagines (sooty moulds), which are always difficult to remove completely during packaging.

This type of damage occurs at the end of the season and adversely affects the presentability of the fruit. It is easily quantifiable in terms of fruit rejected during sorting: as far as the consumption of fresh fruit is concerned, this is a key economic factor.

As for the possibility of using these rejects - which in most cases are sent to canneries for juice extraction - they are only of value if the density of infestation is within a certain limit. Beyond that limit, which was 150 individuals in the case of the California red scale studied in Crete (3), the average weight of the fruit drops, with weight losses of around 20% recorded in the study just mentioned in cases where infestation was very dense. So far very few attempts have been made to quantify the damage caused to citrus trees by scale insects at the start of the season in spring, with the renewal of activity of the various species, most of which are able to thrive on any part of the tree. Limited attempts have, however, been made for U. yanonensis (9) and A. aurantii (3, 5) in the knowledge that each species causes a specific kind of damage. The damage

30

which scales inflict on the plant rather than the fruit itself is not directly quantifiable: there is no premature loss of foliage, and the initial formation of the fruit is unaffected.

Dense swarms usually cause loss of foliage and slower growth. For example, yearly branch growth is inversely proportional to the density of A. aurantii (Fig. 1). The reduction in length of the new growth resulting from heavy infestation by that insect thus ultimately has the effect of reducing the following harvest, since fruit yield is normally conditioned by the volume of the crown.

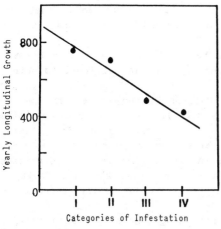

(Fig. 1)

At the end of the spring, when the fruit begins to form, it can be deformed by mobile larvae, e.g. those of the second yearly generation of A. aurantii(4). This phenomenon, already apparent with C. ficus, is particularly noticeable in this case. When the larvae become attached to the fruit, quite severe sclerification occurs, resulting in deep cavities. The fruit stops growing, and is often no bigger than a walnut.

At this time, if the pest reaches a phase of dispersal while the tree is at a receptive stage, the effect on crop yields can be extremely serious economically: not only are quantities reduced by the fact that the growth of the young fruit has been halted, there is also the danger of a deterioration in the quality of the harvest, since there is generally an additional generation in the autumn.

These considerations have been the focus of all research undertaken to develop effective control methods which seek primarily to reduce the size of the spring generation and hence prevent as far as possible larvae becoming attached to the fruit.

In some cases this approach has resulted in:

- attempts to determine tolerance thresholds beyond which measures would have to be taken to avoid all infestation of fruit;
- the use of pheromone traps as a more effective means of detection capable of improving methods of chemical control, which are still indispensable in many cases;
- testing the possibilities of biological control, where parasites, either specific or ecologically well-suited to Mediterranean conditions, are available.

4. OUTLINE OF RESEARCH

4.1 It is difficult to establish appropriate tolerance thresholds owing to the generally explosive nature of population increases among scale insects, especially the diaspids, coupled with the fact that they are rarely distributed homogeneously, but tend rather to form aggregations.

Moreover, the young larvae on hatching are attracted to the fruit in large numbers. This is true of L. beckii, studied at Valencia Late in Corsica (10). By way of comparison, in Crete infestations by A. aurantii of the variety Washington-Navel are also concentrated on fruits, which are the most infested parts of the tree: they are twice as heavily contaminated as last year's leaves and six times as heavily as this year's (3).

This discriminatory behaviour of the mobile larvae means that the threshold beyond which it becomes economically necessary to take action must be set very low.

In the few cases in which this has been studied, it is usually much lower than one individual per leaf, a density which is difficult to detect with the naked eye.

For example, in the case of the California red scale, studied in Israel (20), then some years later in Egypt (22), the presence of one live female per 100 leaves in July/August resulted in 10% of fruit being rejected during sorting, a threshold which is regarded as the upper economic limit. In the conditions prevailing in the Middle East, densities of two or more live females per 100 leaves means that when the fruit is harvested in autumn, scale population densities are not easily reconcilable with market requirements.

In the case of U. yanonensis, 3 hibernating females per 10 leaves, a density which is also hardly perceptible, would be sufficient to give rise to populations which, under Japanese conditions, would require 3 treatments per year to ensure a high-quality yield (25).

The thresholds are quite different, however, for the non-diaspine scales. In France, for example, measures to combat Saissetia oleae do not really become necessary until 10 egg-laying females per 10 cm of branch are observed frequently (11).

In the particular conditions prevailing in the lemon groves of Campania, 5% of fruit infested by mobile stages of P. citri is the economic threshold for intervention (36).

4.2 Thus, for most known species, tolerance thresholds are always exceeded if there is no chemical control after the first signs of contamination become visible. Chemical control is therefore necessary, and in each case insecticides are used when sensitive stages of the insect appear.

Effective treatment is guaranteed by early intervention. This has now become possible since the discovery a few years ago of scale pheromones as a result of work carried out in California on A. aurantii (32). The use of pheromone traps permits rapid detection of low-level infestation on the basis of the number of males caught. Although the two main components of the California red scale pheromone have been identified and synthesized (28,33), these substances are not yet applied universally, and virgin females of A. aurantii are still used in many cases (30). Fifty male California red scales per trap each week is now the lowest threshold beyond which immediate action is taken (31).

Similarly, among the Pseudococcinae, the existence of pheromones in females of Planococcus citri and Pseudococcus calceolariae and the experimental work carried out on both insects (34) already offer new prospects for practical control methods.

Regardless of the species to be treated, these methods always involve one or more applications of organophosphorous substances. Because of their numerous drawbacks, the intensive use of these general purpose and long-lasting products is now frowned upon by most importing countries, which have adopted strict legislation on acceptable levels of persistence of toxic residues. There is moreover the danger that the constantly increasing cost of treatment will rapidly damage the profitability of citrus fruits in a market in which supply is now tending to outstrip demand.

All these factors make it absolutely essential to reduce the volume of toxic products used in citrus groves, without altering the quality of the end product.

4.3 Biological control methods involving the use of exotic entomophagous insects have therefore aroused considerable interest, especially since scales account for half of the 15 or so problems which have been brought under control over the past two decades thanks to the successful introduction of a number of exotic entomophagous insects (15).

The true potential of this approach has been demonstrated by work carried out in the Mediterranean, in particular in connection with the activities of the OILB (7). The aim was to gain a thorough understanding of the ecology of the various pests while assessing the real potential of the indigenous fauna. Observation work was accompanied by the increasing use of imported entomophagous insects in order to assess the limits of their effectiveness on the basis of suitable experimentation.

Of the entomophagous insects used, the various species of the genus Aphytis have been the subject of the most intensive work on the control of diaspids (8).

All are ectophagous parasites which develop at the expense of the young adult females, and sometimes of the nymphs of their hosts. Another common feature is their breeding pattern. They breed at the expense either of a substitute scale represented by Hemiberlesia lataniae of the partheno-genetic strain of Aspidiotus nerii (in particular A. melinus, A.holoxanthus and A. yanonensis) or of their specific host (A. lepidosaphes). All the scales used, with the exception so far of U. yanonensis, will breed on the ripe fruit of Citrullus vulgaris. The adult stages of all the above para-sites are highly sensitive to seasonal climatic factors; they are abundant in spring and autumn, when moderate temperatures combine with relatively high humidity, and decline in summer when excessive dryness can quickly kill off a large proportion of individuals. In Morocco in particular, A. melinus is incapable of surviving in the interior owing to periodic chergui winds (1).

The non-selective insecticides in common use today, moreover, com-pletely cancel out the usefulness of the Aphytis insects by wiping them out in all the plots treated. This happened in recent years with A. chilensis in Crete, where its elimination by chemical spraying resulted in exceptional population explosions of Aspidiotus nerii (2).

33

TABLE II

Use of entomophagous insects: the various possibilities in the Mediterranean region.

Introduction	Features	Effectiveness	Name of Pest
1 single species:			
Aphytis melinus	Introduction possible, dispersal limited	Very good in the Mediterranean / Useful in certain locations	**C. dictyospermi** / **A. aurantii** / **A. nerii**
Aphytis lepidosaphes	Introduction possible, dispersal limited	Good in the Mediterranean	**L. beckii**
Aphytis chilensis *	Very abundant in all untreated areas	Very good if protected	**A. nerii**
Pseudaphycus maculipennis *		Regular	**P. maritimus**
2 complementary species:			
A. holoxanthus + P. smithi	Introduced in the Middle East	Good in Israel	**C. ficus**
A. yanonensis + P. fulvus	Easily introduced / In the process of being introduced in the Mediterranean	Good in Japan	**U. yanonensis**
A. hispanicus + P. pergandei	Introduced in the Mediterranean	Partial	**P. pergandei**

TABLE II (cont.)

Periodic Introduction	Features	Effectiveness	Name of Pest
complex of entomophagous insects:			
Metaphycus helvolus + **M. bartletti** + **Diversinervus elegans**	Introduction possible	Satisfactory if releases are staggered	**S. oleae**
parasite/predator combination:			
Leptomastix dactylopii	Introduction difficult	Regular	
Cryptolaemus montrouzieri	Introduced	Irregular	**P. citri**
Nephus reunioni	Introduction possible	Good	

* Indigenous species

35

To guard against both hazards, which can hinder or even interrupt the introduction process, a basic breeding program must be maintained to preserve stocks. At the same time, modification of spraying programs to safeguard as far as possible the entomophagous insects introduced should also be considered.

If these conditions are met, effective biological control may result from various possible combinations of factors (Table 2), such as:

- the introduction of a single species in the case of a specific parasite or of a species which is ecologically well adapted to local conditions. This applies in particular in the Mediterranean region to A.melinus, whose effectiveness in combatting the red scale C. dictyospermi has proved complete and immediate, although it takes somewhat longer to deal with the California red scale (A. aurantii). The specific parasite, A. lepidosaphes also helps to reduce populations of L. beckii wherever microhymenopteron has been introduced.

- introduction of 2 complementary parasites, the one ectophagous and the other endophagous, which normally attack their hosts at different stages in their development. In the case of U. yanonensis, for example, while the ectophagous A. yanonensis attacks the young females, Physcus fulvus, an endophagous species which cohabits with the first, lays its eggs mainly in the older females of the scale (18). A similar phenomenon occurs in Israel at the expense of C. ficus, in which Aphytis holoxanthus and Pteroptrix smithi combine to reduce populations of the Florida scale (29).

or the periodic introduction of entomophagous insects in the form of either:

- a parasite complex which in the case of the black scale (Saissetia oleae) combines 2 species of the genus Metaphycus (M. helvolus and M. bartletti) with Diversinervus elegans, both of which are microhymenoptera and affect the scale at different stages in its yearly development (26), or

- a combination of two different types of entomophagous insects: a microhymenopteron of the Encyrtidae family (Leptomastix dactylopii) and one of the two species of predatory scales Cryptoloemus montrouszieri Muls. and Nephus reunioni. The combination of the first two entomophagous insects gives good results in combatting P. citri, provided individuals from both species are released in the spring, since they can successfully establish themselves only in very specific locations: Cyprus in the case of L. dactylopii (24) and the Côte d'Azur in that of C. montrouzieri (26). The other species, Nephus reunioni, on the other hand, is a useful entomophagous insect, since it can be successfully introduced.

Thus, considering the sum total of knowledge accumulated to date concerning the scales and the associated entomophagous species, it would seem that rational use of the latter offers real potential for control in certain cases.

5. CONCLUSION

The coordinated studies caried out in recent years by the various producer countries in the Mediterranean region already indicate that there are certain effective methods whereby high quality fruit can be supplied to the various markets.

Biological control introducing natural enemies is now becoming an increasingly widespread method of combatting most known scales.

To follow up the work already done, it is therefore necessary to maintain and improve cooperation with research bodies throughout the world in order to identify as soon as possible entomophagous insects for certain specific cases where none have hitherto been found.

REFERENCES

1. ABBASSI,M.(1977). Recherches sur deux Homoptères fixés des **Citrus, Aonidiella aurantii** MASK. **(Homoptera, Diaspididae)** et **Aleurothrixus floccosus** MASKELL. **(Homoptera, Aleurodidae).** Thèse, Univ. Provence, 119 pp.
2. ALEXANDRAKIS,V.(1979). Contribution à l'étude d'**Aspidiotus nerii** BOUCHE **(Homoptera, Diaspididae)** en Crète. Thèse Univ. Bordeaux I, 117 pp.
3. ALEXANDRAKIS, V. (1980). Essai d'appréciation des dégâts provoqués sur oranger en Crète par la présence d'**Aonidiella aurantii** (MASK.) (Hom. **Diaspididae).** Fruits, 35; 555-560.
4. ALEXANDRAKIS, V. (1983). Données biologiques sur **Aonidiella aurantii** MASK. **(Hom. Diaspididae)** sur agrumes en Crète. Fruits, 38; 831-838.
5. ALEXANDRAKIS, V. and MICHELAKIS, S. (1980). Distribution d'**Aonidiella aurantii** (MASK.)**(Hom. Diaspididae)** en fonction de son emplacement sur l'arbre et de la variété d'agrumes en Crète. Fruits, 35; 639-644.
6. ARGYRIOU, L.C. and IOANNIDES, A.G. (1975). **Coccus aegaeus (Hom. Coccidae)** DE LOTTO, nouvelle espèce de Lécanines des **Citrus** en Grèce. Fruits, 30; 161-162.
7. BENASSY, C. (1982). Groupe de travail O.I.L.B. "Lutte Intégrée en Agrumiculture". C.R. IVe Ass. Gen. O.I.L.B. Antibes, Oct. 81; 122-127.
8. BENASSY, C. (1984). Les problèmes Cochenilles Diaspines chez les **Citrus.** Fruits (à parâitre).
9. BENASSY, C. and PINET, Ch. (1972). Note sur **Unaspis yanonensis** KUW. dans les Alpes-Maritimes. Ann. Zool. Ecol. Anim., 4; 187-212.
10. BENASSY, C., BIANCHI, H. and BRUN, P. (1981). Données préalables à la définition d'un seuil d'intervention chez **Lepidosaphes beckii** NEWM. C.R. Réunion Experts C.E.E. San Giuliano - Siniscola, 4-6/11/1980. Eds. R. Cavalloro et R. Prota, Eur.: 7342; 19-25.
11. BENASSY, C. et al. (1976). Orientation vers la lutte intégrée en agrumiculture dans le Sud-Est de la France. P.H.M. Revue Horticole, 167; 41-48.
12. CASSIN, J. (1984). Comportement des variétés d'Agrumes dans les différentes régions de production. Fruits, 39; 263-276.
13. COMMEAU, J. and SOLA, E. (1964). Une nouvelle Cochenille des Agrumes sur la Côte d'Azur. Phytoma, 16; 49-50.

14. DE BACH, P. (1962). Biological control of the California Red Scale, **Aonidiella aurantii** (MASK.), on **Citrus** around the world. Verh. XI. Int. Kongr. Entomol. Wien, 1960, 2; 749-753.

15. DE BACH, P. (1974). Biological control by natural enemies. Cambridge Univ. Press, 323 pp.

16. EBELING, W. (1959). Subtropical Fruit Pests. University of California Div.of Agricultural Sciences; 436 pp.

17. FABRES, G. (1978). Analyse structurelle et fonctionnelle de la bio-cénose d'un Homoptère (**Lepidosaphes beckii** HOW.)(**Diaspididae**) dans deux types d'habitats agrumicoles de la Nouvelle Calédonie. Thèse Univ. Paris VI; 291 pp.

18. FURUHASHI, K. and NISHINO, M. (1983). Biological control of arrowhead scale, **Unaspis yanonensis**, by parasitic wasps introduced from the People's Republic of China. Entomophaga, 28; 277-286.

19. GERSON, U. (1967). Studies of the chaff scale on **Citrus** in Israël. J.Econ.Entomol., 60; 1145-1151.

20. GRUMBERG, A. (1969). The economic threshold and treatment forecast for the California Red Scale (**Aonidiella aurantii** MASK.) in Citrus groves in Israël. Israël J. Entomol;, 4; 129-137.

21. HAFEZ, M. and SALAMA, H.S. (1969). Biology of **Citrus** purple scale, **Lepidosaphes beckii** NEWM. in Egypt. (**Hemiptera, Homoptera, Coccoidea**). Bull. Soc. Entomol. Egypte, 65; 517-532.

22. HOSNY, M.M., AMIN, A.H. and SAADAMY, G.B. (1972). The damage threshold of the red scale, **Aonidiella aurantii** MASKELL, infesting mandarin trees in Egypt. Z. Ang. Entomol. 71; 286-296.

23. INSERRA, S. (1966). Introduzione ed acclimatazione di due **Aphytis (A. melinus** DE BACH et **A. lingnanensis** COMPERE) parssiti ectofagi di alcune cocciniglie degli agrumi. Tecnica Agricolo, 2; 1-11.

24. KRAMBIAS, A. and KOTSIONIS, A. (1981). Acclimatation de **Leptomastix dactylopii** HOW. à Chypre. Bull. S.R.O.P., 1981, 4; 94-97.

25. MIY AHARA, M. and YAMADA, K. (1969). Relation between frequency of spray necessary to control economically and population density of arrowhead scale. Studies on the improvement of the cooperative control program of the **Citrus** tree pests; 98-100.

26. PANIS, A. (1981). Note sur quelques insectes auxiliaires régulateurs des populations de **Pseudococcinae** et de **Coccidae** (**Homoptera, Coccoidea**) des Agrumes en Provence orientale. Bull. S.R.O.P., 1981, 4; 88-93.

27. PATTI, I. (1976). Rilievi sulla diffusione negli agrumeti siciliani della nuova cocciniglia **Coccus pseudomagnoliarum** (KUW.). Tecnica Agricola, 4; 5-10.

28. ROELOFS, W. et al., (1978). Identification of the California Red Scale sex pheromone. J. Chem. Ecol. 4; 211-224.

29. ROSEN, D. (1965). The Hymenopterous parasites of **Citrus** armored scales in Israël (**Hymenoptera, Chalcidoidea**). Ann.Entomol.Soc.Amer.,58; 388-396.

30. SHAW, G.J. and MORENO, D.S. (1971). Scale detection system for Coachella. Citrog., 56; 67-69.

31. STERNLICHT, M. (1978). Development of a pheromone-based method for control of the California red scale. Inst. Plant. Prot. Beit Dagan Pamphlet, 185: 12.

32. TASHIRO,H. & CHAMBERS,D.L. (1967). Reproduction in the Calif. red scale **Aonidiella aurantii (Homoptera,Diaspididae)**. I. Discovery and extraction of a female sex-pheromone. Ann.Entomol.Soc.Amer.,60; 1166-1170.

33. TASHIRO, H. et al., (1979). Residual activity of a California Red Scale synthetic pheromone component. Env. Entomol., 8; 931-934.

34. TREMBLAY, E. and ROTUNDO, G. (1981). The use of sex pheromones in the control of the **Citrus** scale insects in C.R. Reunion Experts C.C.E. San Giuliano-Siniscola, 4-6/11/1980. Eds. R. Cavalloro et R. Prota. EUR. 7342; 59-66.

35. ÜNCÜER, C. and TUNCYÜREK, M. (1975). Observations sur la biologie et les ennemis naturels de **Coccus pseudomagnoliarum** KUW. dans les vergers d'Agrumes de la région égéenne. Fruits, 30; 255-257.

36. VIGGIANI,G. (1975). Méthode d'estimation des populations de **Planococcus citri** (RISSO) au niveau d'un verger d'agrumes. Fruits, 30; 177-178.

Distribution and density of scale insects (Homoptera, Coccoidea) on citrus-groves in Eastern Sicily and Calabria

S.Longo & A.Russo

Institute of Agricultural Entomology, University of Catania, Italy

Summary

A survey on coccids infesting citrus-groves was made in Eastern Sicily and Ca labria during 1979–80. The most widely distributed species were: Planococcus citri (Risso), Coccus hesperidum L., Saissetia oleae (Oliv.), Ceroplastes rusci L., Parlatoria pergandei (Comst.), Aspidiotus nerii Bouchè and Mytilococcus beckii (New.).
Other species collected in more restricted areas were: Pericerya purchasi (Mask.), Pseudococcus longispinus Targ., Coccus pseudomagnoliarum (Kuw.), Saissetia coffeae (Walk.), Ceroplastes sinensis D.G., Parlatoria ziziphus (Lucas), Mytilococcus gloverii (Pack.), Chrysomphalus dictyospermi (Morg.) and Aonidiella aurantii (Mask.).
On the basis of the infestation-rate, the most important species were: Pl. citri (Citrus mealybug), S. oleae (Black scale), C. rusci (Fig wax scale), P. pergandei (Chaff scale) and M. beckii (Purple scale); with regard to the lemon-trees, A. nerii (Oleander scale) and A. aurantii (California red scale) are particularly injurious.
The most important scales are discussed from the biological and ethological point of view, with particular emphasis on the biological control of Pl. citri by the entomophagous Scymnus (Nephus) reunioni Fursch, Cryptolaemus montrouzieri Muls.(Col. Coccinellidae) and Leptomastidea abnormis (Grlt.), Anagyrus pseudococci (Grlt.), Leptomastix dactylopii (How.) (Hym. Encirtidae).

1. Introduction

Periodic observations on scale fauna of citrus-plants were carried out from 1979 in the context of biological and integrated control programmes against the primary pests of such cultures.

From these studies it emerges that 16 species (among the 20 scales species reported in Italy) occur in Eastern Sicily and in Calabria, where there is more than 70% of Italian citrus industry.

Studies on natural enemies and on the bio-ethology of such scales allowed us to explain some aspects of their population dynamics in the investigated areas.

Planococcus citri (Risso) was particularly studied in relation to the possibility of biological control by means of indigenous and exotic entomophagous.

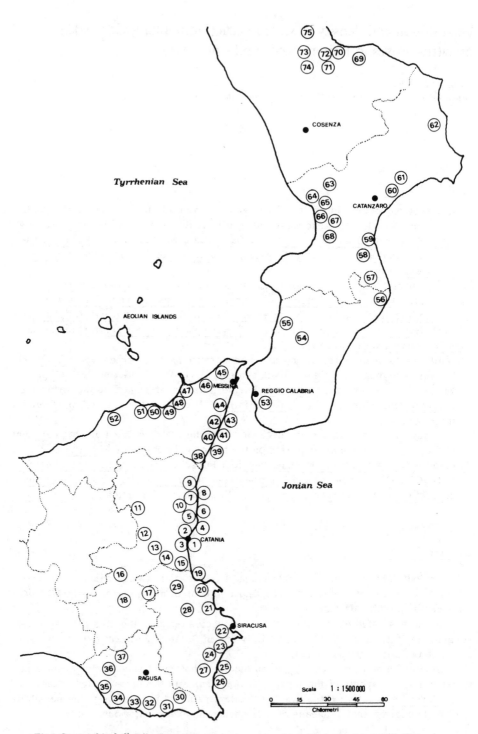

Fig I *Geographical distribution of the citrus groves in which the present investigation was conducted.*

42

2. Material and methods

2.1. Remarks on scale-fauna

With the purpose of giving uniformity to the data on the composition of scale fauna, from June 1984 to March 1985 samples were collected from branches, leaves and fruits in 75 areas of Sicily and Calabria (*) (see Pic. I).

At least in 2 different citrus-groves of each area, 4 to 6 trees were chosen at random in each grove: from these plants at least twenty 2-3 year old twigs were picked off from different cardinal points/sides and at various heights of the trees.

On such samples laboratory observations were carried out on: 100 leaves (upper and under face of the leaves, for a total surface of about 5.000 cm^2), 200 cm of twigs (for an average surface of about 400 cm^2) and 10 fruits (average surface of 1.300 cm^2).

The scales present on the above mentioned surfaces were counted, distinguished into living, dead and parasitized forms and afterwards prepared on slides for specific identification and for distinction of the various biological stages.

The density of each species, expressed as a number of living individuals per cm^2 of surface examined, is indicated in Tab. I, in which only 2 of the density classes considered have been reported. The first includes the species present in the samples in density equal to or higher than 0,1 specimens per cm^2 of surface (symbol ●), the second class includes the species present in density less than 0,1 specimens per cm^2 of surface (symbol o).

Part of the material collected in each area was selected and put in suitable containers, for the emergence of the adult entomophagous.

2.2. Insectary breeding and field distribution of entomophagous of Planococcus citri

The exotic entomophagous Leptomastix dactylopii (How.), Cryptolaemus montrouzieri Muls. and Scymnus (Nephus) reunioni Fursch and the indigenous Leptomastidea abnormis (Grlt.) and Anagyrus pseudococci (Grlt.) were bred in insectaries on pumpkim and on sprouts of potatoes infected with Pl. citri. Inoculative releases of natural enemies were realized from June of each year.

The remarks on the distribution of the scale and on the course of the parasitation were carried out following the sampling methods suggested by the Commission of the European Community (4) (21).

3. Results and discussion

3.1. Composition of the scale-fauna

Among the Monoflebinae, Peryceria purchasi (Mask.) occurs at very low density in almost all the investigated areas but in no case did it form a phytosanitary problem since it was efficaciously controlled biologically by the coccinellid Rodolia cardinalis Muls. (2).

Among the Coccid, the Brown soft scale (Coccus hesperidum L.) also spread over many of areas, is not generally injurious due to the biological activity of its

---

(*) Thanks to Dr. G. Carbone, Director "Osservatorio per le malattie delle piante di Catanzaro" for his kind collaboration during the remarks in Calabria.

43

Tab I *Distribution and density of the scales collected in 75 areas of Sicily and Calabria during* 1983 84

AREAS \ SPECIES	(dens.)	Pericerya purchasi	Planococcus citri	Pseudococcus longispinus	Coccus hesperidum	Coccus pseudomagnoliarum	Saissetia oleae	Saissetia coffeae	Ceroplastes sinensis	Ceroplastes rusci	Parlatoria pergandei	Parlatoria zizyphus	Mytilococcus beckii	Mytilococcus gloverii	Aspidiotus nerii	Chrysomphalus dictyospermi	Aonidiella aurantii
A. SICILY																	
CATANIA																	
1 Catania	o.m.l	○	○		○						○	○			○		
2 CT-Barriera	o						●					○	○				
3 CT-Librino	o		●		○					●	○		○				
4 Acicastello	l		●							○			○			○	○○
5 Acireale	l				○					○						●	●●
6 Pozzillo	l		○	○												●	●●
7 S.G.Montebello	c	○		○			●			○	○				○		
8 Giarre	l	○	○		○		○								○		○
9 Macchia	c	○			○		●	○									
10 S.Venerina	o	○			○		○	○									
11 Adrano	o	○													●		
12 Paterno	o	○	●		○		●			○						○	
13 Motta S.A.	o	○			○		○										
14 Misterbianco	l	○	○		○			○		●				○		○	
15 Torremuzza	o	○	●		○												
16 Ramacca	o				○	●											
17 Scordia	o		●		○		○			○							
18 Mineo	o		●		○	○											
SIRACUSA																	
19 Vaccarizzo	o				○		○			○	○						
20 Villasmundo	c				●		○			○	○						
21 Priolo	o				○					○	○						
22 Siracusa	l	○															●
23 SR Ciane	l	○			○					○				○			
24 Cassibile	o				○					○							
25 Avola	l	○	●							○				○			○
26 Noto Marina	l	○												●			
27 Noto	o	○			○								○				
28 Francofonte	o		●		○						○						
29 Lentini	o		●		○		○				●						
RAGUSA																	
30 Ispica	o.m				○		○			○					○		
31 Pozzallo	o				○										○		
32 Scicli	l				○												
33 Donnalucata	l				○						○				○		
34 S.C.Camerina	l		○				○				○				○		
35 Scoglitti	m		○				○	○			●	○			○		
36 Vittoria	c						●										
37 Roccazzo	o						○										
MESSINA																	
38 Giardini	l				○										○		
39 C.S.Alessio	l		○	○○	○					○	○				○		
40 Roccalumera	l			●			●			○	○					●	
41 Furci Siculo	l		●		○		○			○					○	●	●●
42 Nizza	l			○						○						●	
43 Ali Terme	l		○○	○						○						●	
44 Galati Marina	l		○		○		○			○							
45 Saliceto	l						○			○						○	
46 Spatafora	o		○							○	○				○	○	
47 Milazzo	o									○	○				○		
48 Barcellona	l						○			○	○						
49 Olivarella	c				○		○			○	○				○		
50 Falcone	l		○							○	○					○	
51 Patti	l	○	○		○					○						○	
52 Brolo	l		○		○					○	○					○	

B CALABRIA

REGGIO CALABRIA

53 Reggio Calabria	l																
54 Taurianuova	o																
55 Gioia Tauro	o																
56 Monasterace	l																

CATANZARO

57 Guardavalle o
58 Davoli Marina o
59 Sellia ol
60 Sineri e Crichi o
61 Sineri e Crichi cm
62 Rocca di Neto m
63 Nicastro o
64 Lamezia l
65 Lamezia m
66 Lamezia c
67 S.Pietro l
68 Curinga co

COSENZA

69 Rossano o
70 Corigliano l
71 Corigliano o
72 Corigliano c
73 S.Demetrio o
74 S.Demetrio c
75 Cassano Jonio c

SICILIA	16	22	4	31	2	23	3	3	18	19	3	13	1	21	4	9
CALABRIA	6	6	1	8	–	20	1	1	2	10	–	8	–	4	3	4
TOTALE	22	28	5	39	2	43	4	4	20	29	3	21	1	25	7	13

● density ≥ 0,1 specimen/cm   O density < 0,1 specimen/cm

l lemon   O orange   m mandarin   C clementin

Tab.II DISTRIBUTION OF ENTOMOPHAGOUS OF Planococcus citri (RISSO) IN SICILY

YEAR	PROVINCE	NUMBER OF ENTOMOPHAGOUS DISTRIBUTED				
		N.reunioni	C.montrouzieri	L.dactylopii	A.pseudococci	L.abnormis
1979	CATANIA	–	–	4552	–	–
	SIRACUSA	–	–	130	–	–
	MESSINA	–	–	–	–	–
1980	CATANIA	–	–	5200	–	–
	SIRACUSA	–	–	2800	–	–
	MESSINA	–	–	300	–	–
1981	CATANIA	–	–	300	4800	–
	SIRACUSA	–	–	800	220	–
	MESSINA	–	–	120	–	–
1982	CATANIA	510	115	300	700	800
	SIRACUSA	110	20	300	–	–
	MESSINA	20	20	100	–	–
1983	CATANIA	800	400	7000	–	–
	SIRACUSA	250	170	1900	–	–
	MESSINA	500	200	1500	–	–
1984	CATANIA	950	250	8200	–	–
	SIRACUSA	400	100	1900	–	–
	MESSINA	100	40	1900	–	–

numerous indigenous entomophagous (11). Pullulation of this scale is frequent on single branches or trees, without it usually reaching a generalized diffusion in the citrus-groves.

Citricola scale (Coccus pseudomagnoliarum (Kuw.)), (1) which appeared quite recently in Italy, has a very restricted distribution area and only occasionally reaches population levels that require specific control interventions.

Saissetia oleae (Oliv.) is the most frequently encountered species in the areas under observation and it causes the most relevant damage to the citrus-tree; in 19 biotopes its population density was higher than 0,1 specimen/cm^2 of branches and leaves.

As it has already been evidenced by other A.A. (6) (16) and by personal observations, the Black scale has only one generation per year. Among its entomophagous, the Coccinellid Chilocorus bipustulatus (L.) proved to be the most frequent and active in the observed areas, followed by Exochomus quadripustulatus (L.). Among the parasitoids, the most active are the Pteromalidae Scutellista cyanea Mots. and the Encyrtidae Metaphycus flavus (How.) and M. lounsburyi(How.). The latter is a recent accidental introduction and has rapidly become widespread in many Sicilian citrus-groves; it has reached in some of such groves a parasitism rate of 70%, living on the ovigere and preovigere females of S. oleae (9).

Among the Lecanidae, S. coffeae (Walk.) (Hemispherical scale), that mostly infests ornamental plants, has been found in Sicily and in Calabria on Lemon and Mandarin in citrus-groves of the Jonian coast in low density and with a very irregular presence (10).

Ceroplastes rusci L., a very poliphagous species, occurs in 20 areas. This scale, that makes two generations a year and overwinters as third instar larva , is presently among the most interesting species in citrus-groves for some areas of Sicily. The action of its entomophagous, represented in addition to S. cyanea also by two Hymenopterous species of Tetrastichus,is not able to contain the populations.

The other Wax scale, present in the investigated areas, C. sinensis D.G. has a more generalized diffusion. In the course of the mentioned studies the presence of Ceroplastes japonicus Green, which was reported in 1983, with centres of infection in the metropolitan areas of Venice, Rome, Padua and Pisa, on some ornamental plants and some isolated citrus plants, was not evidenced. The diffusion of this scale, considered in U.S.S.R. and in Japan among the primary pest of various cultivated plants, could represent a serious danger for our citrus-groves and fruit-cultivations (10).

Among the Diaspididae, the Chaff scale (Parlatoria pergandei (Comst.)) is present especially on orange. As a rule it makes 4 generations in a year infesting all the organs; the tick stratifications, that it forms on the branches,frequently determine perishing that often leads to death.

In 6 areas out of 29 its population density was higher than 0,1 specimen cm^2. The presence of Hymenopterous Chalcid Aphytis hispanicus was observed in all the investigated areas.

Black parlatoria (Parlatoria ziziphus (Lucas)) does not make as a rule generalized phytosanitary problems, since its infections are usually limited.

Among the Diaspini Mytilococcini, the Purple scale (Mytilococcus beckii (New.)) spread above all over coastal citrus groves, rarely reaching high population densities; parasitation levels ranging from 10% to 40% have been found by the parasitoid Aphytis lepidosaphes Comp.. The Glover's scale (Mytilococcus gloverii (Pack.)) whose distribution in Italy has for a long time been confined to the Palermo region (18) has recently been found in Calabria (Taurianova) on orange cv "Ovale". Of the Aspidiotini group of Diaspini, the Dictyospermum scale (Chrysomphalus dictyopsermi (Morg.) (13) (19), which some years ago was considered the most harmful of the scale insects in our citrus groves, is today only met with sporadically and in a not seriously infesting form; in one only of the 7 groves where it was found was its density higher than 0.1 sample/cm^2.

Of considerable interest for the lemon are the current infestations by the Oleander scale (Aspidiotus nerii Bouché). In almost all lemon growing areas this scale is present and considerably damages the infested fruits, which remain green in those areas where the scale has become established. The entomophagous have very little effect against it; among these are Cybocephalus rufifrons Reit., Chilocorus bipustulatus (L.) and Lindorus lophantae (Blais.) which are frequent in plants infested by Aspidiotus and other scales, such as Aonidiella aurantii (Mask.).

In Sicily and Calabria the latter scale infests particularly lemon trees and its populations are found on all the vegetative organs, causing a withering of the branches, defoliation and the dropping of fruit, which would in any case have been very depreciated.

In the same Regions the Chalcididae Aphytis melinus De Bach (7) has been introduced for biological control of A. aurantii.. The former, even though producing 12 or 13 generations annually on various armored scale, cannot or do not its own manage to reduce satisfactorily the pest infestations.

3.2. Remarks on the biocenosis of Pl. citri

Of the 4 species of Pseudococcidae reported in Italy, those found in the areas investigated are Planococcus citri (Risso) and Pseudococcus longispinus Targ.. There is the occasional presence of a few specimens of Ps. obscurus Es. on an orange tree growing near a plant of Diospyros kaki L. highly infested by the mealybug. The fourth species, Ps. calceolariae Mask., is present in the Naples area (20) but has not been noted in the investigated biotopes.

Among the mealybugs living on citrus, Pl. citri is the most widespread and harmful species, especially for the lemon and those varieties of oranges that tend to bear fruit in groups (e.g. cv "Moro"). The mealybug produces from 4 to 6 generations yearly, reaching its population density peak in the autumn, when the thermo-hygrometric values are most suitable for its development.

Investigations on the parasites and predators of Pl. citri revealed the presence of the indigenous entomophagous already known for the Mediterranean area (3) (8).

Among the latters the Encyrtid parasitoids Leptomastidea abnormis (Grlt.) and Anagyrus pseudococci (Grlt.) are particularly active above all in the presence of dense populations of scale insects. Associated with them is the predatory action of Neuroptera (genus Sympherobius Ram.), Coleoptera Coccinellidae (i.e. Scymnus includens Kir.) and Diptera (genera Dicrodiplosis Kieffer and Leucopis Meig.). The larvae of these Diptera, mostly in autumn, prey on the eggs and first stage larvae of the Planococcus. As a whole, the indigenous entomophagous while playing an important part in controlling biologically the populations of scale insects do not manage, however, to control the autumn pullulations in those biotopes where particular microenvironmental conditions (e.g. citrus groves with closely planted trees) and the presence of ants hinder their activity, while anhancing the development and dif fusion of the Planococcus (10). In the framework of actions aimed at increasing the population of entomophagous of Pl. citri, recently there have been introduced or reintroduced in Sicily three exotic species which have proved to be particularly effective against scale insects in various citrus groves (22)(17)(12). Of these the Encyrtid L. dactylopii acts effectively against the third instar larvae and young females of the Planococcus, on which it reaches parasitation levels of 60-95%. At the same time, the parasitoid has not become acclimatized in our citrus groves and so considerable numbers of adults must be released annually in the field.

N. reunioni, native from South Africa and now bred in insectaries since 1982 and annually released in groves infested by Pl. citri, seems also unable to live through the winter period, both due to the low temperatures and to the considerable contemporaneous rarefaction of the scale host, which in such period is at its lowest level of annual population.

The other Coccinellidae C. montrouzieri, has been released since 1982 in various citrus groves in Eastern Sicily but has only recently been acclimatized in some biotope where launchings have been effected; here it has managed to survive, preying in the winter period on larvae and eggs of the woolly whitefly (Aleurothrixus floccosus Mask.). The presence of C. montrouzieri has been observed by us also in some Calabrian citrus groves, where, as in the case of Western Sicily (15) and Campania (14), the Coccinellidae seems to have been acclimatized for a longer time.

4. Conclusions
Of the 16 species of scale insects identified in 75 biotopes in the Eastern Sicilian and Calabrian citrus-groves examined in this investigation, presently only 7, Pl. citri, S. oleae, C. rusci, P. pergandei, M. beckii, A. nerii and A. aurantii, are really harmful to citrus, on account of their frequency and population density.

C. hesperidum and P. purchasi are fairly widespread but generally not harmful; Ps. longispinus, S. coffeae, C. sinensis, P. ziziphus and C. dictyospermi are infrequent and of no practical interest; C. pseudomagnoliarum and M. gloverii are localized in restricted areas but cause consistent infestations which are carefully followed up. As regards Pl. citri, its biological control by means of exotic entomophagous has yielded interesting results in those biotopes where the field releasing were swiftly carried out in presence of low populations of the scale.

REFERENCES

1. BARBAGALLO, S. (1974). Notizie sulla presenza in Sicilia di una nuova cocciniglia degli agrumi Coccus pseudomagnoliarum (Kuwana) (Homoptera, Coccoidea). Entomologica, 10: 121-139.
2. BENFATTO, D. and LONGO, S. (1982). Risultati di una prova di lotta integrata contro Icerya purchasi Mask. Atti Giorn. Fitopatol. 1982, 3: 131-138.
3. BODENHEIMER, F.S. (1951). Citrus entomology in the middle east. Dr. Junk, S. Grovenhage, XII + 663 pp..
4. CAVALLORO, R. and PROTA, R. (1982). Integrated control in citrus orchards: sampling methodology and threshold for intervention against the principal phytophagous pests. Proceedings of the E.C. - Experts' Meeting "Integrated control in citrus: comparison of results archivied by appling a standardized methodology". Siniscola Muravera, 43-51.

5. CHAZEAU, J., ETIENNE, J. and FURSCH, H. (1974). Coccinellidae de l'île de la Réunion. Bull.Mus.Nat.Hist.Not., 210: 275-276.
6. DI MARTINO, E. and BENFATTO, D. (1981). Note etologiche sulla Saissetia oleae (Oliv.) in un biotopo olivicolo della Sicilia Orientale. Inf.tore Fitopatol., 12: 61-64.
7. INSERRA, S. (1968). Prove di lotta integrata contro l'Aonidiella aurantii Mask. ed altre cocciniglie degli agrumi in Sicilia. Entomologica, 4: 45-77.
8. LIOTTA, G. and MINEO, G. (1963). Prove di "lotta biologica artificiale" contro lo Pseudococcus citri R. (Cotonello degli agrumi). Boll.Ist.Ent.agr.Oss. Fit.Palermo, 5: 129-142.
9. LONGO, S. (1984). Distribution and density of scale-insects (Homoptera - Coccoidea) on olive trees in eastern Sicily. Proceeding international Joint Meeting "Integrated Pest Control in olive-grove", 1984, in press.
10. LONGO, S. (1984). La difesa degli agrumi: attuali possibilità di controllo biologico integrato delle cocciniglie e aleirodi. Terra e sole, 508: 731-739.
11. LONGO, S. and BENFATTO, D. (1982). Note biologiche su Coccus hesperidum L. (Rhyncota, Coccidae) e risultati di prove di lotta. Atti Giorn. Fitopatol. 1982, 3: 139-146.
12. LONGO, S. and BENFATTO, D. (1982). Utilizzazione del Leptomastix dactylopii How. per la lotta biologica al Cotonello degli agrumi in Sicilia orientale. Inf.tore agrario, 38 (9): 19671-19676.
13. LUPO, V. (1959). Gli insetti dannosi agli agrumi e i mezzi per combatterli. Tecnica agricola, II: 187-203.
14. MAZZONE, P. (1977). Recenti distribuzioni di Cryptolaemus montrouzieri (Muls.) in Campania. Boll.Lab.Ent.agr. Portici, 34: 223-227.
15. MINEO, G. (1966). Sul Cryptolaemus montrouzieri Muls. (Osservazioni morfobiologiche). Boll.Ist.Ent.agr.Oss.Fitopatol.Palermo, 6: 3-47.
16. MINEO, G. (1971). Prime osservazioni sulla dinamica di popolazione della Saissetia oleae (Oliv.) in Sicilia. Boll.Ist.Ent.agr.Oss.Fit.Palermo, 10: 69-80.
17. MINEO, G. and VIGGIANI, G. (1976). Su un esperimento di lotta integrata negli agrumeti in Sicilia. Boll.Lab.Ent.agr.Portici, 33: 219-231.
18. MONASTERO, S. (1954). Morfologia e biologia del Mytilococcus gloverii Packard 1869. Boll.Lab.Ent.agr.Oss.Fit.Palermo, 6: 87-136.
19. SILVESTRI, F. (1935). Le cocciniglie degli agrumi in Italia. Atti I Congresso agrumario tenuto in Palermo, 123-133.
20. VIGGIANI, G. (1977). Lotta guidata contro i fitofagi degli agrumi. Inf.re Fitopatol., 6/7: 39-43.
21. VIGGIANI, G. and BATTAGLIA, D. (1984). Osservazioni su Planococcus citri (Risso) e Pseudococcus calceolariae (Mask.), cocciniglie associate su agrumi in Campania con prove di lotta chimica. Atti giornate fitopatologiche 1984, 2: 363-374.
22) ZINNA, G. (1959). Esperimenti di lotta biologica contro il cotonello degli agrumi (Pseudococcus citri (Risso)) nell'Isola di Procida mediante l'impiego di parassiti esotici, Pauridia peregrina Timb. e Leptomastix dactylopii How.. Boll. Lab.Ent.agr.Portici, 18: 257-284.

Research work supported by CNR Italy. Special grant I.P.R.A. Subproject 1.

Population dynamic for *Aonidiella aurantii* (Mask.) (Homoptera, Diaspididae) and its parasitoids on lemon trees in Sicily*

G.Liotta, A.Agro' & M.C.Perricone

Institute of Agricultural Entomology, University of Palermo, Italy

Summary

Aonidiella aurantii (Mask.) was found mainly on the fruit itself and the branches, but much less frequently on the leaves. The infestation of the fruit did not exhibit any significant differences in relation to exposure to the four points of the compass. The figure of almost 100% for March 1981 dropped to a mean value of less than 20% for May 1983, probably as a result of the combined action of three main factors, viz: a) a mineral oil treatment in the autumn of 1981, b) the high temperatures recorded in June 1982, and c) the parasitic action of two Aphelinidae, Aphytis melinus De Bach and Aphytis chrysomphali (Mercet).

1. INTRODUCTION

Aonidiella aurantii (Mask.) (Citrus red scale) was found in western Sicily in 1968 (Liotta, 1970) and in eastern Sicily in earlier years (Inserra, 1967).

Since the time when it first appeared, the scale has become more widespread, so that today it is thought to appear in almost all the citrus groves of the island, although to a varying extent from one area to another. This paper gives a survey of the trend in the population of the Scale and on the action of the Aphelinidae to contain it.

2. MATERIALS AND METHODS

The observations were carried out between March 1981 and May 1983 in a citrus grove in the Santa Flavia area, on the 'Femminello comune' variety of lemon tree. The trees were about 30 years old, and throughout that whole period were given just one anti-parasite oil-based treatment, in the autumn of 1981. For trends in the phytophage population, we referred to the records of the I.O.B.C. Working Group on citrus red scale and citrus Aleyrodes, which were adopted by the CEC group of experts on the standardization of biotechnical methods of integrated pest control in citrus orchards (1980) with a few modifications as indicated below.

---

* Studies by the CNR Working Group on the integrated control of animal plant pests, No. 259.

2.1. Fruit

Out of 20 trees chosen at random in an orchard covering approximately
2 ha, situated in an area completely covered with citrus orchards, 20
fruits on each tree were observed at monthly intervals, at the rate of 5
fruits for each point of the compass. In total, 400 fruits were monitored
each time.

The fruits were broken down into four classes as follows:

Class	No. of A. aurantii observed
0	0
1	1 - 3
2	4 - 10
3	more than 10

2.2. Branches

At the same rate as for the fruit, four branches were taken, each two
years old, for each of the trees designated. Each branch was 10 cm long,
and one branch was taken for each point of the compass. The surfaces of
these branches were exmamined in the laboratory and the individual numbers
of insects of various ages were recorded separately. In all, 800 cm of
branches were exmained each time, at the rate of 40 cm per tree.

2.3. Leaves

The leaves of the above-mentioned branches were collected, their
surface area was measured, and either the upper or lower surface was
examined under the microscope in the laboratory in order to determine the
incidence of A. aurantii in the various stages.

2.4. Parasitoids

In order to monitor the degree of attack by parasites, at monthly
intervals between September 1981 and May 1983, fruits showing the most
signs of infestation were chosen. These were ripe or nearly so, and were
examined in the laboratory at the rate of 3 areolae per fruit, each
approximately 2 cm^2 in area, to determine the percentage of active
parasites on the phytophages present, which were in various stages of
growth, i.e. in the second stage, young females and male prepupae and pupae
(De Bach and White, 1960; De Bach, 1969; Rosen and De Bach, 1979; Baker,
1976; Orphanides, 1982).

The individual specimens attacked by parasites were isol.ted in petri
dishes which were moistened as required with gypsum and animal carbon, or
in gelatin vessels, so that the adults of the Aphelinidae could leave their
cocoons.

The examination was confined to the fruits, on which infestation was
more constant. On the leaves and branches, because of the forcing technique, the population was sparse and unevenly distributed.

52

3. RESULTS

3.1. Pattern of infestation on fruit

The incidence of infestation decreased between the begining of the observation period (March 1981) and the end (April 1983). The percentage figure dropped from over 90% to an average of approximately 30% in 1982 and 1983.

It must be said that the fruits were considered to be infested if they exhibited specimens of phytophages, regardless of the numbers of these.

A statistical analysis did not reveal any significant differences between the percentage of infestation on fruits exposed to different points of the compass (Figs. 1 - 5). Figure 6 shows the percentage ratio between fruits in the four classes.

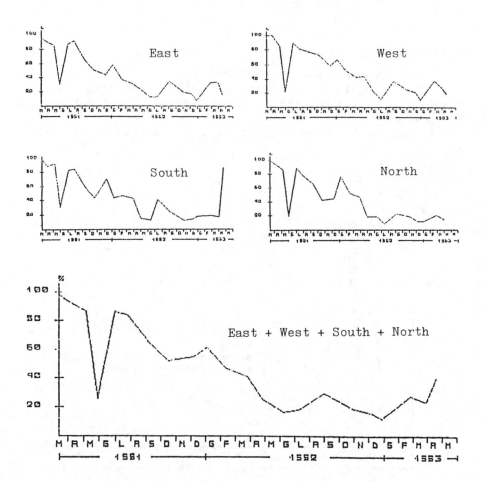

Figs. 1-5: <u>Aonidiella aurantii</u> (Mask.) Percentage of infested fruit for the four points of the compass (1-4) and overall pattern (5).

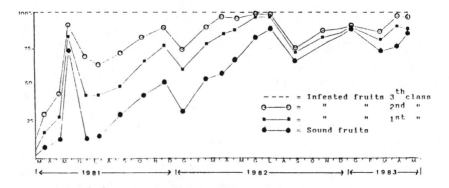

Fig. 6: <u>Aonidiella aurantii</u> (Mask.) Percentage ratio between sound and infested fruits.

3.2. <u>Infestation on branches</u>

On the branches the degree of infestation observed in 1981 was one specimen of <u>A. aurantii</u> per linear cm.   In the next two years this figure
dropped to substantially lower levels, averaging less than 0.05 specimens
per linear cm.

The young in the first stage of growth were always the most numerous
(Fig. 7).

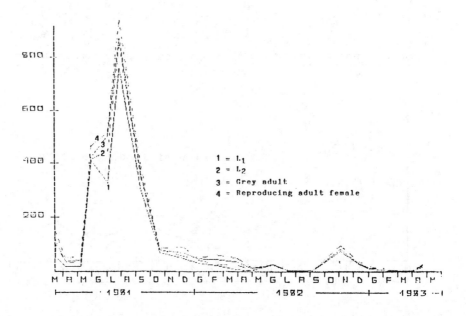

Fig. 7: <u>Aonidiella aurantii</u> (Mask.) Number of specimens in the
various stages observed on 800 linear cms of branches.

3.3. Infestation on leaves

The degree of infestation on the leaves was very low in all cases. However, there were always more specimens of A. aurantii on the upper surfaces
than on the lower.

On 400 cm^2 of leaf surface the maximum number of specimens observed in
1981 was approximately 1/cm^2 on the upper surfaces and approximately
0.1/cm^2 on the lower surfaces. In subsequent years practically no specimens were observed on either the upper or lower surfaces (Figs. 8 - 9).

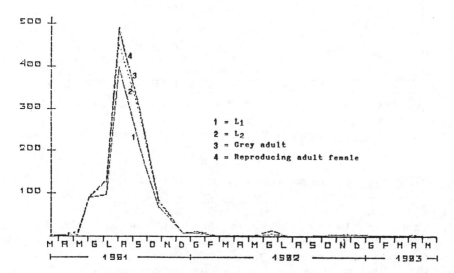

Fig. 8: Aonidiella aurantii (Mask.) Number of specimens in all stages
observed on 400 cm2 of upper surfaces of lemon tree leaves.

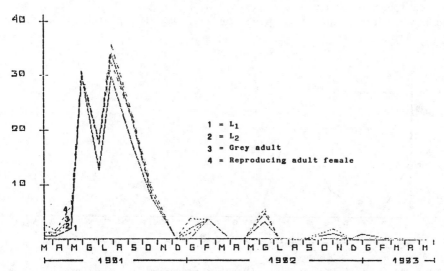

Fig. 9: Aonidiella aurantii (Mask.) Number of specimens in all stages
observed on 400 cm^2 of lower surfaces of lemon tree leaves.

55

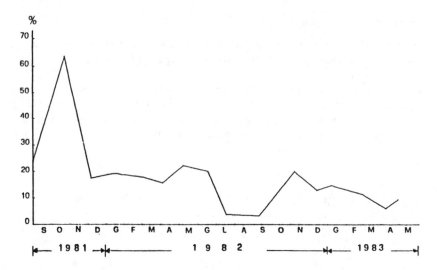

Fig. 10: <u>Aonidiella aurantii</u> (Mask.) Percentage of parasitical action
on available females.

3.4. <u>Parasitoids</u> In the course of the survey two species of ectophage
parasitoids were recorded, viz. <u>Aphytis chrysomphali</u> (Mercet) and <u>Aphytis
melinus</u> (De Bach).

On the available females (young at the second stage and young females)
the degree of initial parasitical action due to the two ectophages together
was very high, exceeding 60% in October 1981 (Fig. 10), whereas in the following years it remained at an average level of approximately 15%, except
for the period July-August 1982, when there was a fall in the percentage of
parasitical action due to the hot sultry weather in June of that year.

On the male prepupae and pupae the percentage of parasitical action
was not constant, which was also the case for the phytophage population
(Fig. 11).

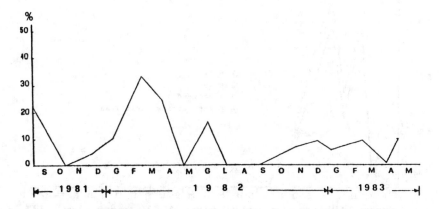

Fig. 11: <u>Aonidiella aurantii</u> (Mask.) Percentage of parasitical action
on male prepupae and pupae.

56

4. CONCLUSIONS

The high degree of infestation recorded on the fruit at the beginning of the survey period (March 1981) should be seen in relation to the fact that the phytophage was new in the area and thus had not encountered any hostile agents to inhibit its development. Subsequently, the effect of the two Aphelinidae, assisted by a mineral-oil treatment (in autumn 1981), brought about a certain equilibrium and stablilized the degree of infestation at around 20%. However, it must be said that the fruits were considered to be infested even if they exhibited only one specimen. In reality, in addition to a reduction in the percentage of infested fruit, there was also a substantial drop in the number of specimens found on each fruit.

No statistically significant differences were found between fruits exposed to the different points of the compass.

In addition, the sparse population observed on the leaves and branches must probably be seen in relation to the adverse conditions of humidity in which these parts of the tree were placed because they were being forced-grown.

In conclusion, the findings are in line with other reports (Atkinson, 1977), i.e. that the red scale was present mainly on the fruits, to a much lesser extent on the branches, and only rarely on the leaves.

With regard to the parasitical action of the Aphelinidae, following a very high initial level (above 60%), this stabilized at around 15%, which is lower than the figure recently recorded in Cyprus (Orphanides, 1982).

The action of each of the two Aphelinidae was not studied separately. However, in general the A. melinus was the more active. This species, which was introduced into Sicily by Inserra in 1966, consequently appears to have found optimum conditions on the island for its acclimatization.

The total level of parasitical action, although lower than that recorded in other regions where citrus fruits are grown, such as Morocco (Abbassi, 1980), South Africa (Atkinson, 1977), Australia (Campbell, 1976), and Cyprus (Orphanides, 1982), is nevertheless substantial.

If to the parasitical action we add the "host-feeding" action of the females (Rosen and De Bach, 1979), we can see that the part played by the Aphelinidae in containing A. aurantii is far from negligible.

REFERENCES

1. ABBASSI M. (1980). Recherche sur deux Homoptères fixés des citrus, Aonidiella aurantii Mask. (Homoptera, Diaspididae) et Aleurothrixus floccosus Maskell (Homoptera, Aleurodidae). Cah. rech. agronom., 35, Rabat.
2. ATKINSON P.R. (1977). Preliminary analysis of a field population of citrus red scale, Aonidiella aurantii (Maskell), and the measurement and expression of stage duration and reproduction for life tables. Bull. Ent. Res., 67; 65-87.

3. BAKER J.L. (1976). Determinants of host selection for species of Aphytis (Hymenoptera: Aphelinidae) parasites of diaspine scales. Hilgardia, 44; 1-25.
4. BARBAGALLO S. and NUCIFORA A. (1981). Sampling methods and economic threshold for the control of lemon pests in Italy. In: Standardization of biotechnical methods of integrated pest control in citrus orchards - S. Giuliano (F) - Siniscola (I) 4-6 Nov. 1980, Eds. R. Cavalloro and R. Prota. EUR/7342; 27-34.
5. DE BACH P. and WHITE E.B. (1960). Commercial mass culture of the California red scale parasite Aphytis Lingnanensis. Bull. Calif. Agric. Expt. Stn. 770; 2-58.
6. DE BACH P. (1969). Biological control of diaspine scale insects on citrus in California. In: H.D. Chapman (Ed.), Proc. Ist. int. Citrus Symp., 2; 801-815.
7. INSERRA S. (1966). Introduzione e acclimatazione di due Aphytis (A. melinus De Bach ed A. lingnanensis Compere) parassiti ectofagi di alcune cocciniglie degli agrumi. Tecn. Agric., 18(2): 176-183.
8. LIOTTA G. (1970). Diffusion des cochenilles des agrumes en Sicile et introduction d'une nouvelle espèce en Sicile occidentale. Al Awamia, 37; 33-38.
9. ORPHANIDES G.M. (1982). Biology of the California red scale, Aonidiella aurantii (Maskell) (Homoptera, Diaspididae), and its seasonal availability for parasitization by Aphytis spp. in Cyprus. Boll. Lab. Ent. agr. Filippo Silvestri - Portici, 39; 203-212.
10. ROSEN D. and DE BACH P. (1979). Species of Aphytis of the World (Hymenoptera: Aphelinidae). W. Junk. BV Publ. - The Hague, Boston, London.

Biological features of *Pseudococcus affinis* (Mask.) (Homoptera, Pseudococcidae) as guidelines of its control in water-sprinkled citrus orchards

A.Panis

INRA, Zoological & Biological Control Station, Laboratory 'E.Biliotti', Valbonne, France

Summary

Pseudococcus affinis (Maskell) is a new pest of French Riviera citriculture, particularly damageous to industry of luxury clementines from 1978
to 1984 . Incomplete control with usual scalicides (methidathion, white
oil) needs a revision of citrus protection plan, on biological grounds
stated in special agronomical context of local citrus plantings . Observations during summer scalicide applications show that cloud-like chemical
product mainly covers leaves, twigs and fruits whereas most of bugs
inhabiting trees are on crownwoods . As a result of biological study, two
behavioural features, i.e. sheltering and crawling, explain partial
inefficacy of sprays and leads sampling plan available for setting up
patterns of within-tree spatial distribution,which can contribute to
improve control of the mealybug and to choose adequate dates of intervention . Repeated applications of insecticides against the new pest, up to
4 per year, are costly and alternative to present artificial rain as
water-supply system, both with modification or replacement of chemical
control programme, is discussed .

1. Introduction

Pseudococcus affinis (Maskell) is known as outbreaking sporadically
or frequently in some subtropical citrus areas . It was stated to occur as
major or minor pest, under wet subtropical climates of China, North Aegean
and Black Sea coasts (1 ; 3) and mediterranean-type climates of California
and Crimea (2 ; 10 ; 7) . According to present status of Pseudococcus
maritimus-malacearum taxonomical complex (4), the citrus mealybug of French
Riviera, hitherto named Pseudococcus maritimus (Ehrhorn) (8 ; 9 ; 6) refers
to P. affinis . The bug might be old established with its habitus
erroneously attributed to Planococcus citri (Risso), but signs in this way
are insufficient .

2. Agronomical context where biology of the mealybug is studied

Economic interest of the pseudococcine appeared from 1978 to 1984 in
localities of Alpes-Maritimes and Var departments, where maritime influence
and local exposure seem to give the highest atmospheric humidities of the
region during summer (more than 65 % R.H. at midday as a july mean) .
Additional favouring factors obviously are summer water-supply and dense
canopy along windscreen or trees far from plantation edges . Except in
Porquerolles Island, off Hyères (Var), where it was observed in 1978 and
1979 on sour orange (bigarade), never scale was found again on this

host plant . It began as minor pest of clementine, then it became the most injurious coccid of this citrus in some farms and it spread to sweet orange, mandarin, lemon, lime, kumquat . All citrus trees inhabited by the pseudococcine were either with dense foliage or dense planting and frequent irrigations from june to october . The most favourable environment which was encountered, with permanent outbreaks on five years, is old clementine-trees in wind-protected situation, with summer water-supply by sprinkling top of the canopy .

Observations on sweet orange, clementine, lemon, lime show that, during all the year, scales are found under small bark-splits and within cracks of stemwood and branchwoods . It is noticeable that bug cannot stay for long time on leaves, except where they touch twig or another leaf . When living on lemon "Eureka" or on some varieties of sweet orange, coccid infests sooner the young fruits than do on clementine . Aspect of lemon skin, navel of some oranges or relatively more appropriate morphology of fruit sepales can explain earlier fruit-infestations . However on clementine, scales shelter first beneath fruit sepales at the end of june and in july and then against sepales . Such preliminar results of observations tend to assume that sheltering is a main component of behaviour of P. affinis on citrus and that chlorophyllous organs (smooth young twigs, leaves and fruits) are as favourable for feeding as woods . Scale localization is obviously influenced by rugosity of epidermal tissues, factor to take into account for sampling plan of the pest .

Even as many mealybug species, difficulties arise when farmer tries to obtain high efficacy of a scalicide on evergreen trees . But maximum chemical expenses seem to be reached in a citrus farm at La Gaude (Alpes-Maritimes), on right bank-terraces of Var river, where main economical end is to produce fruits of high gustative quality (lemon, limes, kumquats, clementine) . Sooty mould induced by the coccid, need fruit brushing and washing, preventing clementines to be saled with two attached leaves (a quality mark of french clementine production) . Moreover, farmers opinion is that scale increases percentage of small fruits, split clementines and discoloured fruit skins . Heavy infestations on fruits are early in autumn till to end of gathering, according to the grower . Before 1978, two to three chemical sprays were needed against the various pests . Then, mealybug infestations were tentatively controlled by four interventions more, using methidathion, white petroleum oil or both as scalicides . From 1978 to 1983, farmer decided to try various dates of applications, with first and last dates respectively as follows : early august-late october ; midmarch-late october, late may-late september . The two others occured between july and september . The best efficacy but incomplete control of pest was reached in 1983, with following dates of sprays against it : 24 th May, 20 th July, 16 th August, 26 th september .

The same grower allocated 15 clementine-trees 15 years old for our experiments, without acaricide and insecticide applications (only keeping bouillie bordelaise in march), from june 1981 to november 1983 . A tree-belt two rows wide surrounded such an experimental plot, preventing pesticide cloud derive when remaining of the 2 hectares clementine-parcel was treated . Uncontrolled plot was two tree-rows far from a northwestern windscreen of cypress . All plantations of the farm are at a southeastern exposure and receive from june to october, an artificial rain sprinkled at 3 metres above ground, on the top of canopy . Except when significant rainfall occurs, water-supply is as follows : 10 mm on the 1st day, 25 mm on the 4th, 10 mm on the 8th, 25 mm on the 11th, 10 mm on the 14th, and so on, so that weakly irrigation amounts 40 mm, all through fruit growth season .

3. Biological features of P. affinis on clementine-trees

Whatever rugosity (i.e. thygmotactism commonly found among mealybugs) may play the main role in bug localization, microclimatic differences may exist between bottom and top of tree-foliage, particularly when rain on canopy causes gradients of temperature and relative humidity within crown, during many hours after it ceases . If moderate to high atmospheric humidity is required, for survival, any data is available for its action and that of temperature on developmental rate of P. affinis living on citrus . It is only known that the insect population of Black Sea Coast can easily survive to negative winter temperatures .

Behavioural features of bug were specified in experimental plot of La Gaude . Combination of temperature and relative humidity gradients within the crown are assumed to be responsible of frequent displacements along stem, branches and twigs of trees, till end of june to early october. Maximum of crawlers occurs some time about noon of summer days . Bugs stay about sunset . Crawlers were unusually numerous in july and august 1981, an exceptional hot summer in France . Their first steps are markedly delayed on days with artificial rain on trees . Except eggs of course, and egg-laying females, all instars are potential crawlers, with third instar larva and female by far the most mobile . Numerous enough of them fall on the ground and stand beneath various shelters, where they may be predated . Few of them shall reach again the trunk . All around the year, even in summer, a lot of bugs shelter in cracks, small hollows and splits of trunk and lowest branchwoods, just as in interstices between pruning cuts of branches and mastic covering cuts, at low and mid-level of the crown . This part of insect population is more steady during summer, than those coloni- zing fruits and twigs . But after moulting, bugs tend to leave their shelter and to enter in another one after a more or less long period of crawling . Even do so the virgin females, climbing upward the canopy when mature, for mating with alate males . Then they go to lay eggs in anfractuosities of barks, in pruning cross-sections, on twigs and fruits in summer and autumn months . As said below, combination of sheltering and crawling conditions seasonal patterns of within-tree spatial distribution .

Generation number is four, may be five a year, overlapping during hotest months confusing life history which would need accurate countings for its better knowledge . Third instar larvae and females are overwin- tering stages . They remain in anfractuosities mainly of B1 and B'1 strata (see legend of fig.1), i.e. trunk and bottom of branchwood . Eggs of first generation appear in april-early may, those of second one between mid-june and early july . In july and august, third and fourth generations overlap. Unfortunately, it lied outside of three years-observations to specify when they begin . Hot weather seems to delay larval development of both generations even as egg hatching of the fourth . Increase of population density which appears on fig.1, between june and august,(1982, 1983) results of low egg-hatching rate for last individuals of the third generation and for whole the fourth ones . Between july and august, decrease of popula- tion density on mid-crown of the trees (see B2 and F1-F3 strata on fig. 1) is obviously linked to active moving of bugs towards more temperate and humid B1 and B'1 strata . If it truly exists, a new egg-laying period about mid-september would constitute beginning of a fifth generation . These eggs will lead to overwintering old larvae and females . It seems that high mortality occurs among crawling scales in august as they delay to stay for feeding, so as among steady bugs inhabiting top of the crown in november, because they are not able to migrate towards wood-shelters .

61

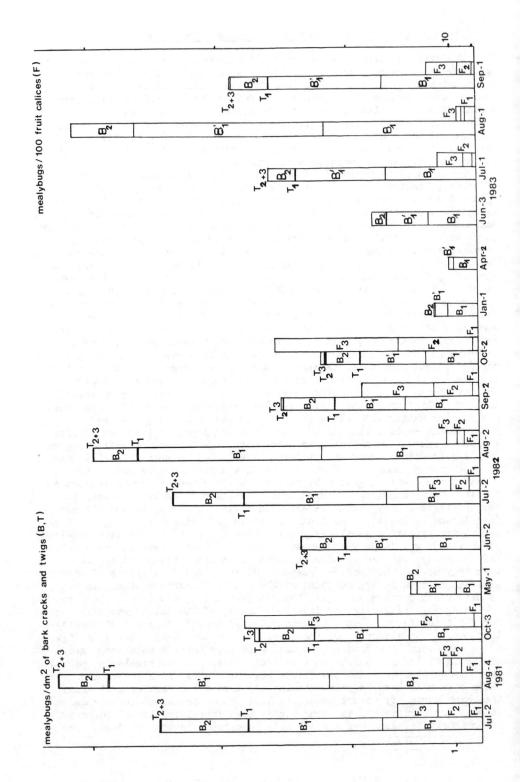

mealybugs / 100 fruit calices (F)

mealybugs / dm² of bark cracks and twigs (B,T)

4. Within-tree spatial distribution of mealybug on clementine-trees

Taking into account behavioural features of P. affinis described above, it is easier to define strata in the tree and a sampling plan. Explanations on strata and sampling plan in the 15-trees experimental plot are included legend of fig.1 . Four quadrats (north, east, south, west) partitioning the nine strata were retained in 1981 and 1982 and do not appear on legend of fig.1 and graphical representation , because any pattern and density differences do occur on studied clementine-trees . Surfaces of leafy twigs are obtained by considering twig as a cylindrical surface and by computing clementine-leaf surface from measurement of length (L) and width (l) of individual leaves, applying Onillon's formula (5) :

$$S = 0.707 \, L^{0.9392} \, x \, 1^{0.9893}$$. Fifteen samples from july 1981 to september 1983 establish seasonal patterns of mealybug distribution as an average over 15 clementine-trees (fig.1) . Such patterns show that scales are very few or missing on twigs all through year, and permanently average to high on wood parts from may to october, so as on fruits, except in august, from july to october (properly, till gathering) .

Figure 1 . Mean within-tree distribution of uncontrolled Pseudococcus affinis population infesting clementine plantation, sprinkled with water during fruit growth (july-october) .

Weak of month is numbered after month .
Individuals of all instars of development (an egg-laying female or living eggs within an egg-sack counted as 1) numbered on non-removed samples of 15 trees .
Each tree partitioned into 9 strata and insect numbers averaged per dm^2 of stratum surface or per 100 calices for fruit strata .
Stem and branchwood bottom with B1 stratum : bark cracks, hollows or splits of the trunk and branchwood fork (up to 0.8-1 metre ; sample unit surface ranged in 0.01 to 0.5 dm^2 as approximate assessment of anfractuosity size .
Crown from the lower part of foliage up to 1 m with,
B'1 stratum : bark cracks and interstices between pruning cuts and their mastic covering, of branchwoods more than 3 cm in diameter ; sample unit surface ranged in 0.01 to 0.1 dm^2 as approximate assessment of anfractuosity size,
T1 stratum : twigs (excluding beared fruits on july-october) ; leaf plus twig surfaces computed from measurements (20 sample unit per tree),
F1 stratum : fruits (before july, scales under calice then against sepales and unfrequently between two fruits or fruit and leaf) ; 20 fruits per tree .
Crown between 1 and 2 m, with,
B2 stratum (as B'1),
T2 stratum (as T1),
F2 stratum (as F1) .
Crown from 2 m to the top with,
T3 stratum (as T1),
F3 stratum (as F1) . Branches more than 3 cm in diameter are insignifiant stratum) .

5. Guidelines for control device of P. affinis

When sprinkler-engine is working, it is observed that whole canopy is easily covered by insecticide cloud, droplets of which penetrate the foliage . However, unless sprayer should be introduced at level of bottom-fork by the worker, as it was done really by farmer cited above (a dangerous practice for health), central woods of the crown remain incompletely covered of chemical . Many scales are not killed, particularly on pruning cuts where mastic is rarely sticking closely to cross sections . Samething occurs within bark cracks situated at downward side of branch fork . Chemical penetration is better in bark cracks of trunk, easier to wet . On the other hand, waxy cover of mealybugs insure them some good protection and white oil, despite effects on quality of clementines, as it has been reported, would increase effectiveness of methidathion . As claimed by farmer, chemical control of P. affinis is time spending, it needs high volume of liquid per hectare and frequent interventions are costly . Pesticide may have a good penetration within wood anfractuosities . As follows of biological observations, instead of canopy covering in march, winter spray localized to crownwoods might be tried .

Sprinkling water on top of the canopy is known as doing gustative quality contribution to lemon, lime, kumquat under climatic conditions of French Riviera, and as improving yield of clementine-trees . Lower level of infestation was reported in citrus plantings where water was sprinkled at ground level . It is suggestive to experiment on yield and quality of fruits with this kind of water-supply, arguing that four additional applications of insecticide in citrus protection programme is costly . Integrated control system fitted to French Riviera citriculture is to be established . Few farmers seem favourable to biological control of mealybugs on citrus . Two entomophagous insects were released in august 1981 on the 15 experimental trees : Nephus reunioni Chazeau, a coccinellid native from Transvaal and cultured at Valbonne Station, and Pseudaphycus maculipennis Mercet, an indigeneous encyrtid very effective in natural control of other populations of P. affinis in the region, collected from momies of the mealybug found in next clementine and lemon plots . Essay is not conclusive though coccinellid had overwintered, it then disappeared during spring 1982 . Despite intensive chemical control, P. maculipennis is maintaining and seems to be increasing from 1981 to 1983 .

REFERENCES

1. CHEN, F. and WONG, F. (1936) . A list of the known fruit insects of China . 1935 Year Book Bur.Ent. Hangchow 5, 82-140 .
2. CLAUSEN, C.P. (1915) . Mealy bugs of citrus trees . Univ. Calif. Publ. Coll. Agric., Agric. Exp. Sta. Berkeley Bull. 258, 19-48 .
3. GOGUIBERIDZE, M.P. (1938) . Cochenilles de la R.S.S. de Géorgie subtropicale humide . Soukhoum, 89 pp. (in russian) .
4. MILLER, D.R., GILL, R.J. and WILLIAMS, D.J. (1984) . Taxonomic analysis of Pseudococcus affinis (Maskell), a senior synonym of Pseudococcus obscurus Essig, and a comparison with Pseudococcus maritimus (Ehrhorn) (Homoptera : Coccoidea : Pseudococcidae) . Proc. Entomol. Soc. Wash. 86(3), 703-713 .
5. ONILLON, J.C., FRANCO, E. et BRUN, P. (1973) . Contribution à l'étude de la dynamique des populations d'homoptères inféodés aux agrumes . Estimation de la surface des feuilles des principales espèces d'agrumes cultivées en Corse . Fruit 28(1), 37-38 .

6. ONILLON, J.C., PANIS, A.et BRUN, P. (1984) . Summary of the studies and works carried out in the framework of the programme on integrated control in citrus fruit groves against Aleyrodes and Lecaninae and Pseudococcinae scale-insects . - in C.E.C. Programme on integrated and biological Control, Final Report 1979/1983, Eds. R. Cavalloro and A. Piavaux, EUR 8689 Bruxelles, 89-101 .

7. OUMNOV, M.P. (1940) . Cochenilles (Coccidae) de quarantaine et autres, en Crimée . Simféropol, 64 pp. (in russian) .

8. PANIS, A. (1980) . Dégâts de Coccidae et Pseudococcidae (Homoptera, Coccoidea) des Citrus en France et effets particuliers de quelques pesticides sur l'entomocénose du verger . Fruits 35(12), 779-782 .

9. PANIS, A. (1981) . Note sur quelques insectes auxiliaires régulateurs des populations de Pseudococcidae et de Coccidae (Homoptera, Coccoidea) des agrumes en Provence orientale . Fruits 36(1), 49-52 .

10. SMITH, H.S. and ARMITAGE, H.M. (1931) . The biological control of mealybugs attacking citrus . Agric. Exp. Sta. Calif. Bull. 509, 1-74 .

The citrus aphids: Behaviour, damages and integrated control

S.Barbagallo & I.Patti
Institute of Agricultural Entomology, University of Catania, Italy

Summary

Citrus in the world are affected by several aphids of different economic importance; the most relevant species are the "brown" and the "black" aphids Toxoptera citricidus (Kirk.) and T. aurantii B.d.F., respectively , the "green
citrus aphid" (Aphis citricola v.d.G.) and the "cotton aphid" (A.gossypii Glov.).
Their geographical distribution, biology, damages (including their ability in the
transmission of citrus tristeza virus),effectiveness of natural enemies, economic thresholds and integrated control measures are briefly exposed.

1. Introduction

Among the citrus pests, aphids represent a group of remarkable economic importance, owing not only to the direct damages, but above all to the indirect ones,
due to the risk of natural spreading of the tristeza virus (CTV), which is one of the
most harmful citrus diseases.

The aphids that affect citrus groves in the world are little less than about twen
ty species (see tab. I), but only a few of them (4 species) have major economic importance.

In the Mediterranean basin about 14 aphid species may be recognized on citrus
but, fortunately, in such area the brown aphid, Toxoptera citricidus (Kirk.),hasn't
been found yet. In fact such aphid is, as it is well known, the most efficient vector
of citrus tristeza virus.

Literature about citrus aphids in the world is supported by several papers, and
the species identification can be tried following the morphological description or
keys available on the single species or groups of them (Essig, 1911; Ebeling, 1959;
Stroyan, 1961; Barbagallo, 1966a; Eastop, 1966; Raychaudhuri, 1980; Patti, 1983).

2. Epidemiology

The development of the aphid colonies on citrus depends strictly on the presen
ce of tender shoots, and consequently the highest density of population is reached,
in the Mediterranean climate, late in spring (May-June). A second period of infesta
tion usually takes place between the end of summer and earl Autumn (September).

Aphids living on citrus are polyphagous species (rarely olygophagous), and
with a prevalent anholocyclic behaviour. Consequently their survival in winter may
happen with the presence of small colonies on occasional tender shoots of the same
citrus trees; but usually they move to other host plants, mostly herbaceous, which
grow in the surrondings of the citrus groves. Thus such alternative plants may play
an important epidemiological role during summer and winter periods, when citrus

Tab. I - Conspectus of aphids recorded on citrus in the world.

Aphid	Life cycle	Range of host plant	Geonemy	Transmission of CTV
A. Main species				
Aphis citricola v.d.G.	anholocyclic & holocyclic	polyphagous(mostly Rosaceae & Compositae)	cosmopolitan	+
Aphis gossypii Glov.	anholocyclic & ?holocyclic	very polyphagous	cosmopolitan	+ +
Toxoptera aurantii (B.d.F.)	anholocyclic	polyphagous	cosmopolitan	+
Toxoptera citricidus (Kirk.)	anholocyclic	mostly Rutaceae	Southern hemisphere & Far East	+ + + +
B. Minor species				
Myzus persicae (Sulz.)	holocyclic & anholocyclic	very polyphagous	cosmopolitan	+
Macrosiphum euphorbiae (Thom.)	anholocyclic	very polyphagous	cosmopolitan	-
Sinomegoura citricola (v.d.G.)	? anholocyclic	polyphagous	South-East Asia & Australia	-
Aulacorthum magnoliae (Ess.& Kuw.)	anholocyclic	polyphagous	South-East Asia & Australia	-
Aulacorthum solani (Kalt.)	anholocyclic	very polyphagous	cosmopolitan	-
Aphis craccivora Koch	anholocyclic	polyphagous (mostly Leguminosae)	cosmopolitan	+
Aphis fabae Scop.	anholocyclic & holocyclic	polyphagous (mostly Leguminosae)	cosmopolitan	-
Toxoptera odinae (v.d.G.)	anholocyclic	polyphagous	South-East Asia & central Africa	-
C. Occasional species				
Rhopalosiphum maidis (Fitch)	anholocyclic	mostly Graminaceae	cosmopolitan	?
Hyadaphis coriandri (Das)	anholocyclic	mostly Umbelliferae	Mediterranean Countries Africa & Asia	-
Aphis nerii B.d.F.	anholocyclic	mostly Asclepiadaceae & Apocynaceae	cosmopolitan	-
Brachycaudus helicrysi (Kalt.)	anholocyclic & holocyclic	polyphagous	cosmopolitan	-
Brachycaudus cardui (L.)	anholocyclic	mostly Compositae	holarctic	-
Pterochloroides persicae (Chol.)	anholocyclic	Rosaceae Prunoideae	Eastern Eurasia	-

plants are not susceptible of infestation, having no tender shoots. The aphid infestation on citrus, therefore, starts with the beginning of the vegetable activity through alates coming from other host plants where they overwintered. Some species, however, may develop an holocycle, like Aphis citricola v.d.G. or Myzus persicae (Sulzer), which overwinter in such cases as eggs on Spiraea and Peach tree, respectively. Anyway citrus represent the "secondary host" in the Mediterranean area as well as in most other citrus countries in the world; but Komazaki et al. (1979) recorded the ovoposition of A. citricola, A. gossypii and T. citricidus on citrus in Japan.

3. Citrus aphids in the Mediterranean area

3.1 Main species

The aphids particularly harmful to citrus in the Mediterranean area belong to three species. The "black citrus aphid", Toxoptera aurantii (B.d.F.) is the most typical one, living on all the species of citrus and their hibrids indifferently. It has a cosmopolitan geonemy and is known all over the citrus growing countries in the world, particularly in the tropical and subtropical areas (Comm.Inst.Ent.,map n.131); its diffusion ranges within the 45° of North and South latitude, then spreading little beyond the citrus growing area. In colder climates it can survive only in protected environments, like greenhouses, inside which virginopare can overcome winter.

The aphid, in fact, is exclusively anholocyclic and evolves only through conti̲nuous generations of virginoparae; amphigonics have never been found in nature. In the Mediterranean district the host range of T. aurantii seems to be rather scanty, while it increases in the tropical areas; a list of its host plants has been related by Essig (1949) and by Bodenheimer (1951).

The optimum for the insect growth is provide by temperatures of about 22°–25°C, with extreme activity values between 7° and 32°C. In the best environmental conditions the post–embryonal development takes 6–7 days, and each adult apterous can, on an average, generate 70 nymphs (Rivnay, 1938). The direct damages of the black aphid involves all the citrus species; on orange, mandarin and similar citrus species damages reach an intermediate incidence among those of the two following aphids (A. citricola and A. gossypii).

The "green citrus aphids", Aphis citricola v.d.G. (= A. spiraecola Patch) is thought to be of nearctic origin and appeared for the first time almost 20 years ago on the citrus groves of the Mediterranean countries; its present diffusion is almost cosmopolitan (Comm.Inst.Ent., map n. 256), though confined to not very wide districts of the various continents. Presently the aphid represents the most harmful species for orange, mandarine and clementine trees, both in Italy and in the other citrus growing Countries of the Mediterranean area; the lemon tree is almost never infested by this species. The green citrus aphid is, however, polyphagous, living on various species of Dicotyledons, above all in warm climates, tropical and subtro̲pical areas, where it usually performs as anholocycle; at constant temperature the aphid development reached the maximum increase rate at about 27°C (Komazaki, 1982). Besides the more frequent anholocyclic behaviour, the aphid can perform a holocycle and in such a case, primary host is represented by Rosacee of the genus Spiraea (Barbagallo, 1966). The discovery of the aphid oviposition on the same citrus trees observed in Japan (Komazaki et al., 1979) is to be considered quite exceptional. The "cotton aphid" Aphis gossypii Glover is a species considered by

69

most Authors permanently anholocyclic all the world over; but still in Japan Komazaki et al. (1979) have pointed out the holocycle, with citrus as primary host. The species has a cosmopolitan geonemy (see Comm.Inst.Ent., map n. 18) and is highly polyphagous; the most typical host plants of this aphid are Cucurbitaceae, Malvaceae, Rutaceae and others. Its infestations on the citrus groves affect the same species damaged by the green aphid, that is orange,clementine, mandarine trees and similar ones, while it is rarely found on the lemon tree. The colonies of the cotton aphid cause limited deformations to the shoots of the citrus infested plants, the leaves of which develop almost regularly, leaving them nearly underformed; however high infestations are harmful to the affected plants because of the loss of sap and nutritional components.

3.2 Minor species

In addition to the previous species, other polyphagous aphids, generally of secondary economic importance, are found on the citrus groves of the Mediterranean basin. Their infestations, even if with different degrees according to the species they belong to, are almost always of moderate extent, limited in time and, except very rare cases, they do not need any control measure. One of the most frequent of such aphids is the polyphagous green peach aphid, Myzus persicae (Sulzer), that colonizes young shoots of different citrus species above all the orange clementine and mandarine, in the early spring period. On these plants it is found in many citrus growing countries of the world; in Italy its presence is occasional and of limited importance, but in Spain it has been lately pointed out (Melia, 1982) as one of the most harmful aphids on mandarine tree.

A similar behaviour is displayed by two other polyphagous species, Macrosiphum euphorbiae (Thomas) and Aulacorthum solani (Kalt.) whose colonies are often found on very tender shoots. Also the two black aphids Aphis craccivora Koch and A. fabae Scop. can infest tender orange shoots and may generate, particularly the former, dense colonies on the under side of the leaves, often mixed with those of other citrus aphids (i.e. A. gossypii and T. aurantii, etc.).

We can seldom find in Italy colonies of two other greenish aphids, Rhopalosiphum maidis (Fitch) and Hyadaphis coriandri (Das), the former of which has been detected in North America citrus groves, too (Essig, 1911, sub Aphis cookii Ess.).

Finally other aphid species have been now and them found on citrus trees by different Authors; one of them is Aphis nerii B.d.F., the yellow oleander aphid, mentioned by Bodenheimer and Swirski (1957) for the Middle East and recorded at the end of the last century also in Riviera Ligure on the orange tree by Ferrari (1872, sub Myzus asclepiadias Pass.). The presence of Brachycaudus helicrysi (Kalt.) on citrus trees is referred only by sporadic records in Spain, India and France, and in this latter Country also B. cardui (L.) has been found (Leclant, 1978).

Hall (1926) mentions also the presence of Pterochloroides persicae (Chol.) on orange tree even if it is hard to expect that this dendrophilus species, linked to Rosacee Prunoidee, can really colonize the citrus plants.

4. Other citrus aphids out of the Mediterranean Countries

Among the exotic aphids harmful to citrus groves, the brown aphid, Toxoptera citricidus (Kirk.), excels for economic importance; it is greatly dangerous not so much for the direct damages it causes, comparable with those of T. aurantii but par ticularly for the indirect ones, owing to its high efficiency as a vector of the citrus tristeza virus. Should it accidentally be introduced into the Mediterranean area, where it is till now absent, it would provoke very harmful effects.

At present the Sahara desert may provide a natural barrier to the spreading of such aphid from the infested African countries towards the Mediterranean area, but it might find other entrance possibilities through the continuous commercial activities, as passive transport on infested plants or part of them, evading the phytosanitary controls.

Presently this aphid is widespread all over the citrus growing countries of the Southern hemisphere, below the cancer tropical line, though in the Far East (North Indian district, China, Korea, Japan) it goes as far as abouth the 40° of north latitude (Comm.Inst.Ent., map n. 132). Its ecological needs (as regards the climatic conditions and the host plants) seem to be compatible with the ones that can be found in the Mediterranean area; hence the necessity to keep it off as long as possible.

T. citricidus has a biological and ethological behaviour (Symes, 1924; Komazaki, 1982) similar to T. aurantii, but is less polyphagous, living mainly on the various species of Citrus and on other plants, mostly of the family Rutaceae. Morphological separation between T. citricidus and T. aurantii does not show any difficulties and be made on the basis of some differences both in apterous morphs and in the alatae ones (Stroyan, 1961; Tao, 1961; Eastop, 1966).

Another species of the same genus, Toxoptera odinae (v.d.G.), has much less economic importance. It is widespread is several countries of the South-East Asia (India, China, Korea, Japan, the Philippines) and lately it has been introduced into central Africa (Burundi); it lives on Rutaceae (Citrus spp.) as well as on other several genera (Anacardium, Coffea, etc.) of different families (Tao, 1961; Remaudiere e Autrique, 1984). Finally, two more species, of exotic aphids found on the citrus trees are the two Macrosiphinae Sinomegoura citricola (v.d.G.) and Aulacorthum magnoliae (Ess.et Kuw.). The former is an Oriental species (Asiatic South-East; China, Japan, Philippines, Formosa, Indonesia, India), but is also present in New Guinea and in Australia (Eastop, 1966); the latter aphid, A. magnoliae, is confined to the Far East: Japan, Korea, Indian region (Anonimo, 1981; Raychaudhuri, 1980). Both species are polyphagous aphids but are frequently found infesting the citrus trees.

5. Damages

The citrus aphids are responsible of direct and indirect damages. The former are highly correlated to the aphid species and to the density of its population, as well as to the species and the age of the infested citrus trees. Damages consist of leaves deformation and a reduced development of the infested shoots; such injuries have a negative effect mostly on the growth of young trees. In addition, the aphid infestation to the flowers and the very young fruits, may cause their drop. In this connection the most harmful aphid, among the three main species present in the Mediterranean area, is A. citricola as it infest clementine, orange and mandarin groves, while damages caused by T. aurantii and A. gossypii are less injurious.

Indirect damages are caused by the excretion of honeydew, on which, sooty-mould fungi rise later. But the greatest threat of the aphid infestation on citrus is undoubtedly represented by their efficience as vector of citrus virus diseases and particularly of the "tristeza" (including the strains "stem pitting" and "seedling yellows"). The most efficient vector of this destructive citrus disease is T. citricidus, the presence of which is responsible of the quick spreading that such virus has had since its appearance in the citrus growing areas of the Southern hemisphe re.

A direct correlation between the spreading of this aphid and the presence of the disease in the different citrus growing countries of the world is clear enough. Nevertheless such aphid is not the only responsible of the transmission of the tristeza virus since other species, among those infesting citrus plants, can transmit the virus (tab. I). A. gossypii, to this concern, has proved to be relatively efficient in transmission test carried out in various citrus growing district (Norman and Grant, 1959; Varma et al., 1965; Roistacher et al., 1984); the same possibility, though of minor extent, has been pointed out for A. citricola and T. aurantii in Florida (Norman and Grant, 1957); M. persicae and A. craccivora, as well as Uroleucon (= Dactynotus jaceae (a species not feeding on citrus) have been also recorded in India (Varma et al., 1965). The different transmission efficiency of such aphids proves to be susceptible of changes in relation to biological (i.e. virus strains and isolates, host plant) and climatic factors (Bar-Joseph et al., 1977; Roi stacher et al., 1984).

As a consequence it can be said that aphid species normally considered as inef ficient vectors may give, in favorable circumstances, positive response to infection.

In Italy, preliminary tests of transmission of a virus strain from Japan by means of four aphids (A. citricola, A. gossypii, A. craccivora, T. aurantii), gave negative results (Cartia et al., 1980); but further studies carried out with A. citricola, A. gossypii, A. fabae, M. persicae and T. aurantii, and an isolate (T4) of CTV from Florida, have pointed out a fairly efficiency as vector (3 positive tests out of 22) of A. citricola (Davino & Patti, 1985).

At last it is useful to note that A. gossypii, M. persicae and T. citricidus are also vectors of the virus which cause the vein enation-woody gall disease of citrus (Wallace, 1975).

6. Effectiveness of natural enemies

The citrus aphids, with reference to the Mediterranean countries and in parti cular to the Italian territory, are often kept under natural biological control by se veral parasites and predators (Barbagallo & Patti, 1982; Patti, 1983).

Relevant efficiency has the Hymenopterous group of the Aphidiinae. Among the eleven species of such parasite wasps recorded in Italy on citrus aphids, the most interesting, for the effectiveness shown, are Lysiphlebus testaceipes (Cr.), L. fabarum (Mash.) and L. confusus Tr. et Ed.; these wasps may reach a degree of para sitism of up to 90-100% on the aphids T. aurantii and A. gossypii (Tremblay et al., 1980). Particularly interesting is L. testaceipes, of nearctic origin, artificially in troduced into Corsica and Mediterranean France (Stary et al., 1977) and spontaneously widespread in the Italian citrus areas, where it has shown a high parasi tic pressure towards the two mentioned citrus aphids. The same aphidiphagous spe cies is also able to parasitize T. citricidus (De Huiza & Ortiz, 1981).

Unfortunately L. testaceipes, or at least the strain by now spread in the Mediterranean area does not succeed in parasitizing A. citricola, and at present this aphid is not biologically controlled by any endophagous species in the Mediterranean (Tremblay & Barbagallo, 1982).

Among the citrus aphid predators, the Coccinellids (several species of Coccinellinae and Scymninae) are the most important, followed by the Syrphids, the Chrysopids and finally by the Cecidomyids.

On the whole the biological control by means of parasites and predators may reach a high degree of effectiveness, to keep the aphid populations at low levels of density; in the Italian citrus groves this case of effective biological control involves the two aphid T. aurantii and A. gossypii. Scarcely controlled, on the contrary, is A. citricola, against which attempts were suggested for the introduction of exotic entomophagous, in order to improve its biological control and avoid chemical sprays, at time necessary to control its population.

7. Economic thresholds and supervised control

Principles of integrated pest management in the citrus agro-ecosystem should be applied to the aphid control as well as to other injurious insects. For such purpose a preliminar valuation of some operative parameters is worth being done; they concern the density of aphid population (i.e. percentage of attacked shoots and extension of the colonies on them), the aphid species, the affected citrus groves (i.e. species involved and age of the host plants), and, finally, the density of entomophagous among the aphid colonies.

The matter is then rather complex, though, generally speaking, the necessity of chemical control in the Mediterranean area is required mainly in the case of the green citrus aphid infestation on young groves and more rarely for other aphid species.

The experts of C.E.C. have suggested the sampling methodology and pointed out the economic thresholds for the principal aphid species infesting the citrus groves in the Mediterranean area (Cavalloro & Prota, 1983). From such suggestions weekly controls kept on the 10% of the trees of the grove are adviced, and the following economic thresholds were proposed:
- for T. aurantii: 25% of infested shoots (a similar level of thresholds may be applied to A. gossypii);
- for A. citricola and M. persicae: 10% of infested shoots for oranges and 5% for clementines.

When the intervention threshold is overcome – and contemporaneously the aphidiphagous insects among the aphid colonies are absent or scarcely present – the application of an aphicide spray is suggested. As to the chemical products the use of a selective and non-persistent aphicide (i.e. pirimicarb, ethiofencarb, ...) should be chosen, in order to reduce the side-effects of the treatment on the beneficial entomofauna.

REFERENCES

1. ANONIMO (1981). Major citrus insect pests in Japan. Laboratory of Entomology (Okitsu Brank), Japan Plant Prot. Ass., Tokyo, 16 pp..
2. BARBAGALLO, S. (1966). Contributo alla conoscenza degli afidi degli agrumi. I – Aphis spiraecola Patch. Boll.Lab.Ent.agr. Portici, 24: 49-83.
3. BARBAGALLO, S. (1966a). L'afidofauna degli agrumi in Sicilia. Entomologica, 2: 201-260.
4. BARBAGALLO, S. and PATTI, I. (1982). Citrus aphids and their entomophagous in Italy. In: Cavalloro R. (Ed.), Aphid antagonists, A.A. Balkema, Rotterdam (1983): 116-119.

5. BAR-JOSEPH, M., RACCAH, B. and LOEBENSTEIN, G. (1977). Evoluation of the main variables that affect citrus tristeza transmission by aphids. Proc. Int.Soc.Citricult., 3: 958–961.
6. BODENHEIMER, F.S. (1951). Citrus Entomology in the Middle East. Dr W. Junk (publ.), The Hague, 663 pp..
7. BODENHEIMER, F.S. and SWIRSKY, E. (1957). The Aphidoidea of the Middle East. Weizmann Sci.Press, Jerusalem, 378 pp..
8. CARTIA, G., BARBAGALLO, S. and CATARA, A. (1980). Lack of spread of citrus tristeza virus by aphids in Sicily. Proc. 8th Conf. I.O.C.V.: 88–90.
9. CAVALLORO, R. and PROTA, R. (1983). Integrated control in citrus orchards: sampling methodology and threshols for intervention against the principal phytophagous pests. Commission European Communities, Brussels-Luxembourg, 63 pp..
10. COMMONWEALTH INSTITUTE OF ENTOMOLOGY. Distribution maps of pests, n. 131: Toxoptera aurantii (Boy.) (1961); n. 132: T. citricidus (Kirk.) (1961); n. 18 (revised): Aphis gossypii Glover (1968); n. 256: A. spiraecola Patch (1969).
11. DAVINO, M. and PATTI, I. (1985). Preliminary results of citrus tristeza virus transmission by aphids in Italy (This volume).
12. DE HUIZA, J.R. and ORTIZ, P.M.S. (1981). Algunos Aphidiinae (Hymenopt.: Braconidae) parasitoides de afidos (Homopt.: Aphididae) en el Perù. Revista Peruana de Entomologia, 23: 129–132. See R.A.E., 70 (1982), n. 3824.
13. EASTOP, V.F. (1966). A taxonomic study of Australian Aphidoidea (Homoptera). Aust.J.Zool., 14: 399–592.
14. EBELING, W. (1959). Subtropical fruit pests. Univ. of California, Los Angeles, 436 pp..
15. ESSIG, E.O. (1911). Aphididae of Southern California. VIII. Plant lice affecting the citrus trees. Pomona J. Ent., 3: 586–603.
16. ESSIG, E.O. (1949). Aphids in relation to quick decline and tristezia of citrus. Pan-Pacific Ent., 25: 13–23.
17. FERRARI, P.M. (1878). Aphididae Liguriae. Ann.Mus.Civ.Storia Nat.Genova, 2: 49–85.
18. HALL, W.J. (1926). Notes on the Aphididae of Egypt. Bull.Min.Agric. Egypt, n. 68: 1–62.
19. KOMAZAKI, S. (1982). Effects of constant temperatures on populations growth of three aphid species, Toxoptera citricidus (Kirkaldy), Aphis citricola van der Goot and Aphis gossypii Glover (Homoptera: Aphididae).
20. KOMAZAKI, S., SAKAGAMI, Y. and KORENAGA, R. (1979). Overwintering of aphids on citrus trees. Japanese J. Appl.Ent. & Zool., 23: 246–250 (in Japanese) (see R.A.E. (1980), n. 5082).
21. LECLANT, F. (1978). Etude bioécologique des aphides de la région méditerranéenne. Implications agronomiques. Acad. de Montpellier, Univ. des Sc. et Tech. du Languedoc. Thèse, 2 voll., 318 + XLII pp..
22. MELIA, A. (1982). Prospeccion de pulgones (Homoptera, Aphidoidea) sobre citricos en Espagna. Bol.Serv. Plagas, 8: 159–168.
23. NORMAN, P.A. and GRANT, T.J. (1957). Transmission of tristeza virus by aphids in Florida. Proc.Fla.Hort.Soc., 69: 38–42.
24. NORMAN, P.A. and GRANT, T.J. (1959). Transmission of T3 severe strain of tristeza virus of citrus by the melon aphid. J.econ.Ent., 52: 632–634.

25. PATTI, I. (1983). Gli afidi degli agrumi. Pubbl. C.N.R. collana del P.F. "Promozione della qualità dell'ambiente", AQ 1/231, Roma, 63 pp..

26. RAYCHAUDHURI D.N. (1980). Aphids of North-East India and Bhutan. The Zoological Society, Calcutta, 521 pp..

27. REMAUDIERE, G. and AUTRIQUE, A. (1984). Toxoptera odinae (van der Goot) puceron asiatique recemment decouvert an Afrique. C.R. Acad.Agr. de France, 70: 379-385.

28. RIVNAY, E. (1938). Factors affecting the fluctuations in the population of Toxoptera aurantii Fonsc. in Palestine. Ann.appl.Biol., 25: 143-154.

29. ROISTACHER, C.N., BAR-JOSEPH, M. and GUMPF, D.J. (1984). Transmission of tristeza and seedling yellows Tristeza Virus by small population of Aphis gossypii. Plant Disease, 68: 494-496.

30. STARY, P., REMAUDIERE, G. and LECLANT,P. (1977). Nouveaux complements sur les aphidiidae (Hym.) de France et leurs hotes. Annls.Soc.ent.Fr. (N.S.), 13: 165-184.

31. STROYAN, H.L.G. (1961). Identification of aphids living on Citrus. FAO Plant Prot. Bull., 9: 45-65.

32. SYMES, C.B. (1924). Notes on the black citrus aphids. Rhodesia Agric.J., 21: 612-626 & 725-737. See R.A.E., (1925), pg. 78-79.

33. TAO, C.C. (1961). Revision of the Genus Toxoptera Koch, 1856 (Homoptera: Aphididae). Qtly. J.Taiwan Mus., 14: 257-260.

34. TREMBLAY, E., BARBAGALLO, S., MICIELI DE BIASE, L., MONACO, R. and ORTU, S. (1980). Composizione dell'entomofauna parassitica vivente a carico degli Afidi degli Agrumi in Italia (Hymenoptera Ichneumonoidea, Homoptera Aphidoidea). Boll.Lab.Ent.Agr.Portici, 37: 209-216.

35. TREMBLAY, E. and BARBAGALLO, S. (1982). Lysiphlebus testaceipes (Cr.) a special case of ecesis in Italy. In Cavalloro R. (Ed.), Aphid antagonists, A. A. Balkema, Rotterdam (1983): 65-68.

36. VARMA, P.M., RAO, D.G. and CAPOOR, S.P. (1965). Transmission of tristeza virus by Aphis craccivora (Koch) and Dactynotus jaceae (L.). Indian J. Ent., 27: 67-71.

37. WALLACE, J.M. (1975). Vein enation - woody gall. In: Description and illustration of virus and virus-like diseases of Citrus. SETCO-IFAC (Ed.), 6 pp..

Seasonal evolution of *Myzus persicae* (Sulz.) (Homoptera, Aphidoidea) with relation to citrus fruit trees

A.Melia
Service for Plant Protection & Phytopathological Inspection, Castellòn de la Plana, Spain

Summary

The seasonal evolution of Myzus persicae (Sulz.)was studied from 1977 to 1982, proving that it exists both in its holocyclical form, alterna- ting host plants, and in its anholocyclical form, resulting in the possible survival of certain clones or species on secondary host plants. This explains the phenomena of resistance observed on citrus fruit trees. A total of 32 host plants – 29 of these being secondary hosts – were identified and listed.
On the peach tree the laying of winter eggs begins in the middle of Oc- tober and ends at the end of December. Hatching starts at the beginning of January and finishes at the beginning of March.
Although the winged insects coming from secondary host plants attack the peach tree in a very limited manner, their offspring is incapable of reproducing.
Three periods of flight have been noted, but it is the winged species of the first flying period (February–March) that attack citrus fruit trees the most.

1. Introduction

Myzus persicae (sulzer) is an aphid that attacks a large range of plants in Spain more than one hundred different kinds – not only causing extensive damage directly but also transmitting numerous viruses.

This is a species that has always been quoted as being typically holo- cyclical, as having the Prunus as its preferred host plant followed by a wide range of very varied plants as secondary hosts. However, several authors have affirmed that this pest can also be anholocyclical, spending the winter as parthenogenetic females on vegetable crops (cruciferae, potatoes, beets, etc.) and weeds (8).

Myzus persicae was first located on citrus fruit trees in 1975 in Cas- tellón (5), later in 1980 in Valencia, Castellón and Alicante (3) and in a study made on citrus fruit trees in Spain, has been identified in Alicante, Almeria, Cadiz, Castellón, Santa Cruz de Tenerife, Sevilla, Tarragona and Valencia, and has a preference for Citrus reticulata (6).

This green fly is Known for the ease which it becomes resistant to in- secticides, a very widespread phenomenon reported in numerous articles pu- blished in U.S.A., Europe, Far East and Australia (7). Given its preference for trees bearing the clementine species – a very common variety – and the increasing inefficiency of products used to treat this pest, the latter could represent a serious problem for our citrus fruit trees.

The various aspects of the biology of this aphid , which help to account for its resistance to insecticides, have been studied in this report.

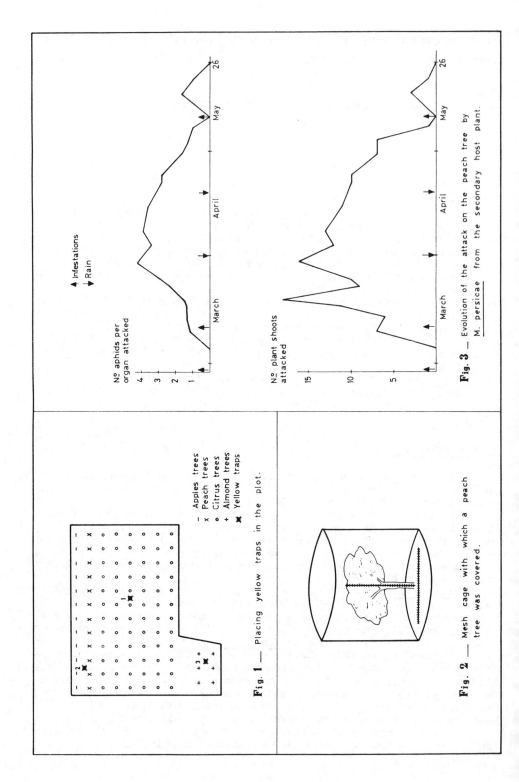

Fig. 1 — Placing yellow traps in the plot.

- Apples trees
x Peach trees
o Citrus trees
+ Almond trees
✹ Yellow traps

Fig. 2 — Mesh cage with which a peach tree was covered.

▲ Infestations
▼ Rain

Nº aphids per organ attacked

Nº plant shoots attacked

Fig. 3 — Evolution of the attack on the peach tree by M. persicae from the secondary host plant.

78

2. Material and methods

2.1. The Winter laying and hatching of eggs.

For several years (1977–82), samples of peach trees – consisting of 10 shoots each 30 cm. long – have been taken, and the weekly evolution of the laying and hatching of eggs in winter observed.

2.2. Host plants

Both cultivated and wild plants have been observed at all times of the year in order to discover whether they had been attacked by M. persicae or not, and if so to what degree.

2.3. Primary host plants infested by specimens from secondary host plants.

At the beginning of September 1981 a peach tree was enclosed in a mesh cage with two zips (fig. 2); one to allow placing the tree, and the other for access. The winter laying had not yet started and, moreover, it was treated with an insecticide (pirimicarb) to ensure that no egg laying would take place on the tree. Tests were carried out to prove this.

In the laboratory, specimens of M. persicae which had been collected from potatoes at the beginning of November 1981, were cultivated on potted Vicia faba. The plant pots were placed inside the cage in order to record the peach tree being infested. For this a count of the parts of the flowers and plants which had been attacked, was taken periodically and compared with another peach tree placed in the open air.

2.4. Study of Migrations.

Three yellow traps, each with a diameter of 26 cm., were placed at a height of 1'20 m. from the ground. One was put amongst citrus trees, another on a peach trees and the third amongst almond trees (Fig. 1). The aphids that were caught were counted and the species identified each week.

3. Results

3.1. Laying and hatching of winter eggs

Table I below gives the results obtained showing the start and finish of both the winter laying and the hatching of eggs on the peach tree.

Table I – The winter laying and incubation of M. persicae eggs.

Year	Laying		Hatching	
	Start	Finish	Start	Finish
1.977	15–X	27–XII	–	–
1.978	31–X	12–XII	10–I	7–III
1.979	20–X	26–XII	9–I	27–II
1.980	14–X	28–XII	2–I	26–II
1.981	23–X	7–XII	7–I	2–III
1.982	–	–	12–I	8–III

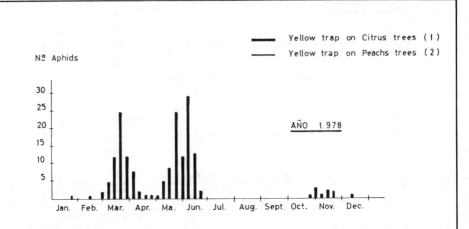

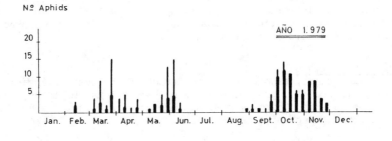

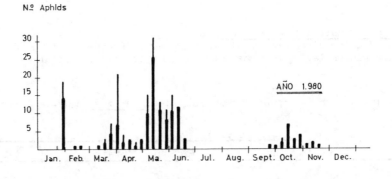

Fig. 4 ___ The trapping of winged M. persicae during the years 1978 to 1980

80

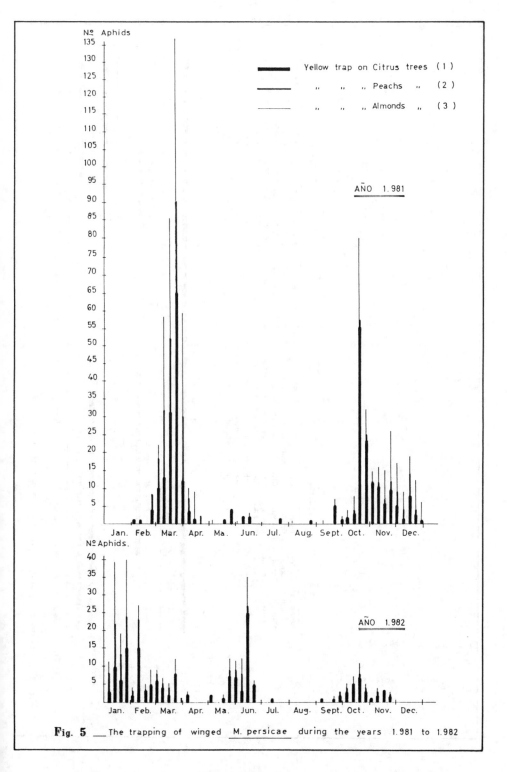

Nº Aphids

Yellow trap on Citrus trees	(1)
„ „ „ Peachs „	(2)
„ „ „ Almonds „	(3)

AÑO 1.981

Jan. Feb. Mar. Apr. Ma. Jun. Jul. Aug. Sept. Oct. Nov. Dec.

Nº Aphids.

AÑO 1.982

Jan. Feb. Mar. Apr. Ma. Jun. Jul. Aug. Sept. Oct. Nov. Dec.

Fig. 5 ___ The trapping of winged M. persicae during the years 1.981 to 1.982

81

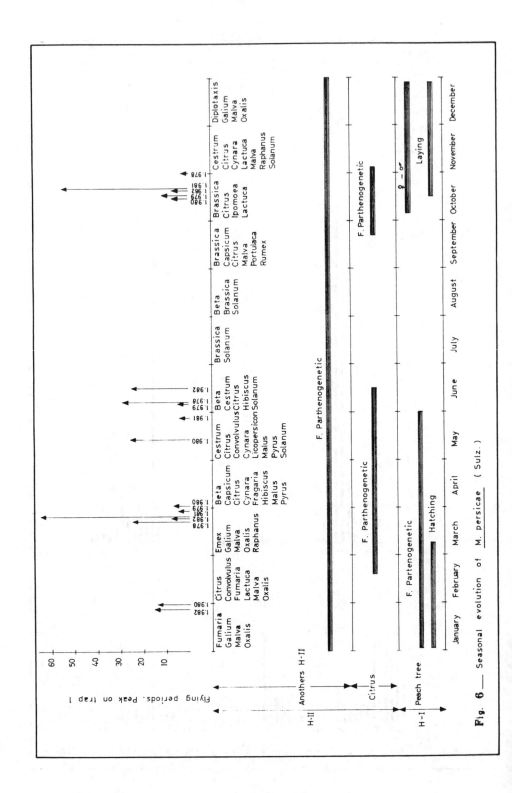

Fig. 6 — Seasonal evolution of _M. persicae_ (Sulz.)

3.2. List of Host Plants

The names of the different host plants on which M. persicae has been found throughout the twelve months of the year, are listed below:

January: Fumaria sepium Boiss et Reuter, Galium aparine L., Malva neglecta Wallr., Oxalis pes-caprae L.

February: Citrus reticulata Blanco, Convolvulus arvensis L., Fumaria sepium Boiss et Reuter, Lactuca sativa L., Malva neglecta Wallr, Oxalis pes-caprae L.

March: Emex spinosa (L.) Campd., Galium aparine L., Malva neglecta Wallr., Oxalis pes-caprae L., Prunus dulcis (Miller) D.A. Webb, Prunus persica (L.) Batsch, Raphanus sativus L.

April: Beta vulgaris L., Capsicum annuum L., Citrus reticulata Blanco, Citrus sinensis Osbeck, Citrus unshiu Marcovitch, Cynara scolymus L., Fragaria ananassa Duchesne, Hibiscus rosasinensis L., Malus domestica Borkh., Prunus avium L., Prunus dulcis (Miller) D.A. Webb., Prunus persica (L.) Batsch, Pyrus communis L.

May: Cestrum nocturnum L., Citrus aurantium L., Citrus limon Burman, Citrus reticulata Blanco, Citrus sinensis Osbeck, Citrus unshiu Marcovitch, Convolvulus arvensis L., Cynara scolymus L., Lycopersicon esculentum Miller, Malus domestica Borkh., Prunus persica (L.) Batsch, Pyrus communis L., Solanum melongena L., Solanum tuberosum L.

June: Beta vulgaris L., Cestrum nocturnum L., Citrus reticulata Blanco, Citrus sinensis Osbeck, Hibiscus rosasinensis L., Solanum tuberosum L.

July: Brassica oleracea L., Solanum melongena L.

August: Beta vulgaris L., Brassica oleracea L., Solanum nigrum L.

September: Brassica oleracea L., Capsicum annuum L., Citrus reticulata Blanco, Citrus sinensis Osbeck, Malva sylvestri L., Portulaca oleracea L., Rumex acetosa L.

October: Brassica oleracea L., Citrus reticulata Blanco, Citrus sinensis Osbeck, Ipomoea baratas (L.) Lam. Lactuca sativa L., Prunus persica (L.) Batsch.

November: Cestrum nocturnum L., Citrus reticulata Blanco, Citrus sinensis Osbeck, Cynara scolymus L., Lactuca sativa L., Malva neglecta Wallr., Prunus persica (L.) Batsch, Raphamus sativus L., Solanum tuberosum L.

December: Diplotaxis erucoides (L.) DC., Galium aparine L., Malva neglecta Wallr., Oxalis pes-caprae L.

3.3. Primary host plant infested by specimens from secondary host plant

On 26th February 1982 two plant pots with large quantities of winged aphids were put into a cage already containing a peach tree. Another two plant pots were put there on the 10th March and on 10th May yet another two were placed inside.

The results of the greenfly count are given in table II.

3.4. Study of migration

The weekly results of the fly catching in the yellow traps are given below (table III).

1. Yellow trap on orange trees.
2. Yellow trap on peach tree.
3. Yellow trap on almond tree.

Table II - Infestation of peach tree by specimens from H-II

Date	Phenology	Nº floral organs attacked	Nº plant shoots attacked	Total nº organs attacked	Nº aphids per organ attacked
26-II-82	B	-	-	-	-
4-III-82	B/C	-	-	-	0'00
9-III-82	B/C/D	7	-	7	1'14
10-III-82	B/C/D				
13-III-82	C/D	6	-	6	1'33
16-III-82	C/D/E	11	-	11	1'36
18-III-82	D/E	18	-	18	1'50
22-III-82	D/E/F	9	-	9	2'22
24-III-82	D/E/F	10	-	10	2'80
29-III-82	F	16	-	16	4'31
3-IV-82	F/G	10	2	12	3'41
7-IV-82	F/G	11	2	13	3'92
14-IV-82	G/H	7	4	11	3'63
21-IV-82	H	4	6	10	2'80
23-IV-82	H/I	3	7	10	2'80
29-IV-82	I	-	7	7	1'71
3-V-82	I	-	7	7	1'28
7-V-82		-	1	1	1'00
10-V-82		-	-	-	0'00
17-V-82		-	3	3	1'66
21-V-82		-	1	1	1'00
26-V-82		-	-	-	0'00

4. Comments and conclusions

It was noted that the period of winter laying was from mid October to the end of December. The hatching period lasts from the beginning of January to the beginning of March.

During the year M. persicae was identified on the following 32 plants; 29 of these are secondary hosts.

Beta vulgaris L.
Brassica oleracea L.
Capsicum annum L.
Cestrum nocturnum L.
Citrus aurantium L.
Citrus limon Burman
Citrus reticulata Blanco
Citrus sinensis Osbeck
Citrus unshiu Marcovitch
Convolvulus arvensis L.
Cynara scolymus L.
Diplotaxis erucoides (L.)DC.
Emex spinosa (L.) Campd.
Fragaria ananassa Duchesne
Fumaria sepium Boiss. et Reuter
Galium aparine L.

Hibiscus rosasinensis L.
Ipomoca batatas (L.) Lam.
Lactuca sativa L.
Lycopersicon esculentum Miller
Malus domestica Borkh
Malva neglecta Wallr.
Malva sylvestri L.
Oxalis pes-caprae L.
Prunus avium L.
Prunus dulcis (Miller) D.A. Webb.
Prunus persica (L.) Batsch
Pyrus communis L.
Raphanus sativus L.
Solanum melongena L.
Solanum nigrum L.
Solanum tuberosum L.

Some years, during the flying period of M. persicae (February—March), winged specimens were observed on the peach tree, producing offspring on small flowers and leaves when the winter eggs were hatching. This has led to the belief that the peach tree could have been infested by specimens from a secondary host plant.

It was reported that in South Africa peach trees might have been infested by swarms of anholocyclical winged aphid (2). In France, however, it was reported that specimens from a secondary host plant can reproduce on peach trees, but are unable to continue reproducing (4). In France the infestations were very late (April—May) and so the tests were carried out earlier (February—March) and also in May, in order to establish whether the vegetative state of the peach tree during the first season was acceptable to M. persicae coming that the invasion that this species produces is very limited and although they could reproduce, the offspring could not continue reproducing. On the peach tree in the open air, however, the level of infestation was high.

It has been proved that three flying periods exist:

a) The first occurs at the end of winter and some years can be divided into two phases. During this stage the peak flying period is the second fortnight of March. In 1.980 and 82 there was another peak period at the end of January; it is necessary to note, however, that the catching of the winged M. persicae continued throughout the winter of 1981—82.
   The first flying period is by winged insects from a secondary host plant as it is at this time when both the hatching of the winter laying and the first infestations are taking place on peach trees are still no winged insects from this host plant itself. This flight is the one which produces infestations in citrus fruit trees.

b) The second period of flight reaches its peak at the end of May or the beginning of June. It is produced basically be specimens from a primary host plant and although it can produce infestations on citrus fruit trees, the latter are very limited as the peak flying period coincides with the time when M. persicae are beginning to cease their attack on citrus fruit trees.

c) The third flying period reaches its peak during mid October. This is the return flight from the secondary host plant to the primary host plant for the egg laying.

After consideration of these facts it can be said that M. persicae exists both in its holocyclica form alternating the host plant (I: peach tree and other Prunus, II: numerous other plants) and in its anholocyclical form with parthenogenetic reproduction all year round, resulting in the possible survival of certain clones or strains on secondary host plants during the winter.

Blackman (1) indicated the existence of clones of M. persicae of 4 different types; holocyclical — with sexual reproduction; anholocyclical with asexual reproduction; androcyclical with the production of some males together with continued parthenogenesis; and intermediate with production of winged specimens in a category between ginopars and virginopars. Although there is no evidence to prove the existence of the androcyclical and intermediate clones it is believed that they do in fact exist.

It can be said that the swarms of M. persicae on citrus fruit trees come from:

- Insects that spend the winter on the same citrus fruit trees. This source is of little importance.

Table III – Winged caught of M. persicae on yellow traps

FECHA	1	2	3	T	FECHA	1	2	3	T	FECHA	1	2	3	T	FECHA	1	2	3	T
3-1-79	-			-	2-1-80	-			-	7-1-81	-			-	5-1-82	3	6	3	11
10-1-79	-			-	8-1-80	-			-	13-1-81	-			-	12-1-82	10	12	17	39
17-1-79	-			-	15-1-80	-			-	20-1-81	-			-	19-1-82	5	7	5	19
23-1-79	1			1	22-1-80	-			-	27-1-81	-			1	26-1-82	15	9	15	40
30-1-79	-			-	29-1-80	14	5		19	3-2-81	1			1	2-2-82	2	1	4	4
6-2-79	-			-	5-2-80	1			1	10-2-81	1			1	9-2-82	15	9	4	27
13-2-79	1	1		1	12-2-80	-			3	17-2-81	-			1	16-2-82	3		2	5
20-2-79	2			-	19-2-80	1			1	24-2-81	4	2	2	8	23-2-82	5	1	3	9
27-2-79	2			2	26-2-80	1			-	3-3-81	10	9	4	22	2-3-82	5	2	1	9
6-3-79	5	3		5	4-3-80	1			4	10-3-81	13	18	27	58	9-3-82	4		2	7
13-3-79	12	6		12	11-3-80	1			9	17-3-81	31	21	33	85	15-3-82	2	2	1	5
20-3-79	25	1		25	18-3-80	2			2	24-3-81	65	25	47	137	23-3-82	8		4	12
27-3-79	12	10		12	25-3-80	4	3		15	31-3-81	12	18	29	59	30-3-82			1	1
3-4-79	8	4		8	1-4-80	7	14		21	7-4-81	3	4	3	10	6-4-82	2	1		3
10-4-79	2	3		2	8-4-80	3	2		4	14-4-81	1	1	7	9	13-4-82	-			-
16-4-79	1	1		1	15-4-80	3	2		5	21-4-81	1	2		2	20-4-82	-			-
25-4-79	1	2		1	22-4-80	1	1		4	28-4-81	-			-	27-4-82	2			2
30-4-79	1	2		1	29-4-80	3			1	5-5-81	-	1		1	4-5-82	-			-
7-5-79	5	1		5	6-5-80	10	5		15	12-5-81	-			1	11-5-82	-		1	2
14-5-79	9	3		9	13-5-80	26	5		31	19-5-81	4			4	18-5-82	1		1	2
22-5-79	25	2		25	20-5-80	11	2		13	26-5-81	-			-	25-5-82	7	2	3	12
29-5-79	12	4		12	27-5-80	8	3		11	2-6-81	2		1	3	1-5-82	7	2	2	11
5-6-79	29	5		29	3-6-80	11	4		15	9-6-81	2		1	3	8-6-82	3	5	4	12
12-5-79	13	10		13	10-5-80	12			12	16-5-81	2		1	1	15-5-82	25	2	3	35
19-5-79	2	2		2	17-5-80	2			3	23-5-81	-			-	22-5-82	5			6
27-5-79	-			-	24-5-80	-			-	30-5-81	-			-	29-5-82	-			-
4-7-79	-			-	1-7-80	-			-	7-7-81	-			-	6-7-82	-			-

86

FECHA	1	2	3	T	FECHA	1	2	3	T	FECHA	1	2	3	T	FECHA	1	2	3	T	FECHA	1	2	3	T
11-7-78	-	-		-	10-7-79	-	-		-	8-7-80	-	-		-	14-7-81	-	-		-	13-7-82	1	-	-	1
18-7-78	-			-	17-7-79	-			-	15-7-80	-			-	21-7-81	-			-	20-7-82	-		-	-
25-7-78	-			-	24-7-79	-			-	22-7-80	-			-	28-7-81	-			-	27-7-82	-		-	-
1-8-78	-			-	31-7-79	-			-	29-7-80	-			-	4-8-81	-			-	3-8-82	-		-	-
8-8-78	-			-	7-8-79	-			-	5-8-80	-			-	11-8-81	-			-	10-8-82	-		-	-
15-8-78	-			-	14-8-79	-			-	12-8-80	-			-	18-8-81	-			-	17-8-82	1		-	1
22-8-78	-			-	21-8-79	1			1	19-8-80	-			-	25-8-81	1			1	24-8-82	-		-	-
29-8-78	-			-	28-8-79	1			1	26-8-80	-			-	1-9-81	-			-	31-8-82	-		-	-
5-9-78	-			-	4-9-79	1	1		1	2-9-80	-			-	8-9-81	-			-	7-9-82	1		-	1
12-9-78	-			-	11-9-79	1	1		1	9-9-80	-			-	15-9-81	-			-	14-9-82	-		-	-
19-9-78	-			-	18-9-79	-	1		1	16-9-80	-			-	22-9-81	5		2	7	21-9-82	1	1	1	2
25-9-78	-			-	25-9-79	3	2		5	23-9-80	-			-	29-9-81	1	-	1	2	28-9-82	2	1	-	3
3-10-78	-			-	2-10-79	10	2		12	30-9-80	1			1	6-10-81	2		2	4	5-10-82	3	1	1	5
10-10-78	-			-	9-10-79	11	3		14	7-10-80	1			1	13-10-81	3	1	4	8	12-10-82	5		2	7
17-10-78	1				15-10-79	11	1		12	14-10-80	2	1		3	20-10-81	55	2	22	79	19-10-82	6	1	3	11
24-10-78	3			3	23-10-79	5	1		6	21-10-80	7			7	27-10-81	23	2	7	32	26-10-82	3	1	1	5
31-10-78	1				30-10-79	5	1		6	28-10-80	3			3	3-11-81	12		3	15	2-11-82	1	-	-	1
7-11-78	2			2	6-11-79	9			9	4-11-80	4			4	10-11-81	11	1	4	15	9-11-82	2	1	-	4
14-11-78	2			2	13-11-79	9			9	11-11-80	1			1	17-11-81	6	1	8	15	15-11-82	3	-	-	3
21-11-78	-				20-11-79	4			4	18-11-80	2			2	24-11-81	10	2	14	26	23-11-82	1	1	-	3
29-11-78	-				27-11-79	2			2	25-11-80	1			1	1-12-81	5	4	8	17	30-11-82	1	-	-	-
5-12-78	1				4-12-79	2			2	2-12-80	-			-	8-12-81	5	3	5	9	7-12-82	-	-	-	-
12-12-78	1			1	11-12-79	-			-	9-12-80	-			-	15-12-81	8	6	5	19	14-12-82	-	-	-	-
19-12-78	-				18-12-79	-			-	15-12-80	-			-	22-12-81	3	6	9	12	21-12-82	-	-	-	-
25-12-78	-				25-12-79	-			-	23-12-80	-			-	29-12-81	1	0	5	6	29-12-82	-	-	-	-
										30-12-80	-			-										

- Winged specimens from the first flying period (February-March) that come from other secondary host plants. This is considered to be the most important of all and because of this, the control of this flying period by yelbw traps or suction could be a good way to predetermine the importance of the attack of this kind of aphid on citrus fruit trees.

- Winged specimens from the second flying period (May-June) that come from the primary host plant. This flying period reaches its peak when the swarms of this aphid start to leave the citrus fruit trees. Because of this, the importance of this flight as a supplier of specimens which contaminate citrus fruit trees is minimal.

That is to say, citrus fruit trees are mainly infested by the clones with anholocyclical forms, which could explain the presence of phenomena of resistance to insecticides. As there is no sexual reproduction and therefore no genetic segregation in their offspring either, the characteristic of resistance, produced by the pressure of insecticide tratment to which the citrus fruit trees and other secondary host plants are continually subjected must be transmitted over the years.

REFERENCES

1. BLACKMAN, R.L., (1.971) Variation in the photoperiodic response within natural populations of Myzus persicae (Sulz.).- Bull. ent. Res. 60, 533-546.

2. DAIBER, C.C. (1.963) Notes on the host plants and winged dispersal of Macrosiphum euphorbiae (Thomas), and Myzus persicae (Sulzer) in South Africa.- J. ent. Soc. S. Afr., 26 (1), 14-33.

3. HERMOSO, A. (1.982) Pugons (Homoptera, Aphidinea) dels citrics del Pais Valenciá.- An. INIA/Ser. Agric., 21 157-174.

4. LECLANT, F., REMAUDIERE, G. (1.970) Elements pour la prese en consideration des aphides dans la lutte intégrée en vergers de pêchers. Entomophaga, 15 (1), 53-81.

5. MELIA, A. (1.980) Investigación del Suborden Aphidinea en la provincia de Castellón sobre plantas de interés agrícola.- Comunicaciones INIA, Serie: Protección Vegetal, nº 12, 176 pp.

6. MELIA, A. (1.982) Prospección de pulgones (Homoptera, Aphidoidea) sobre cítricos en España.- Bol. Serv. Plagas, 8, 159-168.

7. SAWICKI, R.M. (1.981) Les phénomènes de resistance. En "Les pucerons des cultures". Journées d'études et d'information, pp. 83-87, ACTA, Paris, 1.982.

8. VAN EMDEN, H.F., EASTOP, V.F., HUGHES, R.D., WAY, M.J. (1.969) The ecology of Myzus persicae.- Ann. Rev. Entomol., 14, 197-270.

White-flies and psyllids injurious to citrus

S.Barbagallo, S.Longo & C.Rapisarda
Institute of Agricultural Entomology, University of Catania, Italy

Summary

Information is given on the whiteflies and psyllids injurious to citrus, with par
ticular reference to the species occurring in the Mediterranean Region. The behaviour and geographical distribution of the most noxious species are discussed;
their damages, biological and integrated control are also pointed out.

1.1 Introduction

All over the world, citrus groves are injuried by at least one species of whitefly or psyllid. The first of the two groups has a wider importance, since it is re
presented by a very large number of species in the citrus ecosystems, some of
which can become primary pests of such groves. On the contrary, very few psyllids
are presently known to live on citrus.

In the following pages a brief review is made of the main species of both the
groups, including some exotic representatives mainly in relation to the threat of
their introduction into the Mediterranean area.

2.1 Whiteflies

Several species of aleyrodids are recorded on citrus plants in various parts
of the world. The whiteflies hitherto collected on these host plants amount to a total of 55 species, 43 of which belong to the subfamily Aleyrodinae and 12 to the Aley-
rodicinae (32). Nevertheless, not all these species are serious citrus pests, since
many of them are very polyphagous insects and their presence on these orchards
must be considered as occasional.

2.2 Species occurring in the Mediterranean Region

Eight species are actually known to infest citrus in the Mediterranean basin.
Nevertheless, only three of them (Aleurothrixus floccosus (Maskell), Dialeurodes
citri (Ashmead) and Parabemisia myricae (Kuwana)) can be considered serious pests.
The first two species occur in Italy, together with few others of negligible economic importance.

2.2.1 Aleurothrixus floccosus (Maskell)

This is surely the most important whitefly attacking citrus in Italy and one of
the major pests of such groves occurring in the Mediterranean Region, both for its

89

damages and for the difficulties in the control of its infestations.

This species appears to be of neotropical origin, since it is widely distributed in South America, where several of its natural enemies can also be found. From this area, A. floccosus moved northward to the United States and eastward to the Canary Islands and Western Africa. The introduction of this pest in the Western Mediterranean Region is relatively new, since it appeared almost contemporary in Spain and South France towards the end of the sixties (33,34), later invading other Countries of this area, such as Morocco (1), Portugal (27) and Italy. With regard to the latter Country, the insect occurs in many areas, having been recorded in the Ligurian coast (3,34), Sardinia (36), Rome (1981, personal collecting data), Calabria (1983, personal collecting data), Sicily (14). In the latter Region the insect first appeared in the western territories, gradually invading the provinces of Trapani, Agrigento and Palermo. Later on, the occurrence of A. floccosus was recorded in the neighbourhood of Catania (just along the Jonian coast), in February 1983. At only two years' distance, the whitefly is now widespread in all the eastern areas of Sicily. This fact represents a serious danger for the economy of this Region, since about the 90% of the citrus groves of the island is concentrated just in its oriental part.

Infestations by A. floccosus are very easy to recognize, owing to the copious waxy secretion which covers the young stages of the insect, in the lower surface of citrus leaves.

The whitefly is a polyphagous species which is recorded as feeding on a large number of plants, belonging to about 20 genera of several families. Nevertheless, in the Mediterranean area it infests nearly exclusively species of the genus Citrus L..

In our climatic conditions this species has an almost continuous development, showing many generations per year. Its life-cycle undergoes only to a slight slowing down during the warmest and the coldest periods; as a consequence of that, in every season the population of the insect is represented by all its stages. The female oviposits on the underface of young leaves, where the newly hatched larvae can find good feeding conditions. Thus the flushing rhythm of the host plants greatly influences the life-cycle of the whitefly; as a result of the combined action of this rhythm with the climatic conditions, the highest densities of the insect population can be observed in spring and in autumn. Various natural enemies of A. floccosus have been recorded. In addition to some generic predators (like Neuroptera Chrysopidae and Conyopterigidae and Coleoptera Coccinellidae), the two parasitic wasps, Cales noacki How. and Amitus spiniferus (Bréth.), are very important to mention.

2.2.2 Dialeurodes citri (Ashmead)

Among the citrus-feeding whiteflies, D. citri is the most widely distributed in the Mediterranean Region, yet it is not of primary economic importance owing to the rather moderate damages it causes to the infested plants.

It is an oriental species, probably differentiated in the Indian Region. Its present distribution extends to several Continents, since this species occurs in North and South America, Far East and Mediterranean Region. With regards to the latter, D. citri is almost everywhere present, having been recorded in Algeria, Egypt, France, Greece, Israel, Italy, Morocco, Spain, Tunisia, Turkey. In Italy this insect can be found nearly in every region. As to Sicily, infestations by this whitefly have largely regressed during these last years; nevertheless, massive populations still remain in some areas and strong infestations have been recently found by the present Authors in some coastal zones of the province of Messina.

D. citri can be easily recognized from the other common citrus-feeding species (i.e. A. floccosus) for the lack of the waxy secretion on the dorsum of the larvae. It is a polyphagous insect which infests plants belonging to about thirty families. In Italy it mainly lives on citrus but its presence can also be noted on Persimmon and Gardenia.

This citrus whitefly overwinters in the larval stages (mainly as fourth instar). The pupae appear in April and the adults emerge at the end of the latter month and in May. Females oviposit on the underface of the tender leaves and a conspicuous second emergence of adults occurs in July. A third emergence (involving up to 50% of the population) takes place in September-October. Therefore in our regions D. citri shows 2-3 generations per year. In more suitable bi otopes till 5-6 generations per year are reported. Various antagonists of this whitefly are presently known. Among the predators, the Coccinellid Clitostethus arcuatus (Rossi) is the best known. With regard to the parasites, for the biological control of D.citri the specific endophagous Encarsia lahorensis (How.) is of peculiar importance.

2.2.3 Parabemisia myricae (Kuwana)

This is a species of recent introduction in Israel and it is probably destined to spread around such area, thus representing a serious danger for the citrus industry of all the Mediterranean countries.

P. myricae is original of the Far East, where it has been recorded in Japan, Malaya, Taiwan (17, 32). In 1978 the insect was discovered in the citrus groves of California (41) and recently it has been found in Israel (50).

The adults of this whitefly show a dusty blue-grey colour and are smaller than both D. citri and A. floccosus. When freshly deposited, the eggs are yellowish but they darken after one day. The pupa-case is pale yellow-greenish and is characterized by the presence of 13 pairs of rather long tubercled spines along all the margin.

P. myricae is reported to feed on plants of 14 families (32). Its noxious activity on citrus has been recorded in Japan, California, Israel. In the latter Country the insect showed to be particularly injurious to the Avocado groves, too. Very little is presently known on the insect biology. The species has a quite high reproductive rate and has a peculiar bent to the deuterotoky parthenogenesis; the sex-ratio of its populations is particularly favourable for the females. In laboratory experiments under variable temperature of 17.3° C to 21° C and 65 to 100% of R.H., the duration of the entire life-cycle, from egg to adult, was of 21 days (41). Among the antagonists of P. myricae the predator mites Amblyseius rubini (Swirski & Amitai) and A. swirskii Athias-Henriot can be mentioned (50), together with the parasitoids Prospaltella bemisiae Ishii, Encarsia sp. and Eretmocerus sp. (32, 41); their activity may ensure a valuable biological control.

2.2.4 Secondary species

Acaudaleyrodes citri (Priesner & Hosny) is a species of minor importance for citrus, owing to its low reproductive rate. Though never found in Italy, its occurrence has been recorded in several Countries of the Mediterranean Region: Cyprus,

Egypt, Israel (4, 8, 32, 39). Out of this area, the insect is present in Saudi Arabia, Iraq, India, Pakistan and in a large part of Central Africa (32). Adult of this whitefly are easily recognized by their three dark spots on the forewings; the pupacase is black, surrounded by a marginal waxy fringe. A. citri feeds on plants of numerous families and may infest all the grown citrus species (28, 32, 39). Very little is known on the biology of this insect. Its populations always remain at very low levels and serious infestations on citrus (with damages reaching the leaf-drop) are only reported for the upper Egypt (39). Encarsia davidi Vigg. & Mazz., E. lutea (Masi)and Eretmocerus roseni Gerling (15, 42, 49) are parasites of such whitefly.

Bemisia afer (Priesner & Hosny) (= B. hancocki Corbett) is another polyphagous species, which infests citrus but without causing any considerable damages. It seems to be native of the Central Africa and its present geographical distribution includes India, Pakistan and the Mediterranean Region (30, 32, 39). In Italy this whitefly is widely distributed at very low densities. Its biology is almost unknown.

Bemisia tabaci (Gennadius) is a well known, cosmopolitan and widely polyphagous species, whose noxious occurrence on citrus in Israel is recorded by several Authors (4, 8, 13). In Italy, as well as in all other parts of the Mediterranean Region, this species is largely diffused; nevertheless, apart from Palestine, it has never been collected on citrus there.

Dialeurodes kirkaldyi (Kotinsky) is another cosmopolitan species, whose geonemy involves the African, American, Asiatic and Australian Continents (32, 45). In the Mediterranean areas this species occurs in Egypt, Israel, Lebanon, Syria (4, 32, 45). It is not hitherto recorded in Italy. D. kirkaldyi injuries mainly plants of the genus Jasminum L.; Russell(45) reports Citrus sinensis (L.) as a host of this whitefly. In any case this species is of secondary importance as a pest of citrus.

Trialeurodes vaporariorum (Westwood) is a widely polyphagous whitefly, mainly injurious to the protected cultivations. The occurrence of this species on citrus, that was already reported by Russell(44), has been recently recorded in Italy by Arzone & Vidano (3). A heavy infestation by T. vaporariorum on Citrus sinensis (L.) seedlings was also seen by the present Authors from a glasshouse of Eastern Sicily (leg. Dr D. Benfatto). However, these findings must be considered as occasional and the economic importance of this species, as a parasite of citrus, is really negligible.

2.3 Main exotic species of citrus-feeding whiteflies

Among the citrus-feeding whiteflies which have not hitherto appeared in the Mediterranean Region, Aleurocanthus woglumi Ashby is one of the most important. Commonly known as "citrus blackfly", this species is largely distributed over the tropical and subtropical regions of the Asiatic and American Continents; it also occurs in various African Countries (Kenya, Seychelles, South Africa, Tanzania, Uganda) (13, 32, 43). The adults of this species are covered with a characteristic black-bluish waxy secretion; the pupa-cases are black. This whitefly is largely polyphagous but its economic importance is mainly connected with its injuries to citrus. The insect biology is highly variable in relation to the environment conditions; it can normally show from 3 to 6 generations per year. These are frequently overlapped, owing to the rather continuous development of the whitefly in the various seasons, and the adults are thus constantly present all the year through.

A similar species, Aleurocanthus spiniferus (Quaintance) (whose adult is cha-

racterized by the orange tinte of the abdomen) is mainly distributed in Southern and
Eastern Asia; nevertheless it also occurs in some African Countries, Mauritius,
Caroline Islands and U.S.A. (Hawaii). In the regions of Far East, particularly in
Japan, this species is highly injurious to citrus.

In the same Country (Japan) Aleurotuberculatus aucubae (Kuwana) is another
whitefly noxious to citrus.

Finally, Dialeurodes citrifolii (Morgan), commonly known as "cloudy-winged
whitefly" (for the dark spot in the median part of the adult forewings) is widely distributed in Asia (China, India, Japan, Vietnam) and America.

Numerous other species of whiteflies have been recorded on citrus. Neverthe
less they have not any considerable importance both for the limited damages they
cause to these groves and for the little probability they have to be introduced in the
citrus areas of the Mediterranean Region.

2.4 Damages by citrus-feeding whiteflies and control of their infestations

In general the injuries caused by these insects are related to the suction of
sap from the infested leaves. An indirect, but equally important, damage is caused
by the production of honeydew, which promotes the development of sooty-moulds.
In addition to this, the larvae of some whiteflies, such as A. floccosus, produce a
large amount of waxy filaments that, sticking to the honeydew, form a continuous
coat on the underface of the leaves, hardly permeable to the insecticides, giving a
sort of protection to the larvae of the same whitefly. In case of strong infestations,
specific control measures against their populations can be required.

2.4.1 Possibilities of biological control

In various Countries the efficiency of biological control has been studied
against several species of citrus-feeding whiteflies, mainly by means of predators
and parasitoid insects and, secondary, by the use of entomopathogenous fungi.

In the Mediterranean Region, with a particular reference to Italy, the introduction of Encarsia lahorensis (How.), a specific parasitoid of Dialeurodes citri,
is worth being recorded (46). The Aphelinid has been introduced in Western (19) and
Eastern (5, 38) areas of Sicily, where it became well acclimatized. In such areas
of the island, where D. citri represented a serious problem for citrus groves till
few years ago, its populations are presently reduced by the wasp to very low levels.
In addition to E. lahorensis, the considerable predacious activity of Clitostethus
arcuatus (Rossi) against the citrus whitefly is reported (25).

With regard to the other important species (A. floccosus), we can mention the
introductions of its parasites Cales noacki How. and Amitus spiniferus (Bréth.) in
several parts of the Mediterranean basin (2, 9, 20, 26, 35). In Oriental Sicily the
two wasps were introduced in 1983, soon after the discovery of the first areas of
the whitefly infestation. During 1984 a large distribution of C. noacki has been rea
lized in all the territories affected by the pest. The Aphelinid gave everywhere satisfactory responses, reaching in few months parasitization rates varying from 60
to 100%.

In various Countries a satisfactory biological control is also achieved against
Aleurocanthus woglumi by means of Encarsia opulenta (Silv.) and Amitus hesperidum
Silv.. The latter, together with Encarsia smithi (Silv.), has also been succesfully
used for the biological control of Aleurocanthus spiniferus.

With regard to the entomopathogenous fungi, Aschersonia aleyrodis Webber, A. goldiana Lace & Ellis and Aegerita webberi Faw. have been positively used in U.S.A. for the control of both Dialeurodes citri and D. citrifolii (13).

Attempts of bio-technical control were carried out by means of chromo-attrac tive traps against A. woglumi (16) and P. myricae (29).

2.4.2 Chemical control

The activity of natural enemies is not always enough to keep the populations of the whiteflies within the economic threshold levels (47), thus specific chemical treat ments are sometimes necessarily needed. In our Country, various comparative in- secticide trials have been made against D. citri (6, 12, 21, 22, 23, 40) and A.floc- cosus (7, 24, 26). According to the results of these researches, the chemical con- trol of the whiteflies does not show any considerable economic problems, since it can be easily linked to the control of the scales. Winter applications of white mine ral oils (1-2 Kg/hl), eventually mixed with some organophosphorous compounds or with some synthetic pyrethroids, can ensure satisfactory results, if the underface of the leaves and the internal part of the plants are accurately sprayed. For the control of A. floccosus it is necessary to use high pressure nozzles, which help the insecticide compounds to penetrate the waxy masses produced by the insect. In any case it is important to perform the chemical control of these pests respecting the principles of the integrated control methods, by using (when possible) compounds showing a low toxicity for the entomophagous.

3.1 Psyllids

Only two species of this group are known to feed on citrus: Diaphorina citri Kuwayama and Trioza erytreae (Del Guercio). Both of them do not presently occur in the Mediterranean Region; nevertheless the risk of their introduction in this area is to be feared, owing to their injuriousness to the citrus industry.

D. citri is the most widely diffused, since it occurs in almost all the southern regions of Asia, in Saudi Arabia, in Mauritius and Reunion Islands and in Brazil (10, 48). It lives on Rutaceae (genus Citrus L. and Murraya L.) and shows up to 10 overlapping generations per year (37), overwintering in every biological stage. A complete life-cycle, from egg to adult, takes 15-47 days.

The occurrence of T. erytreae involves the African Continent, where it has been recorded in the central and southern regions, the Madagascar, the Mauritius and the Reunion islands (10, 48). Besides living on citrus, this species lives on Clausena anisata, Fagara capensis and Vespris undulata, too. It seems that the main hosts of this species are just C. anisata and V. undulata and that the psyllid subsequently adapted itself to feed on citrus, which are not native from African Re- gions (31). The insect can make up to eight overlapping field generations (11), yet its development is influenced by the flushing rhythm and its survival is greatly con- ditioned by the extremes of weather. T. erytreae lives mainly on the underface of the citrus leaves, where the feeding activity of the nymphs causes the formation of a pit-like depression beneath their body, protruding to the upperface of the leaves.

3.2 Damages by citrus-feeding psyllids and their control

The injuries caused by the two psyllids to citrus plants are connected with the suction of sap and the consequent production of honeydew; T. erytreae also causes

94

the formation of small leaf-galls and, in case of a strong infestation, the upper curling of the leaves. Nevertheless, the main damage is connected with the transmission of the micoplasma-like organism which is supposed to be the cause of the "citrus greening" (18). For this reason the introduction of both D. citri and T. erytreae in regions where they do not presently occur must be avoided, opportunely controlling all the materials coming from infested Countries. The infected material must be destroyed or disinfested with methyl-bromide. When a centre of infestation is found, large amounts of contact insecticides must be seasonably applied.

REFERENCES

1. ABBASSI, M. (1975). Presence ou Maroc d'une nouvelle espéce d'aleurode, Aleurothrixus floccosus Maskell (Homoptera, Aleurodidae). Fruits, 30 (3): 173-176.
2. ABBASSI, M. (1980). Recherche sur deux homoptères fixes des citrus, Aonidiella aurantii Mask. (Homoptera, Diaspididae) et Aleurothrixus floccosus Maskell (Homoptera, Aleurodidae). Cah.Rech.agron., 35: 1-168.
3. ARZONE, A. and VIDANO, C. (1983). Indagini sui parassiti di Aleurothrixus floccosus in Liguria. Informatore fitopatologico, 33 (6): 11-18.
4. AVIDOV, Z. and HARPAZ, I. (1969). Plant pests of Israel. Israel University press, Jerusalem: X + 549 pp..
5. BARBAGALLO, S., LONGO,S. and PATTI, I. (1981). Primi risultati di lotta biologica-integrata in Sicilia orientale contro il Cotonello e il Dialeurode degli Agrumi. Fruits, 36: 115-121.
6. BARBAGALLO, S. and PATTI, I. (1978). Note biologiche ed orientamenti di lotta contro Dialeurodes citri (Ashm.) in Sicilia orientale. Atti giornate fitopatologiche 1978, 1: 237-244.
7. BENFATTO, D. (1982). Risultati di prove preliminari di lotta chimica contro Aleurothrixus floccosus (Mask.) (Hom. Aleyrodidae). Atti giornate fitopatologiche 1982, 3: 111-118.
8. BODENHEIMER, F.S. (1951). Citrus entomology in the Middle East. Uitgeverij Dr W. Junk "S - Gravenhage": XII + 663 pp.
9. CARRERO, J.M. (1979). Contribucion al estudio de la biologia de la "mosca blanca" de los agrios, Aleurothrixus floccosus Mask. en la region valenciana. IV.Parasitismo per Cales noacki How. Anales del Instituto Nacional de Investigaciones Agrarias, Proteccion vegetal, 9: 159-162.
10. CATLING, H.D. (1970). Distribution of the Psyllid vectors of Citrus Greening Disease, with Notes on the Biology and Bionomics of Diaphorina citri. FAO Plant Protection Bulletin, 18 (1): 8-15.
11. CATLING, H.D. (1972). The bionomics of the South African citrus psylla, Trioza erytreae (Del Guercio) (Homoptera: Psyllidae). VI. Final population studies and a discussion of population dynamics. J.ent.Soc.sth. Afr., 35 (2): 235-251.
12. DI MARTINO, E., LANZA, G. and CARUSO, A. (1980). Esperienze di lotta in diversi periodi contro il Dialeurodes citri in Calabria. Atti giornate fitopatologiche 1980, 1: 223-230.
13. EBELING, W. (1959). Subtropical Fruit Pest. Univ.of California, Division of Agr. Sciences: 436 pp..

14. GENDUSO, P. and LIOTTA, G. (1980). Presenza di Aleurothrixus floccosus (Mask.) (Hom. Aleyrodidae) sugli agrumi in Sicilia. Boll.Ist.Ent.agr.Oss.fitopat. Palermo, 10: 205–211.

15. GERLING, D. (1972). Notes on three species of Eretmocerus Haldeman occurring in Israel with a description of a new species. Ent.Ber., 32: 156–161.

16. HART, W. et al. (1978). Development of a trap for citrus blackfly, Aleurocanthus woglumi Ashby. Southwestern Entomologist, 3 (3): 219–225 (R.A.E., 67: 1484).

17. KUWANA, I. (1928). Aleyrodidae or white flies attacking citrus plants in Japan. Sci.Bull.Min.Agric.Forest.Dept., 1: 41–78.

18. LAFLECHE, D. and BUVAT, R. (1972). Mycoplasma – Like Bodies in the Leaves of Orange Trees Infested with Greening Disease. F.A.O. Plant Protection Bulletin, 20 (2): 28–30.

19. LIOTTA, G. (1978). Introduzione in Sicilia della Prospaltella lahorensis How. (Hym. Aphelinidae) parassita specifico del Dialeurodes citri (Ashm.) (Hom. Aleyrodidae). Atti Giornate Fitopatologiche 1978, 1: 231–236.

20. LIOTTA, G. (1982). La mosca bianca fioccosa degli agrumi. Informatore fitopatologico, 32 (12): 11–16.

21. LIOTTA, G. and MANIGLIA, G. (1974). Essais de lutte contre Dialeurodes citri (Ashmead) (Homoptera – Aleyrodidae) sur mandarinier en Sicile. Meded. Fac.Landb.Teusch.Gent., 39: 875–883.

22. LIOTTA, G. and MANIGLIA, G. (1975). Action des huiles blanches contre les stades hibernants de Dialeurodes citri (Ashm.) (Hom. Aleyrodidae) sur citronnier. Meded.Fac.Landb. Rijks.Univ.Gent., 40: 323–327.

23. LIOTTA, G. and MANIGLIA, G. (1980). L'olio minerale a confronto con alcuni piretroidi nella lotta contro Dialeurodes citri (Ashm.) (Hom. Aleyrodidae) su limone. Atti giornate fitopatologiche 1980, 1: 215–222.

24. LIOTTA, G. and MANIGLIA, G. (1982). Confronto dell'efficacia insetticida di alcuni fitofarmaci nei riguardi delle neanidi delle diverse età di Aleurothrixus floccosus (Mask.) (Hom. Aleyrodidae). Atti Giornate fitopatologiche 1982, 3: 119–129.

25. LOI, G. (1978). Osservazioni eco-etologiche sul coleottero coccinellide Scimnino Clitostethus arcuatus (Rossi), predatore di Dialeurodes citri (Ashm.) in Toscana. Frustula Entomologica, 1: 123–145.

26. LONGO, S., PATTI, I. and RAPISARDA, C. (1984). Il controllo dell'Aleurothrixus floccosus (Maskell) negli agrumeti della Sicilia orientale. Atti Gior nate fitopatologiche 1984, 2: 337–346.

27. MAGALHAES, G.S. (1980). Note on the introduction of Aleurothrixus floccosus (Mask.) (Homoptera, Aleurodidae) in South Portugal and its control by Cales noacki How. (Hymenoptera, Aphelinidae). Proc.int.Symp. of IOBC/WPRS 1979, 1: 572–573.

28. MANSOUR, M.M. et al. (1976). Preliminary studies on the abundance of the ci trus fly Aleurotrachelus citri on eight species of citrus trees. Agriculture Research Review, 54 (1): 161–165 (R.A.E., 66: 766).

29. MEYERDIRK, D.E. and MORENO, D.S. (1984). Flight behaviour and color-trap preference of Parabemisia myricae (Kuwana) (Homoptera: Aleyrodidae) in a citrus orchard. Envir. Ent., 13 (1): 167–170.

30. MINEO, G. and VIGGIANI, G. (1975). Sulla presenza di Bemisia citricola Go

mez–Menor (Homoptera – Aleyrodidae) in Italia. Boll.Lab.Ent.agr. "F.Silve-
stri", Portici, 32: 47–51.

31. MORAN, V.C. (1968). The development of the citrus Psylla, Trioza erytreae
(Del Guercio) (Homoptera, Psyllidae), on Citrus limon and four indigenous host
plants. J.ent.Soc.sth.Afr., 31 (2): 391–402.

32. MOUND, L.A. and HALSEY, S.H. (1978). Whitefly of the world. British Museum (Natural History) and John Wiley & Sons, Chichester: 340 pp..

33. ONILLON, J.C. (1969). A propos de la présence en France d'une nouvelle
espéce d'aleurode nuisible aux Citrus, Aleurothrixus floccosus Maskell (Homopt. Aleurodidae). C.r.Acad.Agric.Fr., 55 (13): 937–941.

34. ONILLON, J.C. and ABBASSI, M. (1973). Notes bio–ecologiques sur l'aleuro
de floconneux des agrumes Aleurothrixus floccosus Mask. (Homopt., Aleurodidae) et moyens de lutte. Al–Awamia, 49: 99–117.

35. ONILLON, J.C. and ONILLON, J. (1972). Contribution a l'étude de la dynamique des populations d'Homoptères inféodes aux agrumes. III. Introduction, in
dans les Alpes–Maritimes, de Cales noacki How. (Hymenopt., Aphelinidae),pa
rasite d'Aleurothrixus floccosus Mask. (Homopt., Aleurodidae). C.r.Acad.
Agric.Fr., 58: 365–370.

36. ORTU, S. (1983). Una nuova introduzione in Sardegna: Aleurothrixus floccosus (Mask.) nocivo agli agrumi. In: Cavalloro R. & Prota R. (Ed.). Lotta inte
grata in agrumicoltura: metodologia di campionamento e soglie d'intervento per
i principali fitofagi. C.E.C., Brussels: 27–28.

37. PANDE, Y.D. (1971). Biology of citrus psylla, Diaphorina citri Kuw. (Hemiptera: Psyllidae). Israel J.Ent., 6: 307–311.

38. PATTI, I. and RAPISARDA, C. (1980). Efficacia dell'entomofago Encarsia
lahorensis (How.) nel controllo biologico del Dialeurode degli agrumi. Tecnica
agricola, 32 (5): 291–299.

39. PRIESNER, H. and HOSNY, M. (1934). Contributions to a knowledge of the
white flies (Aleurodidae) of Egypt (III). Bull.Minist.Agric.Egypt tech.scient.
Serv., 145: 1–11.

40. PRIORE, R. and PANDOLFO, F.M. (1972). Prove di lotta chimica contro il
Dialeurodes citri (Ashm.) (Homoptera: Aleurodidae) in Campania negli anni
1971–72. Boll.Lab.Ent.agr. "F.Silvestri", Portici, 30: 139–144.

41. ROSE, M., DE BACH, P. and WOOLLEY, J. (1981). Potential new citrus pest:
Japanese bayberry whitefly. California agriculture, 35 (3–4): 22–24.

42. ROSEN, D. (1966). Notes on the Parasites of Acaudaleyrodes citri (Priesner
& Hosni (Hem.: Aleyrodidae) in Israel. Ent.Ber., 26: 55–59.

43. RUSSELL,L.M. (1962). The citrus blackfly. F.A.O. Pl.Prot.Bull., 10: 36–38.

44. RUSSELL,L.M. (1963). Hosts and distribution of five species of Trialeurodes
(Homoptera: Aleyrodidae). Ann.ent.Soc.America, 56: 149–153.

45. RUSSEL, L.M. (1964). Dialeurodes kirkaldyi (Kotinsky), a whitefly new to the
United States (Homoptera: Aleyrodidae). Fla.Ent., 47: 1–4.

46. VIGGIANI, G. (1976). Sull'introduzione di Prospaltella lahorensis How. per
il controllo biologico di Dialeurodes citri (Ashm.) in Italia. Atti XI Congr.
Naz.It.Entomologia: 375–377.

47. VIGGIANI, G. (1977). Lotta guidata contro i fitofagi degli agrumi. Informato
tore fitopatologico, 27 (6/7): 39–43.

48. VIGGIANI, G. (1980). Rincoti. Agricoltura e ricerca, 3 (12): 73–84.

49. VIGGIANI, G. and MAZZONE, P. (1980). Le specie paleartiche di Encarsia del gruppo lutea Masi (Hym. Aphelinidae), con descrizione di due nuove specie. Boll.Lab.Ent.agr. "F. Silvestri", Portici, 37: 51-57.
50. WYSOKI, M. and COHEN, M. (1983). Mites of the family Phytoseiidae (Acarina, Mesostigmata) as predators of the Japanese bayberry whitefly, Parabemisia myricae Kuwana (Hom., Aleyrodidae). Agronomie, 3(8): 823-825.

Fecundity, survival and life cycle of the citrus white-fly, *Dialeurodes citri* (Ashm.)

J.C.Malausa & E.Franco
INRA, Zoological & Biological Control Station, Laboratory 'E.Biliotti', Valbonne, France

Summary

The aim of the present study is to analyze under controlled laboratory conditions the main components of the rate of increase of the white-fly. The fecundity of about a hundred eggs is maximum between 20° and 25°C and under long photoperiods. The longevity of females reaches 18 days at 20°C and is higher than that of males. The length of the life cycle from egg laying to adult emergence is approximately two months at 22°C and with a photoperiod of 16L/8D, but the variability of the duration of the fourth larval instar can considerably increase the length of this life cycle. At 25°C the minimal length is 45 days. Mortality during the life cycle is very high, ranging up to about 50% of the initial population.
Photoperiods lower than 12L/12D stop the development of the fourth larval instar. This dormancy can be classified between a photoperiodic quiescence and a facultative diapause (= oligopause). From these observations we described some practical applications on the permanent rearing of the **Citrus** white-fly and its entomophagous insects.

1. INTRODUCTION

Although the citrus white-fly has been the subject of several papers dealing with its general biology (Priore, 1969), population dynamics (Onillon et al. 1975), its economic effects and attempts at biological and chemical control (Priore and Pandolfo, 1972; Liotta and Maniglia, 1975; Viggiani and Mazzone, 1978; Makare et al., 1980), precise information on its biotic potential is still fragmentary and generally based on observations under uncontrolled conditions. To provide the basis necessary for understanding and applying an integrated control approach to this major citrus pest, we considered it essential to improve our knowledge of the main factors determining the biotic potential of the white-fly under laboratory conditions and to determine the type of dormancy occuring during its life cycle.

2. EQUIPMENT AND METHODS

The <u>Dialeurodes citri</u> studied came from the experimental citrus groves

TABLE I

Conditions		Number of Experimental Couples		Fecundity (eggs)	Maximum fecundity (eggs)	Daily fecundity (eggs)	Female longevity (days)	Male longevity (days)
Photoperiod	Temp.							
16H	15°	20	m	26.86	69	0.87	13.50	10.20
			Sm	23.37		0.68	13.02	11.47
	20°	80	m	92.04	211	1.93	18.78	8.65
			Sm	52.66		1.15	17.67	10.93
	25°	20	m	86.50	189	2.73	13.95	3.80
			Sm	65.85		1.36	14.33	3.37
	29°	20	m	4.30	12	1.12	4.20	2.95
			Sm	3.86		0.56	1.91	1.70
8H	20°	20	m	8.86	19	1.00	7.65	5.20
			Sm	6.20		0.63	4.33	4.16

Variations in the biotic potential of Dialeurodes citri as a function of temperature and photoperiod.
(m = mean;   Sm = standard deviation)

100

(containing orange and clementine trees) at Valbonne, Alpes-Maritimes.

The fecundity and longevity of D. citri adults were studied as follows: each couple aged less than 24 hours was isolated in a ventilated ovipositor with a volume of approximately one litre containing a young Seville orange tree plant (Citrus aurantium) with tender leaves. Various combinations of temperature and photoperiod were tested (Table 1). The eggs were counted once a week, while adult mortality was noted daily. In calculating total fecundity, we took no account of females which died prematurely for experimental reasons or as a result of injury.

Daily fecundity is the mean daily fecundity of all females which have laid eggs, irrespective of their longevity. The daily fecundity of each female is obtained by relating its total fecundity to its longevity.

For reasons of convenience and reliabilty, and in order to study the duration of development and the mortality of the preimaginal stages of the white-fly, egg-laying had initially to be synchronized. We therefore left several hundred females to lay their eggs on young Seville orange tree seedlings for as short a period as possible. After two hours there were usually enough eggs to start the monitoring. Two kinds of examination were carried out - individual daily monitoring of 30 marked larvae at a temperature of 22°C and with a photoperiod of 12 hours, and general monitoring consisting of twice-weekly counts of the various stages of the white-fly at 22°C and with a photoperiod of 16 hours (initial number of eggs 100). The latter method makes it possible to work out the cumulative duration of development of the various stages and, by subtraction, the mean duration of each stage. Mortality in the various embryonic and larval stages was also noted.

In the studies carried out at the Valbonne grove, conventional methods were used, i.e. counting and monitoring of the various stages of the white-fly on marked leaves, and isolation of the insects in muslin sleeves around the citrus branches. At the same time, and to make it easier to transfer D. citri larvae from the field to the laboratory without adversely affecting their long-term development (an impossible objective when using parts of cut plants), the infested Seville orange seedlings were placed in outside conditions in cages of muslin which protected them from any uncontrolled infestation.

Using this technique, with a temperature of 22°C and a photoperiod of 16 hours, it was possible to reactivate larvae during their winter dormancy.

3. RESULTS AND COMMENTS

3.1. Fecundity (Table 1)

Observed fecundity was very heterogeneous, and the differences in the numbers of eggs laid by different individuals were very marked. Optimum fecundity, with an average of 92 eggs laid per insect, was observed at 20°C with a photoperiod of 16 hours, the absolute maximum being 211 eggs laid by one female. These results do not differ significantly from those obtained at 25°C. Temperatures of between 20° and 25°C therefore appeared to be ideal for D. citri oviposition.

101

Conditions of Observation	Individual (30 individuals) Temperature: 22° - Photoperiod: 12 H			General (100 individuals) Temperature: 22° - Photoperiod: 12 H		
	Mean duration (days)	Extreme durations (days)	Mortality (% of initial population)	Mean Duration (days)	Extreme Cumulative Duration (days)	Mortality (% of initial population)
Egg	15.38	14 – 18	6.67	17.48	14 – 44	0
L1	9.31	8 – 16	3.33	8.23	22 – 51	0
L2	7.77	6 – 10	0	8.77	26 – 65	6.58
L3	9.96	8 – 11	10.00	7.45	33 – 85	11.84
L4	Development of L4 larvae halted (+ 12 months)			42.24	51 – 248 +	25.00
N	–	–	–	6.57	58 – 248 +	5.26
Total	–	–	–	90.74	58 – 248 +	48.68

TABLE II

Duration of development and mortality of preimaginal stages of Dialeurodes citri under controlled conditions.

At extreme temperatures overall fecundity drops dramatically: at 15°C the drop (to 26.86 eggs/female) is due to reduced daily fecundity (0.87 eggs/day) and the reduction in longevity is not significant. However, at 29°C not only are fewer eggs laid (1.12 per day), but there is also a reduction in the longevity of females (4.20 days) which results in a very low total fecundity (4.30 eggs).

Short photoperiods of 8 hours also have a negative influence on the overall and daily oviposition of females.

Daily fecundity, which is at its maximum at 25°C with 2.73 eggs per day, declines steadily with increasing deviation from that temperature. It is difficult to compare these results with the fragmentary existing data, generally obtained under unknown experimental conditions. Targe and Deportes (1953) recorded a fecundity of around 100 eggs per female, which is on a par with the fecundity which we established at 20°C. However, Makar et al. (1980) in India have recorded an average fecundity level of 274 eggs per female. Longevity was very short (6 to 11 days), which resulted in a daily fecundity rate which was much higher than that of our own insects. The fecundity of the whitefly is undoubtedly influenced very greatly by the host plant.

3.2. Longevity (Table 1)

Like fecundity, longevity among adults differs greatly from one individual to another. Longevity is much higher among females than among males. The optimum temperature for the survival of females (18.78 days) is 20°C. Female longevity still exceeds 13 days at temperatures of between 15° and 25°C, but it is substantially reduced by high temperatures and short photoperiods.

All these results confirm the findings of Makar et al. (1980) and Targe and Deportes (1953), who recorded longevities of between 6 and 11 days for adults and 8 to 10 days for females.

3.3. Periods of development of preimaginal stages (Table II)

The duration of development of the embryonic and larval stages was measured at 22°C using two different photoperiods - 12 and 16 hours. Using the first photoperiod it was not possible to go beyond the fourth larval stage, which thus remained blocked for more than 12 months - evidence of a dormancy which is discussed in a subsequent section. The egg incubation periods and the duration of growth of the first three larval stages were the same in the two experiments. Only maximum durations differed - in the case of egg incubation, this was 44 days. According to Priore (1969), the incubation period varies from 16 to 135 days, but under varying conditions. Makar et al. (1980) recorded an average incubation period of 18.5 days, which is very similar to our own results.

At 22°C the minimum duration of development from oviposition to the emergence of the adult was found to be 58 days. However, the L4 larvae stage may be extended by anything up to several months under apparently favourable conditions, thus increasing the duration of the cycle. This phenomenon is confirmed by other writers. Priore (1969) reported preimaginal development lasting from 34 to 228 days. Periods of from 57 to

103

247 days were later observed by the same writer (in Viggiani and Mazzone, 1978) and these figures differ from our own results by only one day. Longer development times were reported by Makar et al. (1980), with a minimum of 93 days at unknown temperatures. Viggiani and Mazzone (1978) observed a cycle of approximately 60 days at a temperature of 25-27°C. Our experiments at 25°C showed that the first adults appeared 45 days after oviposition.

3.4 Mortality of preimaginal stages

By studying larval development in the laboratory at 22°C, we were able to establish the mortality for the various stages (Table II). About 20% of the initial population would appear to die before stage L4. The duration of this stage is highly variable, and mortality increases the longer it continues. 25% of the initial population die before the L4 stage. Overall, the mortality rate between oviposition and emergence of the adult on Seville orange tree seedlings is approximately 50%.

In the conditions prevailing in citrus groves, and leaving aside the effect of antagonistic organisms, the natural mortality rate seems to be much higher still. Examination of 2000 L4 larvae of the hibernating generation showed that only 27.9% developed into adults; the remainder died or disappeared with the heavy fall of leaves at the first rising of the sap.

3.5 Hibernation of larvae

3.5.1 Findings: Observations in the field show that development of the last larval stage ceases in winter. It was, therefore, interesting to investigate the development of D. citri larvae in the laboratory under various conditions in order to shed light on the factors responsible for the onset and end of dormancy.

Experiments to determine the influence of photoperiods on the larval development of the whitefly showed that only long photoperiods of 16 hours were sufficient to produce adults (at 20°). The development of all L4 larvae was halted for at least four months (experiments interrupted) when photoperiods of 8 and 12 hours were used (Table III). This confirms the previous results, which are given in Table II. Short photoperiods are therefore responsible for halting the development of the last larval stage of D. citri. In the laboratory it is easy to avoid dormancy by consistently

Temperature & photoperiod		Duration of preimaginal development
20°C	16 hours	65 to 70 days
	12 hours	Halted at L4 stage (> 4 months)
	8 hours	Halted at L4 stage (> 4 months)

Table III - Duration of preimaginal development (from oviposition to emergence of the first adult) as a function of photoperiod.

maintaining long photoperiods. In nature, besides the effect of autumn temperatures which slow down larval development, the yearly cyclical shortening of the photoperiod invariably induces dormancy in L4 larvae. By determining the conditions in which dormancy is ended, we should be able to establish whether this dormancy is an optional diapause or simply a phase of photoperiodic quiescence.

3.5.2 Reactivation of dormant larvae: At the start of each month between October 1983 and April 1984, two Seville orange seedling plants infested by L4 larvae of D. citri were transferred to the laboratory under conditions which were ideal for reactivation, i.e. 22°C and a photoperiod of 16 hours. Figure 1 shows the rate of emergence of adults on each seedling. The origin of the abscissae is the date on which the seedling was transferred to the laboratory. To improve the reliability of comparisons between the graphs, we marked with an arrow the day on which half of the adults had emerged. This period never exceeded 46 days and did not change during the first two months of dormancy (November and December). However, reactivation periods became shorter thereafter and amounted to only 21 days in the final month of dormancy.

The general pattern of the graphs confirms the previous results. The emergence of reactivated insects is spread widely over the first two months of dormancy (November and December). From January on, there is a tendency towards synchronization.

The cumulative effect of the cold in the first two months could affect the course of dormancy and thus the subsequent development of the larvae when subjected to photoperiods.

3.5.3 Conclusions: The fact that adults can be obtained fairly quickly by reactivating the larvae at the start of dormancy (beginning of November) and that the larvae have not been influenced greatly by factors affecting the diapause, such as cold, does not enable us to describe this phenomenon as a true facultative diapause. Neither does the fact that the period required to produce adults becomes shorter the later the sample larvae are taken in the winter, enable us to describe it as simply photoperiodic quiescence which may be reversed at any time. This intermediate type of dormancy - which in the case of D. citri is conditioned by short photoperiods - corresponds to the oligopause described by Muller (1970), which is generally dependent on photoperiods. In the South of France the halting of the development of L4 larvae due to the effect of photoperiods goes some way towards explaining the small number of generations. In the 1984 season we observed in the Valbonne citrus grove two complete yearly generations and one very partial third generation.

All the hibernating larvae hatch between the beginning of May and the end of June. The larvae from this first brood produce adults from 20 July to 20 October, and all hatch before winter. This second brood produces larvae, the vast majority of which spend the following winter as mature larvae. Only a small proportion of rapidly-developing larvae escape the effect of the shortening photoperiods and produce a few adults between 5 October and the end of October. This small brood is overlapped by that of

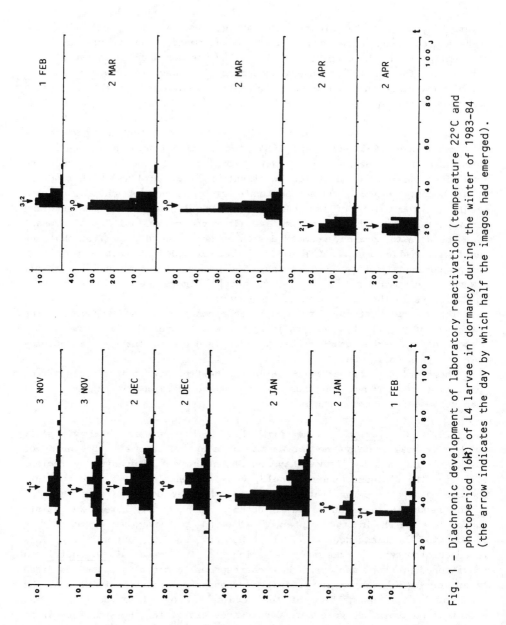

Fig. 1 – Diachronic development of laboratory reactivation (temperature 22°C and photoperiod 16h) of L4 larvae in dormancy during the winter of 1983-84 (the arrow indicates the day by which half the imagos had emerged).

the preceding generation. These observations tally with those of Priore (1969), Onillon et al. (1975) and Patti and Rapisarda (1981), who report two to three generations a year in France and Italy. A larger number of yearly generations is, in our view, highly unlikely in Mediterranean countries.

4. CONCLUSIONS

These laboratory studies provide a clearer picture of the ideal conditions for the development of D. citri. Nevertheless, the influence of the host plant and its fertilization, not considered in this study, clearly has a very important bearing on the whitefly's potential for multiplication.

The fact that larval dormancy can be avoided is a decided advantage for applications involving the permanent breeding of the pest and thus of its various predators. We have been continuously breeding D. citri at Valbonne for two years, with all stages present at any time of the year. Breeding is carried out on the one hand in the laboratory at 22°C and with a photoperiod of 16 hours by rotation and periodic infestation of citrus seedlings, and on the other in an insect-proof citrus conservatory kept at a minimum temperature of 15°C, and with 16 hours of light thanks to a back-up lighting system to extend the natural photoperiod during the short winter days, so as to keep a strain of Encarsia lahorensis and Seville orange seedlings infested with whitefly larvae bearing parasites. It appears preferable to carry out all operations to acclimatize this parasite using citrus seedlings in pots, thus permitting prolonged hatching of the parasites and a better yield of releases than with methods using cut plants (Viagginai, 1980).

REFERENCES

1. LIOTTA, G. and MANIGLIA, G. (1975). Action des huiles blanches contre les stades hibernants de **Dialeurodes citri** (ASHM.) (Hom. Aleyrodidae) sur citronnier. Meded. Fac. Landb. Cijk. Univ. Gent. 40; 323–327.
2. MAKAR, P.V., KHAROLE, V.U. and CHOUDHARI, K.G. (1980). Biology of **Citrus** whitefly (**Dialeurodes citri** ASHM.) and its reaction to the toxicity of a few organo-phosphorous insecticides. Journal of MAU. 5(3); 256–257.
3. MULLER, H.J. (1970). Les différents types de dormance chez les Insectes Nova Acta Leopoldina. 35 (191); 7–25.
4. ONILLON, J.C., ONILLON, J. and BRUN, P. (1975). Contribution à l'étude de la dynamique des populations d'Homoptères inféodés aux Agrumes. II.3. Premières observations sur l'évolution comparée des populations de **Dialeurodes citri** ASHMEAD (Homopt. Aleyrodidae) en Corse et dans le Sud-Est de la France. Fruits, 30 (3); 167–172.
5. PATTI, I. and RAPISARDA, C. (1981). Reperti morfo-biologici sugli Aleirodidi nocivi alle piante coltivate in Italia. Boll.Zool.Agr. Bachic. Ser.II.16; 135–190.

6. PRIORE,R.(1969). Il **Dialeurodes citri** (ASHMEAD) (<u>Homoptera Aleyrodidae</u>) in Campania (Note di morfologia e biologia). Boll.Lab.Ent.Agr.Portici, 27; 287-316.
7. PRIORE, R. and PANDOLFO, F.M. (1972). Prove di lotta chimica contro **Dialeurodes citri** (ASHM.) (<u>Homoptera: Aleyrodidae</u>) in Campania negli anni 1971-72. Boll.Lab.Ent.Agr.Portici. 30; 139-144.
8. TARGE, A. and DEPORTES, L. (1953). L'aleurode des agrumes (**Dialeurodes citri** ASHM.) dans les Alpes-Maritimes. Premiers résultats d'expérimentation de traitements. Phytoma. 44; 9-15.
9. VIGGIANI, G. (1980). Progress toward the integrated control of **Citrus** pests in Italy. Proceedings Intern. Symposium IOBC/WPRS 1980; 293-296.
10. VIGGIANI, G. and MAZZONE, P. (1978). Morfologia, biologia e utilizzazione di **Prospaltella lahorensis** HOW. (<u>Hym. Aphelinidae</u>), parassita esotico introdotto in Italia per la lotta biologica al **Dialeurodes citri** (ASHM.). Boll.Lab.Ent.Agr.Portici. 35; 99-160.

Effect of fertilizer on the population level of the citrus white-fly
Dialeurodes citri (Homoptera, Aleyrodidae)

J.C.Onillon
Zoological Station, CNRA, Versailles, France

P.Brun
INRA, Agronomical Research Station, San Nicolao, France

Janine Onillon & G.Decoux
Laboratory of Biometry, CNRA, Versailles, France

E.Franco
INRA, Zoological & Biological Control Station, Laboratory 'E.Biliotti', Valbonne, France

Summary

To determine the effect of fertilizers containing nitrogen, phosphorous and potassium on the population level of the citrus whitefly, nine combinations of fertilizer were tested on a 2 ha plot of clementine trees. The fertilizer, in combinations of from zero to excess (quadruple) doses of each of the three elements N, P and K, was compared to a control plot receiving a normal dose. Each type of fertilizer was applied each year in a dose proportional to the age of the tree.

One sample per type of exposure, consisting of a leaf–bearing branch of the spring growth, was taken from 216 clementine trees at the centre of the individual subplots.

Variance analysis of the densities of citrus whitefly eggs has revealed sharp differences in the degree of infestation of the trees. The effect of the fertilizer is highly significant, and the plots showing an excess of all the nitrogen and phosphopotassium fertilizer differed from all the others in having very much higher egg densities. The analysis also established that only N_2 and P_2O_5 were involved in the fertilizer effect, with increased doses of phosphorous and then of nitrogen giving rise to an increase in the embryo population of the citrus whitefly.

1. INTRODUCTION

Purely chemical control has been found to have limitations when ap-plied to citrus fruit growing. Although the fact that it is easy to apply and results can be obtained in a short time makes it attractive, sooner or later and depending on voltinism and the peculiar features of the biotic potential of the main pests concerned (mites, whiteflies, diaspid scales insects, lecania and pseudococci), it has serious drawbacks in that it gives rise to resistance, especially among mites, and causes faunal imbalance.

If an integrated approach is to be applied instead, there are certain difficulties inherent in the three main options: the action of indigenous or acclimatized entomophagous insects, the use of selective pesticides below a damage threshold and the effect of techniques.

It is extremely difficult to develop any new techniques with regard to

entomophagous insects, since effective entomophagous insects have been acclimatized in the case of most accidentally introduced pests: <u>Cales noacki</u>
to control <u>Aleurothrixus floccosus</u> (10), <u>Prospaltella Lahorensis</u> to control
<u>Dialeurodes citri</u> (16), <u>Metaphycus bartletti</u> and <u>Metaphycus lounsburyi</u> to
control <u>Saissetia oleae</u> (15), and <u>Aphytis lepidosaphes</u>, a parasite of
<u>Lepidosaphes beckii</u> (4).

The damage thresholds for most of the main citrus fruit pests are at
present being determined (4, 14), and it seems very likely that efforts
will now be directed towards experimentation to verify and apply on a wide
scale (14) methods already worked out in outline form.

However, the third aspect of the integrated approach -the effect of
crop-growing techniques - is a completely uncharted area. The different
techniques applied play a predominant role in the evolution of the phytophagous entomophagous complex, whether because of the <u>size</u> of the crop
concerned, which has a bearing on population levels at the end of winter,
<u>irrigation systems</u>, which in certain cases, as with irrigation by spraying,
give rise to preferential mortality of the mobile stages of the pests
(whiteflies or scale insects) and may encourage the development of mycoses,
and lastly <u>fertilization</u>, which not only modifies the resources offered by
the host plant but also affects the degree of fulfillment of the biotic
potential of the pests.

The effect of fertilization, both directly on the availability of
resources provided by the host plant (per m^3 of leaf surface or in linear
metres of branches) and in its more subtle effects on the modification of
certain parameters connected with the fulfillment of the biotic potential
of predators, is particularly important, since the pest (whitefly, mite,
greenfly or scale insect) at the end of the food chain combines all the biotic and abiotic factors determining the equilibrium and evolution of the
"soil-plant-host-climate-pest" association.

2. EQUIPMENT AND METHODS

2.1 Test site

The work was carried out on a 2.2 ha plot at the Station de Recherches
Agronomiques, Corsica, comprising 864 clementine trees of the variety SRA
63 grafted to Citrange Troyer. Spacing was 6 x 4 m. The plot was planted
in April 1973 to monitor, in accordance with the Station's research
programmes, the influence of fertilizer on homoptera populations, taking
account of the phenological development of the host plant (5), and was also
used for the study to determine the damage thresholds of <u>Dialeurodes citri</u>
and <u>Saissetia oleae</u> (14).

Plot C_4 is bounded on three sides by a wind-break of cyprus trees with
an average height of 7-8 m. The fourth side, facing east, is open to what
was for a long time a "maquis".

2.2 Composition and application of the fertilizer

Three fertilizer doses were used to make it easier to determine any
effect on <u>D. citri</u> population levels:

- "zero" dose, corresponding to a total absence of fertilizer: N_0, P_0 and K_0;
- dose "1" representing normal fertilization: $N_1P_1K_1$ in the following proportions:

     80 kg N_2 /ha/year   )
     20 kg P_2O_5/ha/year)   for adult trees
     40 kg K_2O/ha/year   )

- dose "4", representing excess fertilization: $N_4P_4K_4$ or four times the dose $N_1P_1K_1$:

    320 kg N_2 /ha/year
     80 kg P_2O_5/ha/year
    160 kg K_2O/ha/year.

In addition to the above three balanced fertilizers, six permutations of absence and surplus of each of the elements N, P and K were produced: $N_0P_0K_4$, $N_0P_4K_0$, $N_0P_4K_4$, $N_4P_0K_0$, $N_4P_0K_4$ and $N_4P_4K_0$. For each of these 9 combinations of fertilizer, 6 repeated doses were applied to individual subplots of sixteen clementines trees each. The initial allocation of the combinations to the individual subplots was determined on a random basis.

Table I shows the dosages applied from 1973, when the trees were planted, to 1983, when they were considered to be adult.

Years	% of fertilizer applied
1973	Year of planting
1974	20% of the adult dose
1975	40% of the adult dose
1976	60% of the adult dose
1977	70% of the adult dose
1978	75% of the adult dose
1979	80% of the adult dose
1980	85% of the adult dose
1981	90% of the adult dose
1982	95% of the adult dose
1983	100% of the adult dose

Table I. Application of fertilizer as a function of tree age.

The various fractions of the fertilizer were applied over the year as follows:

1st application: 1/3 nitrogen fertilizer (end of March-beginning of April)
2nd application: 1/3 nitrogen fertilizer (mid-June)
2rd application: 1/3 nitrogen fertilizer (end of July)
4th application: all the phosporous/potassium fertilizer (end of September-
                 mid-October).

2.3 Sampling procedure and data collection

Sample clementine leaves were taken on one single occasion in the first week of July 1983, at the height of the flight phase of the adults of

111

the first generation of D. citri, i.e. between the second and third applications of the nitrogen fertilizer.

From each of the 4 clementine trees in the centre of each of the 54 individual subplots (9 combinations x 6 repeated applications), a leaf-bearing branch of primary sap (i.e. spring) growth was taken, each branch comprising between 4 and 7 leaves.

The total number of eggs (hatched or unhatched) of D. citri on each leaf was divided by the surface of the leaf established on the basis of the formulae (11), permitting the application of the principle of numerical density per unit surface area previously used (12).

3. ANALYSIS OF RESULTS

All the data relating to the average surface area of clementine tree leaves, the number of D. citri eggs per branch and the numerical density of the embryo phase of the whitefly were subjected to variance analysis.

3.1 Influence of fertilizer on average surface of clementine leaves

The effect of fertilizer on the average surface area of clementine leaves was tested by variance analysis (Table II).

Source of variation	Sum of squares	Degrees of freedom	Mean square	F. value
Fertilizer	267.60	8	33.45	1.37 N.S.
Plot	1,099.05	45	24.42	2.57 **
Orientation	880.57	3	293.52	30.86 **
Fertilizer/ Orientation	328.50	24	13.69	1.44 N.S.
Residual	7,443.75	783	9.51	
Total	10,019.47	863		

Table II.   Variance analysis of the surface of a clementine tree leaf

The type of fertilizer used was not found to have any significant influence on the average area of clementine leaves. The combinations of fertilizer tested had no direct effect on the average surface area of leaves on the types of branches tested (4 to 6 leaves). It is possible that fertilizer influences the total number of leaves produced rather than their average area. However, two factors seem highly significant - the effect of the particular plot tested and that of its orientation. There seem, therefore, to be differences between plots, which would suggest that before any supposed action of fertilizer can be tested, it is necessary to examine certain parameters relating to the topography and the history of the plot (previous experiments, soil differences, microclimatic peculiarities due to the presence or proximity of wind-breaks, etc.).

The importance of plot orientation with regard to the average surface area of a leaf is shown in Table III.

112

Orientation	EAST	SOUTH	WEST	NORTH
Surface (cm^2)	11.98	11.54	10.78	13.55

Table III. Average surface area of a clementine
leaf as a function of orientation

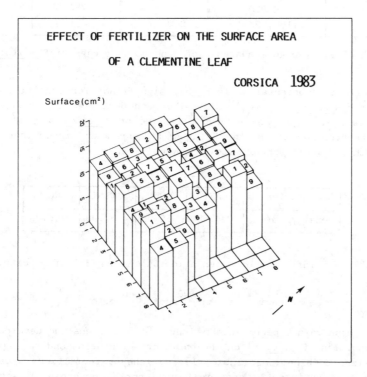

Fig.1. Influence of fertilizer combinations on average surface area of
clementine leaves. (The figures refer to the numbers assigned to
the combinations of fertilizer in Fig. 2)

Among the 216 sample branches, i.e. with a total of approximately
1,000 leaves per exposure, the largest leaves were found to the north, and
the smallest to the west, while there was little difference between the
leaves to the east and south.

3.2 Effect of fertilizer on numerical density of eggs of Dialeurodes citri

Variance analysis, on two levels, of the numerical densities of eggs
of Dialeurodes citri showed that all the factors tested had a significant
effect, as shown in Table IV.

113

Source of variation	Sum of squares of differences	Degrees of freedom	Mean square	F. value
Fertilizer	161,150.6	8	20,143.8	3.05 **
Plot	297,542.0	45	6,612.0	8.59 **
Orientation	69,021.4	3	23,007.1	29.88 **
Fertilizer/ Orientation	30,073.3	24	1,253.1	1.63 *
Residual	602,938.9	783	770.0	
Total	1,160,726.1	863		

Table IV. Variance analysis of densities of eggs of D. citri

The effect of orientation is highly significant for densities of eggs of Dialeurodes citri. If we consider the mean densities for each orientation (Table IV), we see the embryo densities of D. citri per cm^2 of leaf surface observed for the north (56 eggs/cm^2) are much higher than for the east, south and west (approximately 35 eggs/cm^2).

Orientation	EAST	SOUTH	WEST	NORTH
Density/cm^2	34.59	39.41	34.96	56.49

Table V. Influence of orientation on egg density of D. citri

$$D_{east} = D_{south} = D_{west} < D_{north}$$

These mean values become clearer when one considers the development of embryo densities of D. citri on the basis of orientation and taking account of fertilizer combinations (Table VI). Whichever fertilizer is used, egg densities are consistently greater to the north, probably on account of microclimatic conditions favouring egg-laying by adult whiteflies.

The general tendencies illustrated in Table VII can be easily discerned for each orientation, with the lowest densities of less than or just over 20 eggs per cm^2 of leaf surface, regardless of the exposure considered, being observed for fertilizer N° 2 ($N_0P_0K_4$). Fertilizer N° 8 ($N_4P_4K_4$) is associated with very high densities for all orientations, and the difference of around 20 eggs increases to around 50 for the northern orientation, with the closest densities for the other fertilizers being observed in the east, south and west.

The effect of the particular is once again significant (F = 8.59 - Table IV) and it is highly probable that earlier observations concerning the development of leaf surfaces apply equally to Dialeurodes citri which, since it is at the end of the food chain, is affected by all the preceding factors.

N°	Fertilizer	East	South	West	North
1	$N_1 P_1 K_1$	28.87	38.33	26.65	48.98
2	$N_0 P_0 K_4$	17.09	21.54	19.09	23.03
3	$N_0 P_4 K_0$	38.77	33.87	29.92	55.37
4	$N_0 P_4 K_4$	35.25	36.13	39.52	50.29
5	$N_4 P_0 K_0$	32.67	41.33	32.42	49.21
6	$N_4 P_0 K_4$	29.72	38.11	32.24	53.31
7	$N_4 P_4 K_0$	42.51	38.74	37.08	61.59
8	$N_4 P_4 K_4$	55.53	70.04	61.94	114.17
9	$N_0 P_0 K_0$	30.92	36.56	33.82	52.47

Table VI. Densities of _Dialeurodes citri_ eggs per cm^2 of leaf surface as a function of fertilizer and orientation

The fertilizer has a significant effect on densities of the embryo phase of the whitefly.

If we consider the average densities of _D. citri_ eggs as shown in increasing order in Table VII, we see that there are many values which are very close to 35-40 eggs/cm^2. Only fertilizer N° 8 ($N_4P_4K_4$) is associated with significantly higher densities (Fig. 2).

The four highest densities were obtained when using fertilizers with surplus of P_2O_5. This tendency, which was observed in places where the use of P_4 was associated with the highest densities, led us to consider more closely the effect of the individual components of the fertilizers.

N°	2	1	9	6	5	3	4	7	8
Densities per cm^2	20.19	35.71	38.44	38.84	38.91	39.48	40.30	44.98	75.42
Fertiliz.	$N_0P_0K_4$	$N_1P_1K_1$	$N_0P_0K_0$	$N_4P_0K_4$	$N_4P_0K_0$	$N_0P_4K_0$	$N_0P_4K_4$	$N_4P_4K_0$	$N_4P_4K_4$
N	0	1	0	4	4	0	0	4	4
P_2O_5	0	1	0	0	0	4	4	4	4
K_2O	4	1	0	4	0	0	4	0	4

Table VII. Effect of fertilizer on _Dialeurodes citri_ egg densities

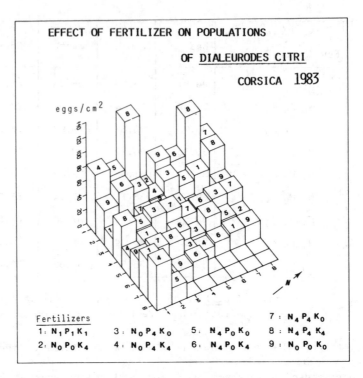

Fig.2.   Effect of fertilizer combinations on embryo
densities of Dialeurodes citri

Origin of of variation	Sum of squares of differences	Degrees of freedom	Mean square	F value
N	45,709.0	1	45,709.0	6.91 *
P_2O_5	49,259.1	1	49,259.1	7.45 **
K_2O	1,786.7	1	1,786.7	0.27 N.S.
Other	64,395.8	5	12,879.2	1.95 N.S.
Plot	297,542.0	45	6,612.0	

Table VIII.   Variance analysis of the fertilizer constituents N, P and K.

	Coefficient	Confidence intervals
N	0.046	± 0.036
P_2O_5	0.197	± 0.146
K_2O	0.019	± 0.073

Table IX.   Regression analysis of doses of fertilizer constituents

Two constituents have a significant effects: P_2O_5 (significant at the 1% level) and nitrogen. The analysis in Table VIII confirms, therefore, the tendencies observed earlier: P_2O_5 provides high densities of whitefly and may well prepare the way for the action of the nitrogen. No effect of K_2O is apparent.

Analysis of the regression coefficients (Table IX) confirms the preceding variance analysis indicating the significant effect of P_2O_5, with a regression coefficient of nearly 0.2.

4. OBSERVATIONS

The influence of fertilizers dynamics on population and on the beha- viour of insects and mites has been studied for numerous species (9). Usually one constituent, most often nitrogen (2,3,6,7,17), is added in varying amounts while another, phosphoric acid, is kept at a constant level (8). Little or no account is taken of the development of the host plant, since very often in the cases studied the host plant completes its cycle within a few months. Citrus fruit trees are a special case owing to the fact that they are evergreen, the average life-span of the leaves being several years, and the foliage is renewed cyclically by the rising sap.

The fact that fertilizer has no effect on the average surface area of clementine leaves is not in itself surprising. The samples taken in fact had only 4 to 6 leaves, whereas the number of leaves on clementine branches ranges from 1 to 15-16 (5), and work in prograss in Corsica shows that the frequency of branches bearing from 1 to n leaves varies markedly depending on the age of the tree and the type of fertilizer used. Fertilizer is not thought to effect the average surface area of a leaf calculated on the basis of an average sample taken from branches with 4 to 6 leaves, but influences the total number of leaves produced at each rising of the sap, taking account of the balance between the different periods when the sap rises (spring, summer and autumn).

The effect of orientation on the average surface area of a clementine leaf is more surprising. Work on growth stimulation of the leaves of Seville orange trees (13) shows that leaves develop and attain their maximum size in 5 to 6 weeks, depending on the temperatures obtaining in May-June. The larger average surface areas observed to the north suggest that a microclimate, characterized by high humidity and reduced sunshine, might have selectively affected leaf growth two months earlier. No definite conclusions can be drawn, since no microclimatic data were recorded.

If we now consider the pest Dialeurodes citri and the development of embryo densities, we see that the effect of orientation is highly sig- nificant and that very high densities are to be found on leaves facing north. Since the average surface area of leaves is greater there, this means that there are far more eggs on such leaves. This phenomenon, coupled with the fact that the citrus fruit trees most severely infested by D. citri are normally situated on low ground and near streams, supports the theory suggesting the existence of a special microclimate in the north which is characterized by high humidity and which seems to have a marked influence (Table VI), irrespective of the fertilizer used.

Fertilizer has a significant effect on densities of the embryonic phase of D. citri, and only the combination $N_4P_4K_4$ (Table VII) differs from eight others. This increase in D. citri populations when, the N, P and K contents increase from 0 to four times the normal dose, is very close to the observations made with regard to populations of Bemisia tabaci on Manihot esculenta (8), since the total absence of fertilizer resulted in the lowest densities of whitefly larvae. In that particular case all combinations, whether the dose was 0 or surplus, contained equal proportions of N, P and K (1 : 1 : 1). Similarly, plots where N and K were used in the proportion 1 : 1 (with P constant) showed a lesser degree of infestation by B. tabaci than those where N and K were combined in the ratio 1 : 1.5 - 1 : 2 and 1 : 3; a combination of N and K in the ration 1 : 2 - 1 : 4 leads to an increase in the larval population of Bemisia (8). The same author points out that a high potassium dose (100 - 150 kg/ha) affects the development of B. tabaci.

In the clementine tree D. citri relationship, N, P, and K were combined in the ratio 4 : 1 : 2 to produce a normal balanced or surplus fertilizer. There is therefore over-fertilization and no deficiency of one of the constituents. It is interesting to note that the lowest densities of D. citri eggs were found using the fertilizer $N_0P_0K_4$, i.e. with an excess of potassium (Table VII). Potassium at a high dose compared with N and P would appear to affect larval development (8), as well as egg-laying by the adult whitefly, which confirms the results obtained (1) for Aleurodes brassicae.

The highest densities of whitefly eggs correspond to the four combinations of fertilizers containing not only P_2O_5 but also additional N, which is confirmed by analyses of Tables VIII and IX. Admittedly, we can only speak of tendencies, since fertilizer N° 8 differs significantly from the others, but we cannot discount the theory that P_2O_5 has a special role to play, since - let us not forget - it was applied nine months earlier. Further analyses of the distribution of D. citri egg densities per type of fertilizer and allowance for climatic factors (temperature and rainfall) should provide a clearer picture of the effect of fertilization on the development of D. citri.

5. CONCLUSION

In applying integrated methods to citrus fruit groves, many very different factors are likely to play a part. The action of auxiliary factors has been largely clarified, and the study of exchanges of entomophagous insects between cultivated and natural environments is just beginning. As far as insecticides are concerned, a few papers have been published on the local effect of certain active substances, but a procedure for assessing mortality in the field as a result of general or specific treatments has yet to be established.

Certain parameters, however, have yet to be clarified - the effect of different varieties of the plant and, in particular, the influence of crop-growing techniques (fertilization, irrigation, pruning, etc.) which can, owing to the physiology and phenology of the host plant, favour or prevent a balance being established between pests and useful insects. These diverse factors, which are all the more difficult to assess since they act through a host plant whose functioning is as yet little understood, must be considered when drawing up plant protection measures to be applied to citrus fruit trees.

REFERENCES

1. AL-HOUTY, W. (1979). Some ecological studies on the cabbage whitefly, Aleyrodes brassicae (WALKER). Hemiptera-Homoptera-Aleyrodidae. thesis, Bath University U.K.: 192 pp.
2. ARANT, F.S. and JONES, C.M. (1951). Influence of lime and nitrogenous fertilizers on populations of greenbugs infesting oats. J. Econ. Entom., 44 : 121-122.
3. ARCHER, T.L., ONKEN, A.B., MATHESON, R.L. and BYNUM, Jr. E.D., (1982). Nitrogen Fertilizer Influence on Greenbug (Homptera: Aphididae). Dynamics and Damage to Sorghum. J. Econ. Entomol., 75 :695-698.
4. BENASSY, C., (1977). Note sur l'acclimatation en France d'Aphytis lepidosaphes COMP. (Hym., Aphelinidae) parasite de Lepidosaphes beckii NEWM. Fruits, 32(6) : 432-437.
5. BRUN, P. and ONILLON, J.C. (1978). Dynamique du végétal et estimation des populations de ravageurs inféodés aux Citrus. Fruits, 33(12): 807-810.
6. CARRILLO, L.R., and MUNDACA, B.N. (1976). Efecto del nitrogeno en las poblaciones, progenie y velocidad de desarrollo de los afidos Metopolophium dirhodum y Sitobion avenae en trigo (Triticum aestivum L.) Agro. Sur. 4(1) : 15-20.
7. JACKSON, P.R. and HUNTER, P.E. (1983). Effects of Nitrogen Fertilizer Level on Development and Populations of the Pecan Leaf Scorch Mite (Acari : Tetranychidae). J. Econ. Entomol., 76 : 432-435.
8. LAL, S.S. (1981). An ecological study of the whitefly Bemisia tabaci (GENN.) population on Cassava. Pestology, 5(1), 11-20.
9. McNEILLL, S. and SOUTHNOOD, T.R.E. (1978). The role of nitrogen in the development of insect plant relationship : pp. 77-98. In J.B. Harbone (ed.) Biochemical aspects of plant and animal coevolution. Academic Press, Inc. New York, 435 pp.
10. ONILLON, J.C. and ONILLON, Janine. (1972). Contribution à l'étude de la dynamique des populations d'Homoptères inféodés aux Agrumes. III. Introduction dans les Alpes-Maritimes de Cales noacki HOW (Hymenopt., Aphelinidae) parasite d'Aleurothrixus floccosus MASK. (Homopt., Aleyrodidae) C.R. Acad. Agric. France, 58(6) : 365-370.
11. ONILLON, J.C., FRANCO, E. and BRUN, P. (1973). Contribution à l'étude de la dynamique des populations d'Homoptères inféodés aux Agrumes. I.2. Estimation de la surface des feuilles des principales espèces d'Agrumes cultivées en Corse, Fruits, 28(1) : 37-38.
12. ONILLON, J.C., ONILLON Janine and BRUN, P. (1975). Contribution à l'étude de la dynamique des populations d'Homoptères inféodés aux Agrumes. II.3. Premières observations sur l'évolution comparée des

populations de <u>Dialeurodes citri</u> ASHM. en Corse et dans le Sud Est de la France. Fruits, 30(3) : 167-172.

13. ONILLON, J.C., ONILLON Janine, FRANCO, E. and RODOLPHE, F. (1978). Contribution à l'étude de la dynamiques des populations d'Homoptères inféodés aux Agrumes. I.3. Stimulation de la croissance foliaire des éléments de la frondaison du bigaradier, <u>Citrus aurantium</u>. Ann. Zool. Ecol. anim., 10(2): 303-314.

14. ONILLON, J.C., PANIS, A., BRUN, P., ONILLON Janine, FRANCO, E. and MARRO, J.P. (1981). Recherches préalables à l'estimation du seuil de nuisibilité de <u>Dialeurodes citri</u> et de <u>Saissetia oleae</u> en vergers de clémentiniers. In "Standardisation de méthodologie biotechniques sur la lutte intégrée en agrumiculture". San Giuliano-Siniscola, 4-6 nov. 1980: 147-161.

15. PANIS, A., (1981). Note sur quelques insectes auxiliaires régulateurs des populations de <u>Pseudococcidae</u> et de <u>Coccidae</u> (<u>Homoptera</u>, <u>Coccoidea</u>) des agrumes en Provence Orientale. Fruits, 36(1) ; 49-52.

16. VIGGIANI, G and MAZZONE, P.(1976). Introdotta in Italia la <u>Prospaltella lahorensis</u> HOW. per il controllo biologico di <u>Dialeurodes citri</u>. ASHM., Inf. Fitop., 27 : 5-7.

17. PUTTASWAMY and CHANNABASAVANNA, G.P. (1982). Influence of Nitrogen Fertilization of the Host Plant on the Population of <u>Tetranychus ludeni</u> (Acari, <u>Tetranychidae</u>). Indian J. Acar., 6 : 64-71.

Thrips on citrus-groves

S.Longo

Institute of Agricultural Entomology, University of Catania, Italy

Summary

About forty species of Thrips live on citrus in the world; nevertheless only few of them – mostly in the sub-order Terebrantia – actively infest the citrus groves: Heliothrips haemorroidalis Bouchè, Thrips tabaci Lind., Scirtothrips aurantii Faure, S. citri (Moulton), S. dorsalis Hood. The Greenhouse Thrips(H. haemorroidalis) is the commonest species damaging blossoms, fruits and leaves of citrus in Italy.

1. Introduction

Of the 40 or so species of Thysanoptera recorded on citrus in various parts of the world (1), (4), (5), (7), (8), (9), (10), (11), (12), (14), 15), (18), about ten are considered predators of other harmful arthropods. Of the phytophagous species at least 20 are incidental, in that they are present only sporadically and in small numbers on citrus. The remaining sap feeding species are all polyphagous and are not closely associated with citrus; therefore their pullulation on citrus host plants is generally sporadic and occasional and mainly occurs in spring and autumn when the thermo-hygrometric conditions are most favorable for their development.

2.1 Harmful species in Italy and in the Mediterranean area

Of the widest spread species, the cosmopolitan and polyphagous Heliothrips haemorroidalis Bouché is the most frequently found in the Mediterranean area and the one best studied from the point of view of the damage it causes to the fruit and leaves of Citrus. The Greenhouse Thrips is a parthenogenetic species tending to develop homodynamically: in fact, it goes through numerous generations a year living on various host plants, wild and cultivated, and in particular on Viburnum (Viburnum tinus L.). Normally in the South of Italy it has 3 or 4 generations a year according to the thermo-hygrometric conditions which are optimal at values of around 28°C and 85% R.H. (19). The insect overwinters in various postembryonic stages but mainly in the adult one; it is frequently found on citrus from the month of May on the underside of leaves and on young fruit. In the summer there are only thin populations of it, whereas in autumn, and above all on the lemon, there may be a recrudescence of the infestations (9).

Another species occasionally considered harmful to citrus in the Mediterranean area and America is Thrips tabaci Lind. (the Onion Thrips) reported in Sicily (Di Martino, pers. comm.), North Africa (7), Palestine (4) and Cyprus; in the latter country it is the commonest species on citrus flowers together with Taeniothrips meridionalis Priesn., Haplothrips andresi Priesn. and H. flavicintus Karny (12).

T. tabaci is an amphigonic species (even if the male is infrequent, above all outside the Mediterranean basin) (20).

It has from 4 to 5 annual generations and overwinters inthe adult stage in the ground. It causes the most serious damages to various vegetable plants (onions, tomatoes and potatoes) above all as a transmitting agent of fungi and viruses (20).

Another species indicated as harmful to citrus in North Africa, and especially Algeria, is Thrips major Uzel (6). Together with T. meridionalis, this species in Italy causes serious rusty alterations to the skin and scars on nectarine fruit (20). Numerous species of Thrips are frequent on citrus flowers, but their presence is not always to be correlated with damage to the fruit. In 1941 Melis reported the presence of Holarthrothrips tenuicornis Bagn. on citrus flowers in Catania; we have observed other species, Thrips urticae Fabr. and T. alni Uzel, with a certain frequency on lemon flowers in some Sicilian groves.

2.2 Exotic species

Among the exotic species several Scirtothrips have been reported as harmful to citrus. Of these, S. aurantii Faure (the South African Citrus Thrips) is also found in Egypt (18) as well as in South Africa, Rhodesia and the Sudan; S. dorsalis Hood is very widespread and has been reported as harmful to fruit of Satsuma, in Japan (1). The third species, S. citri Moulton is only known in California and in neighboring parts of Mexico (18). In Asia, Africa, Australia and North America Thrips hawaiiensis Morg. is widespread and occasionally infests citrus (8).

3. Control of Thrips

3.1 Economic importance and damage

Among the parasitic entomofauna of citrus, phytophagous Thysanoptera have a place of only secondary importance, owing to the rare frequency of their infestations, which sometimes are yet heavy enough, to cause damage to fruit production and to require specific phytosanitary measures.

Generally only a small percentage of the fruit shows typical alterations caused by Thrips. In many cases, alterations to the fruit displaying the more or less exten sive suberization attribued to Thrips are, instead, of a different origin and are caused for example a rubbing of leaves of twigs against the fruit exposed to the wind (12). Alterations to the fruit epicarp of a brownish–rusty or silvery–ashen color, coriaceous and cracked with a symptomatology that could fit those due to Thrips are attributed to mite infestations (Eriophydae, Tarsonemidae, Tenuipalpidae). As regards the damage caused by H. haemorrhoidalis an agent of the so–called "white rust" on citrus, it appears on the fruit, on the underside of the leaf and sporadically on young twuigs. These alterations are a consequence of the punctures made by the insect with its maxilla and mandible for sucking food; on the underside of the leaf near the rib, the Greenhouse Thrips in addition to the punctures for sucking makes oviposition marks with its ovipositor; mainly on those parts tiny dark spots can be observed which are the excrement of the insect.

As a result of the devitalization of the affected cells, the surface at those points at first displays widespread alteration of a whitish metal color; then a corky layer, sometimes having a more or less characteristically scaly appearance, is formed on the fruit. In case of a strong infestation the fruits become unmarketable and a considerable defoliation can occur, too (8).

3.2 Biotic and abiotic factors of control

In all its stages the development of the Thrips is closely conditioned by tempe rature and hygrometric factors. In particular the eggs laid by the female in the leaf mesophyll, having an incubation period from 15 to 44 days are sensitive to va riations in the turgor of the host plant and are devitalized for a high percentage by a marked withering of the leaves. The species which hibernate in the ground in the adult of nymph stage suffer considerable mortality from rains and irrigation.

Numerous groups of entomophagous have been reported in connection with phy tophagous Thrips; among the most common are various predators species of the sa me order belonging to the genera Aeolothrips, Frankliniella, Franklinothrips and Haplothrips (13). The larvae of A. fasciatus L. prey on preimaginal stages of H. haemorrhoidalis as well as on other insects and acari (16). A. fasciatus   and A. gloriousus are reported in Cyprus in connection with T. tabaci and other phyto phagous Thrips (12). Other more or less occasional entomophagous are in the or ders of the Rhynchota (Nabis and Orius), the Neuroptera (Chrysopa sp.), the Or thoptera (Oecanthus), the Diptera (Syrphus) and Coleoptera (Coccinella, Hippoda mia, Scymnus)(13). As regards S. aurantii which pupates on the ground, some spe cies of the terricolous acari predators (Cunaxa, Cunaxoides, Bdella, Anystis, Gaeolaelaps and Hypoaspis) play a considerable role in controlling its populations (17).

3.3 Chemical control

Normally in citrus groves where anticoccid treatments are carried out in win ter and/or summer, it is rare to find the development of infestations needing speci fic control measures.

The sampling methodology suggested by the experts of the Commission of the European Community in 1982 (6) to establish the amount of H. haemorrhoidalis pre sent, recommends the examination of 4 shoots per tree and the study of 20 fruits per tree over 5-10% of the plants in a citrus grove during spring and autumn, the latter season specifically in the case of lemon ,. The presence of 2-3% of fruit infe sted and 5-10% of leaves is considered to be the economic threshold for such spe cies.

As regards the active agents utilized in integrated pest control in citrus gro wing, it should be remembered that the activated white mineral oils used against coccids are effective against the Greenhouse Thrips. When the populations of this Thrips tend to cross the above-mentioned thresholds, treatment with organophosphou rus compounds at the petals fall may considerable reduces precocious alterations of the fruit. Malathion, among the different chemical compounds, offers the greatest guarantee, above all in view of its limited. secondary effects on the useful entomo fauna. Other organophosphorous tried out in South Africa against S. aurantii are Triazophos and Phantoato (3) and Abate (10).

123

Tab. 1 — Thrips recorded on Citrus in the world.

Species	behaviour and/or economic import.	Countries or Regions	References
TEREBRANTIA			
Aeolothrips deserticola Pr.	P	Israel	(4)
Aeolothrips fasciatus L.	P	Italy,Cyprus	(12)(15)
Aeolothrips gloriosus Bagn.	P	Cyprus	(12)
Caliothrips fasciatus (Pergande)	N	Italy,California,Hawaii	(10)(13)
Chaetanophothrips orchidii (Moulton)	U	Florida	(10)
Chaetonophothrips signipennis (Bagn.)	U	Australia	(10)
Frankliniella cephalica (Crowford)	P P	Florida	(10)
Frankliniella cubensis Hood	P P	Cuba,Puerto Rico	(10)
Frankliniella gossypiana Hood	P P	California	(10)
Frankliniella insularis Franklin	P P	Trop.North and South America	(10)
Frankliniella rodeos Moulton	P P	Brazil	(10)
Frankliniella tritici (Fitch)	P P	Central America	(10)
Franklinothrips myrmicaeformis Jan	P	Israel	(4)
Heliothrips haemorrhoidalis Bouché	H P	America,Australia,Mediter. India	(2)(4)(7)(8)(9)(10)
Heterothrips moreirai (Moulton)	U	Brazil	(10)
Holarthrothrips tenuicornis Bagn.	N	Italy	(14)
Isochaetothrips striatus Hood	U	Brazil	(10)
Kakothrips robustus Uzel	N	Israel	(4)
Limothrips cerealium Hal.	N	Israel	(4)
Odontothrips karnyi rivnayi Pr.	N	Israel	(4)
Selenothrips rubrocinctus (Giard.)	U	Florida	(10)
Scirtothrips aurantii Faure	H P	Africa	(10)(18)
Scirtothrips citri (Moulton)	H P	California,Mexico	(10)(18)
Scirtothrips dorsalis Hood	H P	Japan	(1)
Thrips alni Uzel	N	Italy	pers.ob..
Thrips hawaiiensis (Morg.)	H P	Africa,Asia,Australia,America	(8)
Thrips major Uzel	H P	Israel,Algeria	(4)(5)
Thrips tabaci Lind.	H P	California,Chile,Italy,Israel	(4)(10)(16)
Thrips urticae Fabr.	N	Italy	pers.ob.
Taeniothrips meridionalis Priesn.	N	Cyprus,Israel	(4)(12)
Taeniothrips discolor Karny	N	Israel	(4)
Taeniothrips frici (Uzel)	U	Iran	(10)
TUBULIFERA			
Haplothrips andresi Pr.	P	Cyprus,Israel	(4)(12)
Haplothrips flavicintus (Kary)	P	Cyprus	(12)
Haplothrips tritici Kurdj	P	Israel	(4)
Haplothrips gowdey Frankl.	P P	Israel	(4)
Haplothrips cypriotes Pr.	P P	Israel	(4)
Haplothrips spp.	P P	Israel	(4)
Hoplandrothrips sp.	N	Israel	(4)
Karnyothrips flavipes Jon.	P P	Israel	(4)

P = predator; PP = probably predator; HP = harmful phytophagous; N = phytophagous of negligible economic importance; U = unknown behaviour.

124

REFERENCES

1. ANONIMO (1981). Major Citrus Insect Pest in Japan. Japan Plant Protection Association Tokyo: 1-16.
2. AVIDOV, Z. and HARPAZ, J. (1969). Plant Pest of Israel. Ierusalem Israel Univ. Press.: 1-549.
3. BEDFORD, E.C.G. (1976). The present status of the integrated control of Citrus pests in South Africa. Citrus & Subtropical Fruit Research Institute. Nelspruit. South Africa: 203-215.
4. BODENHEIMER, F.S. (1951). Citrus Entomology in the Middle East S. Gravenhage-Groningen: 1-662.
5. BOURNIER, A. (1963). Un nouveau déprédateur des agrumes en Afrique du Nord: Thrips major Uzel. Rev.Path.veg. 42, 2: 119-125.
6. CAVALLORO, R. and PROTA, R. (1982). Integrated control in citrus orchards sampling methodology and threshold for intervention against the principal phytophagous pest. Proceedings of the E.C. – Experts Meeting. Integrated control in citrus comparison of results archivied by appling a standardized methodology, 43-51.
7. CHAPOT, H. and DELUCCHI, V.L. (1964). Maladies, troubles et ravageurs des agrumes au Maroc. Inst.Nat.Rech.Agron.Rabat, 1-339.
8. COMMONWEALTH INSTITUTE OF ENTOMOLOGY (1983). Distribution Maps of Pests. Ser. A Map. N° 431 (Revised), London.
9. DI MARTINO, E. (1956). La ruggine bianca dei Limoni. Rivista di Agrumicoltura, 1: 395-399.
10. EBELING, W. (1959). Subtropical Fruit Pest. Univ. of California Division of Agr. Sciences: 1-436.
11. GEORGALA, M.B. (1974). Thrips control: slow progress whith new treatment. Citrus and Sub-Trop. Fruit Journal N° 484, 6: 8-13.
12. IORDANOU, N.C. and IOANNOU, Y.M. (1979). Effect of thrips and wind on the development of epidermal injuries on citrus fruits in Cyprus. Techn.Bull. Agric.Res.Inst., Min.of Agr. and Nat.Res., 26: 1-12.
13. LEWIS, T. (1973). Thrips, their biology, ecology and economic importance. Academic press. London and New York, 1-349.
14. MELIS, A. (1941). Tisanotteri italiani VII. Genus Holarthrothrips. Redia, 27: 25-43.
15. MELIS, A. (1960). I Tisanotteri italiani (II). Redia, 45: 185-329.
16. MELIS, A. (1961). I Tisanotteri italiani (III). Redia, 46: 331-530.
17. MILNE, D.L. (1977). Biological control of citrus thrips, Scirtothrips aurantii: what are the prospects? Citrus and Subtropical Fruit Journal, 528: 14-16.
18. MOUND, L.A. and PALMER, J.M. (1981). Identification, distribution and host-plants of the pest species of Scirtothrips (Thysanoptera: Thripidae). Bull.ent. Res., 71: 467-479.
19. RIVNAY, E. (1935). Ecological studies of the greenhouse thrips: Heliothrips haemorrhoidalis, in Palestine. Bull.Ent.Res., 26: 267-278.
20. TREMBLAY, E. (1981). Entomologia applicata, 2: 1-310.

125

On citrus trees Lepidoptera and particularly on those inhabiting the Mediterranean area

G.Mineo

Institute of Agricultural Entomology, University of Palermo, Italy

Summary

The Lepidoptera -key and minor pests- of Citrus in the world are
treated with particular mention of the species attacking the citrus-
orchards in the Mediterranean Basin. For many of them bio-ethological
accounts and their economic importance are also reported.

Among the insects infecting citrus trees in the various citrus areas
in the world we find a remarkable amount of species of Lepidotera and among
these a large number seems to be dangerous in an adult stage.

Apart from some species steadily living on Citrus spp., the others are
highly poliphagous an many of them have a limited relationship with citrus
trees; moreover their settlement may be primary or secondary.

Ephestia cautella Walker, Criptoblabes gnidiella Millière and
Ectomyelois ceratoniae Zeller belong to this second group: their presence
on citrus trees is related to Pseudococcid's contemporary one. Larvae of
these moths scratch the bark of the twigs, the epidermis of leaves and
sometimes the epicarp of fruits. On these they can even provoke some holes
or little tunnels in the first layers of the mesocarp.

Bodenheimer and Klein (1934) observed damages similar to those caused
by Ephestia vapidella Mannerheim for C. gnidiella in Israel.

This latter moth according to Avidov and Gothilf (1960) has 5-6 generations with a x ♀ number of eggs of 150-284 in Israel. Life span of
adults varies from about one week to about one month in autumn.

According to Delucchi (1964, in Chapot-Delucchi) there are 5 generations in Morocco, but a maximal number of x ♀ eggs of 98.

Eggs are laid near the cottonlike mass of Pseudococcids where we can
find up to 15 of them, according to the above mentioned authors.

In southern Italy C. gnidiella may have 3 or 4 generations (Silvestri,
1943; Liotta e Mineo, 1964).

Female fecundity observed in Sicily has been much inferior to the one
noticed by the previous authors (15-25 eggs). E. cautella and E. ceratoniae may be living together with C. gnidiella; their phenology and habits
are similar to the former (Mineo, 1964; Delucchi, 1964; Agenjo, 1959; Avidov e Gothilf, 1960; Limon de la Oliva et al.,1972; Gerini, 1977).

The species whose attacks to citrus trees happen in a primary form
belong to various groups. Among these one of most abundant is that of the
Tortricids, whose larvae feed both on shoots which they cover with silky
threads and on fruits. Among these, Cacoecimorpha pronubana (Huebner) and

Archips rosanus L. are present in the mediterranean area, C. citrinella Schir in Japan, C. epicyrta Meyrick and C. pensilia Meyr. In India, C. isocyrta Meyr. in Malaysia, Eulia citrina (Fiernald), Platynota stultana Barnes et Busck in the U.S.A. (Silvestri, 1943), Adoxophyes cyrtosema Mey., A. orana Fischer von Röslerstamm, Archips micaceana var. compacta (Nietner) and Homona coffearia Mey. in China.

C. pronubana is quite well known in the Mediterranea area and its relationship with citrus trees has been studied by Delucchi and Merle (1962) and by Limon de la Oliva et al. (1972). In Morocco where it is particularly dangerous the two above mentioned authors report the following data. From January to March the newly emerged larvae infest sprouts where they first eat the epidermis of the leaves, being hidden beneath a sort of nest made of silken threads; they also bind the little leaves one to the other through such threads; larvae of the first stage more seldom behave as miners.

From April to July one can notice the attack on fruits where the newly emerged larvae settle in both under the calyx and in the contact point between two fruits or better between a leaf and a fruit.

Whatever their place of settlement is, they feed on the escarps, being well protected with nests of silken threads.

Infested fruits keep on enlarging, but healings got on the eaten parts disfigure them so much that they are no longer marketable.

When the attack extends towards ripe fruits (October-November) and there is no further healing, these become rotten and fall.

Still in Morocco, where C. pronubana would accomplish 5 generations, this tortricid infests all the species of citrus trees cultivated there, and the percentage of infested fruits may reach 90%.

The other above mentioned species of Tortricids have ethological behaviours very similar to those mentioned for C. pronubana. On these literature is very scarce. The same thing happens with Archips rosanus present in a few Mediterranean citrus areas (Sicily and Campania). It seems to accomplish one generation per year.

Another interesting group of species is given by Noctuids and Papilionids which eat the leaves of citrus trees. These latter ones can be found above all in the genius Papilio. Among these Papilio demoleus (L.) is spread in China, India, Indonesia, P. xuthus L. is present in Japan and Haway; P. demodocus Esp., P. thoas L. and P. anchisiades capys Huebener are known in tropical and subtropical areas in Northern and Southern America and in Africa (P. demodocus), P. machaon sphyrus Huebner is present in Israel.

P. demoleus is considered in China, India and Indonesia a key insect. Roa (1969) Mishra & Pandey (1965) have studied its biology in Kanpur, where the Papilionid infests mostly young plants. According to the observations of the above mentioned authors, the species overwinters at chrysalis stage; every female may oviposite 40-183 eggs on leaves; 5-6 generations may follow one another in a year. Also Badawi (1969) observes P. demoleus dangerous especially for citrus trees nurseries in Sudan. This Author studies its biological cycle in laboratory, noticing that larvae pass through four moults before reaching maturity and that duration of crysalis stage becamed remarkably longer in the winter months up to over three months.

Moreover Badawi (1981), in another study carried out in field and in laboratory in Saudi Arabia, observes the accomplishing of eight generations per year; the average leaf surface eaten by each larva is of about 284 cm^2.

Yunus and Munir (1972) study the susceptibility of 18 species and varieties of Citrus and one of Poncirus infesting by P. demoleus. On C. aurantium and C. limon larvae of Papilionid develops more quickly (12,5 days). The above mentioned authors report that 4-6 larvae are sufficient in field to eliminate the leaves of little plants of Poncirus 32-64 cm high.

As far as P. xuthus is concerned, Suzuki et al. (1976) have carried out trials on the space distribution of eggs on citrus plantations in field; Nishida (1977) examines the effect of some substances contained in host plants stimulating oviposition; Hirose et al. (1980) study its population dymanics in various citrus orchards; they observe that the species accomplishes 4 generations per year in Japan and that the space distribution of eggs is relatively high, as already observed by Suzuki et al. (1976), moreover the stability of population in every generation is kept by parasitoids: Trichogramma spp. and Pteromalus puparum (L.) whose generations are faster than the host's. Bodenheimer (1951) reports on P. machaon sphyrus. This species also noticed in the citrus orchards in Cyprus and in Israel, accomplishes 3 generations per year, with an overwintering at the crysalis stage, as it happens for nearly all the above mentioned species. Up to now, its presence in the citrus orchards of the Mediterranean area is merely decorative, as it has been observed by Bodenheimer.

Noctuids may be dangerous either at larval stage or at adult stage. Among these, Roa (1969) quotes Achea janata (L.), A. tirrhaca (Cramer), A. catella Guenée, A. echo (Walker), A. lineardi (Boisduval), A. sordida (Walker), Tiracola plagiata (Walker) and perhaps, according to Edwards (1978) also A. argilla Swinh.

A. argilla is present in Australia; A. janata and A. tirrhaca are spread in India; but the first one is present also in Burma, Sri Lanka, Africa and Australia; while the other species of Achaea are at the moment known only in Zimbabwe; T. plagiata is known in China and many other regions of Southern Asia (Malaysia, New Guinea, Papua, Indonesia, Sri Lanka and India). Moreover similar habits have in China the following species: Oraesia emarginata F., O. excavata Butler, Adris tyrannus (Guenée), Maenas salaminia (F.), Lagoptera dotata (F.), Achaea triphaemoides (Wlk.), Parallelia arctotaemia (Guenée), Metopta rectifasciata (Menestriés), Speiredomia retorta (L.), Mocis nudata (F.).

Adults of all the above mentioned species pierce oranges to extract their juices.

Noctuids dangerous at the larval stage are the following species: Laphigma exigua Huebner, Chloridea obsoleta F., C. nubigera Herrich-Schaeffer, Scotia segetum Schff., Trichoplusia Hb., Prodenia litura L., they are all present in the Mediterranean area and like the Papilionids, they may be considered as occasionally eating leaves of nurseries and greenhouses, even if a carpophagous diet has been noticed in some species.

129

Of these P. litura and L. exigua have been more studied by Bodenheimer (1951) in their relationship with citrus trees.

P. litura may accomplish up to 8 generations per year in Israel, but economical damages are caused only by the populations which develop in July and August as the populations present in May and September are controlled (up to 99%) by high temperatures caused by hot desert winds. Temperatures above 40°C (even for few hours) are lethal for young larvae and eggs. Already after 29°C development shows a remarkable slowing down while mortality increases considerably.

Temperatures near zero are also lethal for the insect apart from the individual which are at crysalis stage. L. exigua shows habitus and biology similar to P. litura. Bodenheimer (1951) observes eight generations per year of this Noctuid in Israel, with an overwintering in the larval stage. Its biological cycle becames faster when temperature goes above 28°C, conditions adverse to the development of the insect begin appearing. As regards R.H., it seems certain that high percentage of humidity together with low as high temperatures are infavourable to the development of this Noctuid. Number of eggs x ♀ is between 300-500, with a maxium of 1700.

The species may accomplish up to 18 generations per year in Java. Euxoa segetum seems to accomplish 3 generations per year in Israel, but as it happens for Phytometrani, damages to nurseries, usually noticed in April-May, are very rare.

The same happens for C. peltigera, C. nubigera and C. obsoleta. Larvae of the latter species may eat also the skin of ripe fruits. High infestations to citrus groves in South Africa, especially in spring, are caused by it.

Secondary Lepidopterous pests of citrus trees are: Gymnoscelis pumilata Huebner, Ephestia vapidella Mannerheim, Phyllocnistis citrella Stainton, Charaxes jasius L., they are all present in various citrus area of Palaearctic region.

Diet of G. pumilata on citrus trees is anthophagous, the most recent news about such crops are given by Delucchi (1964) in Morocco and by Mineo (1964) in Sicily. The species may accomplish 6 generations per year in Sicily.

Its damages to flowers are economically irrelevant.

First news about E. vapidella have been by Bodenheimer and Klein (1934) who found it in Israel living on cambium of root stocks corresponding to the grafts when they are still protected by proper bandages.

Afterwards it has been noticed by Delucchi (1964) in Morocco on Citrus spp., living both on Citrus spp., and on not specified citrus areas of Middle East. Biology of this species should be better pointed out. Phyllocnistis citrella is present in the Mediterranean area (Israel, Cyprus) and in the Middle East (Irak), Eastern, tropical and temperate Asia, with neighbouring isles, but it has also been introduced in Australia (Silvestri, 1943) and Southern Africa (Wiltshire, 1944; Bodenheimer, 1951).

Eggs are deposited on younges sprouts (axis and inferior page of the little leaves).

Larvae behave as leaf-miners, becaming crysalis in the final part of miners after building a proper cell. Infested sprouts grow with difficulty, curl up and appear as if they had been attacked by colonies of Aphids.

According to Silvestri (1943) in the tropical regions or in nurseries, the species reproduces itself the whole year around; it overwinters as an adult in Japan; it may accomplish even more than 6 generations per year. Roa (1969) referring to Indian citrus orchards, mentions it occasionally and affirms that it may often become a key-pest.

Ch. jasius is recorded as living on young citrus plants by Morris (quoted by Bodenheimer, 1951).

Many records of Cossids living on sections and trunks of citrus trees are quoted: Cossus cossus L. and Zeuzera pyrina L. in Italy by De Stefani (1904), by Silvestri (1943) and by Costantino (1937); Xyleutes strigilatus (Felder) in Brasil, X. punctifer Hamps in Western Indies; other species of Indarbela Fletcher are reported by Silvestri (1951) in tropical Asia.

Plants mostly damaged by larvae of these Cossids are young, they may become completely destroyed specially when trunks are damaged.

List of Lepidotera species will most probably be longer than the list here given. We want still to consider two Plutellids, the first one is Prays endocarpus Meyrick known, according to Silvestri (1943) only in Philippines where it damages Citrus spp. fruits; the second one is Prays citri Millière spread in all the Mediterranean citrus areas when it develops mainly on lemon and citron trees. Laboratory and field trials have been carried out in Italy, Portugal and Israel on the biology of this species.

As regards organs on which it may develop, P. citri behaves almost like P. oleae in the olive tree. Its development cycles are mainly in-fluenced by temperatures and in one year, in field, 14 generations have been noticed.

Though the species is present the whole year in all its developmental stages, the biggest damages are on the re-flourishing varieties of lemon, on late spring flowerings and on late summer flowerings which produce "verdelli".

Observations on the trend of moths' flight in field have been carried out by Istituti di Entomologia Agraria of Palermo and Catania using pheromon traps.

Tests by researchers of the two Institutes agree on the fact that the technique of "forzatura" highly retains moth populations. Such an effect is probably due to the period of "secca" which lemon orcharda undergo, dur-ing which the Plutellid does not find receptive organs for its development on plants.

REFERENCES

1. AGENJO, R. (1959). La polilla de las garrofas, plaga actual de las narajas (Lep. Phycit.). Graellsia, 17:7-17.
2. AVIDOV, Z. and GOTHILF, S. (1960). Observations on the honeyden moth (Cryptoblabes gnidiella Millière) in Israel. I. Biology, phenology and economic importance. Ktavim, 10:109-124.

3. BADAWI, A. (1969). Biological studies on Papilio demoleus L. a pest of citrus trees in the Sudan (Lepidotera: Papilionidae). Bulletin de la Société Entomologique d'Egypte, 52:397-402.

4. BADAWIN, A. (1981). Studies on some aspects of the biology and ecology of the citrus butterfly Papilio demoleus L. in Saudi Arabia (Papilionidae, Lepidoptera). Zeitschrift für Angewandte Entomologie, 91: 286-292.

5. BODENHEIMER, S.F. (1951). Citrus Entomology in the Middle East. 'S-Gravenhage, W. Junk, 663 pp.

6. BODENHEIMER, S.F. and KLEIN, Z.H. (1934). On some moths injurious to citrus trees in Palestine. Hadar, 7:8-10.

7. CHAPOT, H. and DELUCCHI, L.V. (1964). Maladies, troubles et ravageurs des agrumes au Maroc. INRA, Rabat, 339 pp.

8. COSTANTINO, G. (1937). Il Cossus cossus L. dannoso agli agrumi. Annali della R. Stazione sperimentale di agrumicoltura e frutticoltura in Acireale, 14:199-203.

9. DELUCCHI, L.V. and MERLE, L. (1962). La tordeuse de l'oeillet Cacoecia pronubana Huebner (Lepidoptera, Tortricidae) ravageur peu connu des agrumes au Maroc. Al Awamia, 3:79-86.

10. DE STEFANI, T. (1904). Il rodilegno nei limoni. Nuovi annali di agricoltura Siciliana. 15 (1):7-13.

11. EDWARDS, E.D. (1978). A review of the genus Achaea Huebner in Australia (Lepidoptera: Noctuidae). Journal of Australian Entomological Society., 17:329-340.

12. GERINI, V. (1977). Contributo alla conoscenza dei principali insetti presenti sugli agrumi a Cipro. Rivista di agricoltura subtropicale e tropicale, 71:149-159.

13. HIROSE, Y., SUZUKI, Y., TAKAGI, M., HIEHATA, K., YAMASAKI, M., KIMOTO, H., YAMANAKA, M., IGA, M., YAMAGUCHI, K. (1980). Population dynamics of the citrus swallowtail, Papilio xuthus Linné (Lepidoptera: Papilionidae): mechanism stabilizing its numbers. Researches on Population Ecology, 21:260-285.

14. LIMON DE LA OLIVA, F., BLASCO PASQUAL, J., VICENTE LOPEZ, S., VERNIERE FERNANDEZ, C. (1972). Ciclos biologicos de algunas plagas y enfermedades del naranjo. Boletin Informativo de Plagas, 98:19-40.

15. LIOTTA, G. and MINEO, G. (1964). La Cryptoblabes gnidiella Mill. o tignola rigata degli agrumi (Lep. Pyralidae) Osservazioni biologiche in Sicilia. Bollettino dell'Istituto di Entomologia Agraria e dell'Osservatorio di Fitopatologia di Palermo, 5:155-172.

16. MINEO, G. (1964a). L'Ephestia cautella Walk (Tignola dei fichi secchi) infesta anche gli agrumi (Osservazioni biologiche). Ibidem, 5:1-23.

17. MINEO, G. (1964b). Osservazioni sulla biologia della Gymnoscelis pumilata Hb. (Lepidoptera: Geometridae). Ibidem, 5:173-192.

18. MISHRA, S.C. and PANDEY, N.D. (1965). Some observations on the biology of Papilio demoleus Linn. (Papilionidae, Lepidoptera). Labdev J. Sci. Technolog., 3:142-143.

19. NISHIDA, R. (1977). Oviposition stimulants of some papilionid butterflies contained in their host plants. Botyu-Kagaku. 42:133-140.

20. ROA, V.P. (1969). India as a source of natural enemies of pests of citrus. Proceedings 1st International Citrus Symposium. 2:785-792.
21. SILVESTRI, F. (1943). Compendio di Entomologia applicata. II. Tip. Della Torre, 699 pp.
22. SUZUKI, Y., YAMAGUCHI, K., IGA, M., HIROSE, Y., KIMOTO, H. (1976). Spatial distribution of the eggs of Papilio xuthus Linné (Lepidoptera, Papilionidae) in a citrus grove. Japanese Journal of Applied Entomology and Zoology, 20:177-183.
23. YUNUS, M. and MUNIR, M. (1972). Host plants and host preference of lemon butterfly, Papilio demoleus Linn., caterpillars. Pakistan Journal of Zoology, 4:231-232.

Tephritid pests in citriculture

G.Delrio

Institute of Agricultural Entomology, University of Sassari, Italy

Summary

Various fruit flies (Tephritidae) attack citrus fruit in the most important fruit growing areas in the world. These are tropical and subtropical species, broad-range exploiters of pulpy fruits, multivoltine
and characterized by a great capacity for movement and by a high reproductive potential. Ceratitis capitata (Wied.) is the most important
species, because of its large distribution area, high poliphagous behaviour and its ability to adapt to different climatic and agricultural conditions. C. capitata population densities in the Mediterranean basin are largely determined by climatic factors and by the sequence and availability of host fruits, being highest in areas of
mixed fruit cultivation and in subtropical climates. The seasonal
fluctuation of this species shows a characteristic trend with a minimum population in winter, a maximum at the end of summer and an aggregation in citrus groves in autumn. Fruit flies do not appear to be
well adapted to development in citrus fruit where eggs and larvae suffer high mortality because of toxicity of essential oils in the peel.
The biological and ecological characteristics and an extremely low
economic threshold of fruit flies attacking citrus determine the ways
in which populations may be manipulated for pest management purposes.

1. INTRODUCTION

Fruit flies (Tephritidae) represent a constant threat for citrus cultivations in many parts of the world. One or more species attack fruit in the
most important citrus growing areas and only the U.S.A. (Florida and California) is free from these pests because quarantines have held accidental
introduction to a minimum and the infestations were detected early and
eradicated at great expense. In the infested zones the economic damage is
considerable whether in direct losses, in field control costs or in the
need to fumigate fruit from quarantined areas. Fruit fly control with insecticides disrupts, furthermore, the citrus biocenose   and makes the biological control against other citrus pests less efficient.

Numerous tephritid species have been found to infest citrus , but only
a few are considered important, either because of their larger distribution
area or their constant attacks on citrus fruits (25,70). Among these, the
following are still confined almost entirely to their countries of origin:
the Mexican fruit fly, Anastrepha ludens (Loew), the South American fruit

135

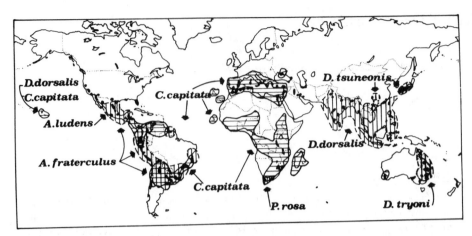

Fig. 1. Distribution areas of principal Tephritid pests and citrus cultivation zones (black marks).

fly, A. fraterculus (Wied.), the Queensland fruit fly, Dacus tryoni (Frogg)
the Chinese and Japanese orange fruit fly, D. tsuneonis (Miyake) and the
Natal fruit fly, Pterandrus rosa (Karsch). The Oriental fruit fly, Dacus
dorsalis Handel, from South East Asia has accidentally been introduced to
Hawaii and the Mediterranean fruit fly, Ceratitis capitata (Wied.), is the
most notorious and widespread of fruit flies attacking citrus (Fig. 1).
Other tephritids such as the West Indian fruit fly, A. mombinpraeoptans
Sein, the Caribbean fruit fly, A. suspensa (Loew), and Dacus zonatus
(Saunders) have been found to infest citrus fruit, but are considered less
important than the species indicated above.

2. BIOLOGY AND ECOLOGY OF FRUIT FLIES

The principal tephritids which damage citrus fruits are tropical and
subtropical species, multivoltine and broad-range exploiters of fleshy
fruit. They attack a great variety of hosts (C. capitata more than 200, D.
dorsalis and D. tryoni more than 150) (11,30), often more preferred and
susceptible than the citrus. These alternative hosts contribute to maintaining populations at a high level regardless of the maturity or availability of susceptible varieties of citrus. The most important cultivated hosts
are citrus, apricot, peach, coffee, mango, papaya, guava, loquat, fig, persimmon, pear, etc. Also many wild hosts contribute in certain areas to maintaining fruit fly populations at a high level. For example, A. ludens frequently attacks the yellow chapote (Sargentia greggii), a wild citrus which
grows extensively in north eastern Mexico (32), A. fraterculus particularly
infests various species of Eugenia and Spondias in Brazil (44), D. dorsalis
in Hawaii develops mainly in wild guava (Psidium guajava) (35, 73). C. capitata develops in great numbers in the fruit of the argan tree (Argania spinosa), a relic of tropical vegetation that covers a considerable area in
Morocco (62) and also in the fruit of the prickly pear (Opuntia ficus-indi-
ca) in some Mediterranean areas.

Adult fruit flies are long-lived and can, in certain conditions, survive for more than a year. They feed primarely on honeydew secreted by homo-
pterous insects, and also on glandular secretions of plants, nectar and sap
exuding from fruit, leaf or stem injuries. A diet rich in aminoacids is
necessary for reaching sexual maturity and it has been recently suggested
that an important source of protein is represented by the phyllosphere bacteria, which are eaten avidly by the flies (6, 23, 34).

The adults are strong flyers and possess a great capacity for dispersal, which can allow them to cover tens of kilometers. Long distance
flights may take place between emergence from the soil and the onset of
sexual maturity, from and to overwintering sites and particularly after the
disappearance of fruit which the population has been utilizing for oviposition (6, 7).

To locate potential hosts, fruit fly females are believed to respond
to plant-associated stimuli perceptible from a distance (e.g., attractive
fruit volatiles; plant color, form and size; form, size and color of individual fruits, which together elicit attraction after fly arrival on host
plant) and stimuli perceptible upon direct contact (shape, size, color, surface structure, chemical factors which elicit attempts of boring and oviposition)(54).

Eggs are deposited singly (A. fraterculus, A. ludens) or in clusters
(C. capitata, D. tryoni, D. dorsalis and D. tsuneonis) in a puncture made
by the ovipositor just below the surface of the fruit (5, 11, 30, 50). One
important feature of some species (e.g., D. dorsalis) is that they show a
pronounced tendency to oviposite in old lesions, and in particular,oviposi-
tion holes made by previous females (11). The females of some species (A.
fraterculus, C. capitata) mark their oviposition sites with pheromones
which deter repeated egg-laying (55, 56), but no evidence was found for the
existence of deterrence in other species of tephritids investigated (D.
tryoni, D. dorsalis) (7). However, the avoidance of fruit that already contain larvae, has been shown to occur with ovipositing females of D. tryoni
(30).

The females possess high fecundity and the maximum number of eggs laid
per female varies from about a thousand for C. capitata to over 3000 for A.
ludens and D. dorsalis (11). Fruit flies are characterized by a rapid development of the immature stages determined by the host fruit conditions and
the temperature. The duration of the larval development is in fact, strongly
influenced by the species and variety of the host fruit and by its physical
structure and chemical composition (11). The development of the immature
stages can occur at constant temperatures, between 11–13°C and 32–33°C, but
survival is possible in a much wider range (3, 6, 8, 48, 49, 53).
Temperatures slightly above 0°C lasting several days are sufficient to kill
eggs and larvae in the fruit (2). Resistance to low temperatures is particularly high in pupae and adults, that can withstand temperatures below
0°C and resist for a few hours even at -4°C (19, 48). A few hours at 45°C
seems to be the upper limit, regardless of the stage of development, but
populations are likely to suffer high mortality in all immature stages if
daily maximums persist in the region of 36–38°C for more than a week (19).
The adult has a heat tolerance which is similar to that of the immature

137

stages, but they can escape high temperatures taking refuge in cooler micro-climates, such as the under surface of shaded leaves. Temperature and relative humidity strongly influence adult activity: the preoviposition period is extended with the lowering of temperature, while mating and egg laying normally occur above 15-16°C and the optimum for egg production is situated between 25 and 30°C (6, 11, 19).

The various species differ slightly in their reaction to climatic factors; for example, under the same conditions the duration of a C. capitata generation was shorter than that of the D. dorsalis (49) and, furthermore, the first species seems more adapt to temperate climates than the second. In Hawaii in fact, where the two species coexist, C. capitata is more widespread in the cooler temperatures of the higher regions, while D. dorsalis is better adapted to the lower tropical areas (35). These different reactions to climate, together with competition between the species could explain, partially, the lack of C. capitata colonization in the south east asiatic which is the distribution area of D. dorsalis. Also in South America, C. capitata occurs in higher frequencies in the temperate regions of Brazil and Argentina, while A. fraterculus is more common in tropical zones (44).

Egg-laying occurs in the field from the beginning of spring to late autumn, provided there are fruit hosts available, and the development cycle, which in ideal situations, lasts less than a month, can be extended in cold temperatures for 2-3 months. Wintering is normally passed at the adult stage, but can occur in the larval stage in the fruit and the pupal stage in the ground. Fruit flies show a distinct seasonal pattern of abundance, determined by the climatic conditions and the availability of fruit hosts. In the temperate zone the fly populations grow typically in the summer, reaching a peak in autumn and declining rapidly in the colder season. In the tropical zones the seasonal fluctuation seems to depend instead on the availability of fruit suitable for infestation. When one or more hosts are available for the whole year the population remains always at a high level; when the host sequence is interrupted for a certain period, the population falls to a low level, but can start immediately a new and important attack on another host that has become available (24, 32, 37, 43).

The factors responsible for long-term trends in populations are very poorly understood. Most species of fruit flies are attacked in their areas of origin by a complex of hymenopter parasites, which normally exist at quite low densities even in the wild host and rarely influence infestation rates in cultivated hosts. The mature larvae in the ground and the pupae, furthermore, are attacked by a broad spectrum of soil insects (ants, Staphilinids, Carabids, etc). The availability and the sequence of susceptible hosts, together with abiotic factors, seem to be more important than the natural enemies in determining the abundance of the fruit flies. Moisture in particular seems of primary importance at least in various areas characterized by very dry seasons (6, 7, 8).

3. CERATITIS CAPITATA IN THE MEDITERRANEAN AREA

The Medfly is highly plastic in its ability to adapt to different ecological conditions and occurs in areas of diverse climates, agriculture,

138

and host plants. In fact, the Medfly was widely distributed in Asia, Africa, Central America, Australia and Hawaii and has been established for some time in the Mediterranean area where it constitutes one of the most important pests in fruit groves.

Numerous studies have been carried out on individual ecology (reactions to climatic factors, requirements for adult and larval food, movement, reproductive behaviour), but knowledge of the ecology of populations (phenology, population dynamics, reproduction strategy, genetic constitution,etc) is far less complete. Generally, only the infestations of the different hosts during the year and the adult seasonal fluctuations recorded by means of traps, are well known in various Mediterranean countries, while the factors responsible for long-term trends in populations have been poorly covered.

3.1 Ecology of individuals

Climate plays a fundamental role as the determining factor of Medfly abundance, influencing development, behaviour of individuals and survival, limiting therefore the geographical distribution of the insect. The temperature, together with the host fruit, acts upon the development rate and determines the number of generations produced by the Medfly in each area.

The relationship between constant temperatures and developmental rate has been clearly defined and the threshold of development was found to be in the neighbourhood of 10-12°C. The threshold of reproduction is on the contrary considerably higher situated at around 16°C for ovarian maturation and oviposition, and about 16-18°C for mating. The availability of temperature thresholds and thermal constants appropriate to each life stage, makes possible the development of models for estimating the life cycle of the Medfly under various climatic conditions (8, 16, 19, 21, 61).

The duration of a generation in the Mediterranean zone varies from about 20 days in the summer to 2-3 months in the winter. The number of generations produced in a year by the Medfly is around 6-7 in the areas of typically Mediterranean climate (e.g., Algeria, Southern Italy), but is reduced to only 2-3 in more northern areas (e.g., Central Italy) (29, 45).

Tolerance to thermal extremes is very high in the different stages of the Medfly. Temperatures of 40°C for a few hours represent the upper limit for eggs and larvae; the pupae are more resistant and can tolerate temperatures of 46°C for several hours, while adults die after only a few minutes of exposure at 40°C (19). No Medfly eggs or larvae survive exposure at 4-7 °C for 7 weeks, 0.5-4°C for 3 weeks, 0°C for 2 weeks or -2.5°C for 1 week (2, 46). Adults and pupae are more resistant to low temperature and survive for a few hours even at temperatures of -4, -6°C (9, 19).

Medfly distribution is limited by winter climatic conditions: average monthly temperatures lower than 2°C for 1-3 months consecutively or under 10°C for 4 months, prevent insect settlement in the locality concerned (8). Therefore in many areas of temperate climate the summer infestations must be attributed to recurrent invasions, because the winter conditions do not permit the overwintering of the insect.

Medfly do not present diapause, but only a slowing down of development at low temperatures. Overwintering can occur at any stage; in winter, adults can only be found in areas with mild climates, while moving gradually north there is a progressive tendency for the species to pass the winter as larvae or pupae. Overwintering adults tend to aggregate on evergreen plants (e.g. citrus) while larvae, in the Mediterranean basin, develop during the winter above all on prickly pear and sour orange, but also on other secondary hosts of minor importance (8, 10, 13, 18).

Moisture together with temperature, influences pupal survival as well as adult longevity and fecundity. In particular, saturation of the soil occuring during seasons of heavy rainfall, can cause high mortality. Relative humidity in the air, together with high temperatures, is considered the determining factor for population abundance in dry climatic regions, for example Israel, where dry winds decimate the adults. Models based on average monthly temperatures and humidity (ecoclimograms) have been proposed to explain Medfly population abundance. Recent studies have demonstrated an optimum longevity and fecundity between 60% and 75% R.H. and a minor tolerance to extremes of relative humidity (8, 19).

The host range of the Medfly includes more than 250 species of fruit and vegetables. In the Mediterranean basin the principal host sequence as from spring is represented by sour orange, grapefruit, late varieties of sweet orange, loquat, apricot, peach, fig, pear, apple, persimmon, prickly pear, clementine and early varieties of sweet orange. There also exists, however, a series of alternative hosts, such as guavas, mangos, plums,dates pomegranates, peppers, Lycium subglobosum, Aberia caffra, etc, which can be of great local importance.

Females locate potential hosts by chemical and visual cues and are particularly attracted to the more aromatic fruits (stone fruit, guava, citrus approaching ripeness. Nevertheless, in periods of host shortage, immature fruit and fruit with a high mechanical resistance to oviposition penetration (e.g. apple) can also be attacked.

The nature of the hosts directly influences the egg-larval mortality, the rate of development and the quality of adults. Substances present in some fruit, e.g. apple and persimmon tannins and citrus essential oils, kill high percentages of eggs and newly-hatched larvae. Development is normally faster in fruit in the course of maturation and rich in sugars (e.g. figs, peaches) and is generally slower in unripe fruit with compact pulp (e.g. apple) or acid fruit (e.g. citrus). Adult body size, longevity and fecundity in Medfly is affected both by the nature of fruit used as the larval host and by the presence of larval conspecifics within the same fruit. Fecundity results higher in females that have developed, for example,on stone fruits as compared to citrus or prickly pear (12, 17, 20, 28, 39, 51, 58).

The larval competition that is to be found sometimes in overinfested fruit in situations of host shortage determines the development of small pupae and adults. The female, nevertheless, releases after egg-laying a marking pheromone which deters repeated oviposition and therefore limits larval competition in nature (39, 56).

The Medfly possess a great capacity for dispersion and can fly for several kilometers even over the sea or among non-host plants (7, 64). The gradual trap catches of the first adults in the year in Spain also suggeste migrations tending from coastal zones towards the interior (60). Short distance movements are, without doubt, particularly influenced by the search for food and host plants as well as by sexual activity. The attraction exercised by host plants must be particularly important since adults appear often suddenly in a grove coinciding with the fruit maturation period.

3.2 Ecology of populations

Population densities of the Medfly in different fruit-growing areas depend on the type of cultivation (specialized or mixed, species and variety), on crop sanitation practices and on climate factors. Population densities

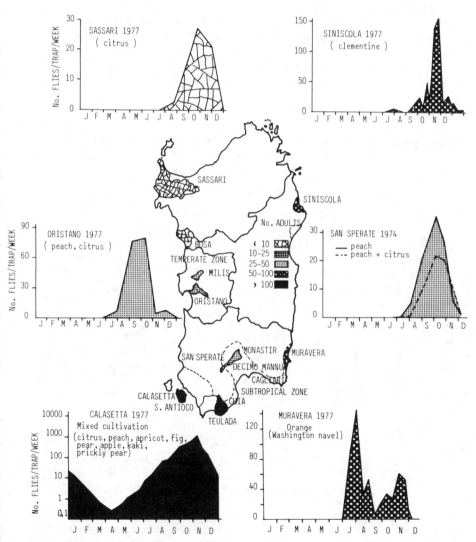

Fig. 2. Monthly fluctuations and Medfly population density in fruit growing temperate and subtropical zones of Sardinia.

141

are generally highest in subtropical climates and in areas of mixed fruit cultivation, where adults can be found throughout the year. Populations are, on the contrary, reduced in temperate climates with lower winter and spring temperatures and in extensive monoculture areas, in which the host sequence is from time to time interrupted for a certain period (Fig. 2)(22).

Medfly populations are relatively instable even in the most favourable areas and show a definite preference for certain plants which offer better ecological conditions for adult life and a substratum more suitable for oviposition. Plants that have maturing fruit attract particularly adults, populations grow following infestation but disperse after the harvest. Nevertheless, low adult populations are to be found also in plants without fruit and non-host plants.

The seasonal fluctuation in the Mediterranean basin shows a characteristic trend determined by the climate, and host sequence, availability and suitability. The populations are reduced to a minimum in winter because of unfavourable climatic conditions. In spring the Medfly begins to reproduce, but growth rate is low due to the shortage of susceptible hosts and because the temperature is still not optimum; survival is however high and adults show a great capacity for dispersion due to the scattered distribution of host plants. In summer the climate and larval food are optimum and the populations reach their highest growth rate. Fruit hosts are highly suitable for development and determine a low larval mortality and a high adult qual-

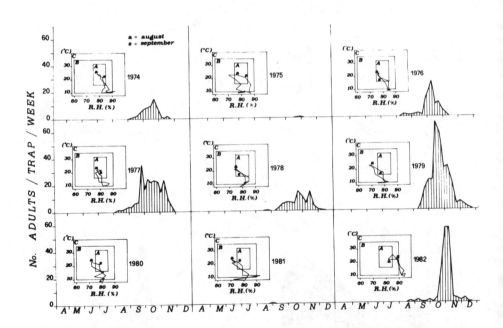

Fig. 3. Medfly population variations in Sardinia and climatograms with Bodenheimer climatic zones (Sassari 1974-1982; A= optimal conditions, B= favourable conditions, C= unfavourable conditions).

ity. At the end of summer populations reach maximum density, but susceptible hosts become rarer leading to an increase in larval competition and species dispersion. In autumn medfly aggregate on citrus and can produce strong infestations, on susceptible varieties, especially if temperatures remain high.

Medfly fluctuations in the course of a year can be very strong. Densities, estimated by the mark-recapture method, resulted in Tunisia as 10-100 flies/ha during the unfavourable season and as 5000 in summer; in the island of Procida (Italy) as 8 flies/ha in spring and as 15,000 in summer and in a Sardinian citrus area as 4 flies/ha in August and as above 6,000 at the population peak in November (13, 68).

Medfly populations show strong annual variations in accordance with climatic conditions. In warm climates (e.g. Israel) very dry summers result in low densities, but in more temperate areas populations can fall strongly when cold and wet weather extends into spring (Fig. 3).

4. RESISTANCE OF CITRUS FRUIT TO FRUIT FLY DAMAGE

Although fruit flies constitute a genuine threat for citrus growers, they do not seem to be highly adapted for development in citrus fruits. In fact the majority of the species, with the exception perhaps of D. tsuneonis, is subject in the citrus to high egg and young larval mortality. The duration of larval development in the citrus results, furthermore, lengthened and the quality of developed adults inferior in respect to many other attacked fruits (1, 4, 8, 20, 38, 39, 51, 59).

The ability of fruit flies to successfully attack citrus fruits proved to be dependent upon fruit species and cultivar, with grapefruit and clementines being most susceptible and lemons and sour lime almost completely resistant. The orange shows a susceptibility to tephritid attack variable according to the cultivars. Rind thickness and texture, density and volume of the oil glands and a high concentration of linalool in relation to limonene in the peel oil seem to play important roles in the relative immunity of citrus fruit. The eggs can be drowned by the gum secreted by the developing fruit and eggs and young larvae can be killed by essential oils in the flavedo region of the peel. The larvae can also meet resistance to penetration by very compact albedo (8, 33, 52).

The citrus resistance to the development of the juvenile stages of the fruit flies seems moreover bound to the degree of fruit ripening, resulting highest in the unripe fruits, both for the higher number of essential oil glands per unit of flavedo surface and for the greater thickness and compacteness of the albedo. Ripe and over-ripe citrus are more susceptible to attack, up to the point that fly development is possible also in over-ripe lemons.

Notwithstanding their temporary resistance the citrus can be subject to serious damage. The punctures in the fruit provoke an anticipated colouring and a premature fall. The bacteria which are introduced by the fly during egg-laying initiate fruit rot, and the puncture and subsequent larval gallery open the way for moulds (particularly Penicillium spp) and other organisms that increase decay. Fruit can furthermore depreciate economically due to considerable blemishes caused by simple puncturing.

Evaluation of economic losses caused by fruit flies is made very difficult due to the many varying factors, such as: fluctuations in abundance of
flies, the variety of types of damage and the different cultural situations
However, there have been reported for example: infestations of A. ludens in
unsprayed grapefruit groves in Belize between 15% and 60% according to the
period of harvest; loss of citrus fruits of 6.7% in Taiwan caused by D. dorsalis (37, 41). Economic losses due to C. capitata attacks were estimated:
in Central America as about 28% for sweet oranges, more than 50% for mandarines and 24% for grapefruit (37); in Spain as 1-2.5% for citrus fruits in
1960-65 but in which no collective control measures were taken (40); in
Turkey about 3.5-6.2% in 1961-1965 (71); in Sardinia 0.7% for oranges and
up to 46% for clementine in 1977.

5. STRATEGY FOR FRUIT FLY CONTROL

Biological and ecological characteristics of the fruit flies attacking
citrus determine the ways in which populations may be manipulated for pest
management purposes. These characteristics indicate that the species are
strong r-selected with a high capacity for increase, high mobility and
searching capacity and few natural enemies. When chosing methods of control
we must also bear in mind that these insects have an extremely low economic
threshold particularly in situations where fruit is grown for export to
uninfested areas, or where there is a demand for a perfect product.

Pest management methods in current use against tephritids throughout
the world include: control methods for protection of individual orchards,
suppression measures aimed at a substantial segment of a very large population with the object of reducing the level of fruit infestation to below
the economic threshold; eradication aimed at the destruction of every individual in a breeding population.

Control methods are based, above all, on the use of systemic insecticide or of bait sprays with protein hydrolysate or more specific male attractants. The first of these techniques is effective against the fruit
flies but has the disavantage that these control method react harshly
against beneficial insects. Bait sprays, although more selective, do not always result efficient on small groves due to the continual risk of adult immigration from untreated areas. Suppression procedures are based on protein
bait sprays applied as a cover spray or as a spot spray on large areas by
aerial or ground application. These techniques are really efficient and are
now used against most species of tephritid pests on citrus throughout the
world. Localized treatments allow the insecticide load of the environment
to be reduced to a very low level, but are not completely without impact on
non-target organisms (26, 31). Biological control can also be used to maintain low fruit fly populations. Numerous attempts at introducing hymenopterous parasites have been successful only in a few cases however and have
rarely been spectacular. A considerable level of success has been achieved
only in Hawaii against C. capitata and D. dorsalis with the introduction of
opiine braconids (particularly Biosteres arisanus (Sonan), B. longicaudatus
Ashm. and B. tryoni (Cam.)) (14, 15, 35).

The major procedures available for eradication are: protein hydrolysat bait sprays, male annihilation using specific male attractants and the sterile insect method (SIT). Efficient male attractants are available for the most important species (e.g. trimedlure for C. capitata; methyl eugenol for D. dorsalis; cue-lure for D. tryoni) and are widely used in detection systems. Spectacular results have been obtained with the male annihilation method, eradicating an isolated population of D. dorsalis by uniform distribution of methyl eugenol-toxicant formulations (69). Other attractants do not seem, however, as efficient as methyl eugenol and the mass-trapping against C. capitata males gave unconclusive results. Numerous experiments have shown that SIT is very efficient and highly suitable for fruit fly control. Many technical and scientific problems that limited the application of the method have now been resolved and large-scale programmes are now being carried out in various countries. The Medfly has recently been eradicated by SIT from Mexico and large areas in northern Guatemala (36).

REFERENCES

1. BACK, E.A. and PEMBERTON, C.E. (1915). Susceptibility of citrous fruits to the attack of the Mediterranean fruit fly. J. Agric. Res. 3: 311-330.
2. BACK, E.A. and PEMBERTON, C.E. (1916). Effect of cold-storage temperatures upon the Mediterranean fruit fly. J. Agric. Res. 5: 657-666.
3. BAKER, E.W. (1945). Studies on the Mexican fruit fly known as Anastrepha fraterculus. J. Econ. Entomol. 38: 95-100.
4. BAKER, A.C., STONE, W.E., PLUMMER, C.C. and McPHAIL, M. (1944). A review of studies on the Mexican fruit fly and related Mexican species. USDA Misc. Publ. 531: 155 pp.
5. BARROS, M.D., NOVAES, M. and MALAVASI, A. (1983). Estudos do comportamento de oviposicao de Anastrepha fraterculus (Wiedmann, 1830) (Diptera, Tephritidae) em condicoes naturais e de laboratorio. An. Soc. Entomol. Brasil 12: 243-247.
6. BATEMAN, M.A. (1972). The ecology of fruit flies. Ann. Rev. Entomol. 17: 493-518.
7. BATEMAN, M.A. (1976). Fruit flies. In: Delucchi V.L. Studies in biological control. Cambridge University Press, 1976, p. 11-49.
8. BODENHEIMER, F.S. (1951). Citrus entomology in the Middle East. W. Junk Pub. Co., The Hague, 661pp.
9. CHEIKH, M. (1969). Etude des effets de quelques températures basses sur les imagos de la Mouche Méditerranéenne des fruits Ceratitis capitata Wied. (Dipteres-Trypetidae). Ann. Inst. Nat. Rech. Agr. Tunisie 42:1-80.
10. CHEIKH, M., BEN SALAH, H. and HOWELL, J.F. (1972). Les différents stades de la mouche méditerranéenne des fruits Ceratitis capitata Wied. pendant l'hiver au nord de la Tunisie. Inst. Nat. Rech. agron. Tunisie, Doc. tech., 64: 1-9.
11. CHRISTENSON, L.D. and FOOTE, R.H. (1960). Biology of fruit flies. Ann. Rev. Entomol. 5: 171-192.
12. CIRIO, U. (1970). Influence de l'alimentation larvaire sur la longevité et la fécondité de Ceratitis capitata Wied. Bol. Asoc. Nac. Ingen. Agr. 209: 501-502.
13. CIRIO, U., DE MURTAS, I., GOZZO, S. and ENKERLIN, D. (1972). Preliminary ecological observations of Ceratitis capitata Wied. on the island of

Procida with an attempt to control the species using the sterile-male technique. Boll. Lab. Entomol. Agr. "F. Silvestri", Portici 30: 175-188.

14. CLAUSEN, C.P. (1956). Biological control of Fruit Flies. J. econ. Entomol. 49: 176-178.

15. CLAUSEN, C.P. (1978). Tephritidae (Trypetidae, Tripaneidae), pp. 320-335 In C. P. Clausen (ed.), Introduced parasites and predators of arthropod pests and weeds: a world review. U. S. Dep. Agric. Handb. No. 40.

16. CROVETTI, A., CONTI, B. and DELRIO, G. (1984). Effect of abiotic factors on Ceratitis capitata (Wied.) (Diptera Tephritidae). 2. Pupal development under constant temperatures (in print).

17. DEBOUZIE, D. (1977). Effect of initial population size on Ceratitis productivity under limited food conditions. Ann. Zool. Ecol. anim. 9: 367-381.

18. DELANOUE, P. and SORIA, F. (1962). Les fruits de l'oranger amer (Citrus Bigaradia Risso) riserve dangereuse en Tunisie de mouche des fruits (Ceratitis capitata Wied.) Ann. Inst. Nat. Rech. Agr. Tunisie 35:187-202.

19. DELRIO, G. Effetti dei fattori climatici sullo sviluppo, sulla mortalità e sulla fecondità di Ceratitis capitata Wied. (unpublished).

20. DELRIO, G. and BASHAR, M.A. Effects of hosts on the Mediterranean fruit fly, Ceratitis capitata (Wied.). Redia LXVIII (in print).

21. DELRIO, G., CONTI, B. and CROVETTI, A. (1984). Effect of abiotic factors on Ceratitis capitata (Wied.) (Diptera Tephritidae). 1. Egg development under constant temperatures (in print).

22. DELRIO, G., CONTINI, C., LUCIANO, P. and PROTA, R. (1977). Osservazioni preliminari sulle popolazioni della mosca mediterranea della frutta (Ceratitis capitata Wied.) in Sardegna. 5émes Journées circum- méditerranéennes Phytiat. Phytopharm., Rabat, 15-20 Mai 1977: 628-634.

23. DREW, R.A.I., COURTICE, A.C. and TEAKLE, D.S. (1983). Bacteria as natural source of food for adult fruit flies (Diptera: Tephritidae). Oecologia (Berlin) 60: 279-284.

24. DREW, R.A.I and HOOPER, G.H.S. (1983). Population studies of fruit flies (Diptera: Tephritidae) in South-East Queensland. Oecologia (Berlin) 56: 153-159.

25. EBELING, W. (1959). Subtropical fruit pests. Univ. Calif. Press 407 pp.

26. EHLER, L.E. and ENDICOTT, P.C. (1984). Effect of malathion bait sprays on biological control of insect pests of olive, citrus, and walnut. Hilgardia 52: 47 pp.

27. FERON, M. (1962). L'instinct de reproduction chez la mouche méditerranéenne des fruits Ceratitis capitata (Wied.) (Dipt. Trypetidae). Comportement sexuel. Comportement de ponte. Rev. Pathol. Vég. Entomol. Agric. France 41: 1-129.

28. FIMIANI, P. (1972). Fertilità delle uova in Ceratitis capitata Wied. (Diptera Trypetidae). Boll. Lab. Entomol. Agr. "F. Silvestri" Portici 34: 150-163.

29. FIMIANI, P. and TRANFAGLIA, A. (1972). Influenza delle condizioni climatiche sull'attività moltiplicativa della Ceratitis capitata Wied. Ann. Fac. Sc. Agr. Portici 6: 3-12.

30. FLETCHER, B.S. and BATEMAN, M.A. (1983). Combating the fruit fly prob-

lems in Australia: The current situation and future prospects. CEC/IOBC Symposium/Athens/Nov. 1982: 555-563.

31. GARY, N.E. and MUSSEN, E.C. (1984). Impact of Mediterranean fruit fly malathion bait spray on honey bees. Environ. Entomol. 13: 711-717.

32. GONZALES-HERNANDEZ, A. and TEJADA, L.O. (1979). Fluctuacion de la poblacion de Anastrepha ludens (Loew) y de sus enemigos naturales en Sargentia greggii S. Watts. Folia Entomologica mexicana 41: 49-60.

33. GREANY, P.D., STYER, S.L., DAVIS, P.L., SHAW, P.E. and CHAMBERS, D.L. (1983). Biochemical resistance of citrus to fruit flies. Demonstration and elucidation of resistance to the Caribbean fruit fly, Anastrepha suspensa. Ent. exp. appl. 34: 40-50.

34. HAGEN, K.S. (1958). Honeydew as an adult fruit fly diet affecting reproduction. Proc. X Int. Congr. Ent., Montreal 1956: 25-30.

35. HARAMOTO, F.H. and BESS, H.A. (1970). Recent studies on the abundance of the Oriental and Mediterranean fruit flies and the status of their parasites. Proc. Hawaiian Entomol. Soc. XX: 551-566.

36. HENDRICHS, J., ORTIZ, G., LIEDO, P. and SCHWARZ, A. (1983). Six years of successful medfly program in Mexico and Guatemala. CEC/IOBC Symposium /Athens/Nov. 1982: 353-365.

37. HOUSTON, W.W.K. (1981). Fluctuations in numbers and the significance of the sex ratio of the Mexican fruit fly, Anastrepha ludens caught in Mc Phail traps. Ent. exp. appl. 30: 140-150.

38. IBRAHIM, A.G., RAHMAN, M.D.A. (1982). Laboratory studies of the effects of selected tropical fruits on the larvae of Dacus dorsalis Hendel. Pertanika 5: 90-94.

39. INGLESFIELD, C. (1981). Larval hosts, adult body size and population quality in Ceratitis capitata Wied. : A laboratory study. "Studi Sassaresi" Ann. Fac. Agr. Univ. Sassari XXVIII: 25-38.

40. KOPPELBERG, B. and CRAMER, H.H. (1969). Control of Mediterranean fruit fly Ceratitis capitata Wied. in Spain. Methods and economic importance. Pflanzenschutz-Nachrichten Bayer 22: 157-167.

41. LEE, L.W. and CHANG, K.H. (1979). Integrated control of plant protection in citrus-integrated control of the oriental fruit flies in Taiwan. Proc. Int. Symp. IOBC/WPRS on Integrated Control in Agriculture and Forestry, Wien, 8-12 Oct. 1979: 568-571.

42. LIU, Y.C. (1983). Population studies on the oriental fruit fly, Dacus dorsalis Hendel, in Central Taiwan. CEC/IOBC Symposium/Athens/Nov. 1982: 62-67.

43. MALAVASI, A. and MORGANTE, J.S. (1981). Adult and larval population fluctuation of Anastrepha fraterculus and its relationship to host availability. Environ. Entomol. 10: 275-278.

44. MALAVASI, A., MORGANTE, J.S. and ZUCCHI, R.A. (1980). Biologia de"Moscas -das-frutas" (Diptera, Tephritidae). I. Lista de hospedeiros e ocorrencia. Rev. Brasil. Biol. 40: 9-16.

45. MARTIN, H. (1953). Contribution à l'étude de la mouche de fruits, Ceratitis capitata Wied., dans la région d'Alger (1949-1951). Rev. Pathol. Vég. Ent. Agric. France 32: 209-246.

46. MASON, A.C. and Mc BRIDE, O.C. (1934). Effect of low temperatures on the Mediterranean fruit fly in infested fruit. J. Econ. Entomol. 27: 897-902.

147

47. MEATS, A. (1983). Critical periods for developmental acclimation to cold in the Queensland fruit fly, Dacus tryoni. J. Insect Physiol. 29: 943-946.

48. MEATS, A. (1984). Thermal constraints to successful development of the Queensland fruit fly in regimes of constant and fluctuating temperature. Entomol. exp. appl. 36: 55-59.

49. MESSENGER, P.S. and FLITTERS, N.E. (1954). Bioclimatic studies of three species of fruit flies in Hawaii. J. Econ. Entomol. 47: 756-765.

50. MIYAKE, T. (1919). Studies on the fruit flies of Japan. I. Japanese orange fly. Bull. Imp. Central Agr. Expt. Station 2: 87-165.

51. MOURIKIS, P. (1965). Data concerning the development of the immature stages of the Mediterranean fruit fly (Ceratitis capitata (Wiedemann) (Diptera: Trypetidae)) on different host-fruits and on artificial media under laboratory conditions. Annls. Inst. Phytopath. Benaki 7: 59-105.

52. ORTU, S. (1982). Osservazioni sulla resistenza temporanea dei frutti di alcune specie di agrumi agli attacchi della Ceratitis capitata Wied. in Sardegna. "Studi Sassaresi" Ann. Fac. Agr. Univ. Sassari XXIX: 159-172.

53. PRESCOTT, J.A. and BARANOWSKI, R.H. (1971). Effects of temperature on the immature stages of Anastrepha suspensa (Diptera: Tephritidae). Florida Entomologist 54: 297-303.

54. PROKOPY, R.J. (1983). Tephritid relationship with plants. CEC/IOBC Symposium/ Athens/ Nov. 1982: 230-259.

55. PROKOPY, R.J., MALAVASI, A. and MORGANTE, J.S. (1982). Oviposition deterring pheromone in Anastrepha fraterculus flies. J. Chem. Ecol. 8:763

56. PROKOPY, R.J., ZIEGLER, J.R. and WONG, T.T.Y. (1978). Deterrence of repeated oviposition by fruit-marking pheromone in Ceratitis capitata. J. Chem. Ecol. 4: 55-63.

57. RHODE, R.H. (1975). A medfly eradication proposal for Central America. In: Controlling fruit flies by the sterile-insect technique.Vienna FAO/ IAEA

58. RIVNAY, E. (1951). The Mediterranean fruit fly in Israel. Bull. Ent. Res. 41: 321-341.

59. RIVNAY, E., NADEL, M. and LITTAUER, F. (1941). The reaction of the orange fruit to the autumn attack of the Mediterranean fruit fly and its economic status. Hadar 11: 16 pp.

60. ROS, J.P. (1975). Control genetico contra C. capitata Wied. por el metodo de insetos esteriles. Trabajos realizados en Espana (1969-1973). In "Sterility Principle for Insect Control 1974" IAEA-SM-186/59: 57-92.

61. ROS, J.P. (1983). Importance of ecological studies for application S.I.T. against Ceratitis capitata Wied. CEC/IOBC Symposium/ Athens/ Nov. 1982: 68-73.

62. SACANTANIS, K. (1957). La forêt d'arganiers, le plus grand foyer de Ceratitis capitata Wied. connu au monde. Boll. Lab. Ent. Agr. "F. Silvestri" XV: 1-53.

63. SELIM, O.F. (1967). Studies on the Mediterranean fruit fly Ceratitis capitata Wied., in U.A.R. Bull. Soc. ent. Egypte LI: 315-341.

64. SEVERIN, H. and HARTUNG, W. (1912). The flight of two thousand marked male Mediterranean fruit flies (Ceratitis capitata W.). Ann. Ent. Soc. Amer. 5: 400-407.

65. SCIROCCHI, A. and CIRIO, U. (1976). Ricerche ecologiche per un programma di lotta contro la Ceratitis capitata Wied. nella Piana di Fondi. Boll. Lab. Ent. Agr. "F. Silvestri" Portici XXXIII: 113–127.

66. SORIA, F. (1962). Epoque d'infestation des divers hôtes de la mouche des fruits (Ceratitis capitata Wied.) dans le nord de la Tunisie. Ann. Inst. Nat. Rech. Agr. Tunisie 35: 48–50.

67. SORIA, F. (1962). Plantes-hôtes secondaires de Ceratitis capitata Wied. en Tunisie. Ann. Inst. Nat. Rech. Agr. Tunisie 35: 51–72.

68. SORIA, F. (1963). A radioisotope study on population density and dispersal of Ceratitis capitata in Tunisia. In "Radiation and radioisotopes applied to insects of agricultural importance." Proc. IAEA Symp. Athens 22–26 April 1963: 357–362.

69. STEINER, L.F. (1969). Control and eradication of fruit flies on citrus. Proc. First Int. Citrus Symposium 2: 881–887.

70. TALHOUK, A.S. (1975). Citrus pests throughout the world. In:CIBA-GEIGY Tech. Monogr. No. 4, Basle, Switzerland: 21–23.

71. TUNCYUREK, M. (1972). Recent advances in bio-ecology of Ceratitis capitata Wied. in Turkey. Bull. EPPO 6: 77–81.

72. VARGAS, R.I., HARRIS, E.J. and NISHIDA, T. (1983). Distribution and seasonal occurrence of Ceratitis capitata (Wiedemann) (Diptera: Tephritidae) on the Island of Kauai in the Hawaiian Islands. Environ. Entomol. 12: 303–310.

73. VARGAS, R.I., NISHIDA, J. and BEARDSLEY, J.W. (1983). Distribution and abundance of Dacus dorsalis (Diptera: Tephritidae) in native and exotic forest areas on Kauai. Environ. Entomol. 12: 1185–1189.

Demographic parameters of the Mediterranean fruit fly at various laboratory conditions

A. Gil
Center for Information Sciences, Superior Council for Scientific Research, Madrid, Spain

M. Muñiz
Institute of Edaphology and Plant Biology, Superior Council for Scientific Research, Madrid, Spain

Summary

Demographic parameters and their standard errors have been estimated for females of Ceratitis capitata (Wied.) at 31,28,25,22 and 19ºC.
    Highest mortalities were obtained during the larval stage and, generally, the lowest mortalities ocurred at 22ºC. Intrinsic rate of increase (r) varied from 0,103 at 19ºC to 0,237 at 31ºC. The longest mean generation time (T) ranged from 22,97 days at 31ºC to 55,57 days at 19ºC. The highest net reproductive rate (R_0 = 308,72 days) was obtained at 19ºC.
    A new mathematical methodology to estimate demographic parame - ters and their standard errors is reported.

1. Introduction

During the last ten years it has been pointed out the significance of biological studies on the species of economic importance in order to adqui re a basic scientific knowledge of the insect pest (5). So, it will possible to apply the different programmes and techniques with an adequate level of efficiency.
    Ceratitis capitata (Wied.) is a highly destructive pest in the medite rranean basin . In Spain the direct and indirect damage is nearly 1% of the citrus fruit production. However, as Carey pointed out, little work has been done on the demography of this species (3). Rössler used the net reproduc- tive rate and the generation time for determinating the quality of lab-rea red fruit flies (13). In other cases, more extensive studies have been done, but regarding one specific temperature (3,4,15); however, errors of such a parameters, a very important topic in statistical works, were not included.
    Using data from earlier studies on daily reproductive activity of Ceratitis at various experimental conditions (12), we report here a new ma- thematical methodology and some results on the demography of this insect. In this manner, we collaborate to increase the basic knowledge of this ma- jor pest in many tropical and subtropical countries.

2. Material and methods

2.1. Laboratory methodology

In the before mentioned studies, one day old males and females of the Medfly were removed from a very well adapted population to lab rearing conditions and then were introduced in specially designed cages with adult food and water,at the rate of one couple per cage (9,10). Each cross was carried out in 20 replicates consisting of 1 male and female each for all environmental conditions of this work.

Eggs were collected once daily and transferred to a Petri dish in the bottom of which,a black filter paper was placed in order to maintain an optimal level of moisture. Neonata larvae were daily seeded in vials (5 cm dia x 9 cm.long) containing 5 grs of a new larval diet (1,11) and provided with a porous stopper to get an adequate aeration.

The following data were daily collected: Oviposition,egg hatch,larval development time.pupal and adult production,mortality of eggs,larvae,pupae and adults,sex segregation of the newly emerged adults and longevity of fe males. In this study food was unlimited and there were no mortality factors other than phisiological ones.

2 2. Mathematical methodology

For the sake of simplicity.we assume that a lot of individuals increases or decreases with an intrinsic rate (r) of increase,positive or negative. The deterministic model is

$$N_t = N_o \cdot e^{rT}$$

where N_t is the number of individuals at time $\underline{t}$. We are also assumed that adult female populations of Ceratitis have reached a stable age distribution; so,if a such population is divided into classes,according their age, each of them has its own growing law,but the proportion between the number of individuals of two different classes is a constant.

Now,let B_t the number of emergencies per unit of time in the (0,t) interval. The birth rate in the population is

$$\lambda = B_t/N_t$$

Furthermore,if p_x is the probability of surviving to age $\underline{x}$,the proportion of individuals,aged between x and x+1 in the stable population is

$$c_x = \frac{B_{t-x} p_x}{N_t} = \frac{\lambda p_x N_{t-x}}{N_t} = \lambda p_x e^{-rx}$$

Thus, $\sum_x c_x = 1$ and then $\lambda = 1/\sum_x p_x e^{-rx}$ (1)

The death rate is defined by

$$\mu = \lambda - r \qquad (2)$$

Also,if m_x is the mean number of female offspring produced in a unit of time by a female aged $\underline{x}$,then

$$B_t = \sum_x N_t c_x m_x = \lambda N_t \sum_x p_x m_x e^{-rx}, \text{ so that}$$

$$\sum_x p_x m_x e^{-rx} = (1/\lambda) \cdot (B_t/N_t) \quad \text{or}$$

$$\sum_x p_x m_x e^{-rx} = 1 \text{ (Lotka equation)} \quad (3)$$

As Birch pointed out (2),usual methods of calculation of r may be found in Dublin and Lotka (6) or Lotka (7). However,these conventional calculations don't offer adequate solutions for high values of this parameter. We have preferred to start from the following initial value:

$$r_o = \frac{1}{n} \sum_t \ln \frac{N_{t+1}}{N_t}$$

where the sum is extended to a sufficient number of days $(n = 3\omega)$; ω is the extinguished time. So, it will possible to estimate the standard errors of r, as we do later.

From (3), we derived the recurrent equation:

$$r_{i+1} = \ln\left[\sum_x p_x m_x e^{-r_i(x-1)}\right]$$

The process is finished when one of the following conditions results:

a) The absolute value of two consecutive terms, r_i and r_{i+1}, is less than 5.10^{-6}, in which case we take $r = r_i$.

b) $|r_i| \geqslant 5.10^{-6}$ and sign $r_i \neq$ sign r_0. In this case $r = r_0$.

c) If the number of iterations is greater than 50, then $r = r_{50}$.

We have considered that $\underline{r}$ is the basic parameter, so that, both, the mean generation time (T) and the net reproductive rate (R_0) are functions of $\underline{r}$. The former is defined as

$$T = \frac{TF + TG}{2} \tag{4}$$

where

$$TF = \sum_x x p_x m_x / \sum_x p_x m_x \quad \text{and}$$

$$TG = \sum_x x p_x m_x e^{-rx} / \sum_x p_x m_x e^{-rx}$$

Last expresions are the age of females at time in which they produce female offspring.

Net reproductive rate is the factor by which one generation is multiplied to obtain the next one:

$$R_0 = N_{t+1}/N_t = e^{rT} \tag{5}$$

In particular, for T=1, expression (5) is known as finite rate of increase.

The method of calculating r_0 allows us to estimate the standard error (SE) of r:

$$SE(r) = \sqrt{\frac{Var(r)}{n}}$$

Taking derivates in (4) and (5) with respect to r, we obtained the standard error of T and R_0:

$$SE(T) = dT/dr \; SE(r); \quad SE(R_0) = R_0 \left| T + r(dT/dr) \right| SE(r)$$

where:

$$dT/dr = \frac{1}{2 \sum_x p_x m_x e^{-rx}} \left[\frac{(\sum_x x p_x m_x e^{-rx})^2}{\sum_x p_x m_x e^{-rx}} - \sum_x x^2 p_x m_x e^{-rx} \right]$$

In like manner, from (1) and (2):

$$SE(\lambda) = \lambda^2 \sum_x x p_x e^{-rx} \cdot SE(r)$$

$$SE(\mu) = \left| \lambda^2 \sum_x x p_x e^{-rx} - 1 \right| SE(r)$$

A programme named TYVI was written in FORTRAN IV language to estimate r, T, R_0, λ and μ, as well as their standard errors. For this purpose, it is necessary to know the daily emergence rate and mortality in the (x,x+1) interval.

3. Results and discussion

Highest mortalities of 54, 42, 40, 33 and 40%, for 31, 28, 25, 22 and 19°C, respectively, were obtained during the larval stage (Table I). Lowest mortalities ocurred during the pupal stage, except for 19°C; At this temperature, egg mortality was 48% and larval mortality was 40%.

Table I. Life table of Ceratitis capitata (Wied.). (12 : 12 hrs L : D)

Experimental conditions	x	l_x	d_x	$100q_x$	s_x	Gen.mort. (%)
T= 31 ± 1ºC RH= 73 ± 3%	Egg	1179	302	25,61	25,61	
	Larvae	877	476	54,28	65,99	
	Pupae	401	60	14,96	71,08	
	Adult	341				71,08
T= 28 ± 1ºC RH= 67 ± 2%	Egg	1340	269	20,07	20,07	
	Larvae	1071	452	42,20	53,81	
	Pupae	619	24	3,88	55,60	
	Adult	595				55,60
T= 25 ± 1ºC RH= 57 ± 4%	Egg	1731	538	31,08	31,08	
	Larvae	1193	475	39,82	58,52	
	Pupae	718	16	2,23	59,45	
	Adult	702				59,45
T= 22 ± 1ºC RH= 67 ± 4%	Egg	1831	494	26,98	26,98	
	Larvae	1337	447	33,43	51,39	
	Pupae	890	11	1,24	51,99	
	Adult	879				51,99
T= 19 ± 1ºC RH= 78 ± 3%	Egg	1777	855	48,11	48,11	
	Larvae	922	365	39,59	68,66	
	Pupae	557	37	6,64	70,74	
	Adult	520				70,74

x = Stage of life cycle; l= Number living at beginning of stage in the x column; d= Number dying within stage in the x column; $100q_x$= Percentage mortality of the stage in the x column; s_x= Percentage mortality of the generation after stage in the x column. Gen.mort.= Generation mortality. Symbols given by Micinski et al.(8).

Generation mortality (egg to adult) varied from 71% at 31ºC and 19ºC to 52% at 22ºC. Note that, in general, the lowest mortalities were obtained at 22ºC. At this temperature and 67% RH we obtained the optimal reproductive parameters of this insect (12). In this Table, the number living are referred to the mean total production in each stage and are related to the whole fertility period of females.

Table II shows the values of the various demographic parameters and their standar errors. The highest intrinsic rate of increase was obtained at 31ºC with the shortest mean generation time, but the shortest net reproductive rate. The highest net reproductive rate was obtained at 19ºC.

Values of these parameters at 25ºC and 57% RH are similar to those reported by Rössler (13) and Carey (3) with data taken from Shoukry and Hafez (14). This fact is a result of the different methodology and lab-rearing.

Table II. Demographic parameters of Ceratitis capitata (Wied.) at various experimental conditions.

(12 : 12 hrs. L : D regime; Mean ± SE)

Experimental conditions	Intrinsic rate of birth (λ)	Intr. rate of death (μ)	Intr. rate of increase (r)	Finite rate of increase (e^r)	Mean generation time (T)	Net reproductive rate (R_0)
T= 31 ± 1ºC RH= 73 ± 3%	0,272 ± 0,015	0,035 ± 0,003	0,237 ± 0,012	1,27 ± 0,015	22,97 ± 0,35	233,15 ± 61,82
T= 28 ± 1ºC RH= 67 ± 2%	0,262 ± 0,013	0,034 ± 0,002	0,228 ± 0,011	1,26 ± 0,014	25,12 ± 0,43	306,69 ± 81,89
T= 25 ± 1ºC RH= 67 ± 4%	0,236 ± 0,015	0,025 ± 0,003	0,211 ± 0,012	1,24 ± 0,015	26,64 ± 0,42	277,59 ± 88,99
T= 22 ± 1ºC RH= 67 ± 4%	0,169 ± 0,007	0,015 ± 0,001	0,154 ± 0,006	1,17 ± 0,007	36,34 ± 0,50	274,01 ± 61,90
T= 19 ± 1ºC RH= 78 ± 3%	0,109 ± 0,004	0,005 ± 0,0004	0,103 ± 0,004	1,11 ± 0,004	55,57 ± 0,62	308,72 ± 60,42

REFERENCES

1. ANDRES,MªP. and MUÑIZ,M. (1984). Desarrollo de una nueva dieta larvaria para Ceratitis capitata (Wied.). Bol.Serv.Plagas,10:85-116.
2. BIRCH.L.C. (1948). The intrinsic rate of natural increase of an insect population. J.Anim.Ecol.,16:15-26.
3. CAREY,J.R. (1982). Demography and population dynamics of the Mediterranean fruit fly. Ecol.Model.,16:125-150.
4. CAREY,J.R. (1983). A life table examination of growth rate and age structure trade-offs in Mediterranean fruit fly populations. In Fruit Flies of Economic Importance (Ed.R.CAVALLORO).Proc.of the CEC/IOBC International Symposium.Athens.(Greece) ; 16-19 Nov.,1982:315-320.
5. CAVALLORO,R. (1983). Fruit Flies of Economic Importance. Proc.of the CEC/IOBC International Symposium.Athens.(Greece);16-19 Nov., 1982.642 pp.
6. DUBLIN,L.I. and LOTKA,A.J. (1925). On the true rate of natural increase as exemplified by the population of the United States,1920.J.Amer.Stat. Ass.,20:305-339.
7. LOTKA,A.J. (1939). Théory analytique des associations biologiques.Deuxiéme Partie. Analyse démographique avec application particuliére á l'espéce humaine. Actualités Sci.Industr.,780:1-149.
8. MICINSKI,S.,BOETHEL,D. and BOUDREAUX,H.B. (1981). Life tables and intrinsic rates of increase of the Pecan leaf scorch mite. J.Econ.Entomol.
9. MUÑIZ,M. (1979). Técnica para la evaluación de puesta en Ceratitis capitata (Wied.). Graellsia,31:277-292.
10. MUÑIZ,M. (1985). Studies on a rapid adaptation of the Mediterranean fruit fly. In Fruit Flies of Economic Importance. Proc.of the CEC/IOBC International Symposium. Hamburg,Aug.1984 (In press).
11. MUÑIZ,M. and ANDRES,MªP. (1983). Investigaciones básicas para la inclusión de Hansenula anomala como aporte proteico en la dieta larvaria de Ceratitis capitata (Wied.). Graellsia,39:165-174.
12. MUÑIZ,M. and GIL,A. (1984). Desarrollo y reproducción de Ceratitis capitata (Wied.) en condiciones artificiales. Bol.Serv.Plagas.Fuera de serie,nº 2:139 pp.
13. RÖSSLER,Y. (1975). Reproductive differences between laboratory-reared and field-collected populations of the Mediterranean fruitfly,Ceratitis capitata. Ann.Entomol.Soc.Am.,vol.68,nº6:987-991
14. SHOUKRY,A. and HAFEZ,M. (1979). Studies on the biology of the biology of the Mediterranean fruit,Ceratitis capitata .Ent.exp.appl.,26:33-39.
15. VARGAS,R.I.,MISHITA,D. and NISHIDA,T. (1984). Life history and demographic parameters of three laboratory-reared Tephritids. Ann.Entom.Soc. Am.,77:651-656.

Importance of old males in the reproductive activity of the Mediterranean fruit fly

M.Muñiz & A.Navas

Institute of Edaphology & Plant Biology, Superior Council for Scientific Research, Madrid, Spain

Summary

Experiments have been carried out in Spain in order to study effects of male age on the reproductive activity of Ceratitis capitata(Wied.) using single pairs. When old males were crossed with young and old virgin females.a very short pre-mating period of about 1 minute was observed,all females were efficiently inseminated and a high daily fecundity and fertility rate was obtained. When neonata larvae,from these crosses,were daily seeded in vials containing a new larval diet. larval development time,pupal production,pupal quality and adult emer gence were similar to those obtained from crosses between sexually ma tures males and females emerged at the same time. Daily oviposition of virgin females was lesser than the mated females one.

Males mated with young and old virgin females,after they mated with the same age females,lived longer;the life span of virgin female les was longer than the mated females one.

1. Introduction

The aim of this study was to point out the importance of basic know-ledge on the reproductive characters of this species,a highly destructive pest of citrus and various other fruit crops in Spain. Generally,laborato-ry researches on insects of economic importance are made on the basis of a population consisting of males and females that are confined together from the day of emergence. However,in order to the efficiency of using several techniques of insect pest management,it is necessary to consider the im - pact of newly emerged females in a population of old males and females. This is a very important requeriment for using SIT programmes.

2. Material and methods

One day old males and females were removed on 7 January 1985 from a laboratory population,that has been reared for over 20 years,and then in-troduced into specially designed cages (4,5). Each cross was carried out in 10 replicates consisting of 1 male and female each. The individuals of these crosses are referred in the text as "Control".

Eggs were collected once daily and then transferred to a Petri dish to observe hatching levels. When the fecundity or fertility of a female was null during 5 or more days,this female was removed from its cage;at the sa-

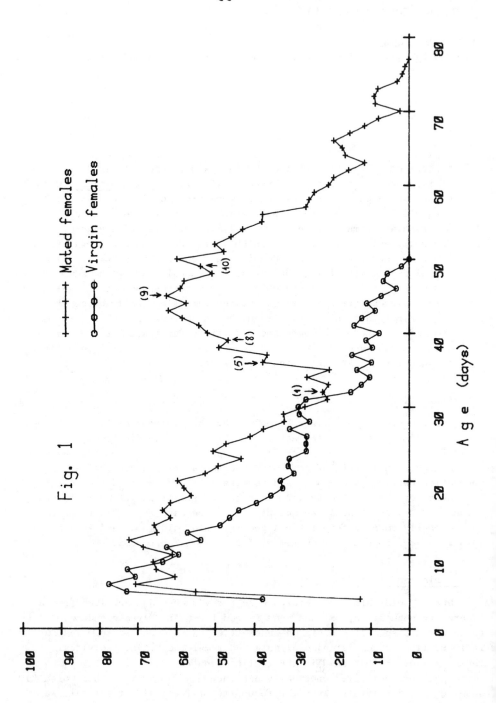

Number of eggs / female

Fig. 1

Mated females
Virgin females

Age (days)

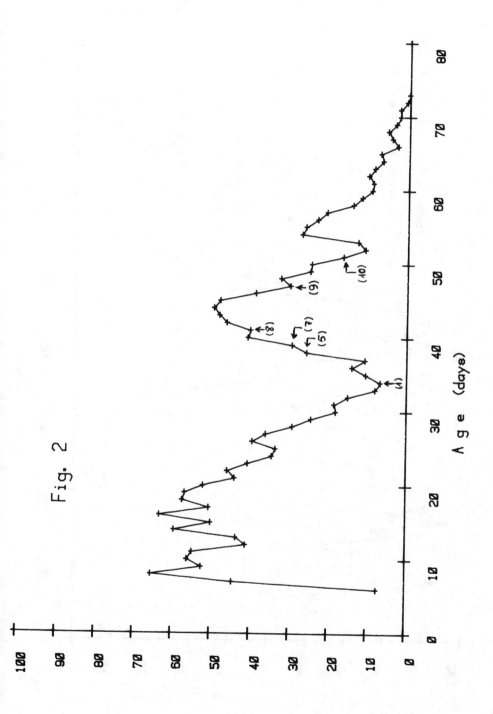

Fig. 2

Number of larvae / female

A g e (days)

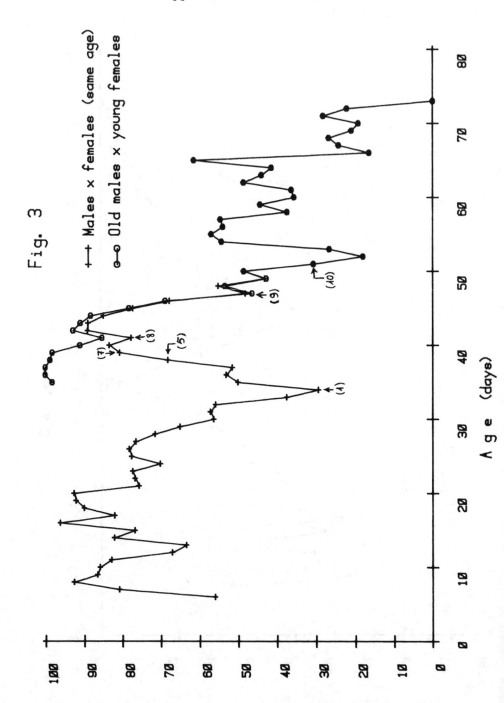

Fig. 3

+—+ Males × females (same age)
o—o Old males × young females

Egg hatch (%)

A g e (days)

me time,a new female was also removed from a virgin females population (1 February 1985) and then was introduced into that cage to mate with the same sexually mature male. So,it was possible to study the variation of the age specific fecundity and fertility obtained from the new and "Control" crosses. Each sterile female was confined in another individual cage to record longevity data. Also,one day old virgin females were introduced into the same cages to collect daily oviposition and logevity;10 replicates were carried out.

On the other hand,80 neonata larvae were daily seeded throughout the females'logevity period,in glass vials containing 5 gr.of a new larval diet (1,6) in order to collect the following data: Larval development time,pupation,pupal weight,pupal density,adult emergence and sex segregation of the newly emerged adults. A Zeiss standard microscope with camera lucida was used to calibrate the size of pupae and to estimate their volume (3,7).

Conditions during the experiments were 26 $\pm$ 1ºC,60 $\pm$ 1% RH and 12:12 hrs L:D regime (1.900 lux).

3. Results and discussion

The variation of the age specific fecundity of mated and virgin females throughout their age is represented in Fig.1;figures in parenthesis are the number of virgin females introduced in the "Control" population with respect to a day of the adult life. It can be noted that the daily oviposition rate of virgin females was lesser than the mated females one,and the preoviposition period was 3 days in the both virgin and mated females,in agreement with Feron (2).

When plotting age specific fertility throughout the female life span, it could be observed that one day after the first 5 days old virgin female was introduced in an individual cage to mate with a 32 days old "Control" male,the daily average of larvae/female began to rise with a high rate of increase (Fig.2). This fact was also observed when the percentage of egg hatch was plotted (Fig.3).The total fertility period was almost duplicated in relation to the "Control" one;that indicates the high insemination power of the old males.

Reciprocal crosses (old mated females X young virgin males)didn't give these results because of the existence of atrophic and hipertrophic ovaries in the "Control" females (3).

When a virgin female and a "Control" male were confined together to mate,a very short pre-mating period was observed (about 1 minute),irrespective the male age;in various cages this period was significantly shorter (about 20 sec.).

When neonata larvae,produced by old males X young females were daily seeded in the larval diet,a good pupal characters were obtained (Table I). Weights,mean volume and densities were adequate for the experimental conditions of this work. In earlier studies,similar results were obtained at 25ºC and 57 R.H.

Table II shows various reproductive patterns obtained form crosses between "Control" males and young females. In the before mentioned earlier studies,data of 7,61 days,60,22% and 95,53% were reported for larval development time,pupation and adult emergence,respectively (7).

Table I. Pupal characters of _Ceratitis_ _capitata_ (Wied.)

Pupal age (days)	Weiht/pupa (mgr.)	Volume (mm^3)	Density (gr/cm^3)
1	11,2487 ± 0,033		1,0778 ± 0,0028
2	10,3517 ± 0,031		0,9919 ± 0,0025
3	10,1186 ± 0,029		0,9695 ± 0,0023
4	9,9365 ± 0,025		0,9521 ± 0,0020
5	9,8292 ± 0,025	10,44 ± 0,05	0,9418 ± 0,0027
6	9,7363 ± 0,025		0,9329 ± 0,0027
7	9,6529 ± 0,023		0,9249 ± 0,0028
8	9,5137 ± 0,017		0,9116 ± 0,0026
9	9,3929 ± 0,018		0,8998 ± 0,0028

Data based on 57 replicates; 5 pupae/replicate. Mean ± S.E.

Table II. Reproductive patterns of _Ceratitis_ _capitata_ (Wied.)

Larval development time (days)	Pupation (%)	Emergence (%)	Sex segregation (%) Males	Females
7,31 ± 0.10	63,88 ± 6,53	83,30 ± 4,72	47,35 ± 2,5	52,65 ± 2,5

Data based on 22 replicates. Mean ± S.E.

With regard to the average of longevity,the life span of the "Control" males was significantly longer than the virgin females one (t = 4,5734; p< 0,001;18 d.f) and the virgin females lived more than the "Control" fema-les (t = 2,9225; p< 0,01;18 d.f). These values are showed in Table III.

Table III. Longevity data of _Ceratitis_ _capitata_ (Wied.). Mean ± S.E.

No. pairs	"Control"males	"Control"females	Virgin females
10	73,90 ± 6,20	30,20 ± 2,80	42,30 ± 3,05

These results can be explained by the existence of a higher phisiolo-gical degeneration in the "Control" females because of a higher sexual ac-tivity in relation to the virgin ones. On the other hand,the longer life span of males has sense from the stand point of the theory of natural selec-tion because of their reproductive ability after sterility of the same age females (8).

162

REFERENCES

1. ANDRES,MªP. and MUÑIZ,M. (1984). Desarrollo de una nueva dieta larvaria para Ceratitis capitata (Wied.). Bol.Serv.Plagas.10:85-116.
2. FÈRON,M. (1962). L'instinct de reproduction chez la mouche méditerranéenne des fruits,Ceratitis capitata Wied.(Dipt.Trypetidae).Comportement sexuel-comportement de ponte.Rev.Path.Veg.Ent.Agr.Fr..t.XLI,129 pp.
3. MUNGUIRA,M.L. MUÑIZ,M. and SALOM,F. (1984). Efectos del 5-fluoruracilo sobre la evolución del volúmen ovárico en Ceratitis capitata (Wied.). Bol.Serv.Plagas.10:43-57.
4. MUÑIZ,M. (1979). Técnica para la evaluación de puesta en Ceratitis capitata (Wied.). Graellsia.31:277-292.
5. MUÑIZ,M. (1985). Studies on a rapid adaptation of the Mediterranean fruit fly. In Fruit flies of Economic Importance.Proc.of the CEC/WPRS International Symposium.Hamburg.Aug.1984 (In press.)
6. MUÑIZ,M. and ANDRES,MªP. (1983). Investigaciones básicas para la inclusión de Hansenula anomala como aporte proteico en la dieta larvaria de Ceratitis capitata (Wied.). Graellsia.39:165-174.
7. MUÑIZ,M. and GIL,A. (1984). Desarrollo y reproducción de Ceratitis capitata (Wied.) en condiciones artificiales. Bol.Serv.Plagas.Fuera de serie,nº2:139 pp.
8. MUÑIZ,M. and MARTIN,J. (1985). Postreproductive periods and mortality in Ceratitis capitata (Wied.) (In press).

Beetle pests living on citrus in Italy*

D.Benfatto
Citrus Experimental Institute, Acireale, Italy

S.Longo
Institute of Agricultural Entomology, University of Catania, Italy

Summary

Twenty-two species of phytophagous beetles living on citrus in Italy are reported; they belong to the families Scarabeidae, Bostrychidae, Nitidulidae, Cerambycidae, Cucurlionidae, Alleculidae and Chrysomelidae. But only few species (i.e. weevils) among such groups are economically important and require
control measures.Notes are reported on Trichoferus griseus Fabr. (Col. Cerambycidae), new species on citrus in Mediterranean area, and Apate monachus
Fabr. (Col. Bostrychidae) infesting lemon trees in Sicily.

1. Introduction

There are many species of insects living on citrus, belonging to several orders. Among them only few species may be considered key-phytophagous of such
groves and against them it is usually necessary to apply control.
Some species of beetles, that only occasionally damage citrus, must be considered of secondary importance. In this paper the phytophagous species already
known in Italy and those found during recent studies carried on from 1979 to 1985
in the main Sicilian citrus-growing areas, are reported.

2. Involved species

Twenty-one species of phytophagous beetles are reported to feed on citrus in
Italy. They belong to the families Alleculidae, Bostrychidae, Chrysomelidae, Curculionidae, Nitidulidae and Scarabeidae (1), (2), (5), (9), (12), (18), (20), (24). To
the above species Trichoferus griseus Fabr. (Col. Cerambycidae), must be added
a new species on citrus for the Mediterranean area, recently found on lemon trees,
in Sicily. Table 1 shows the involved species. First we shall discuss the more frequently present ones, that may be of economical interest.

2.1. Cucurlionidae

Among all the beetles of the genus Otiorrhynchus, the best known species, O.
cribricollis Gyll. caused the most serious damage and had the greatest economical
effect. In western Sicily it attacks different kinds of erbaceous plants and trees,
both wild and cultivated.

---

* Research conducted as part of the Ministry of Agriculture and Forestry
project "Development and improvement of industrial fruit-growing, early
fruit growing and citrus fruit growing", Publication n° 161.

Tab. 1 - Beetle pests living on citrus in Italy.

Species	Citrus-growing area	references
ALLECULIDAE		
Omophlus lepturoides Fabr.	?	(10)
BOSTRYCHIDAE		
Apate monachus Fabr.	Sardinia	(9)
	Sicily	pers.obs.
Synoxylon sexdentatum Oliv.	Sicily	(5)
Xylomedes coronata Mars.	Sicily	(5)
CERAMBYCIDAE		
Trichoferus griseus Fabr.	Sicily	pers.obs.
CHRYSOMELIDAE		
Aphtona nigriceps Redtb.	Sicily	(20)
Crepidotera ventralis Illg.	Sicily	(20)
" impressa Fabr.	Sicily	(20)
Longitarsus brunneus Duft.	Sicily	(20)
" tabidus Fabr.	Sicily	(20)
CURCULIONIDAE		
Otiorrhinchus armatus Bohm.	Sicily	(12)
" aurifer Bohm.	Sardinia	(13)
" cribricollis Gyll.	Sicily	(1)
" rhacusensis Germ. siculus Stierl.	Sicily	(7)
NITIDULIDAE		
Carpophilus hemipterus L.	Sicily	(18)
" mutilatus Eric.	Sicily	pers.obs.
SCARABEIDAE		
Cetonia aurata (L.)	Sicily	(20)
Oxythyrea funesta Poda	Sicily	(20)
Pentodon punctatus Vill.	Sicily	pers.obs.
Rhizotrogus rugifrons Burm.	Sardinia	(24)
Tropinota hirta Poda	Sicily	(20)
" squalida Scoop.	Sicily	(20)

Its damages were first recorded more than one hundred years ago (1). The damages caused by the feeding activity of polyphagous and nocturnal adults, are characterised by the eating of the margins of the leaves, primarily evident in Spring and in Autumn.

The larvae feed on the roots, their injuries being particularly serious in the nursery, where this infestation may cause great damage. Similar injuries caused by the species O. aurifer Boh. have been recorded in Sardinia (13) and by the species O. armatus Boh. and O. rhacusensis Germ. var. siculus Stierl. in Sicily (12). They have few known natural enemies, although it is known that the adult O. cribricollis is predated by an ascaride (17).

166

Particular attention must be paid to the Otiorrhynchus spp. infestations only in seedlings and in top-grafts, when their population increases.When control measures against these insects become necessary, a satisfactory result may be obtained by applying to the ground granular organic phosphorous insecticides or by sprayng the trees during the period in which many adults are present. Another satisfactory control method for the adult, which is unable to fly and must climb up the trunk to reach the leaves, is to put a collar of synthetic glue which impeds its climbing.The larvae may be microbiologically controlled, by means of entomopathogenic fungi (Beauveria bronniartii, Metarhizium anisopliae and Paecilomyces fumoso-roseus) in the ground. Good results against preimaginal stages are also obtained by the entomophagous nematode Neoaplectana carpocapsae (Weism.) (19).

2.2 Scarabeidae
During the flowering period of citrus trees, adults of numerous species of Scarabeidae are discontinuously present . The commonest in the observed areas are the species Tropinota hirta Poda, T. squalida Scop. e Oxythyrea funesta Poda. They are thought to be responsible for microlesions in the ovary of the flowers,pre sumably made accidently by the anterior legs. These lesions evolve in large cork-like scars in the mature fruit. The presence of damage, mainly seen on oranges and mandarins, is extremely variable from year to year and may be present in one part of the zone and not the next. Usually it is not of such great extent that control is necessary. In Sardinia the species Rhizotrogus rugifrons Burm. has recently been observed feeding on the leaves and twigs of citrus trees (23). Young orange and mandarin trees occasionally have erosion of the roots caused by Pentodon punc-tatus Vill.; these erosions have been observed in a citrus grove of Western Sicily also containing peach and plum trees.

2.3 Bostrychidae
During this study attacks by Apate monachus Fabr. (*) were observed in a le-mon grove along the sea of Western Sicily. The infestations were limited to a few trees made weak by the presence of gummosis cancer of the branches (Botryosphae-ria ribis Gross and Dugg. pycnidic form Dothiorella ribis (Fuck.) Sacc.. Feeding bores, excavated by the nocturnal adults were observed in the large branches of the trees and in a greater number in the branches and twigs od diameter of 2-3 cm, and always of an age superior than two years.

The same damages have been observed in various plants including citrus in Sardinia (9) (22).

The ethology of preimmaginal stages is not well documented. In Israel, whe-re the species has been found also on citrus, the larvae feed on dead wood (3).

Therefore, as yet, this insect has not been found to have an economical ef-fect because its sporadic presence is limited to plants that are already failing from other causes.

Other boring Bostrychidae living on citrus are Synoxylon sexdentatum Oliv. and Xylomedes coronata Mars.. Wich have also caused in South-Western Sicily spo radic damages (5); on the contrary, in Maroc the latter species has caused serious damage (11).

(*) The classification of the species was confirmed by Dr R. Madge of Commonwe-alth Institute of Entomology, with many thanks.

Pentodon punctatus Vill larva.

Apate monachus Fabr. adult.

Feeding bore excavated by A. monachus adult on a lemon branche.

Trichoferus griseus Fabr. adult.

Otiorrhynchus cribricollis Gyll., adult feeding on a citrus leave.

Otiorrhynchus cribricollis Gyll. larvae.

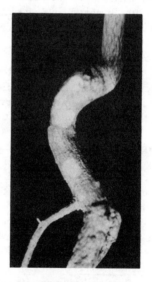

Erosions caused by O.cri-
bricollis larvae on a ci-
trus seedling.

Tropinota sp. adults on citrus flowers.

2.4 Nitidulidae

Two species: Carpophilus hemipterus L. (18) and C. mutilatus Ericson have frequently been found in the decayng fruits of orange, mandarin and clementine in which there are lesions in the skin by various origin (i.e. rain and hail rubbing bores made from larvae of Cryptoblabes gnidiella Mill. in infested fruits by Planococcus citri (Risso); as these species live on citrus fruit already damaged, their presence is of secondary importance, though they cause serious damage on Peach (23) and Pomegranate (21).

2.5 Alleculidae and Chrysomelidae

These two families have historically been responsible for some damage but now their presence does not have any impact. Adults of Omophlus lepturoides Fabr. occasionally damage various erbacious plants and trees, and are reported to live and feed on flowers of citrus trees (10).

Among the Chrysomelidae, Aphtona nigriceps Redtb., Crepidotera ventralis Illg., C. impressa Fabr., Longitarsus brunneus Duft. and L. tabidus Fabr. have been named to be the cause of erosion to the leaves of citrus trees in Western-Sicily (20). However, during our observations, we have occasionally found the presence of various species of Chrysomelidae without noticing any damages.

2.6 Cerambycidae

From the branches of lemon trees taken in a citrus grove of South Western Sicily (Ragusa, september 1983) and put in a rearing cage, we obtained specimen of Trichoferus griseus Fabr., a poliphagous species known from several Countries of the Mediterranean area, living on other plants but not on citrus.

This species has been studied in France and in Algeria(4) in relationship to damages caused to Ficus carica L., Ceratonia siliqua L. and Morus sp.. The adults lay their eggs on various plants including Pistacia lentiscus L., Nerium oleander L., Quercus mirbeckii(4), Acer sp., Eucaliptus sp. and Taxus baccata L.. In Southern France and in North Africa the species has only one generation per year, with adults emerging in August.

Also this species is not of particular threat for citrus; owing to its limited attacks to trees that are already in bad health from other causes (Gummosis cancer).

3. Conclusions

Beetles living on citrus are of secondary importance in relation to their control.

Among twenty-two species recorded in Italy, O. cribricollis and some kindred species, in restrict areas, may sometimes increase so much that require control measures but in the young trees and in the nurseries only.

The injuries to the flowers caused by Scarabeidae so as those caused to the branches by xilophagous species of Bostrychidae and Cerambycidae, are casual and limited.

REFERENCES

1. ALFONSO, F. (1890). Nuovo parassita degli agrumi. Nuovi Annali di agricoltura siciliana, 1: 196-199.

2. ANONIMO (1941). L'Epicometis (Tropinota) hirta, Poda e l'Oxythyrea funesta, Poda: due coleotteri dannosi alla vite, agli alberi da frutto, agli agrumi e ad alcune piante da fiori. Bollettini della R. stazione sperimentale di frutticoltura e di agrumicoltura di Acireale, 68: 1–5.

3. AVIDOV, Z. and HARPAZ, J. (1969). Plant pest of Israel. Jerusalem, Israel Un. Press. 549.

4. BALACHOWSKY, A.S. (1963). Entomologie appliquée a l'agriculture, 2: 878–900.

5. BARBAGALLO, S. (1980). Insetti in Gli Agrumi. REDA: 232–233.

6. BARGAGLI, P. (1878). La flora delle Altiche in Europa. Boll.Soc.Ent.It. 10: 43.

7. BERLESE, A. (1924). Manuale sugli insetti nocivi alle piante coltivate, campestri, orticole ed i loro parassiti. Ed.Tip.M.Ricci, Firenze: 93.

8. BODENHEIMER, F.S. (1951). Citrus entomology in the Middle East. S - Gra venhage Groningen: 662.

9. BOSELLI, F. (1959). Apate monachus Fabr. var. rufiventris Lucas (Col. Bostrychidae), nuovo parassita del mandorlo, del pesco, del limone e dell'arancio in Sardegna. Annali della sperimentazione agraria, 13 (suppl. al n° 1): XXXI-XLVI.

10. CECCONI, G. (1924). Manuale di entomologia forestale. Tip. del Seminario, Padova: 680.

11. CHAPOT, H. and DELUCCHI, V.L. (1964). Maladies, troubles et ravageurs des agrumes au Maroc. Ist.Nat.Rech.Agron. Rabat: 339.

12. COSTANTINO, G. (1941). Danni accidentalmente causati dallo Otiorrhinchus armatus Boh. a due specie del genere citrus: Citrus medica L. (Cedro rugoso, Cedro vozza vozza) e Citrus limonia Osb.. Annali della R. stazione sperimentale di frutticoltura e di agrumicoltura di Acireale, 16: 7–10.

13. CROVETTI, D. (1971). Segnalazione di danni causati dallo Otiorrhynchus aurifer Bohemann (Coleoptera, Curculionidae, Otiorrhynchinae) agli agrumi in Sardegna. Assoc.prov.Dot.Sc.Agr. Sassari, quaderno, 5: 13.

14. DI MARTINO, E. (1956). L'Otiorrhynchus cribricollis Gyll. dannoso agli agru mi della Sicilia orientale. Riv. di Agrumicoltura, 1: 239–251.

15. DI MARTINO, E. (1970). I parassiti animali degli agrumi in Tutto sugli agrumi. Ist.tec.e prop.agr. Roma: 171–172.

16. FERRON, P. and MARCHAL, M. (1983). L'Oziorrinco parassita delle piante ornamentali. In Claimer informa, 8 (6): 171–175.

17. GRANDI, G. (1913). Gli stadi postembrionali di un Coleottero (Othiorrhynchus cribricollis Gyll.) a riproduzione partenogenetica ciclica irregolare. Boll. Lab.Zool.Gen.e Agr. Portici, 7: 72–90.

18. LO GIUDICE, V. and LANZA, G. (1973). Presenza di Sherapheleuchus entomophagus su frutti di arancio. Ann.Ist.Sper.Agrum., 6: 5–8.

19. LOZZIA, G.C. and BIRAGHI, C. (1983). Curculionidi del genere Otiorrhynchus Germ. e loro controllo nei vivai del Nord Italia. La difesa delle piante, 5: 269–280.

20. MINA' PALUMBO, F. (1883). I nemici degli Esperidi. L'agric.ital., 106–107: 392–405.

21. NUZZACI, G. (1968). Danni da Carpophilus mutilatus Erichs. a frutti di melo grano. Entomologica, 4: 167–173.

22. PROTA, R. (1963). Osservazioni su alcuni danni causati in Sardegna da Apate monachus Fabr. (Coleoptera Bostrychidae). Studi sassaresi, 11: 77–88.

23. TREMBLAY, E., ESPINOSA, B. and BALDINI, C. (1984). Dannosità dei carpofili (Coleoptera Nitidulidae) alle pesche in Campania. Informatore fitopatologico, 10: 43–45.

24. ZANARDI, D., LOCHE, P. and FRESU, B. (1979). Rhizotrogus rugifrons Burm., nuovo parassita degli agrumi sardi. Infor. agrario, 14: 5427–5428.

Session 2
Other pests

Chairman: E.Di Martino

Address of the session chairman

E.Di Martino
Citrus Experimental Institute, Acireale, Italy

The reports and communications which we will hear in this session would appear, in their ampience and depth, to thoroughly cover the theme of citrus phytophages.

In point of fact, some decades ago the treatment of insects would have included, at least in Italy, all the enemies of cultivation, but harmful acarids were little taken note of then and even a scholar as important as W. Ebeling was surprised when it was announced, toward the end of World War II, that damage caused by a red spider had been found in citrus groves in the Sorrentine peninsula.

Personally, I can say that since 1950 in Italy I have noted the appearance, accompanied by related economic damage, of 7 species of acarids, including 2 species of Tetranychidae, 2 species of Tenuipalpidae, 1 species of Tarsonemidae and 2 species of Eriophyidae.

Not only harmful acarids, but also benign ones will be discussed, and particular stress will be laid on the useful ones for the contribution that they can make, together with known predatory insects, in the containment of damaging species.

Other phytophages, such as nematodes, do exist, although these belong to another class. In fact, in recent decades a new discipline has arisen - nematology - which among other things studies the relationship between these insects and the root systems of citrus trees. Thus these themes will also be discussed.

For the sake of organization, the topic of weeds in citrus groves has been included in this session - the role which they may assume in their cohabitation with citrus trees in various situations will be examined, as well as their direct influence on cultivation techniques.

A first list of the mites in citrus orchards in Italy[1]

V.Vacante & A.Nucifora
Institute of Agricultural Entomology, University of Catania, Italy

Summary

The results are presented of an investigation on the mite fauna on citrus in Italy.
The presence of 64 species has been observed belonging to 17 families.On these
species 10 are new for the Italian fauna. The phytophagous species (12 in all)
are fewer in number than the predators (28 in all) on than those with a varied
feeding habits (27 in all).

1. Introduction

With the purpose of contributing to the knowledge of the mites in the citrus
agroecosystem in Italy we began in 1982 to survey and examine material collected
from the epigean part of the plants. The investigation has covered the majority of
citrus species as well as the varieties of each of them in Southern Italy and the
islands of Sicily and Sardinia. So far the investigation has been limited to the leaves, branches and fruit on which the phytophagous species of agricultural interest
and their symbiotic antagonists live. Other investigations on species living the
bark and the soil under the canopies have now been begun and will be the subject
of subsequent pubblications. The work carried out so far has led to the identification of 64 species belonging to 17 families, 10 of the former being new for the Italian fauna. The list we give contains brief notes on their feeding habits and on the
role they presumably play in their ecosystem. More detailed information about the
knowledge so far available on their biology, as well as the clues leading to their
recognition and to the determination of the families and species listed in this work
will be the subject of subsequent papers presently in preparation. The purpose is
to facilitate the work of those who wish to follow up our study in greater depth and
help us to identify the presence of other species which may extend the list given here. In our opinion this list should not be considered complete yet and further identification has still to be undertaken. The phytophagous species belonging to the families of the Tetranychidae, Tenuipalpidae, Eriophyidae and Tarsonemidae are relatively few in number; more numerous is the host of predators belonging to the families of the Phytoseiidae, Ascidae, Stigmaeidae, Hemisarcoptidae, Cheyletidae and

[1] This study is part of the interuniversity program financed by the Italian Ministry of Education and aimed at identifying the mite fauna and its symbionts in
the main agricultural cultures in the Mediterranean area. This particular research is being performed jointly with the Institutes of Agrarian Entomology of
the Universities of Palermo and Bari.

Cunaxidae. A group of 26 species (Oribatulidae, Micreremidae, Ceratozetidae, Tydeidae, Acaridae, Glycyphagidae) is characterized by a various feeding habits (mycophagous, microphytophagous, saprophagous, coprophagous or necrophagous). The latter can be considered scavengers of the plants and they are often present is considerable numbers, prevailing on the whole over all the other phytophagous and predatory species. For a large part they are represented by tydeids, for some species of which there is controversy as to whether they are phytophagous and if so to what extent. In our opinion much damage, which they do not cause, is attributed to them.

2. The list of the species found

The majority of the phytoseiid mites we list had already been identified in Sicily by Ragusa and Swirski (1976) and by Ragusa (1977).

Research on the biology of these predators must still for a large part be carried out; that is of considerable importance because of the determining role that their plays in controlling the tetranychids and other phytophagi (thrips). In this context, of particular interest is the action of Phytoseiulus persimilis Athias-Henriot against Tetranychus urticae Koch and of Amblyseius stipulatus Athias-Henriot against Panonychus citri (Mc Gregor). We have been able to identify their role during four years of integrated control experiments carried out in lemon orchards in Sicily (Nucifora, 1983); it is such as to exclude the use of any sort of acaricide against the two phytophagous mites mentioned above. The only hemisarcoptid so far found (Hemisarcoptes malus (Shimer)) was seen to prey on Aspidiotus nerii Bouché, Mytilococcus beckii (Newan), Parlatoria pergandei Comstock and Aonidiella aurantii (Maskel); it has been observed that the predator can control up to 50% of the females of M. beckii present in the field. The two species included in the Cunaxidae family are certainly predators, but we have been unable to establish the size of their role in the equilibrium of their environment. At the moment the same consideration applies to the cheyletids. In the Stigmaeidae family the presence of Zetzellia graeciana Gonzalez and Z. mali (Ewing) is of considerable importance as they prey on the eggs and mobile forms of the tetranychids and eriophyids. Concerning the other species, there is insufficient knowledge and it seems that they may be indirectly useful by feeding on fungi, saprophytes and detritus. In the averall economy of the ecosystem they help toward the survival and multiplication of various predatory mites for which they may be a suitable nutritional substrate.

Family Tetranychidae Donnadieu

Bryobia praetiosa Koch

1836. Bryobia praetiosa Koch, in Koch, Deutschlands Crustaceen, Myriapoden und Arachniden, Ein Beitrag zur Deutschen Fauna, Regensburg, fasc.1, n.8.

A phytophagous species of herbaceous plants growing beneath the canopy; this mite is rarely found on citrus to which, besides, it causes no damage. In this study it was collected on lemon leaves in Sicily.

Aplonobia histricina (Berlese)

1910. Tetranychopsis histricina Berlese, in Berlese, Redia, 6: 243.

A phytophagous species of the Oxalis genus of plants; the mobile forms of this mite are sporadically found on <u>Citrus</u> trees. The masses of eggs of this mite deposi͟ ted in spring and hatching in autumn on the trunks of citrus trees are, instead, frequent. The tetranychid cause no damage to crops. This species has been found in Sicily.

<u>Petrobia tunisiae</u> Manson

1964. <u>Petrobia tunisiae</u> Manson, in Manson, Acarologia, <u>6</u> (1): 73.

A polyphagous species of herbaceous plants beneath the canopies. This mite is rarely found on <u>Citrus</u> and causes it no damage. It has been found on lemon leaves in Sicily.

<u>Panonychus citri</u> (Mc Gregor)

1916. <u>Tetranychus citri</u> Mc Gregor, in Mc Gregor, Ann.ent.Soc.Am., <u>9</u>: 28.

A phytophagous species injurious to all species of <u>Citrus.</u> This mite is present in all the citrus cultivation areas of Italy.

<u>Tetranychus urticae</u> Koch

1836. <u>Tetranychus urticae</u> Koch, in Koch, Deutschlands Crustaceen, Myriapoden und Arachniden, Ein Beitrag zur Deutschen Fauna, Regensburg, fasc.1 n.10.

A phytophagous species injurious to all <u>Citrus.</u> The mite is present in all the citrus cultivation areas of Italy.

Family <u>Tenuipalpidae</u> Berlese

<u>Brevipalpus cuneatus</u> (Canestrini and Fanzago)

1876. <u>Caligonus cuneatus</u> Canestrini and Fanzago, in Canestrini and Fanzago, Atti Accad.scient.veneto–trent.–istriana, <u>5</u>: 136.

A phytophagous species reported on lemon in Sicily in 1903 by Cavara and Mol͟ lica. It was never found during this study.

<u>Brevipalpus californicus</u> (Banks)

1904. <u>Tenuipalpus californicus</u> Banks, in Banks, J.N.Y.Ent.Soc., <u>12</u>: 55.

A phytophagous species jniurious to all species of <u>Citrus;</u> the mite is present in the main areas of citrus cultivation in Italy.

<u>Brevipalpus obovatus</u> Donnadieu

1875. <u>Brevipalpus obovatus</u> Donnadieu, in Donnadieu. Annls Soc.linn.Lyon, <u>12</u>: 116.

A phytophagous species collected on leaves and fruit of lemon and orange in Si͟ cily, Calabria and Sardinia.

<u>Brevipalpus phoenicis</u> (Geijskes)

1939. <u>Tenuipalpus phoenicis</u> Geijskes,in Geijskes,Beitrage zur Kenntnis der Europaischen Spinnmilben(Acari,Tetranychidae),mit Besonderer Berucksichtigung der Niederlandischen Arten,Veenman and Zonen,Wageningen, 42 (4): 23.

A species injurious to all species of <u>Citrus;</u> this mite is present in the main areas of citrus cultivation in Italy.

Family Oribatulidae Thor

Siculobata sicula (Berlese)

1892. Oppia tibialis (Nicolet) var. sicula, Berlese, in Berlese, Acari, Myriopoda et Scorpiones hucusque in Italia reperta, Padova, fasc. 64, n° 1.

A mycophagous or saprophagous species common on all species of Citrus in Sicily, Calabria and Apulia. The mite is found, often in large numbers, on the shady parts of the larger branches, on the leaves and on the fruit.

Family Micreremidae Grandjean

1931. Micreremus gracilior Willmann, in Willmann, Arch.Hydrobiol., 23: 373.

This is a new species for the Italian fauna. In its feeding habits it is probably mycophagous or saprophagous. It was found on leaves and twigs of lemon and orange in Sicily and Calabria.

Family Ceratozetidae Jacot

Humerobates rostrolamellatus Grandjean

1936. Humerobates rostrolamellatus Grandjean, in Grandjean, Annls Soc.ent.Fr., 105: 77.

A mycophagous or saprophagous species found on the leaves, fruit and branches of orange, lemon and clementine in Sicily, Calabria and Apulia.

Trichoribates angustatus Mihelcic

1957. Trichoribates angustatus Mihelcic, in Mihelcic, Zool.Anz., 159: 102.

This is a new species for the Italian fauna. Nutritionally it is probably mycophagous or saprophagous. It was found on twigs of orange and lemon in Sicily, Calabria and Apulia.

Family Tydeidae Kramer

Pronematus ubiquitus (Mc Gregor)

1932. Tydeus ubiquitus Mc Gregor, in Mc Gregor, Proc.ent.Soc.Wash., 34: 62.

A species present on all Citrus in the main citrus cultivation areas in Italy. Its feedings habits are little known; probably it is mycophagous, saprophagous or coprophagous.

Triophtydeus triophthalmus (Oudemans)

1929. Tydeus triophthalmus Oudemans, in Oudemans, Ent.Ber.,Amst., 7: 479.

A species found in all the citrus cultivation areas investigated. Little is known of its feeding habits.

Tydeus ferulus (Baker)

1944. Lorryia ferulus Baker, in Baker, An.Inst.Biol.Univ.Mex., 15(1): 217.

A species found on the leaves and fruit of lemon, orange, mandarin and clementine in Sicily. Nutritionally it is probably heterogeneous (predator, corophagous, mycophagous, saprophagous).

Tydeus reticulatus Oudemans

1928. *Tydeus reticulatus* Oudemans, in Oudemans, Ent.Ber.Amst., 7: 380.

A new species for the Italian fauna. Its feeding habits are unknown. It was collected on lemon leaves in Sicily, Calabria and Apulia.

Tydeus formosa (Cooreman)

1958. *Lorryia formosa* Cooreman, in Cooreman, Bull.Inst.r.Sci.nat.Belg., 34 (8): 7.

A mycophagous species present in all the Italian citrus cultivation areas.

Tydeus teresae (Carmona) Comb. n. (2)

1970. *Lorryia teresae* Carmona, in Carmona, Acarologia, 12 (2): 310.

A new species for the Italian fauna. During this study we collected some few examples on leaves of lemon in Calabria. Its feeding habits are unknown.

Tydeus australensis Baker

1968. *Tydeus australensis* Baker, in Baker, Ann.ent.Soc.Am., 63 (1): 168.

This mite was found in Sicily on orange leaves. Its feeding habits are unknown.

Orthotydeus californicus (Banks)

1904. *Teytranychoides californicus* Banks, in Banks, N.Y.Ent.Soc., 12: 54.

A species present on all Citrus in the Italian citrus cultivation areas. Little is known about its feeding habits. Probably it is heterogeneous (predator, mycophagous, coprophagous, saprophagous).

Orthotydeus foliorum (Schrank)

1776. *Acarus foliorum* Schrank, in Schrank, Beytrage zur Naturgeschichte, Augsburg: 33.

This species, too, is widespread on all Citrus in the main areas of citrus cultivation in Italy. Little is known of its feeding habits.

Orthotydeus kochi (Oudemans) Comb. n.

1928. *Tydeus kochi* Oudemans, in Oudemans, Ent.Ber.,Amst., 7 (164): 337.

A new species for the Italian fauna. It was found in Sicily on orange leaves. Its feeding habits are unknown.

Orthotydeus caudatus (Dugés) Comb. n.

1834. *Tetranychus caudatus* Dugés, in Dugés, Annls Sci.nat.,ser.2,1(Zool.):29.

---

(2) The tydeids dealt with in this paper have been classified according to André's (1980) revision.

181

This species has been collected only in Eastern Sicily on orange leaves. Its feeding habits are unknown.

Family Phytoseiidae Berlese

Seiulus amaliae Ragusa and Swirski

1976. Seiulus amaliae Ragusa and Swirski, in Ragusa and Swirski, Redia, 59:183.

This mite was found in Sicily on orange and lemon leaves. It is a predator.

Typhlodromus exhilaratus Ragusa

1977. Typhlodromus exhilaratus Ragusa, in Ragusa, Acarologia, 18 (3): 380.

This mite was found in Sicily on the leaves and fruit of all Citrus. It is a predator.

Typhlodromus rhenanoides Athias-Henriot

1960. Typhlodromus rhenanoides Athias-Henriot, in Athias-Henriot, Bull.Soc. Hist.nat.Afr.N., 51: 85.

A predator found on the leaves and fruit of all Citrus in Sicily.

Typhlodromus cryptus Athias-Henriot

1960. Typhlodromus cryptus Athias-Henriot, in Athias-Henriot, Bull.Soc.Hist. nat.Afr.N., 51: 89.

A predator species collected on the leaves of lemon, orange and mandarin in Sicily.

Typhlodromus athenas Swirski and Ragusa

1976. Typhlodromus athenas Swirski and Ragusa, in Swirski and Ragusa, Phytopa-rasitica, 4 (2): 111.

A predator collected on orange and lemon leaves in Sicily.

Phytoseius finitimus Ribaga

1902. Phytoseius finitimus Ribaga, in Ribaga, Riv.Patol.Veg., Padova, 10: 178.

A predator collected on orange and lemon leaves in Sicily, Calabria and the Naples area.

Iphiseius degenerans (Berlese)

1889. Seius degenerans Berlese, in Berlese, Acari, Myriopoda et Scorpiones hu-cusque in Italia reperta, Padova, fasc. 54, n° 9.

A predatory species collected on leaves and fruit of lemon, orange and mandarin in all the Italian citrus cultivation areas.

Phytoseiulus persimilis Athias-Henriot

1957. Phytoseiulus persimilis Athias-Henriot, in Athias-Henriot, Bull.Soc.Hist. nat.Afr.N., 48: 347.

A predatory species collected on leaves and fruit of lemon and orange in Sicily and Calabria.

Amblyseius aberrans (Oudemans)

1930. Typhlodromus aberrans Oudemans, in Oudemans, Ent.Ber.,Amst., 8: 48.

A predator collected on lemon and orange leaves in Sicily and Calabria.

Amblyseius messor Wainstein

1960. Amblyseius messor Wainstein, in Wainstein, Zool.Zh., 39 (5): 688.

This species was not found during the investigation. In Sicily it was reportes on lemons by Ragusa (op.cit.). A predator.

Amblyseius barkeri (Huges)

1948. Neoseiulus barkeri Hughes, in Hughes, The mites associated with stored food products, Tech.Bull.Minist.Agric., London: 142.

A predator collected on lemon leaves in Sicily, Calabria and Sardinia.

Amblyseius stipulatus Athias-Henriot

1960. Amblyseius stipulatus Athias-Henriot, in Athias-Henriot, Acarologia, 2:294.

A predator widely spread over all the citrus cultivation areas in Italy on all species of Citrus.

Amblyseius largoensis (Muma)

1955. Amblyseiopsis largoensis Muma, in Muma, Ann.ent.Soc.Am., 48: 266.

During this investigation we never found this species. It was reported by Vig giani in 1982. A predator.

Amblyseius potentillae (Garman)

1958. Amblyseiopsis potentillae Garman,in Garman, Ann.ent.Soc.Am., 51: 76.

A predatory species collected on leaves and fruit of lemon, orange and mandarin in Sicily, Calabria and the Naples area.

Amblyseius californicus (Mc Gregor)

1954. Typhlodromus californicus Mc Gregor, in Mc Gregor, Bul.Sth.Calif.Acad. Sci., 53: 89.

A new species for the Italian fauna. This predator was collected on lemon lea ves in Sicily.

Amblyseius swirski Athias-Henriot

1962. Amblyseius swirski Athias-Henriot, in Athias-Henriot, Annls Ec.natn.Agric. Alger, 3: 1.

A predatory species collected on leaves of lemon in Sicily.

Family Ascidae Voigts and Oudemans

Proctolaelaps pygmaeus (Müller)

1859. Gamasus pygmaeus Müller, in Müller, Lotas, 9: 26.

A predator present in all Italian citrus cultivation areas.

183

Family Stigmaeidae Oudemans

Ledermuelleriopsis plumosus Willmann

1951. Ledermulleriopsis plumosus Willmann, in Willmann, Stit. Zungsb.der Osterr.
Akad.Wissensch.,mathem.natuzw.Kl.,Abt I, 160 (1–2): 140.

A new species for the Italian fauna. Almost certainly a predator.

Agistemus collyerae Gonzalez

1963. Agistemus collyerae Gonzalez, in Gonzalez, Acarologia, 5 (3): 349.

This species, too, is new to Italy. Probably it is a predator.

Zetzellia mali (Ewing)

1917. Caligonus mali Ewing, in Ewing, J.econ.Ent., 10: 499.

This species was collected inside buds and on leaves and fruit of lemons, and
on orange leaves in Sicily, Calabria and Sardinia. It is a predator.

Zetzellia graeciana Gonzalez

1965. Zetzellia graeciana Gonzalez, in Gonzalez, Univ.Calif.Publs Ent., 41: 22.

This species was found inside buds and on the leaves of lemon, orange, manda_
rin and clementine in Sicily, Calabria and Sardinia. It is a predator.

Family Hemisarcoptidae Oudemans

Hemisarcoptes malus (Shimer)

1868. Acarus malus Shimer, in Shimer, Trans.Amer.ent.Soc., 1: 368.

A predator species of Diaspididae so far found only in Sicily.

Family Cheyletidae Leach

Cheletogenes ornatus (Canestrini and Fanzago)

1876. Cheyletus ornatus Canestrini and Fanzago, in Canestrini and Fanzago, Atti
Accad.scient.veneto–trent.–istriana, 5: 106.

A predator found inside the buds and on leaves of lemon, orange, mandarin
and clementine in Sicily and Calabria.

Cheletomimus minutus Soliman

1977. Cheletomimus minutus Soliman, in Soliman, Dt.ent.Z., 24 (3): 209.

A predator species found on orange leaves in Sicily.

Cheletomimus berlesei(Oudemans)

1904. Cheletes berlesei Oudemans, in Oudemans, Ent.Ber.,Amst., 1 (7): 154.

A predator found inside buds and on lemon leaves in Sicily and Calabria.

Eutogenes citri Gerson

1967. Eutogenes citri Gerson, in Gerson, Acarologia, 9 (2): 363.

184

A predator found on lemon leaves in Sicily.

Family Acaridae Ewing and Nesbitt

Tyrophagus tropicus Robertson

1959. Tyrophagus tropicus Robertson, in Robertson, Aust.J.Zool., 7: 156.

This species, too, is new to Italian fauna. It was found on lemon fruit in Sicily. Little is known of its feeding habits.

Tyrophagus putrescentiae (Schrank)

1781. Acarus putrescentiae Schrank, in Schrank, Enumeratio Insectorum Austriae indigenorum, Augsburg: 521.

A species found on leaves and fruit of lemon and orange in Sicily and Calabria. Presumably it is mycophagous or saprophagous.

Thyreophagus entomophagus (Laboulbene)

1852. Acarus entomophagus Laboulbene, in Laboulbene, Annls Soc.ent.Fr.: 54.

A species found on leaves and fruit of orange and lemon in Sicily and Calabria. Presumably this mite is mycophagous, saprophagous, necrophagous or coprophagous in its feeding habits.

Thyreophagus corticalis (Michael)

1885. Tyroglyphus corticalis Michael, in Michael, Jl R.microsc.Soc., 2 (27–31): 885.

A species found inside buds of lemon in Sicily. Probably its feeding habits are analogous to that of the congener already described.

Family Glycyphagidae Berlese

Glycyphagus domesticus (De Geer)

1778. Acarus domesticus De Geer, in De Geer, Mem.Hist.Ins., 7 (87), n° 1.

A species found on leaves of orange and lemon in Sicily. Its feeding habits are probably analogous to that of the species mentioned above.

Family Eriophyidae Nalepa

Aceria sheldoni (Ewing)

1937. Eriophyes sheldoni Ewing, in Ewing, Proc.ent.Soc.Wash., 39 (7): 193.

A phytophagous species widely spread in all the citrus cultivation areas of Italy.

Aculops pelekassi (Keifer)

1959. Aculus pelekassi Keifer, in Keifer, Bur.Ent.Calif.Dep.Agric., n° 1.

A phytophagous species present in all the Italian citrus cultivation areas.

Family Saproglyphidae Oudemans

Calvolia hebeclinii (Sicher)

1899. Glycyphagus hebeclinii Sicher, in Canestrini, Prospetto dell'Acarofauna Italiana, 8, 971, t. 99.

A species found on leaves and fruit of orange and lemon in Sicily. Its feeding habits are probably mycophagous or saprophagous.

Family Tarsonemidae Kramer

Polyphagotarsonemus latus (Banks)

1904. Tarsonemus latus Banks, in Banks, Proc.U.S.natn.Mus., 32 (1553): 615.

A phytophagous species collected in Sicily.

Fungitarsonemus monasterii (Lombardini) Comb. n. (3)

1959. Hemitarsonemus monasterii Lombardini, in Lombardini, Boll.Ist.Ent.agr. Oss.Fitopath.Palermo,3: 164.

This mite has reported on lemon in Sicily by Lombardini (1959). It was never found during the present investigation. Its feeding habits are unknown.

Tarsonemus unguis Ewing

1939. Tarsonemus unguis Ewing, in Ewing, Tech.Bull.U.S.Dep.Agric., 653: 30.

A species sporadically found on lemon leaves in Sicily. Unknown feeding habits.

Tarsonemus aurantii Oudemans

1927. Tarsonemus aurantii Oudemans,in Oudemans,Tijdschr.voor Ent.,70: 34.

This is the mite reported by Lombardini (op.cit.) on lemon in the Palermo area when he listed it as Tarsonemus floricolus Canestrini and Fanzago. During the present investigation we found it several times on the leaves and fruit of lemon and orange in Sicily and Calabria. Its feeding habits are of the mycophagous kind.

Tarsonemus smithi Ewing

1939. Tarsonemus smithi Ewing, in Ewing, Tech.Bull.U.S.Dep.Agric., 653: 18.

A mycophagous species found sporadically on leaves of orange and lemon in Sicily.

Tarsonemus waitei Banks

1912. Tarsonemus waitei Banks, in Banks, Proc.ent.Soc.Wash., 14: 96.

A mycophagous species frequently found on orange and lemon leaves in Sicily.

Tarsonemus bakeri Ewing

(3) The tarsonemids described in this paper have been classified according to Beer and Nucifora's (1965) revision.

186

1939. <u>Tarsontmus bakeri</u> Ewing, in Ewing, Tech.Bull.U.S.Dep.Agric., <u>653</u>: 20.

A species sporadically found on lemon and orange leaves in Sicily. The species is new for the Italian fauna. Unknown feeding habits.

Family <u>Cunaxidae</u> Thor

<u>Cunaxa capreolus</u> (Berlese)

1890. <u>Scirus capreolus</u> Berlese, in Berlese, Acari, Myriopoda et Scorpiones hucusque in Italia reperta, Padova, fasc. 57, n° 9.

A predator sporadically found on lemon leaves in Sicily.

<u>Cunaxa setirostris</u> (Hermann)

1804. <u>Scirus setirostris</u> Hermann, in Hermann, Memoire Apterologique, Hammer, Strasbourg: 62.

A predator found sporadically on lemon leaves and twigs in Sicily.

3. <u>Conclusions</u>

We do not wish to considerer the research reported here and the results obtained so much from the strictly faunistic point of view as rather from the technical agrarian one. The purpose of the study is to lead us to an understanding of the facts and phenomena connected with the use of the various techniques of integrated and biological control. It is seen that by respecting certain equilibria we are able to keep dreaded key–phytophagous infestations below their threshold level; furthermore we should be warned against the recurrent habit of advising – as one commonly finds is done even at highly respectable levels – insecticide or acaricide treatments that lead to useful species being made rare, and to the dismantling of those equilibria had been set up by Nature to our advantage.

REFERENCES

1. ANDRE, H.M. (1980). A generic revision of the family Tydeidae (Acari: Actinedida). IV Generic descriptions, keys and conclusions. Bull.Ann.Soc.r.belge Ent., 116, pp. 103–168.
2. BEER, R.E. and NUCIFORA, A. (1965). Revisione dei generi della famiglia Tarsonemidae (Acarina). Boll.Zool.agr.Bach., (ser.II), 7, pp. 19–43.
3. CAVARA, F. and MOLLICA, N. (1903). Intorno alla "Ruggine bianca dei limoni". Comunicazione fatta alla Accademia Gioenia di Scienze Naturali, Catania, (ser. IV), 17, pp. 1–25.
4. LOMBARDINI, G. (1959). Acari nuovi. XXXVII. Boll.Ist.Ent.agr.Oss.Fitopath.Palermo, 3, pp. 163–168.
5. NUCIFORA, A. (1983). Integrated Pest Control in Lemon Groves in Sicily: Fire Years of Demonstrative Tests and Present Feasibilities of Transferring Results. C.E.C. Programme on Integrated and Biological Control, Final Report 1979/1983, pp. 129–146.
6. RAGUSA, S. and SWIRSKI, E. (1976). Notes on predacious mites of Italy, with a description of two new species and of an unknown male (<u>Acarina</u>: <u>Phytoseiidae</u>). Redia, 59, pp. 179–196.

7. RAGUSA, S. (1977). Notes on phytoseiid mites in Sicily with a description of a new species of <u>Typhlodromus</u> (Acarina: Mesostigmata). Acarologia, 18 (3), pp. 379-392.
8. VIGGIANI, G. (1982). Effetti collaterali di fitofarmaci su Acari Fitoseidi degli agrumi. Atti Giornate Fitopatologiche, 3, pp. 171-179.

Studies on citrus red mite in Sardinia*

G.Delrio
Institute of Agricultural Entomology, University of Sassari, Italy

Summary

Population fluctuations of Panonychus citri (Mc Greg.) and the effects
of climatic factors, natural enemies and of pesticide treatments on
the population of citrus red mite were studied intensively from 1979
to 1984 in Sardinian citrus groves. Seasonal citrus red mite population
cycles occurred with maximum population densities in spring and autumn,
depending on the seasonal growth cycles of the tree and the direct effects of weather on the mites (cold in winter and hot and dry periods
in summer). In the same orchard, grapefruit and lemon trees harboured
the higher mite populations in respect to orange and mandarine trees.
The most effective predators in approximate order of importance were:
phytoseiids (predominantly, Amblyseius stipulatus A.-H.), coniopterygids (Conwentzia psociformis (Curt.)), Stethorus punctillum (Weise)
and other several polyphagous coccinellids, Chrysoperla carnea (Steph.)
A virus disease was detected in late spring, coinciding with high mite
density. High autumn infestations of citrus red mite were noted due to
summer treatments against scales with organophosphates, which eliminate the major part of the predators.

1. INTRODUCTION

A recent investigation conducted on Sardinian citrus pests revealed,
as well as numerous insects, also some species of phytophagous mites (6).
Amongst these, the eriophyoids Eriophyes sheldoni Ewing and Aculops pelekassi (K.) cause damage in a few areas, the tenuipalpid Brevipalpus californicus (Banks) is found in low numbers in some traditional citrus groves and
the tetranychid Tetranychus urticae Koch is considered an important pest
which is controlled with acaricides, particularly in the south-eastern
citrus growing area. Some growers also treat the tydeid Tydeus formosa (Cooreman) with acaricides wrongly judging this mite to be harmful. T. formosa,
however, is commonly found with scales on the citrus where it feeds on honeydew and sooty mold, carrying out a useful role as a sanitizing agent (14).
Panonychus citri (Mc Greg.) was accidentally introduced into central-
western Sardinia in 1978, so greatly complicating the citrus pest situation.
As it was known that the high population levels of this mite can seriously
damage the production and compromise the quality of the fruit, research into
population dynamics, natural enemies and control methods was begun in 1979.

---

*Studies of the C.N.R. Working Group for the Integrated Control of Plant
pests.

2. MATERIALS AND METHODS

Studies were carried out in some citrus groves in central and southern areas of Sardinia, with a warm temperate climate. The average temperatures in July and August are 32°C and the maximum can reach to peaks above 40°C; relative humidity in summer can often fall below 30%. Rainfall is about 700 mm and results as being concentrated in autumn and winter with high intensity of precipitation (mm/hour).

During 1979 and 1980, 5 neighbouring plots each of 0.3 ha located in a 9 ha citrus grove near Oristano, planted with "Eureka" lemons, "Marsh" grapefruit, "Washington navel" oranges, clementines and "Avana" mandarines were chosen for study. The whole grove was not sprayed during the test years and the plots consisted of 150 trees each with a tree spacing of 3.5 X5.5 m. The tree age was 10 years with height varying from 2.5 to 3.0 m. During 1979 to 1981 citrus red mite studies were also conducted in 3 plots of 0.5 ha each, situated in a grove of 19 ha near Simaxis, consisting of "Washington navel" and "Tarocco" oranges. The trees were 17 years old with a spacing between trees of 6X5 m. The grove was sprayed 2-3 times during the summer with white oil and parathion to keep the scales (particularly Planococcus citri Risso) under control. Some plots were left untreated for comparison. A square frame (0.25 m^2) was used weekly to measure the young growth flushes. Citrus red mite and predators from each plot were sampled weekly by collecting 50 leaves 1.5-2.0 m above the ground from various quadrants of 5 trees. Each leaf sample was placed in a separate plastic bag and taken to the laboratory where they were all examined under the stereomicroscope. The numbers of citrus red mite eggs, larvae, nymphs, adult males and females and predators were counted for each leaf, for determining age-class structure. When available a sample of mites was mounted in Hoyer's medium and examined microscopically under polarized light to determine virus infection (19).

In 1982-84 the mite and predators populations were observed in 4 plots each of 0.5 ha, situated in a 40 ha grove near Oristano, consisting of "S. Chiara" lemons and "Washington navel" and "Tarocco" oranges. The trees were 20 year olds and spaced 6X5 m apart. Some plots were treated with white oil and metidathion and others were left untreated for comparison. Citrus red mite and predator populations were counted weekly in situ with the naked eye and a handlens on 100 mature leaves chosen at random per plot.

3. RESULTS AND DISCUSSION

3.1 Seasonal abundance of Panonychus citri

P. citri reproduces on citrus in Sardinia during the whole year. Its population, however, presents seasonal fluctuations with maximum density during the spring and autumn months.

In all groves examined from 1979 to 1981 the densities of the mite were low in the winter and spring periods, falling lower in summer, beginning to rise again from the third week of August and reaching a maximum in September-November (Fig.1).

In 1982 populations resulted as being high in some groves for the whole

190

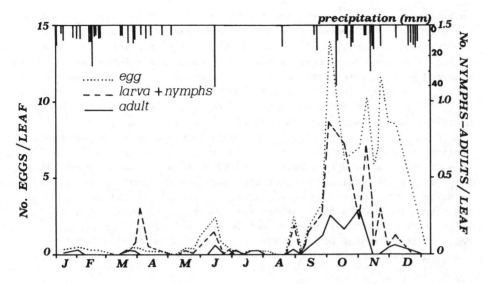

Fig. 1. Fluctuation of citrus red mite population on grape-fruit (Oristano, 1979).

of spring (7-22 mites/leaf), but falling in June, until almost disappearing in July and August. They began rising again in autumn (Fig. 2).

In the winter months of 1979-81 the population of P. citri consisted mostly of eggs and adults; larvae and nymphs began to appear in noticeable numbers only in March and April, coinciding with the first flush growth of the citrus. On grape-fruit, in 1979, for example, larvae and nymphs showed two population maximums in April and June. In the summer months the mite was represented only by a few adults and eggs. Egg laying began again at the end of August on the foliage matured by the second flush growth and reached the maximum density in October-December (Fig. 1).

The analysis of the age-class structure of the population demonstrates that the eggs are more numerous during the whole year than larvae, nymphs and adults. This indicates that the natural mortality factors affect the immature stages in a stronger way than the eggs. The prevalence of eggs over the other stages of the mite has also been found in Japan (21).

3.2 Effects of climatic factors on populations of Panonychus citri

Climatic factors affect the citrus red mite populations whether by slowing down or by accelerating the development cycle or acting as a mortality factor.

It was possible to calculate the theoretic number of generations that the mite can develop in Sardinia on the basis of the relationship between development and temperature which is well known for this mite (5, 12, 22). and it resulted as about 9 generations.

The reproduction of P. citri is interrupted at temperatures lower than 8°C and higher than 32°C (6, 22). The highest fecundity is reached at 24°C (22). Egg hatching is negatively affected both by high temperatures linked with low air humidity and by low temperatures linked with high humidity (8,16).

Unfavourable climatic conditions for the development and reproduction of the mite occur in Sardinia in winter (December–February) and summer (July and August). In fact, in winter the development of the mite slows down due to the low temperature (average monthly temperatures resulted in the years examined as being: in December 11.8°C, January 7.5°C and February 8.0°C). In July and August the daily maximum temperatures normally resulted above 30°C with points of 40°C and these temperatures, linked with the low air humidity, often lower than 20%, caused high mortality in the field.

Rain also reduces the P. citri populations depending on the hourly precipitation intensity and the size of the drops. The mites can in fact be completely eliminated by a rainfall of 80 mm in a day and by winds above 8 m/sec (20). In Sardinia, violent downpours are frequent at the beginning of autumn, when storms can occur bringing from 100 to 150 mm of rain in a few hours. During the observations years, precipitations of above 30 mm in a few hours washed away eggs and motile stages in June 1979 and May 1980. However, the rains of October 1979 resulted of greater importance as they practically eliminated the mites for the whole month.

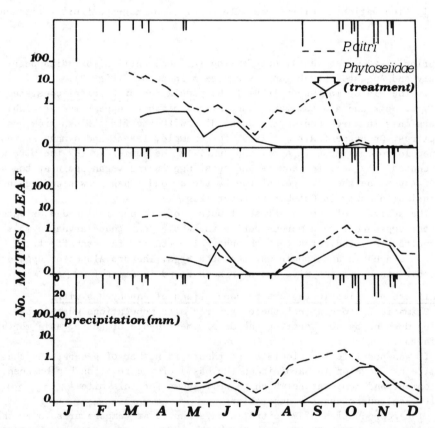

Fig. 2. Population fluctuations of citrus red mite (motile stages) and its main predators, phytoseiid mites, from direct counts in 3 plots of oranges (Oristano, 1982).

3.3 Effects of the host plant on populations of Panonychus citri

Citrus in Sardinia presented three growth cycles during the observation years: in spring, summer and autumn. The spring flush started in early March and continued up to half way through April, that is until the beginning of blossoming. Another new growth (which was the heaviest flush of the year 1979) was produced during June and July; the much lighter third growth flush occurred in September-October.

The first growth flush comes under the attack of the mite in May-June, when the leaves have reached their final size. After a summer slowing down in the reproduction, the mite then goes on to infest the matured leaves of the second growth flush at the end of August. It is during this same period that the mite move on to the small orange fruit (4-5 cm in diameter) where they grow in greater numbers than the leaves. The lack of new foliage as from November reduces the population development of P. citri. The mite is usually found on the fruit in winter and so the most part of the population is eliminated by the harvest.

Flushing rythm is an important biotic factor regulating population of P. citri. The mite shows a preference for medium aged leaves on which they show a lower mortality and major fecundity compared to that of the mature and very new leaves (9, 15). This preference together with climatic factors explains the seasonal cycle of the mite observed in the field. In fact the two maximum populations of the mite occur in coincidence with the spring and summer-autumn flushes. The minimum populations in winter and summer are probably due to the combination of the effects of ageing foliage and nega- tive climatic conditions. The resulting feeding injury in the event of strong mite density, furthermore, determines a fall in population with a tendancy for the mite to migrate. Seasonal cycles linked with the vegetation have been observed in various citrus areas of the world (3, 7, 10, 11).

Small variations in the flushing cycle of different citrus species can determine changes in the mite populations seasonal cycle. However, observa- tions made on the same orchard have clearly shown higher mite densities dur- ing the whole year, per unit of surface foliage, on grapefruit and lemon, than on orange, clementine and mandarin (Tab.I).

Tab. I. Numbers of citrus red mites on leaves of 5 citrus species (Oristano, 1979; average numbers of mites or eggs per 100 cm^2; samples of 100 mature leaves per citrus species).

MONTH	GRAPEFRUIT	LEMON	ORANGE	CLEMENTINE	MANDARIN
MAY	1.56	2.80	0.07	0.18	–
JUN	6.91	3.24	0.92	–	1.05
JUL	0.54	3.78	4.31	0.36	0.66
AUG	6.91	2.70	0.92	0.45	2.11
SEP	10.42	0.27	0.40	0.27	0.66
OCT	40.55	9.70	12.28	0.72	–
NOV	32.40	10.16	3.30	8.64	–
DEC	26.24	2.70	0.92	3.04	–

3.4 Natural enemies of Panonychus citri

Many species of insects and predacious mites (thysanopters, chrysopids, coniopterygids, coccinellids, staphylinids, cecidomyids, phytoseiids and stigmeids) have been reported in various citrus growing areas of the world feeding on the citrus red mite (2, 3, 4, 7, 17). Most of these are incidental feeders, which depend on other sources for their principal supply of food, but a number depend principally on the citrus red mite. Furthermore, a virus disease has been found commonly infesting the mite in California(13)

In Sardinia, high citrus red mite populations resulted associated with the coniopterygids Conwentzia psociformis (Curtis) and Semidalis aleyrodiformis Steph., the chrysopid Chrysoperla carnea (Steph.) and the coccinellid Stethorus punctillum Weise. Other polyphagous coccinellids, particularly Clitostethus arcuatus Rossi, common on citrus, can also prey on citrus red mite (1). These insect predators can be found in the citrus groves, above all in spring and summer months. The maximum yellow sticky trap catches of coniopterygids are in fact in June and September, while Stethorus punctillum has two peaks in June and August-September (Fig. 3). Severe citrus red mite infestations concentrated on single trees have been eliminated on some occasions just by a high concentration of predators (particularly Conwentzia psociformis and Stethorus punctillum).

A virus disease has also been noted at Simaxis in the presence of high mite density (more than 20 mite/leaf) in late spring. The disease does not however seem common and it wasn't possible to define the importance of the presence of the virus in the reduction of the mite populations.

The most important natural enemies of the citrus red mite in Sardinia are, however, the predacious mites. This is due to their constant presence during the whole year, and because of their abundance in the untreated

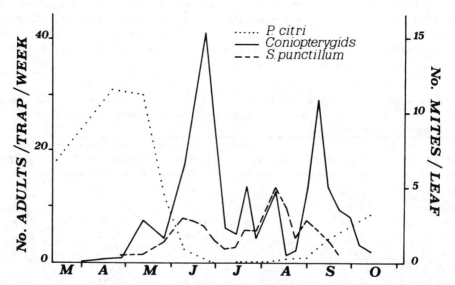

Fig. 3. Captures of some citrus red mite predators by yellow sticky traps (Oristano, 1982).

citrus groves. Four species of Phytoseiids (Amblyseius stipulatus A.-H., Iphiseius degenerans (Berl.), Phytoseiulus persimilis A.-H. and Typhlodromus rhenanoides A.-H.) and one species of Stigmeid (Zetzellia graeciana Gonzal.) were found in the citrus groves. In Sardinia, as in other parts of the Mediterranean area, the most common and numerous Phytoseiid is A. stipulatus. This mite can survive even in the absence of P. citri, feeding above all on Tydeid Tydeus formosa (Cooreman), as well as pollen and honeydew.

Phytoseiids, without doubt, represent one of the most important control factors of P. citri populations, being a density-dependent factor of mortality. In fact the Phytoseiid populations follow that of the P. citri. In 1982, the high phytophagous mite densities in spring were reduced by the Phytoseiids, that in some plots, in June, numerically outgrew the prey. After the summer interruption of population reproduction, the P. citri density was however, higher again in autumn, but at the same time the predacious mite density also grew (Fig. 2).

3.5 Effects of pesticid treatments on Panonychus citri populations

The importance that Phytoseiids have in keeping the P. citri populations under control is evident when the predators are destroyed by insecticide and acaricide treatments. In 1979, for example, the Phytoseiid densities resulted about the same level as the low P. citri populations. The grower carried out an unnecessary treatment using both insecticide and acaricide; this brought about the almost total disappearance of the predacious mites and the P. citri population built up rapidly in an extraordinary way (Fig. 4).

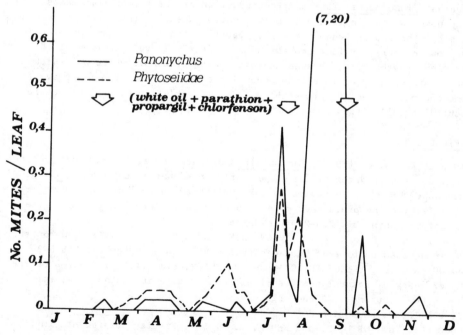

Fig. 4. Population fluctuation of citrus red mite (motile stages) and its main predators, phytoseiid mites, in a treated orange grove (Simaxis, 1979)

195

In many groves the citrus red mite population explosions were caused by summer treatments with insecticides (organophosphates and pyrethroids) carried out to control scales and the medfly. The repeated treatments certainly cause the disappearance of Phytoseiids, but an effect of pesticides on citrus physiology and also a direct effect on P. citri fecundity and survival cannot be excluded.

4. CONCLUSIONS

Before the introduction of Panonychus citri, the pest management of Sardinian citrus groves was centred on the control of Planococcus citri and other scales with white oil and organophosphates in the case of oranges and in the control with dimethoate against the medfly for the clementine.

On some orchards treated in summer with organophosphates, the Panonychus citri (and also Tetranychus urticae) have become increasingly important up to the point of causing defoliations of varying intensity and the consequent use of acaricides by growers. Major reasons for this increase of importance appear to be the reduction of predaceous species by extensive pesticide applications, particularly predaceous phytoseiid mites. The most widely distributed and generally the most abundant phytoseiid mite that feeds on citrus red mite in Sardinia is Amblyseius stipulatus. This species as well as the other mite predators are absent in many orchards where pesticides have been regularly applied.

The studies conducted in Sardinia have shown that the citrus red mite populations can be reduced to below the damage threshold in orchard where integrated control techniques are applied. The use of biological control against Planococcus citri and the reduction of the summer treatments with organophosphates, obtained by only treating when the intervention threshold has been passed and when the early stages of the mealybug are present in maximum numbers, in fact brought about a fairly large reduction of the mite problem. Furthermore, the use against scales of simple white oil which is less harmful to predators than other insecticides, and of bait-sprays for the control of medfly, bring about a considerable improvement of natural control of all citrus pests.

REFERENCES

1. AGEKYAN, N. G. (1977). Clitostethus arcuatus Rossi (Coleoptera Coccinellidae)- a predator of citrus whitefly in Adzharia. Ent. Obzr. 56: 31-33
2. ANONYMOUS (1978). Studies on the integrated control of the citrus red mite with the predaceous mite as a principal controlling agent. Acta Entomol. Sin. 21: 260-270 (in Chinese)
3. CATLING, H.D., LEE, S.C., MOON, D.K. and KIM, H.S. (1977). Towards the integrated control of korean citrus pests. Entomophaga 22: 335-343
4. DE BACH, P., FLESCHNER, C.A. and DIETRICK, E.J. (1950). Studies of the efficacy of natural enemies of citrus red mite in Southern California. J. Econ. Entomol. 43: 807-819
5. DELRIO, G. unpublished
6. DELRIO, G., ORTU, S. and PROTA, R. (1982). Fitofagi di recente introduzione nelle colture agrumicole della Sardegna. Ann. Fac. Agr. Univ. Sassari XXVIII: 57-64

7. GARCIA-MARI, F., SANTABALLA, E., FERRAGUT, F., MARZAL, P., COLOMER, P. and COSTA, J. (1983). El acaro rojo Panonychus citri (Mc Gregor): Incidencia en la problematica fitosanitaria de nuestros agrios. Bol. Serv. Plagas 9: 191-218

8. FUKUDA, J. and SHINKAJI, N. (1954). Experimental studies on the development of citrus red mite (Tetranychus citri Mc Gregor) on the influence of temperature and relative humidity upon the development of the eggs. Bull. Hort. Div. Nat. Tokai-kinki Agr. Exp. Sta. 2: 160-171 (in Japan.)

9. HENDERSON, F. and HOLLOWAY, J.K. (1942). Influence of leaf age and feeding injury on the citrus red mite. J. Econ. Entomol. 35: 683-686

10. JEPPSON, L.R., COMPLIN, J.O. and JESSER, M.J. (1961). Factors influencing citrus red mite populations on navel oranges and scheduling of acaricide applications in Southern California. J. Econ. Ent. 54: 55-60

11. KEETCH, D.P. (1971). Ecology of the citrus red mite Panonychus citri (Mc Gregor) (Acarina: Tetranychidae) in South Africa. I. The seasonal abundance of P. citri in an orchard under natural control. J. Entomol. Soc. S. Africa 34: 63-72

12. KEETCH, D.P. (1971). Ecology of the citrus red mite Panonychus citri (Mc Gregor) (Acarina: Tetranychidae) in South Africa. II. The influence of temperature and relative humidity on the development and life cycle. J. Entomol. Soc. S. Africa 34: 103-118

13. MC MURTRY, J.A., SHAW, J.G. and JOHNSON, H.G. (1979). Citrus red mite populations in relation to virus disease and predaceous mites in Southern California. Env. Entomol. 8: 160-164

14. MENDEL, Z. and GERSON, U. (1982). Is the mite Lorryia formosa Cooreman (Prostigmata, Tydeidae) a sanitizing agent in citrus groves? Acta Oecol. Oecol. Applic. 3: 47-51

15. MIJUSKOVIC, M. (1974). Influence de l'âge et du degré d'épuisement des feuilles d'agrumes sur le développment de Panonychus citri Mc Gregor. Ann. Zool. Ecol. anim. 6: 551-560

16. MUNGER, F. (1963). Factors affecting growth and multiplication of the citrus red mite Panonychus citri. Ann. Ent. Soc. Amer. 56: 867-874

17. NAKAO, S., NOHARA, K. and ONO, T. (1972). Fundamental study on the integrated control of citrus red mite in the summer orange grove.Mushi 46:27

18. SHINKAJI, N. (1959). Studies on seasonal fluctuation of citrus red mite Panonychus citri (Mc Gregor). Bull. Hort. St. Nat. Tokai-kinki Agr. Exp. Sta. 5: 143-166 (in Japanese)

19. SMITH, K.M., CRESSMAN, A.W. (1962). Birefrigent crystals in virus-diseased citrus red mites. J. Insect Pathol. 4: 229-236

20. TANAKA, M. and INOUE, A. (1962). Studies on forecasting of citrus red mite occurrence. (IV. Influence of rainfall to mite reproduction. Analysis on the effect of artificial rainfall). Proc. Assoc. Pl. Prot. Kyushu 8: 18-20 (in Japanese)

21. YASUDA, M. (1980). Changes in the distribution pattern, sex ratio and age-class structure during population growth of the citrus red mite Panonychus citri (Mc Gregor). Jap. J. Appl. Ent. Zool. 15: 447-457

22. YASUDA, M. (1982). Jap. J. Appl. Ent. Zool. 26: 52-57

197

Research on the acarofauna of citrus-groves in Greece with the aim of instituting an integrated control programme for *Tetranychus urticae* (Koch) and *Panonychus ulmi* (Koch)

P.Souliotis
'Benaki' Phytopathological Institute, Kiphissia, Greece

Summary

Research on the acarofauna of citrus groves has demonstrated the presence of various phytophagous species including three Tenuipalpidae - Hystripalpus lewisi (Ewing), Brevipalpus phoenicis (Geijskes) and Cenopalpus pulcher (Can. and Franz.), three Eriophyidae - Eriophyes sheldoni (Ewing), Aculops pelekassi (Keifer) and Phyllocoptruta oleivora (Keifer), three Tetranychidae - Tetranychus urticae (Koch), Tetranychus cinnabarinus (Bois) and Panonychus ulmi (Koch) - as well as one Tarsonemida - Polyphagotarsonemus latus (Banks) - and five predatory Phytoseiidae species, of which Amblyseius aberrans (Oudemans), Amblyseius finlandicus (Oudemans) and Phytoseius plumifer (Can. and Franz.) represent the typical predators on mites in Greek citrus groves. Of the plant-feeding species, Tetranychus urticae, Panonychus ulmi, Aculops pelekassi and Phyllocoptruta oleivora can cause significant economic damage, sometimes destroying the entire crop where they are present in very large numbers. While Tenuipalpidae are widespread, populations are low and they cause no damage to citrus fruit.

INTRODUCTION

Over the last decade, studies of the health of Greek citrus groves clearly demonstrate that mites present a growing threat to crops and one that is no less alarming than that posed by insects.

Research carried out between 1982 and 1984 on citrus mites in various citrus-growing areas has made it possible to identify the phytophagous species, as well as their predators and parasites. These include other Prostigmata species and some Astigmata and Cryptostigmata mites. It was also possible to determine the distribution, biology and any risk they posed to crops, so that a reasearch programme could then be drawn up to study the relationships between predators and phytophagous species, the natural factors determining their populations, and any side-effects of the anti-parasite preparations employed. In the future, the results of this research will allow the introduction of a programme of integrated pest management in citrus groves.

DESCRIPTION OF PHYTOPHAGOUS SPECIES

Tenuipalpidae

The species <u>Hystripalpus lewisi</u> (Ewing), <u>Brevipalpus phoenicis</u> (Geijskes) and <u>Cenopalpus pulcher</u> (Can. and Franz.) are small, red, polyphagous mites widespread throughout the various citrus-growing regions of Greece. They have been found on branches, leaves, and fruit. Their life cycle is similar, wintering during the female stage, mostly in cracks in the bark and branches. Typically they cause rotten patches on fruit and chlorotic stains on leaves. Populations are small, and it is only in southern Greece (Laconia) that much heavier infestations have been found. Damage has never been observed. In other countries as well, only sporadic and minor damage has been recorded (1). Since we do not have a good understanding of the factors liable to limit an excessive growth in the population of these phytophagous species, it is advisable not to underestimate their presence.

Eriophyidae

The tiny ochre-coloured mite <u>Eriophyes sheldoni</u> (Ewing), known as the "wonder mite", has been found in all Greek citrus-growing regions. It shelters in the buds, flower buds, and the angle formed where the stem enters the fruit. Wherever it feeds, <u>E. sheldoni</u> damages the shoots, leaves, flowers and fruit. The shape of leaves and flowers is affected, and the fruit has a misshapen appearance. In Greece, this mite causes damage only in lemons, with the damage being significant only during years in which there are especially large populations. In such cases, quite high crop losses can be encountered in groves not treated during the spring/summer period (beginning of June) with specific acaricides (Akar or Kelthane).

<u>Aculops pelekassi</u> (Keifer) and <u>Phyllocoptruta oleivora</u> (Keifer) are species that have been found in the citrus regions of north-western Greece (Preveza, Arta), as well as in the southern zones (Corinth, Laconia, Crete). Each year, more than five generations appear. The effects of these eriophyds have been observed only on fruit, with no damage to buds or leaves being recorded. They have been found on all cultivated citrus varieties. In southern Greece, these eriophyds are also to be found in all stages of development during the winter, when the weather is mild. They attack the epicarp of the fruit, which turns a characteristic rusty leather-brown or silver colour, later adopting a leathery, spotted or reticulated appearance. Over recent years, these phytophagous species have been responsible for major economic damage. Up to 60% of citrus crops can be lost unless the trees are protected by specific acaracides during the summer (beginning of June) and autumn (mid-September to mid-November).

Tetranychidae

The tetranychidi, <u>Tetranychus urticae</u> (Koch) and <u>Panonychus ulmi</u> (Koch) are our major citrus pests. They are to be found in all zones with

200

citrus plantations and produce 7 - 8 generations a year. In general, the symptoms of Tetranychidi attack are the yellowing, drying and fall of the leaves where infestations are severe. In the case of lemons and oranges, leaves have characteristic pimples, pale on top and dark underneath, while the fruit is marked by leathery spots, either spread over the fruit or concentrated in the apical area. In recent years, these mites have caused major economic damage, especially where chemical control of insect pests makes use of organo-phosphoric anti-parasite preparations. The results achieved by such treatment are now being examined. Pest control relies entirely on chemical means (specific acaracides), and more than two applications are carried out each year. Tetranychus cinnabarinus (Bois) is a mite which occurs only in north-western Greece (Preveza), and damage is not economically significant at the moment.

Predators

We have found five species of predatory Phitoseiidi: Amblyseius aberrans (Oudemans), Amblyseius finlandicus (Oudemans), Phytoseius plumifer (Can. and Franz.), Amblyseius meridonalis (Berlese) and Amblyseius citri (Meyer and Ryke). Of these, Amblyseius aberrans, A. finlandicus and Phytoseius plumifer are to be found in quite high numbers in all Greek citrus groves and comprise the typical predators of citrus mites (2). Furthermore, some species have been collected that belong to the Stigmeidae (Prostigmata) and Acaridae (Astigmata), and these are still being classified, as well as Cryptostigmata and Humerobates rostrolamelatus, which feeds on decomposing organic and animal matter.

Finally, there are quite major populations of the predator Stethorus punctillum (Weise), especially in citrus plantations infested with T. urticae and P. ulmi.

REFERENCES

1. JEPPSON, L.R., KEIFER, H.H. and BAKER, E.W. (1975). Mites injurious to economic plants, University of California Press, pp. 614.

2. PAPAIOANNOU-SOULIOTIS, P. (1981). Predacious mites (phytoseiidae) observed on various plants in Greece. Annal Inst. Phytopath. Benaki, (N.S.), 13; 36-58.

Laboratory observations on the biology of *Tarsonemus waitei* Banks

A.Nucifora & V.Vacante
Institute of Agricultural Entomology, University of Catania, Italy

Summary

The results are reported of a preliminary laboratory investigation into the feeding habits of the peach bud mite (Tarsonemus waitei Banks). This mite was successfully reared at 20° and 30° C and with R.H. between 60% and 70% on Alternaria spp. Rearing trials on lemon leaves according to Munger proved negative. The results of this first phase of the research allow us to exclude that the species can live and reproduce on lemon leaves. The presumed phytophagy of the mite on buds, flowers and young fruit still remains to be proved.

1. Introduction

Within the genus Tarsonemus Canestrini and Fanzago a group of species can be identified, presently known as the Tarsonemus waitei Banks group which are distinguished from other congeneric species by the following characteristics: "On the larva and adult of both sexes, the spinelike seta is absent dorsoproximally near the solenidion, and only 3 setiform setae are present, on tarsus II; 1 pair of scapular setae is usually markedly elongated; and the pharynx is well sclerotized, with somewhat horseshoe-shaped lateral walls, and with a conspicuous pair of glandlike structures posteriorly. The adult of both sexes usually lacks the smaller of the two tibial solenidia on leg I. On the adult male, the anterior margins of apodemes III and IV are weakened and not clearly connected with each other;on leg IV,the femorogenu lacks a flange and is usually at least twice as long as wide, and the tibia is not fused with the tarsus. On the adult female, the tegula is enlarged, the lateral plates usually do not overlap each other medially beneath the tegula, the posteromedian apodeme on the prodorsum is usually well developed and trifid, and the poststernal apodeme is sometimes reduced to a variable extent" (Lindquist, 1978).

Presently the group includes 5 species, 2 of which (T. waitei and T. bakeri Ewing) live on citrus trees in Italy (Vacante and Nucifora, 1985). The peach bud mite is the more common species on Citrus; it is found on leaves, fruit and twigs and in particular those covered with souty-mold. The feeding habits on this tarsonemid is insufficiently known. Beer (1954) reported having reared the species for various generations on fungus cultivations in the laboratory. Mc Gregor (1956) considered it phytophagous and responsible for some deformation of the leaves and fruit of Citrus in California. Suski (1972) found that the mite lived and reproduced in the laboratory on fungus cultures. Jeppson et al. (1975) reported a personal communication of Klimker in which he said that in Marocco T. waitei caused alterations to the epicarp of Citrus fruits. Lindquist (op.cit.) writes on this matter: "T. waitei

Table I – Results of rearing trials of Tarsonemus waitei Banks on Alternaria spp. at 20°and 30°C and with R.H. between 60% and 70%.

Temperature and R.H.	Sex	No.of specimens studied	Avg.development time (in days) of various biological stages			Duration of biological cycle (egg-adult) in days			Period of preoviposition in days	No.of eggs/d	Sex ratio
			egg	larvae	chrysalis	$\bar{X}$	MS	S			
20° C 60–70%	♂	8	4,7	2,6	2,0	9,2	0,27	0,60	—	—	37%
	♀	8	4,7	2,6	3,1	10,3	0,20	0,33	2,1	0,9	63% (117 specimens examined)
30° C 60–70%	♂	8	3,0	1,7	1,7	6,6	0,15	0,19	—	—	27%
	♀	8	3,8	1,7	1,7	7,2	0,13	0,14	2,0	1,9	73% (93 specimens examined)

$\bar{X}$ = average; MS = mean standard error; S = variance.

204

possibly is a facultatively phytophagous species (in addition to being probably primarily a fungivore) that may cause foliar deformities similar to those caused by
eriphyoid mites". The above literature references show that in the present state of
our knowledge the feeding habits and the role of this species within the citrus agro
ecosystem are not precisely known. If one accepts the hypotheses of McGregor (op.
cit.) and of Jeppson et al. (op.cit.), a part of some typical rust anomalies found in
our environment on the epicarp of orange, lemon, mandarin and clementine fruit
could be ascribed to the action of this mite. These considerations led to our investigation with the purpose of having a better knowledge of the feeding habits of the
mite and its bioethology on Citrus. In this paper we report the first results on the
experimental rearing of the mite in the laboratory.

2. Materials and methods

The mite was fed on Alternaria spp. developed on CZAPECK substrate (1) in
5-cm-diam. Petri capsules. The trials were carried out at 20° and 30°C and with
an R.H. of between 60% and 70%. Collateral trials were carried out rearing the
mite on young lemon leaves at 20°C and with R.H. between 60% and 70%, according
to Munger (1942). In all cases the trials were in reduced daylight. Each trial examined 8 specimens, the females coming from a laboratory rearing where the individuals had been collected in the field on mold-covered oranges. Periodic observations were made every 4 hours noting the sequence of stages. At the end of every
trial all the specimens (parents and offspring) were identified and studied morphologically in the light of the recent redescription of the group proposed by Lindquist
(op.cit.).

3. Results

.a) Biological aspects

The investigation allowed the following biological aspects of the mite to be
established:
1) In none of the 8 experimental cases was it possible to rear the tarsonemid on lemon leaves. None of the females isolated showed evidence of being able to reproduce easily on this vegetable substrate. Only 4 of the females laid eggs in the first
24 h of isolation, and only 1 egg each. The other 4 laid no eggs over the whole dura
tion of the trial (7 d).
2) The specimens reared in the two experimental conditions on Alternaria spp. sho
wed that they could easily reproduce on this nutritional substrate, even if displaying
a certain variability between one an other. Table I gives a synthetic summary of so
me of the data emerging from the first phase of this investigation.

b) Ethological aspects

The species has separate sexes (Banks, 1912). In the laboratory the two batches of trials in question displayed a varying sex ratio: 37% males (out of 117 speci
mens examined) at 20°C and 27% males (out of 93 specimens examined) at 30°C. The
species has arrhenotokous partenogenesis; the females lay unfertilized eggs from
which only male offspring emerge, whereas female offspring develop from fertilized

(1) The Authors would like to thank Dr G. Magnano di S.Lio of the Institute of Vegetable Pathology, Catania University, for his kind suggestions.

eggs. Mating is between adult males and young virgin females and lasts from a minu
te or two up to 15–20 minutes. A female can mate with more than on male. The mating mechanisms are similar to those described by Nucifora (1963) for Polyphagotarsonemus latus (Banks). In the few cases examined it has, however, been observed that only rarely does the male transport the chrysalis, binding it to himself
with the help of his genital papilla. Normally he waits impatiently, even for several hours, for the chrysalis to be transformed into a female and become available
for mating, but he does not move it from where it is. In this waiting phase several
males (up to 6 or 7) can be observed around a single chrysalis. They appear exci-
ted and often enact a series of movements that seem simulations of a fight for con-
quering the awaited female; two or more males have been seen to fight even for several minutes (3–7 min), attaching themselves with their fourth pair of legs, that
they move in a pincer–like way. We have had the impression that the contendents are
not hurt by this fighting. The male who manages to have the better of the others is
the one who mates first with the female. He helps the chrysalis free itself on the
cocoon; the female prepares herself motionless to receive the male. During the mating the couple are often disturbed by other males who in some cases manage to interrupt the copulation. The female already fertilized by a first male is generally
not very prepared for a second mating; in the few cases when one occurs, however,
it lasts from a few seconds to a minute or two. We have also observed two cases
where males attempted to mate with a newly deposited egg and another case of two
males in a clear attitude of simulated copulation. Oviposition occurs in a time varying from a few seconds to a minute or two. Generally the female does not leave
the place of oviposition. The eggs are laid separately and rest on the mycelium hyphae. The neonate larvae and the males are the most mobile of the biological stages.

4. Conclusions

The investigation has allowed us to establish that populations of T. waitei collected on citrus plants in Sicily reproduce without any difficulty on fungus substrate
such as Alternaria spp. and exclusively on that. There have been no cases of the
mite developing on a nutritional substrate of lemon leaves. The results of this study
confirm the findings of Beer (op. cit.) and Sucki (op. cit), who have successfully
rear this species on not precisely defined fungi. Concerning the presumed phytopha
gous characteristics mentioned by McGregor (op.cit.) and by Jeppson et al. (op.cit.)
and Lindquist (op.cit.), we have been able to establish that this mite does not live
on the foliar tissues of lemon. It is still a matter of doubt whether the tarsonemid
can infest buds, flowers and young newly–formed fruit of citrus. Periodic observations carried out on numerous buds of lemon in Eastern Sicily have, however, shown
that even if it is present on old leaves the mite is only seldom found inside buds, and,
in those rare cases when it has been found, the buds had already been infested by
Aceria sheldoni (Ewing). Probably the buds are only a place of shelter for the tarsonemid. Further investigations are needed to establish whether it may infest buds,
flowers and fruit, both in the laboratory and in the field.

REFERENCES

1. BANKS, N. (1912). New American mites. Proc.ent.Soc.Wash., 14: 96–99.
2. BEER, R.E. (1954). A revision of the Tarsonemidae of the Western Hemisphere (Order Acarina). Kans.Univ.Sci.Bull., 36: 1091–1387.

3. JEPPSON, L.R., KEIFER, H.H. and BAKER, E.W. (1975). Mites injurious to economic plants. Univ.Calif.Press, Berkeley, 614 pp.
4. LINDQUIST, E.E. (1978). On the synonymy of Tarsonemus waitei Banks, T. setifer Ewing, and T. bakeri Ewing, with redescription of species (Acari: Tarsonemidae). Can.Ent., 110: 1023-1048.
5. MC GREGOR, E.A. (1956). The mites of citrus trees in southern California. Mem.S.Calif.Acad.Sci., 3, 42 pp.
6. MUNGER, F. (1942). A method for rearing Citrus Thrips in the laboratory. J. ec.Ent., 35: 373-374.
7. NUCIFORA, A. (1963). Osservazioni sulla riproduzione di Hemitarsonemus latus (Banks), (Acarina, Tarsonemidae). Atti Accad.naz.ital.Ent., Bologna, 10: 142-153.
8. SUSKI, Z.W. (1972). Tarsonemid mites on apple trees in Poland. X. Laboratory studies on the biology of certain mite species of the family Tarsonemidae (Acarina, Heterostigmata). Zesz.Probl.Post.Nauk Roln., 129: 111-137.
9. VACANTE, V. and NUCIFORA, A. (1985). A first list of the mites in Citrus orchards in Italy. Proc.Integrated Pest Control in Citrus-groves, Acireale, 26-29 March (in press.).

Citrus nematodes

V.Lo Giudice

Citrus Experimental Institute, Acireale, Italy

There is a vast amount of literature on parasitic citrus nematodes in all major citrus growing countries. An association between nematodes and citrus plants would seem to have been first detected by Neal in 1889, when Heterodera radicicola (or Meloidogyne sp.) was established as a citrus root parasite in Florida.

Tylenchulus semipenetrans Cobb was first recorded in the County of Los Angeles (California) by J.R. Hodges in 1912. Cobb described this species and organized an international inquiry from which it resulted that this parasite was widespread throughout the world. He used this example of a parasitic plant nematode to draw attention to the scientific and economic importance of nematodes in agriculture and set up the first laboratory for nematology.

Since then many studies have been carried out with the aim of showing that there are many nematodes associated with citrus plants and that they interfere in various ways and to different degrees with the development of the root system and the plant's growth.

The species of nematodes known to affect citrus plants totalled eight in 1949, 28 in 1959 and 189 belonging to 39 genera in 1968. This number has meanwhile increased to 209 species belonging to 42 different genera of which only a few have been shown to be pathogenic to citrus plants.

Of these the most important is without doubt Tylenchulus semipenetrans Cobb. It is also the most widespread, whereas others are found in specific environments.

I shall therefore concentrate in particular on this nematode and consider the others in less detail.

Tylenchulus semipenetrans Cobb

This is a sedentary semi-endoparasitic nematode distributed throughout the citrus fruit growing areas. The results of surveys published show that between 70 and 80% of all citrus plants in the world may be considered infested to varying degrees with Tylenchulus semipenetrans. Unlike many other nematodes it is clearly sexually dimorphic. When mature the female is sacciform during the development of the ovaries while the male continues to be vermiform.

Since the sexes are distinct, reproduction is amphigonic, although parthenogenesis also occurs. Gametogenesis is meiotic (n=5).

The females attach themselves to the radicle and secrete from their excretive pore a mucilaginous substance to which particles of soil become attached, so as to give the radicles a coral-like appearance. The mucil-aginous substance and the soil attached thereto act as a protective shield against predators and other natural enemies. This material is used by the female to lay her eggs, from which individual nematodes of both sexes emerge. A female generally lays 75 to 100 eggs and the development cycle from egg to egg takes 6 to 8 weeks at a temperature of 25°C. Under laboratory conditions the larvae emerge in 12-14 days after undergoing an initial molt inside the egg (pre-parasitic larva). The male larvae undergo three more molts over the next 7-10 days. They do not become attached to or feed from the roots and cannot therfore be considered parasites. The female larvae attach themselves to the roots and undergo a further three molts. The larvae are spread by water, soil and the infected plants. In soil they spread very slowly and it is estimated that the nematode on average moves 4-6 mm per month.

T. semipenetrans can survive for a number of years in ground not under citrus plants. Larvae can be found in ground kept for 2 1/2 years at a temperature of 15°C while at 33°C the larvae survive for no more than 2 1/2 months. In water at 27°C larvae can survive for 128 days, but their infec-tiousness, breathing rate and food reserves are reduced. From the research carried out the optimum conservation temperature have been established as 10°C.

Since T. semipenetrans performs its biological cycle on the under-ground part of the citrus plant, its reproduction and viability are directly affected by the physical conditions of the soil.

T. semipenetrans can reproduce in a wide range of soil types, but prefers certain textures. The reproduction processes are slower in coarsely textured soil and faster in soil of a fine texture. The highest rates of reproduction and the greatest reduction in root growth on infected plants occur in medium-mixture soils well-stocked with organic substances, while the opposite occurs in sandy soils.

Temperature also affects the activity of T. semipenetrans. Soil temp-eratures between 25°C and 30°C are most suitable for infestation, growth and reproduction while at 20°C and 35°C infestation is reduced. The effect of the water content of the soil differs from one soil type to another. T. semipenetrans is very sensitive to drought conditions. Reproduction is generally helped by high water content. However, the optimum water content is identical for both nematode and plant.

A correlative of ground humidity is aeration. Where the roots of citrus plants suffer reduced oxygenation, the reproduction of T. semi-penetrans also slows. Ground containing high levels of organic substances promote infestation and rapid population growth. Conversely, the addition of large quantities of organic substances in the form of peat (2:1 by volume) inhibits reproduction, probably as an indirect result of a reduction in the pH value.

The pH value affects the T. semipenetrans population. Optimum pH values vary from 6 to 7.5. Below pH 5 and above pH 8 there is a constant reduction in reproduction, although high population levels have been recorded at pH 8-8.5.

Although T. semipenetrans is a parasite acting exclusively on citrus plants, it nevertheless enjoys an extremely wide range of hosts. At least 29 types of citrus, 21 types of citrus hybrids, 11 groups of the Rutaceae family and 6 non-citrus genera are known hosts.

Biotypes

Like other species of nematodes, T. semipenetrans also has various biotypes, a knowledge of which is important for the study of stock resistance and an understanding of any difference in response to treatment with nematocides.

At present 4 T.s. biotypes are known: the Poncirus biotype, which corresponds to the C_3 biotype in California, and reproduces on citrus spp., Poncirus trifoliata and vines, but not on olive trees; the citrus biotype, which corresponds to biotypes C_1, C_2 and C_4 in California and reproduces on Poncirus trifoliata, Citrus spp., citrange Carrizo and Troyer, olive, vines and persimmon; the Mediterranean biotype, which is very similar to the latter but which does not reproduce on olive trees, and the "grass" biotype which reproduces only on Andropogon rhizomatus, found only in Florida.

Host-parasite relationships

T. semipenetrans establishes itself on the root system of all the commercially used citrus stocks, although the degree of vulnerability to parasite attack varies. Although the attack is manifest at root level, signs of significant damage to other parts of the plant are evident.

The symptoms are a general reduction in the growth and vigour of the plant, which may cause a 30% reduction in root volume, a 50% reduction in the fresh weight of the epigaeal organs, a 35% reduction in trunk diameter and a 43% reduction in height, as well as a 50% reduction in the plant's development. Citrus seedlings may suffer a growth reduction of between 10 and 60%.

In Israel decline symptoms are associated with an infestation level on the order of 40,000 larvae per 10 grams of root. Beyond this threshold the decline symptoms become increasingly serious.

The seriously infested plants decline within 3-5 years, but where infestation reaches high levels the population tends to decrease - only to increase again once the plant develops a new root system, which in turn is attacked according to a cycle marking the plant's decline. It would seem that the decline of the plant is affected more by the destruction of the root system than by the reduced uptake of minerals. The function of the nematode in this cyclical decline is not fully known.

There is evidence, however, that seedlings infested by T.s. have lower levels of copper, zinc and manganese in their leaves and absorb more sodium from the ground where the latter is available in substantial quantities.

The physiological functions of the roots apparently undergo changes, with the result that infested plants show less reaction to adverse and stress-inducing environmental factors.

Most of the larvae penetrate the cortical layer of 4-5 week old roots,

but older roots may also be affected. Second-stage larvae penetrate the
cortical layers of the root, causing necrosis and creating a 'feed zone',
which affects between 6 and 10 cortical cells, known as the 'host cells'.
These discharge their contents, thereby creating a cavity accommodating the
head of the nematode. The walls of the host cell thicken and the nucleus
and nucleolus become enlarged and devoid of vacuoles. The adjacent cells
remain unaffected and show no signs of hypertrophy or hyperplasia. These
feed zones are invaded by secondary microorganisms which cause the root to
disintegrate and eventually lead to its destruction. The necrosis caused
by the enzymatic hydrolysis of the starch and protein is blocked by the
inhibitors produced by the plant itself to protect it against the enzymes
produced by the nematode.

Defence mechanisms

Of the plants which are resistant to T.s., Poncirus trifoliata is the
most important. It is easily crossed with most of the citrus family and is
used as a source of resistance to develop new stock plants. The defence
mechanism of Poncirus trifoliata involves three elements: a) a hyper-
sensitive cell reaction to the nematodes' feeding activities at hypodermic
level, b) the formation of periderm from the wound in the cortical layer
in response to feeding activities and c) a possible toxic effect produced
by substances secreted by the plant when nematodes are present which slows
their development at the initial stages. These resistance factors are not
transferable.

A similar retardant morphological effect was also seen on soil-
cultivated Severinia buxifolia, while no such effect is evident when agar
is used as a culture medium.

In two-membered plants the graft apparently influences the development
and reproduction of the nematode while the stock controls the availability
of roots for feeding, penetration and development.

Relationship between the T.s. and other parasites

Citrus plants infested with nematodes are often attacked by fungi,
which find a hold in the lesions caused by nematodes. The composite damage
caused by the nematodes and fungi is often more serious as a result of the
synergetic action of the two parasites. It is still unclear why nematodes
promote fungus attack. The mechanical lesions are apparently not the sole
cause and there are grounds for assuming that physiological changes in the
vegetable tissue are brought about by secretions from the esophagal glands
of the nematode.

In citrus plants infested with T.s. the plants are more likely to be
attacked by Fusarium oxysporum and F. solani. The joint fungus-nematode
action is considered to be the cause of citrus plant decay in Arizona.
Similar results have been obtained in connection with other research.

The effects of mycorhizal fungi were also studied. The influence on
the growth of seedling citrus lemons (rough lemons) of Glomus mosseae, a
mycorhizal fungus of the vesicular-arbuscular variety, in the presence of

T.s. was studied under glass. When C. lemon seedlings infested by T.s. were transplanted to soil containing G. mosseae, they developed better when the mycorhizal fungus was present than in its absence.

The same experiment was carried out with Radopholus similis in the presence of G. etunicatus. The growth of the rough lemon seedlings was significantly better where the fungus had taken hold than where it had not or where they were infested with R. similis. The stimulating effect of the fungus on the growth of the seedlings was inhibited by R. similis. When the seedlings were inocculated with G. etunicatus and/or R. similis, the depressant effect of R. similis on the growth of the seedlings was much less in evidence when the micorhizal fungus was also established. Some of the nematodes, belonging mainly to the genera Xiphinema, Longidorus and Tricodorus species are known to be virus carriers.

Although some species of nematode are known to attack citrus plants of the genera indicated above, no scientific data are available which prove the existence of nematodes which transmit viruses or virus-type diseases among citrus plants, although some attempts have been made.

Other nematodes

The various types of nematodes which attack citrus plants differ in shape, size, biological cycle, damage caused, distribution and economic impact. Experiments have shown that, apart from T.s., the following group of nematodes is pathogenic to citrus plants: four migratory endoparasites, one sedentary endoparasite and eight migratory ectoparasites.

The migratory endoparasites, which are the most important in terms of their economic impact, reproduce in the root tissue of the host plants, where they complete their biological cycle in cavities which they produce in the inter- and intra-cellular spaces of the cortical tissue. The wounds they cause inside the roots trigger a weak attempt at growth by the plants and leads to reduced rootlet mass.

In surveys carried out in Florida, Pratylenchus brachyurus was found on citrus plants in 90% of the citrus groves, frequently together with populations of R. similis. The infective forms are the female and her larvae, while the male is rare or absent. Reproduction occurs in the cortical layer of the rootlets in the course of one year. Where infection of the Seville orange tree (C. aurantium) and the rough lemon (C. limon) occurs, the result may be a growth reduction of 35% compared with a healthy seedling. Infestation rarely exceeds 2,000 specimens per gram of root, which would suggest that in spite of their distribution citrus plants are not ideal hosts. Young plants are the most seriously affected and damage decreases gradually as the plant develops, since the population density of the nematodes becomes inversly proportional to the tolerance limit. The physical properties of the soil or other environmental factors have no decisive effect on P. brachyurus and it manages to survive even under adverse environmental conditions.

Pratylenchus coffeae is more pathogenic than Pratylenchus brachyurus and also more widespread. At a temperature of 25°C this nematode completes its biological cycle in one month or less. All the active stages are capa-

213

ble of penetrating the cortical layers of the rootlets, where they create cavities and necrosis in the cells, but do not give rise to hyperplasia or other cell reactions. Reproduction is amphigonous and populations levels increase rapidly, possibly up to 10,000 specimens per gram of root. The high reproduction rate, both in coarse and fine soils, coupled with a preference for attacks on almost all citrus stock plants make this a particularly dangerous species.

The ability to adapt to different soil types and a wide range of hosts (125 types of citrus hybrids and associated types are known hosts) help it to spread and heighten the threat it poses for citrus plants.

Pratylenchus vulnus, which is associated with citrus plants in California has also been found on young citrus trees in nurseries in Italy, where it had previously only been recorded on olive trees. In cross-innoculation tests, P.V. from citrus trees were found capable of parasitizing olive trees and vice versa. All the active stages can attack the cortical tissue of the rootlets, causing lesions and large cavities. A population of 1,000 specimens per gram of root can slow growth and reduce the dry weight of Seville orange seedlings cultivated in fine or sandy soils.

Unlike P. brachyurus, the increase in population density of P. vulnus on growing plants exceeds the tolerance thresholds of the host plants. The range of host plants in the citrus family is limited, but the widespread occurrence of olive trees in areas where citrus fruit is grown makes P.V. a potentially dangerous new nematode for this crop.

Radopholus similis is found on citrus trees only in Florida, where it causes diseases known as "spreading decline." Following detailed inspections of nurseries and speedy action to control its spread, it has been limited to less than 1% of the citrus growing area. The infective stages are accounted for by the female while the male, owing to its smaller stylet and its primitive esophagous is unable to enter the tissues of the host plant other than by passing through the lesions caused by the parasitic forms. At a temperature of 25°C it completes its biological cycle in 20 days, feeding and reproducing inside the cortical tissue of the rootlets. It is also able to inflict damage on the central cylinder with consequent changes to the phloem.

The endodermis and pericycle also undergo changes. The pericycle shows signs of hyperplastic and tumorous formations. The affected tissues exude a gummy secretion which gathers on the surfaces of the lesions and turns brown due to oxidation. Under favourable conditions on a suitable host the average population density is between 500 and 1,000 specimens per gram of root. Population increases rapidly and may reach a high level three months after innoculation.

Fungi, bacteria and other microorganisms complete the destruction of the tissues damaged by R.s. and cause further injury to the infested plants. Where no host is available R.s. cannot survive for more than six months and as it is unable to survive on the decayed root it will die unless it migrates to new roots. Under favourable conditions the infested area spreads naturally and within a space of one year may cover some 15 m. Well-drained and sandy soil promotes reproduction. Most of the stock plants are affected by only a few of the thousands of different species

214

whereas cultivars and proven hybrids were shown to be resistant and able to tolerate R.s. In Florida two biotypes are known, one which attacks citrus species and banana trees (Musa spp.) and another which attacks only banana trees.

The plants affected show dessication of small branches and twigs, signs of malnutrition and smaller fruit.

Among the sedentary endoparasitic nematodes five species of the genus Meloidogyne have been found to affect citrus plants.

Of these only M. javanica, which is occasionally encountered in California, Israel and Italy, is worth mentioning as the citrus family is not a normal host of this nematode, which has problems completing its biological cycle on the roots of citrus plants.

The second-stage larvae form gall-like swellings, but are unable to complete their development. Nevertheless in California the ability of M. javanica to reproduce on Seville orange roots, Troyer citrange and Cleopatra tangerines has been established.

The eight species of migratory ectoparasitic nematodes recognized as pathogens for citrus plants are relevant only in specific areas.

Belonolaimus longicaudatus and Trichodorus christiei have been detected in Florida in the rhizosphere of decayed citrus plants. Both attack the tips of the rootlets of citrus plants. B. longicaudatus causes a swelling, distortion and occasionally the lesion of the root tips of grapefruit seedlings and a reduction in the growth of rough lemon seedlings. No other citrus hosts are known.

T. christiei, like the T. porosus recorded in California, restrict their activities to the meristematic region of the roots, thereby reducing root length.

Hemicycliophora arenaria is known in California to damage citrus plants grafted on rough lemons (C. limon). Other hosts are C. aurantifolia, Severinia buxifolia and the Cleopatra mandarin. The parasitic activities of the nematode are concentrated on the epidermic and cortical cells of the root tips, on which it causes the formation of gall-like swellings, thereby increasing cell division. This nematode is typical of the desert regions of southern California.

Paratrichodorus lobatus causes reduced growth, swelling of root tips and some leaf chlorosis in the affected plant.

Xiphinema brevicolle and X. index were shown in pathogenicity tests in Israel to reduce the growth of Seville orange seedlings by 44 and 46% respectively.

X. basiri has also been recorded on bitter lemon roots in the Sudan.

CONCLUSION

The depressant effects of nematodes on citrus plants, in particular that of T.s., are frequently masked and not readily identifiable. Genetically vigorous or well-nourished plants in favourable environmental conditions can still provide economically acceptable crops, but these nevertheless fall short of the potential yield. Action against nematodes has in some cases increased the yield of apparently healthy and productive plants by between 20 and 50%.

215

Current research is showing that suitable management of groves infested with nematodes can help to improve the development and productivity of citrus plants. However, it should be remembered that any management should take account of the soil type, the stock plant and the agricultural practices in use. Any action to control nematodes must also avoid causing any stress to the citrus plant itself.

It therefore seems indispensible that before any form of control is applied there should be a clear understanding of the biology and ecology of nematodes and their relationships with both the plants and other organisms.

REFERENCES

1. BAINES, R.C. et al. (1960). Susceptibility of some species and varieties of citrus and some other rutaceous plants to the citrus nematode. Plant Dis. Reptr., 44:281-285.
2. BAINES, R.C. et al. (1978). Nematodes attacking citrus. in: W. Reuther (ed.) The Citrus Industry, IV:321-345. Univ. Calif. Div. Agr. Sci.
3. COHN, E. (1972) . Nematode diseases of Citrus. in: J.M. Webster (ed.) Economic Nematology: 215-244. Academic Press, London.
4. INSERRA, R.N. (1977). I nematodi parassiti degli agrumi e relativi mezzi di lotta. Italia Agr., 114 (3): 90-97.
5. INSERRA, R.N. e VOLVAS, N. (1977). Nematodes other than Tylenchulus semipenetrans Cobb pathogenic to citrus. Int. Soc. Citriculture, 3: 826-831.
6. INSERRA, R.N. et al. (1980). A classification of Tylenchulus semipenetrans biotypes. J. Nematol., 12:283-287.
7. LO GIUDICE, V. (1981). Present status of citrus nematode control in the mediterranean area. Proc. Int. Soc. Citriculture, 1:384-387.
8. MEAGHER, J.W. (1969). Nematodes as a factor in citrus production in Australia. in: Chapman H.R. (ed.) Proc. 1st Int. Citrus Symp., 2:999-1006.
9. O'BANNON, J.H. et al. (1975). Bibliography of nematodes of citrus. A.R.S.-S-U.S.D.A.A.:1-41.
10. O'BANNON, J.H. et al. (1979). Bibliography of nematodes of citrus. Supplement. Sci. Educ. Adm., U.S.D.A.:1-11.
11. REYNOLDS, H.G. e O'BANNON, J.H. (1958). The citrus nematode and its control on living citrus in Arizona. Plant Dis. Reptr., 42:1288-1292.
12. STOKES, D.E. (1969). Andropogon rhizomatus parasitized by a strain of Tylenchulus semipenetrans not parasitic to four citrus rootstocks. Plant Dis. Reptr., 53:882-885.
13. STOKES, D.E. (1977). Possibility of nematode transmission of citrus viruses. Proc. Int. Citrus Congress, Florida, 3:831-833.
14. VAN GUNDY, S.D. (1958). The life history of the citrus nematode Tylenchulus semipenetrans Cobb. Nematologica, 3 (4):283-294.
15. VAN GUNDY, S.D. e KIRKPATRIK, T.D. (1964). Nature of resistance in certain citrus rootstocks to the citrus nematode. Phytopathology, 54:419-427.
16. VILARDEBO, A. e LUC, M. (1961). Le "Slow decline" des citrus dû au nematode Tylenchulus semipenetrans Cobb. Fruits, 16 (9):445-454.

Nematodes of citrus-groves in the Spanish Levante
Ecological study focused to their control

A.Bello, A.Navas & C.Belart

Institute of Edaphology & Plant Biology, Superior Council for Scientific Research, Madrid, Spain

Summary

454 soil samples from 85 localities in "La Plana" Region have been studied in order to determine the influence of the plant parasitic nematodes in citrus groves; 66 species from 33 different genera of agricultural interest have been found with an irregular frequency. The absence of some proved citrus parasitic nematodes is pointed out.

The relationships between these nematodes and different factors as soil conditions, plant age, vigour, variety, nutritional state, weeds presence and sampling season have been studied. The interest of knowing the influence of these factors on the dynamics of nematodes populations, as a basis for planning an integrated control, is remarked.

1. Introduction

This contribution is part of a work about the phytonematological problems in Spain (2) and its main porpuse is to prove (to lay the foundation of) the ecological hipothesis that allows us to recognize which environmental parameters have the highest influence on distribution and development of parasitic and pathogenous species in the Spanish citrus groves.

To carry out this study we have chosen one of the most representative areas for the citrus in our country, that we can consider the Northen border of citrus distribution (Fig.1), where this crop reaches tradition of more than a century and where the action of the environmental factors can be more manifest due to being the limit of this crop in Spain.

2. Material and methods

454 soil samples have been studied distributed according to crop intensity (Fig.1). To know the influence of environment on the citrus nematofaune, two kinds of statistical analysis have been carried out: the first to define species associations and their ecological characteristics and the second one, autoecological, upon <u>Tylenchulus semipenetrans</u>, for it is the most frequent and abundant species on this crop.

To characterize the associated nematofaune to citrus groves we have used the factorial analysis of correspondence (5), which simplify the

217

Fig. 1: Nematodes of citrus groves in the Spanish Levante. Ecological study focused on their control (A. Bello, A. Navas and C. Belart).

218

inicial data structure to produce a plan according some variation senses.
We have also analyzed the relationship of T.semipenetrans infestation
levels with the variables by means of ecological profiles (6) and, after
that, we comfirmed the results by means of x^2 analysis (total x^2 to see
the association level and partial x^2 to see the positive or negative of
that association).

The variables that we have studied are grouped as follow: Physical
variables: pH, organic matter, sand, silt and samples moisture: sampling
seasons: Spring, Summer, Autumn and Winter; host characteristics:Variety,
vigour (5 classes), deficiencies (Mg,Zn and Fe); T.semipenetrans infesta
tion levels: Less of 1.000 individuals/Kg; 1.000- 5.000 ind./Kg and more
than 5.000 ind./Kg and species: the species that appeared in lower frequency have been omitted from the total, and only 43 have been taken in
consideration.

3.   Results and discussion

3.1 Nematodes found
     From the 66 species belonging to 33 different genera (3) we found in
roots only 8 genera whose frequency, besides T.semipenetrans were what
appear in Table I. The most important species from the agricultural
point of view are: Tylenchulus semipenetrans, Helicotylenchus digonicus,
Zygotylenchus guevarai, Paratylenchus microdorus, Rotylenchulus borealis,
Paratrophurus loofi, Pratylenchus minyus, P.thornei and Xiphinema brevicolle.

Table I. Frequency of T.semipenetrans and genera found in root.

Nematodes	Frequency (%)
T.semipentrans	95,8
Aphelenchus sp.	49,0
Criconemoides spp.	4,2
Aphelenchoides spp.	1,5
Helicotylenchus spp.	1,2
Boleodorus spp.	0,7
Tylenchus spp.	0,7
Rotylenchulus spp.	0,5
Tylenchorhynchus spp.	0,2

We have to. remark that from the total number of species known as pathogenous in citrus crops (7) and according with (1) nor Belonolaimus longicaudatus, Hemicycliophora arenaria, Hoplolaimus indicus, Radopholus similis and Trichodorus christiei, whose distribution is recorded in areas of
U.S.A. (Florida and California) and Union India, neither Pratylenchus bra
chyurus, P.coffeae and P.vulnus, cosmopolitan pathogenous in fruit trees
in general (7), have appeared in this study.

219

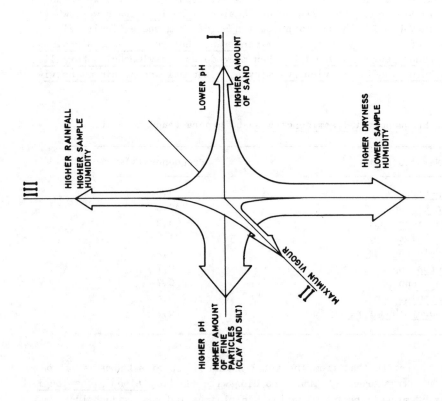

Figures 2 & 3: Nematodes of citrus groves in the Spanish Levante. Ecological study focused on their control (A. Bello, A. Navas and C. Belart).

3.2. Nematofaune characteristics(*)

After suppressing the less frequent of them we have considered the presence of 43 species: Aglenchus agricola (1), A. bryophilus (2), A. costatus (3), Aphelenchoides spp. (4), Aphelenchus avenae (5), Basiria spp. (6), Boleodorus thylactus (7), Crossonema multisquamatum (8), Ditylenchus spp. (9), Helicotylenchus digonicus (10), H.dihystera (11), H.crenacauda (12), H.erythrinae (13), H.vulgaris (14), Helicotylenchus spp. (15), Hemicriconemoides cocophilus (16), Macroposthonia antipolitana (17), M.curvata (18), M.xenoplax (19), M.sphaerocephala (20), Merlinius brevidens (21), Merlinius spp. (22), Miculenchus spp (23), Neopsilenchus magnidens (24), Nothotylenchus affinis (25), Paratylenchus microdorus (26), Pratylenchus minyus (27), P.thornei (28), Pseudhalenchus anchilisposomus (29), Psilenchus cf. iranicus (30), Quinisulcius acti (31), Rotylenchulus borealis (32) Trophurus sculptus (33), Tylenchorhynchus goffarti (34), Tylenchorhynchus spp. (35), Tylenchus ditissimus (36), T. filiformis (37), T. helenae (38), T.plattensis (39), T. thornei (40), T.vicinus (41), Tylenchus spp. (42), and Zygotylenchus guevarai (43).

The formation of factorial axes is basically made up of physical variables and Autumn with contribution of two varieties and vigour 4 for axis I. Silt, Autumn, vigour 1 and minimum infestation level of T. semipenetrans as well as the nematodes M. antipolitana, Miculenchus spp. and T. ditissimus, for axis II and finally, the season variables, varieties, vigour (and scantiness), samples moisture and the most of nematodes species for axis III.

The main variation directions of the factorial space are showed in Fig. 2. It can be noted a textural gradient (Axis I), a little vigour variation (Axis II) and a fundamentally climatic and moisture gradient which explain the maximum species vatiation (Axis III).

Samples, species and variables disposition according to their projection in Axes II-III (Fig. 3) allow us to recognize four main groupes,three of them (A-B, C and E) distributed along axis III and one of them (D) more conditioned by axis II. According projection on I-III axes (Fig. 4) it is possible to recognize separately A and B groupes, being these axes that have a greater descriptive importance, because allow us to define the ecological characteristics of the groupes.

Group A: Samples with more sand and humidity. There are two associations of species; one of them appears very close to Berna and Nules varieties, weeds presence and Zn scantiness. The other one appeares associated to highest values of sand and no specific variety. Valencia Late variety and vigour 5 (declining) seems to be common characteristics to both associations.

Group B: Is defined by 15 species associated to Spring, to 3 var - ieties (Satsuma, Navelina and Sanguina) and in a strong relationship with vigour 2, that means a Mg deficiency. This group is in the most heavy soils and a relationship

(*) Figures in brackets are the number of species in Fig. 4

221

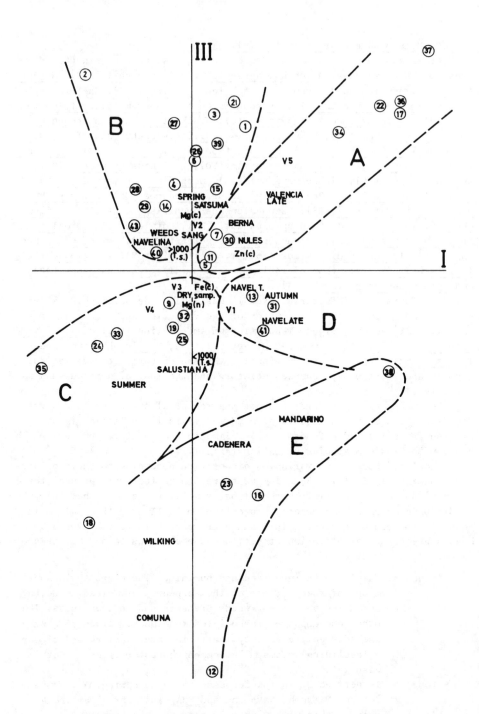

Fig. 4: Nematodes of citrus groves in the Spanish Levante. Ecological study focused on their control (A. Bello, A. Navas and C. Belart).

between weeds and high infestation levels of T.semipene-
trans is remarkable.

Group C: Is defined by 7 species. These species are associated to 3
and 4 vigour and it has tendency to appear in plants with
Fe deficiency and less abundance of T.semipenetrans.

Group D: This groupe express the association for Autumn, maximum
plant vigour, Clementino variety and presence of three
species, being also associated to soils with medium average of sand.

Group E: Appears along axis III, it is in relationship to dryness,
and orientated to highest percentage of sand. It is defined
by the varieties Comuna, Wilking, Cadenera and Mandarino in
association with four species.

3.3 Relationship of T.semipenetrans population with variables

On the following tables are displayed summarily the results of the
relationship of T.semipenetrans infestation levels with plant and soil
characteristics and sampling season.

Table II. T.semipenetrans - Plant characteristics

	x^2		Population levels indv./Kg	Association
AGE (3 classes)	9.15†	< 8 years	< 1.000	(+)*
		8 - 50 "	> 5.000	(-)†
		> 50 "	1.000- 5.000	(+)*
VARIETY (15)	52.33**	NULES	1.000- 5.000	(+)*
		SANGUINA	1.000- 5.000	(+)**
		NAVEL W.	> 5.000	(+)*†
		NAVEL T.	< 1.000	(+)*†
		VALENCIA L.	> 1.000	(+)*
VIGOUR	6.91 (NS)			

Table II shows us the most important relationships with plant characteristics, appearing significative association to host age and variety.
It has not appeared any statistical association to vigour which means that
the trees with healthy appearance to have high infestation levels.

As for age, it is clear association of the younger plants to lowest
infestation, while the highest nematode abundance has not any relationship to the oldest plants (4). This lack of relationship between the
highest abundance and any age class can be due to the wide variation
rank of medium class and so, could be possible to define exactly (having

Table III T.semipenetrans - Soil characteristics

	x^2		Population levels indv./Kg.	Association
SAMPLE MOISTURE	$12.01^{†}$	VERY DRY	< 1.000	(+)*
		SEMI-HUMID HUMID	> 5.000	(+)*
TEXTURE	31.03^{***}	HEAVY	1.000 - 5.000	(+)**
		LOAMY	> 5.000	(+)*
pH (5 classes)	18.40^{**}	> 8	< 1.000	(+)**
		7.5 - 8	< 1.000	(-)*
		> 8	> 5.000	(-)**
		7.5 - 8	> 5.000	(+)***
ORGANIC MATTER	1.33 (NS)			

Table IV. T.semipenetrans - Sampling Season

		Population levels indv./Kg	Association
$x^2 = 34.58^{**}$	SPRING	< 1.000	(-)***
	SUMMER	> 1.000	(+)***
	SPRING	> 5.000	(+)***
	SUMMER	> 5.000	(-)***

in account the relationship of young old plants) to which age, more or less, is associated.

From a total of 15 varieties only those represented seem condition T.semipenetrans abundance. That, firstable, is unespected because most of them are graften on the same root-stock (C.aurantium). The explanation could be found in the association between root-stock and variety (i.e. compatibility or not compatibility), which causes physiological alterations and consequently more or less development of roots system (8).

224

Table IIIsummarizes the results of the relationship between T.semipe-
netrans population and soil characteristics, being significative the influence of sample moisture, texture and pH and not the organic matter levels.

For the sample moisture, there is a clear association, being the
lowest levels of infestation in very dry samples. So, samples distribution according humidity classes allows to recognize, at least, that the
highest population are in association to highest moisture levels (11).

Association to texture is very significative, according to total x^2
value. We have grouped the 11 classes of soil textures in two groups,
heavy and loamy. We confirm the association of medium level infestation
to heavy texture and the maximum level infestation to loam (4): this means
probably, there is a tendency to increase the specimens number according
sand increase and to decrease of silt content, because we could prove that
the content of clay in the zone's soils is almost constant.

For the pH, we take in account the opposed associations found in the
maximum and minimum levels on the intervals $< 7.5-8$ and >8; it can be said
that above a threshold ($\simeq 8$), the pH increment has a negative effect in
the"citrus nematode"populations, so it can be considered that its optimal
development is between a 7.5 - 8 pH values (9 and 11).

Finally, the percentage of organic matter does not influence the
relative abundance of this species, although others authors consider that
it can do it (10).

The relationship of infestation levels and sampling season (Table
IV) is influenced by the difference between two extreme pluviometric
season in the zone, as we observed in the general analysis, because there
is a positive association of the highest populations to Spring and the lowest one to Summer, being indifferent the association among any level of
T.semipenetrans and Autumn-Winter.

4.   Conclusions

As a general conclusion, a good ecological characterization of nematodes is an important part of integrated pest control because:
1. On defining and typifying the faunistic association tendences, together
   with the knowledge of biogeographical characteristics allow us to prevent
   the introduction of patogenous species which are not in our country or in
   the studied area.
2. To define sampling season according to the population dynamic.
3. To predict where can appear problems according to soil and plant characteristic.
4. To predit the more resistant varieties in relation to plant vigour, age
   and states of deficiency.

So, being the nematodes a primordial element in the citrus groves
pathological chain, because fungi, virus and bacteria find in their
action easiness to the further development; the adecuate control of them,
could be the basis to integrated control of illness and perhaps pests if
we take into account that there are pesticides which can act not only
against nematodes but also against insects and mites.

225

ACKNOWLEDGEMENTS

The help of Dr.M Arias is very appreciated. We wish to thank Dr.A.Gil his guidance in computing and Visitacion Alvira her technical assistance.

REFERENCES

1. BELLO,A. y LABORDA,E.(1974). Caracteristicas biogeográficas de cinco especies de nematodos parásitos de los agrios no encontrados en España. Las Ciencias, 39:50-52.
2. BELLO,A.,LABORDA,E. y ALVIRA,P.(1973). Estudios realizados en España sobre los nematodos de los agrios.Bol.R.Soc.Esp.Hist.Nat.(Biol.),71:17-59
3. BELLO,A.,NAVAS,A. y BELART,C.(1985).Los nematodos de los cítricos. Su interés agronómico para la Comunidad Valenciana. 186 pp. (In press).
4. COHN,E.,MINZ,G. and MONSELISE,S.P.(1965). The distribution, ecology and pathogenicity of the citrus nematode in Israel.Israel J. agric. Res., 15:187-200.
5. GIL CRIADO,A.(1978). Métodos de análisis multivariante en ecología. Aplicaciones a una comunidad herbácea heterogenea. Tesis doctoral.Fac. Ciencias Univ. Sevilla, 271 pp.
6. GODRON,M.(1965). Les principaux types de profils ecologiques. CEPE.L. Emberger. 8 pp. Montpellier.
7. INSERRA,R.N.and VOVLAS,N.(1977).Nematodes other than Tylenchulus semipenetrans Cobb pathogenic to citrus.Proc. International Society of Citriculture, 3:826-831.
8. KIRKPATRICK,J.D. and VAN GUNDY,S.D.(1966). Scion and rootstock as factors in the development of citrus nematode populations. Phytopathology, 56:438-441.
9. MACARON,J.(1972). Contribution to the study of plant-parasitic nematode Tylenchulus semipenetrans Cobb 1913 (nematoda-Tylenchida).Ph.D.Thesis, University of Science and Technology, Languedoc. Montpellier. 190 pp.
10.O´BANNON,J.H.(1968). The influence of an organic soil amendment on infectivity and reproduction of Tylenchulus semipenetrans on two citrus rootstocks. Phythopatology, 58:597-601.
11.VAN GUNDY,S.D. and MARTIN,J.P.(1962). Soil texture, pH and moisture effects on the development of citrus nematode Tylenchulus semipenetrans Phytopathology, 52:31.

Weeds in citrus-groves

V.Lo Giudice
Citrus Experimental Institute, Acireale, Italy

G.Maugeri
Botanic Institute, University of Catania, Italy

Summary

This paper discusses the beneficial and harmful effects of weeds, their biology and their effects on the yield of citrus fruit trees. In particular it shows that a good knowledge of the role and the biology of weeds is required for weed control to be applied correctly.

1.1 INTRODUCTION

In citrus fruit growing, weed control is a serious problem because of the amount of labour and energy it requires and its effects on plant growth. Consequently it would appear useful to analyse the part played by weeds in order to determine what action to take involving the soil and the fruit trees themselves.

To define a weed might appear simple, given the large number of definitions in existence, but in reality each weed may have its own definition with regard to both its harmful and beneficial effects.

We know that weeds can be propagated by means of one or more of the following: seeds, tubers, rhizomes, stolons, bulbs or bulblets. At the same time, however, we need to know what ambient conditions, such as temperature, water content of the soil, light, etc., encourage growth and development through the year.

1.2 BIOLOGY OF WEEDS

The environment of weeds in citrus groves is influenced not only by the soil and the macro-climate, but also by the protection given by tree foliage and the constant repetition from year to year of cultivating operations, which are both facets of human action. All this contributes to selecting the weeds from amongst the members of the flora of a region which could potentially take root among the fruit trees.

The total accompanying flora (3) is determined by a process of "sociological" selection (25) of the plants, which in the course of time find ideal growing conditions in fruit groves, as well as by geographical and ecological selections.

However, weeds in citrus groves may also display a variety of physiognomical features under the same environmental conditions.

The flora generally consists of some widespread plants together with a fairly large number of other less frequent species. In particular environmental conditions and as a result of human action (tilling, fertilizing, irrigating, weeding, etc.) it is possible for facies or populations of practically a single species to establish themselves, to the virtual exclusion of other species (35). A typical example is the Oxalis pes-caprae, which may be observed in citrus groves in the winter period.

A knowledge of the biology of weeds in citrus groves and the related problems cannot be confined to taxonomy alone - which is not always as accurate as it should be - or to general information on the distribution and frequency of occurrence of the species which affect the crop. It is necessary to survey various aspects of the biology, not forgetting the origin of the species and their corology, biological forms, and their requirements in terms of heat, water and nutrition (nitrophilous, alo-philous and calciphilous). It is also useful to know which of the weeds present in the citrus grove also occur in natural or man-made environments. This detailed knowledge is important in that some weeds, e.g. Amaranthus spp., are found exclusively on cultivated lands, whereas others may also be found in non-cultivated environments, e.g. Bromus tectorum, or natural environments, e.g. Stellaria neglecta, the presence or absence of which is linked to various factors. In addition, a knowledge of the bio-rhythms of weeds at the point when they are adapting to agronomical practices is particularly useful.

Research carried out on weeds in Sicilian citrus groves (33, 35, 36, 49) representative of the Mediterranean environment has shown that the vegetation is usually luxuriant throughout the year, but two distinct differences emerge, linked to specific periods.

The two types of vegetation, winter/spring and summer/autumn (35, 36) which in some respects contrast with each other, change from one to the other in the course of the year. The former reaches its peak in the coldest month of the year and the latter in the hottest. The transition from winter/spring to summer/autumn vegetation occurs between May and June and the transition from summer/autumn to winter/spring vegetation between November and December (Fig. 1).

The seasonal variation in the vegetation (33) is linked to changes in the species occurring over the course of the year. Depending on their time of appearance, weeds in citrus groves may be divided into three groups as shown in Table 1.

The presence of different species at different times of the year is linked to the capacity and power of their seeds to germinate and their requirements in terms of periods of heat and light for flowering (33).

The winter/spring species have high germination rates at relatively low temperatures (between 10° and 1 C° for Stellaria neglecta).

They do not require low temperatures to complete the process of flowering and for the most part they need long periods of daylight.

In the case of summer/autumn species with maximum germination rates at temperatures of between 30° and 40° C, such as Amaranthus graecizans, the flowering accelerates in the absence of low temperatures and in general they need short periods of daylight.

The species which are found for most of the year show less marked requirements both for germination from their seeds, e.g. Sonchus oleraceus, and for completion of the flowering process (Fig. 2).

228

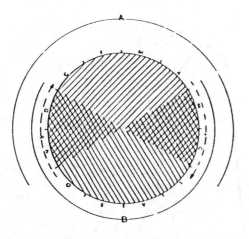

Fig. 1 – Diagram of the alternation between 2 types of vegetation in the course of the year in Sicilian citrus groves (the arrows indicate transitions):

A: winter/spring   a: transition from A to B
B: summer/autumn   b: transition from B to A

Table 1 – Seasonal occurrence of a number of weeds in Sicilian citrus fruit groves

PLANTS FOUND IN THE WINTER/SPRING PERIOD:

Bromus spp.   Oxalis pes-caprae
Calendula arvensis   Poa annua
Chrysanthemum coronarium   Stellaria media
Chrysanthemum segetum   Stellaria neglecta
Lolium rigidum

PLANTS FOUND IN THE SUMMER/AUTUMN PERIOD:

Amaranthus spp.   Echinochloa colona
Chenopodium album   Portulaca oleracea
Cyperus rotundus   Setaria spp.
Digitaria sanguinalis

PLANTS FOUND MOST OF THE YEAR:

Convolvolus arvensis   Solanum nigrum
Cynodon dactylon   Sonchus oleraceus
Parietaria diffusa   Urtica urens

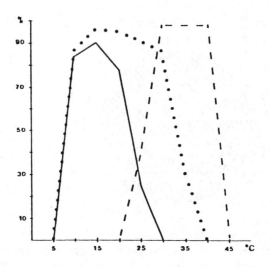

Fig. 2 - Percentage seed germination:

        —— Stellaria neglecta (winter/spring)
        --- Amaranthus graecizans (summer/autumn)
        ... Sonchus oleraceus (found all the year)

(F. Luciani and G. Maugeri - not published)

1.3 <u>FUNCTION OF WEEDS</u>

It is known that weeds have both beneficial and harmful effects on other organisms in the agro-ecosystem (25).

Of the adverse effects commonly considered significant and for which weed-control is carried out we should mention the following:

- removal of water for their development (8,21,53);
- competition for nutrients, in particular nitrogen (4,6,21,22,53);
- secretion of toxic substances (allelopathic phenomena) (13,27);
- interference with the light in the young citrus plants (53);
- provision of hosts for a number of insects and mites (20,30);
- reduction of aeration directly below tree foliage.
- increased risk of freezing through irradiation(2,12,15,20,21,31,50,51)
- interference with cultivation (53).

Among the beneficial effects, which often are not evaluated exactly, we have:

- protection against vertical and horizontal erosion of the soil (53);
- removal of water by transpiration in soil with an excessive water content;
- improvement of surface drainage of the soil (17,26);

- production of organic substances (12,34,38,43,48,53) and their maintenance;
- fixation of atmospheric nitrogen through leguminous plants (5), particularly in soil with a low nitrogen content (9);
- reduction in the leaching of elements;
- distribution in depth of nutrients, in particular phosphorus and potassium (43,12)
- protection against the battering and compacting action of rain;
- facilitation of the passage of mechanical implements;
- reduction in damage by Phytophtorae and snails (30);
- changes in the microflora (10);
- reduction of the development of roots on the surface and hence protection against extremes of temperature (44).

These features of weeds are evaluated in general with an eye to the effect of their presence or absence on the soil, the development and yield of the citrus plants and the quality of the fruit. Many of these effects on citrus plants have been documented, but the findings are very divergent owing to the fact that the standard of reporting varies according to the grove surveryed (21).

1.4 WEEDS AND SOIL

Ever since the introduction of non-tillage of the soil, which involves the total elimination of weeds from the land, the results of the surveys conducted in large areas have shown a considerable improvement in the physical condition of the soil in groves managed in this way (15,32, 39).

Increased control of either annual or perennial weeds generally results in an increase in the water content of the soil even under different systems of irrigation. This effect is greater when the weeds are completely eliminated and less when they are left undisturbed (17, 21). This is important, because there is a close positive correlation between the water content of the soil and plant-growth and the characteristics of the fruit, e.g. size, total soluble solids, and the ratio between total soluble solids and acidity.

The presence of a grassy mulch produces a high percentage of water-stable aggregates, promotes more rapid infiltration (12, 17, 26) and reduces loss through water leaching (Tab. 2). However, it may also be the case, particularly in fine soil, that in a soil which is not tilled and covered with a green mulch and mowed regularly, the percentage of water is higher than in land which is tilled and weeded regularly, although the water consumption is greater. It is thought that this is due to better infiltration of the water and consequently greater build-up, which gives optimum physical conditions (26). Graminaceous plants appear to be more effective than annual leguminous plants in improving water infiltration (56).

The temperature of the soil also varies. Land which is free of weeds warms up more quickly in spring and cools more slowly in autumn than land which is tilled or is permanently infested with weeds (55). Weeds increase

Table 2 - Effect of winter mulch on water infiltration

MULCH	AVERAGE WATER INFILTRATION (cm)	
	Without Mulch	With Mulch
Ammonium sulphate	34.50	46.50
Sodium nitrate	36.00	45.00
Calcium nitrate	69.00	82.50

(E.R. Parker and W.W. Jones, 1951 - adapted)

the amount of organic substances in the soil, as stated above, and consequently improve its structure and fertility (12).

1.5 WEEDS AND PLANTS

Increased weed control also leads to bigger plants. The plants where weeds are totally controlled are larger than those where weeds are allowed to develop. In the case of 100% control, particularly where Cynodon dactylon is the main weed, the trunk circumference is greater (10,12,21) (Fig. 3),the leaf area larger and the shoots more numerous and longer than on weed-infested land (Tab. 3), whereas these effects are less pronounced where weed control is only partial (21). However, in many trials no evidence of increased trunk development has been found even after long periods of weed control (47).

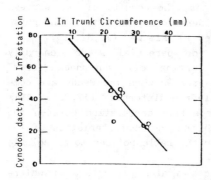

Fig. 3

Relation between weed infestation (Cynodon dactylon) and increase in trunk circumference.

(R. Goren and S.P. Monselise, 1969)

Table 3 - Effect of cultivation system on the growth of the grapefruit Redblush cultivar grafted onto a sour orange tree on the basis of trunk diameter measurements.

Cultivation System	Trunk diameter (cm)		
	1964	1965	1966
Untilled (Chemical treatment)	8.70a	13.60a	18.20a
Tilled	6.00b	10.40b	15.50b
Weed-infested	4.95b	7.70b	12.10c

The figures followed by the various letters are significant at the 5% level on the Duncan test (R.F. Leyden, 1969).

The development of the roots is also dependent on the amount of weed vegetation present (Tab. 4). In a comparison between roots of plants treated with green manure on tilled soil and on soil with no weeds (untilled), the latter showed greater root development (24).

Table 4 - Effect on orange tree rootlets of non-tillage the land.

Depth of Soil (cm)	Weight of Rootlets (gr/plant)		Distribution of Rootlets (%)	
	tilled	untilled	tilled	untilled
0 - 15	126	1,480	2.3	14.5
15 - 30	1,386	3,960	25.3	38.7
30 - 60	1,719	2,430	31.4	23.7
60 - 150	2,250	2,367	41.0	23.1
0 - 150	5,481	10,237	100.0	100.0

(Adapted from M.H. Kimball et al, 1951)

The water potential of the leaves, which is a measure of the water stress, increases or becomes less negative when weeds are eliminated. The chlorophyl content is not affected by the presence of annual weeds, but is reduced when Cynodon dactylon is present. The level of nitrogen in the leaves is reduced by the presence either of annual weeds Bermuda grass (4, 22). The photosynthesis process is affected by the presence of either annual or perennial weeds (21).

Complete control (Fig. 4, Tab. 5)- using weed-killers - of weeds which affect citrus groves gives increased production (4,10,11,14,15,16,17,22, 23,29,39,40,41, 52,54). These effects are observed both in citrus groves where weeds are periodically eliminated and in those where they are completely eliminated (21). This effect on production is proportional to the degree of control achieved (14, 50, 52). However, increased production generally means a reduction in the size of the fruit, as there is a known inverse correlation between. The total elimination of weeds in comparison with green manure may also improve the size and the juice content of the fruit, but at the same time reduce the total soluble solids, acidity and vitamin C content without altering the ratio between total soluble solids and acidity (6).

In some cases the elimination of weeds caused earlier ripening of the fruit (42). In general no definite conclusions can be drawn regarding the quality of the fruit (18, 37, 52). The presence of weeds which reduce yield affects the quality of the fruit (17), increasing the size (16,17,21,22,23) the total soluble solids and acidity without altering the ratio between the total soluble solids and acidity, while the effect on the juice content may be variable (21) or improved it (16,17,23). However, an increase in size accompanied by improved quality of the fruit have also been reported in the absence of weeds (14).

233

Table 5 - Effect of the cultivation system on the yield of the Redblush grapefruit cultivar on sour orange tree in the initial years of production.

Cultivation System	Yield	(Kg/plant)
	1966	1967
Untilled (chemical treatment)	95	107
Tilled	10	74
Weed-infested	4	13

(R.F. Leyden, R.F., 1969)

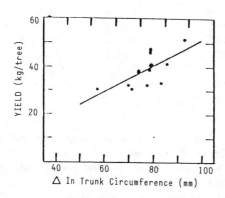

Fig. 4
Ratio between trunk circumference growth and yield from cultivar Shamouti orange tree after 5 years-old on sweet lime.

(R. Goren and S.P. Monselise, 1969)

In general it is the perennial weeds which have more harmful effects on yield. For example, in the presence of Cynodon dactylon there may be a reduction in yield because of both the competition for nitrogen and water, and the effects of allelopathy (10, 13).

1.6 CONCLUSIONS

The choice between constantly maintaining a natural covering of vegetation, sowing winter or summer green manure, or eliminating weeds periodically as part of the tilling process or permanently by means of weed killers, is not easy in view of the large number of variables which may be encountered in the various citrus fruit growing areas of the world. Thus the choice depends on the physical and chemical properties of the soil, annual rainfall, the irrigation facilities, the lay of the soil, the type of flora and other of local factors. Many detailed surveys are being conducted into the effects of weeds in citrus groves either with regard to plant yield or in connection with the various biotic and abiotic factors.

It should be noted that it is in the interests of ecological balance to allow weeds to grow and their biology should be studied with a view to ascertaining any beneficial effects so that the weeds can be eliminated by the most suitable methods depending on the individual situation, if they prove to have no beneficial effects. Lastly, beneficial effects should be weighed against the harmful effects, such as reduced growth and yield, in a long-term view which takes account of the fertility of the soil.

REFERENCES

1. AHARONI, M., HILLER, W. and PATT, J. (1969). Effect of drainage water from the root medium of Bermuda gras (Cynodon dactylon) and of annual weeds on the development of citrus seedlings. Israel J. Agric. Res. 19:1.
2. BEAR, W.H. (1969). Car bare soil contribute to grove cold protection? Citrus and Vegetable Magazine, 7:12-13 and 21.
3. BRAUN-BLANQUET, J. (1964). Pflanzensoziologie, Grundzuge des Vegeta-tionskunde. Ed. Springer, Vienna.
4. CASAMAYOR, R., GARCIA, R. and ARIAS, N. (1981). Chemical weed control and its influence on the production of "Valencia late" oranges (Citrus sinensis). Proc. Int. Soc. Citriculture: 496-498.
5. CHAPMAN, H.D., LIEBIG, G.F. and RAYNER, D.S. (1949). A lysimeter investigation of nitrogen gains and losses under various systems of cover cropping and fertilization and a discussion of error sources. Hilgardia, 19:57-128.
6. DANTUR, N.C., CASANOVA, M.R. and GARGIULLO, C. (1977). Efecto de la applicacion de herbicida sobre la fertilidad del suelo y el estado nutricional del limonero. Rev. Ind. y Agric. de Tucuman, 54 (1):1-8.
7 DI MARTINO, A. and RAIMONDO, F.M. (1979). Biological and chorological survey of the sicilian flora. Webbia, 34 (1):209-235.
8. GERARD, C.J., BLOODWORTH, M.E. and COWLEY, W.R. (1959). Effect of grasses and mulches on soil moisture losses and soil temperature changes in the lower Rio Grande Valley of Texas. J. Rio Grande Valley Hort. Soc., 13:97-101.
9. GIOBEL, G. (1926). The relation of soil nitrogen to nodule development and fixation of nitrogen by certain legumes. New Jersey Agr. Expt. Sta. Bul. 436:1-125.
10. GOREN, R. and MONSELISE, S.P. (1969). Studies on the utilization of herbicides in citrus orchards in Israel as related to efficiency. Proc. First Int. Citrus Symp., 1:483-492.
11. HERHOLDT, J.A. (1969). Weed control in South african citrus orchards. Proc. First Int. Citrus Symp. 1:499-502.
12. HILGEMAN, R.H. (1959). Response of Marsh Grapefruit to different methods of soil tillage in the Salt River Valley of Arizona. J. Rio Grande Valley Hort. Soc., 13:9-17.
13. HOROWITZ, M. (1973). Competitive effects of three perennials weeds. Cynodon dactylon (L.) Pers. Cyperus rotundus L. and Sorghum halepense (L.) Pers. on young citrus. J. Hort. Sci., 48:135-147.
14. JACKSON, L.K. (1970). The death of the plow. Herbicides - the total picture. Citrus Indus., 51 (2):24 and 27.
15. JOHNSTON, J.C. and SULLIVAN, W. (1949). Eliminating tillage in citrus soil management. Calif. Agr. Ext. Circ., 150:1-16.
16. JONES, W.W., CREE, C.B. and EMBLETON, T.W. (1961). Some effects of nitrogen sources and coltural practices on water intake by soil in a "Washington" navel orange orchard and on fruit production, size and quality. Proc. Amer. Soc. Hort. Sci., 77: 146-154.

235

17. JONES, W.W. and EMBLETON, T.W. (1973). Soils, soil management and cover crops. In: Reuther W. (ed.). The Citrus Industry, vol. III, Univ. of Calif. div. Agr. Sci., Berkeley, California.

18. JORDAN, L.S., DAY, B.E. and RUSSEL, R.C. (1969). Herbicides in citrus trees and soils. Proc. First Int. Citrus Symp., 1:463-466.

19. JORDAN, L.S. and DAY, B.E. (1973). Weed control in citrus. In: Reuther W. (ed.). The Citrus Industry, vol.III, Univ. of Calif. Div. Agr. Sci., Berkeley, California.

20. JORDAN, L.S. (1978). Benefits and problems of herbicide use in citriculture. Proc. Int. Soc. Citriculture, 209-214.

21. JORDAN, L.S. (1981). Weeds effect citrus growth, physiology, yield, fruit quality. Proc. Int. Soc. Citriculture, 481-483.

22. JORDAN, L.S. and RUSSEL, R.C. (1981). Weed management improves yield and quality of "Valencia" oranges. Hortscience, 16 (6):785.

23. KIMBALL, M.H., WALLACE, A. and MULLER, R.T. (1950). Changes in soil and citrus root characteristics with non-tillage. Calif. Citrog., 35:409, 432-433.

24. KIMBALL, M.H. et al. (1951). Non tillage without cover cropping in a California citrus orchard. Proc. Amer. Soc. Hort. Sci., 58:141-145.

25. LANGFORD, A.N. and MURRAY, F.B. (1969). Integration, identity and stability in the plant association. Advanced Ecol. Res., 6:83-135.

26. LEYDEN, R.F. (1959). Relation of cultural practices to soil moisture depletion in a young citrus orchard. J. Rio Grande Valley Hort. Soc., 13:102-107.

27. LEYDEN, R.F. and ROHRBAUGH, P.W. (1963). Protection of citrus trees from cold damage. Proc. Amer. Soc. Hort. Sci., 83:344-351.

28. LEYDEN, R.F. (1964). Chemical weed control with no tillage: a cultural practice for Texas. Citrus orchard. J. Rio Grande Valley Hort. Soc., 18:25-28.

29. LEYDEN, R.F. (1969). Development of the herbicide program in Texas. Proc. First Int. Citrus Symp., 1:473-477.

30. LO GIUDICE, V. (1982). Un programma di diserbo chimico per l'agrumeto. Inf. Fitop., XXXII (12):17-24.

31. LO GIUDICE, V. (1983). Diserbo chimico e abbassamenti termici. Terra e vita, XXIV (9):44-45.

32. LOMBARD, T.A. (1951). Does non-tillage increase Valencia orange yields. Calif. Citrog., 36 (5):182.

33. LUCIANI, F. and MAUGERI, G. (1984). Recherches sur la périodicité des mauvaises herbes des cultures siciliennes. Proc. EWRS 3rd Symp. on Weed Problem in the Mediterranean area: 437-446.

34. MAUGERI, G. and LO GIUDICE, V. (1975). Funzione delle erbe infestanti negli agrumeti. Produzione di biomassa erbacea. Ann. Ist. Sperim. Agrum., VI-VII: 291-302.

35. MAUGERI, G. (1979). La vegetazione antropogena della Sicilia. Primo quadro sintetico. Boll. Acc. Gioneia Catania, S. 4, 13 (10):137-159.

36. MAUGERI, G. (1979). La vegetazione infestante gli agrumeti dell'Etna. Not. Soc. It. Fitosoc., 15:45-56.

37. MC CARTHY, C.D. et al. (1960). Effect of monuron on lemon fruit quality. Calif. Citrog., 45:174.

38. MERTZ, W.M. (1918). Green manure crops in southern California. Univ. Calif. Agr. Expt. Sta. Bull., 292:1-31.

39. MOORE, P.W. (1944). Oil sprays for weed control in untilled citrus orchard. Calif. Citrog., 29:246,251.

40. MOORE, E.C. (1944). Nontillage weed control by oil spray. Calif. Citrog., 30:53.

41. MOORE, E.C. (1945). Nontillage weed spray program in Tulare County. Calif. Citrog., 30:289-291.

42. MOORE, E.C. (1947). Non-tillage report from Tulare County. Calif. Citrog., 32 (10):435.

43. MORETTINI, A. (1963). La concimazione fosfo-potassica delle colture arboree. L'Italia Agr., 100 (8):739-754.

44. MORLAT, R. (1981). Effets comparés de deux techniques d'entretien du sol sur l'enracinement de la vigne et sur le milieu édaphique. Agronomie, 1 (10):887-896.

45. NORRIS, R.F. (1982). Interactions between weeds and other pests in the agro-ecosystems. In: J.L. Hatfield and I.J. Thomason (ed.), Bio-meteorology in integrated pest management. Academic Press, New York: 343-406.

46. PARKER, E.R. and JONES, W.W. (1951). Effects of fertilizers upon the yields, size and quality of orange fruits. Univ. Calif. Agr. Expt. Sta. Bul., 722:1-57.

47. PARSONS, J.W. (1984). Green manuring. Outlook in Agriculture, 13 (1):20-23.

48. PATT, J. (1959). Experiments in non-tillage of citrus (in the coastal zone of Israel). Livre du IV Congrès International de l'Agrumiculture méditerranèenne: 219-230.

49. PRATT, P.F. et al. (1959). Chemical changes in an irrigated soil during 28 years of differential fertilization. Hilgardia, 28:381-420.

50. RAIMONDO, F.M. et al. (1979). Aspetti stagionali e caratteri bio-corologici della vegetazione infestante gli agrumeti del Palermi-tano. Not. Soc. It. Fitosoc., 15:159-170.

51. RYAN, G.F. and KRETCHMAN, D.W. (1962). Chemical and weed control in beddes groves. Citrus Ind., 43 (12):25.

52. RYAN, G.F. (1966). Evolution of substituted uracil herbicides for use in citrus. Proc. Fla. State Hort. Soc., 79:30-36.

53. RYAN, G.F. (1969). The use of chemicals for weed control in Florida citrus. Proc. First Int. Citrus Symp., 1:467-472.

54. SUZUKI KUNIHIKO (1981). Weeds in citrus orchard and their control in Japan. Proc. Int. Soc. Citriculture: 489-492.

55. TURPIN, J.W. et al. (1970). Herbicide management in citrus orchard. Agr. Gaz. N.S. Wals., 81 (2):50-58.

56. WEERTS, P.G. and CARY, P. (1978). Effects of soil management on soil temperatures in an orange orchard. Proc. Int. Soc. Citriculture: 221-226.

57. WILLIAMS, W.A. and DONEEN, L.D. (1960). Field studies with green manure and crop residues on irrigated soils. Soil Sci. Soc. Amer. Proc., 24:58-61.

Session 3
Diseases

Chairman: C.C.Thanassoulopoulos

Address of the session chairman

C.C.Thanassoulopoulos
Regional Direction of Agriculture, Patras, Greece

This is Session three, the final Session of the first part of our meeting. The object here will be diseases caused by fungi, bacteria, viruses and other agents.

The insects and other pests which we dealt with in the two previous sessions are extremely interesting agents causing serious damages in citrus groves.

The diseases we are going to discuss have a twofold interest:

1) they can destroy the tree itself and

2) they can destroy the entomologists' work by exterminating the trees.

Fungal and bacterial diseases of citrus in the Mediterranean region

M.Salerno
Department of Plant Pathology, University of Bari, Italy

G.Cutuli
Plant Protection Observatory, Acireale, Italy

Summary

Some peculiar features of the common fungal and bacterial diseases of
the Mediterranean region are described.
Beside the bacterial disease due to Pseudomonas syringae pv. syringae,
the fungal diseases treated are: root rots (Armillaria mellea, Rosellinia necatrix, Phytophthora spp.), Phytophthora gummosis (Phytophthora spp.) brown rot (Phytophthora spp.) gum diseases (Botryosphaeria ribis, Phomopsis citri, Hendersonula toruloidea, Diplodia
natalensis), anthracnose (Colletotrichum gloeosporioides) and mal
secco (Phoma tracheiphila). This last diseases is exclusive of the
Mediterranean basin.

1.1 Introduction

The climate in most of the citrus-growing areas of the Mediterranean
region is characterized by adequate rainfall during autumn and winter, and
by a hot, virtually rainless summer. Under such conditions the fungal and
bacterial diseases of citrus, other than mal secco, root rots and Phytophthora gummosis, do not constitute a major problem.
    In this paper, some peculiar features of each of the common fungal
and bacterial diseases of citrus in the Mediterranean region are described.

1.2 Root rots

Independently of the etiology, trees affected by root rot are declined,
with little and chlorotic leaves, untimely shedding; the spring growth is
stunted and the canopy progressively reduced, because of withered twigs
and branches. Root symptoms are quite specific of the different causal
agents.
    The mushroom root rot caused by Armillaria mellea is a very insidious
disease, since symptoms on the above ground parts of the tree are very late,
when the fungus has already become well established on the roots. Luckily
the disease is quite infrequent in the citrus orchards of the Mediterranean
region; at present it seems very serious only in Spain, in the Valencia
area, and in a less severe form in Sicily, in the volcanic loose soil on
the slopes of Mount Etna, where grapes were formely grown (5). Sour orange
rootstock, still widespread in the Mediterranean citriculture, has a fairly

good resistance, but it is non so for other rootstocks, as the Citranges.

The root rot due to Rosellinia necatrix is not present in Italy and rarely reported in the other Mediterranean countries (2). The rootstock Sour orange is susceptible.

Phytophthora root rot is a problem wherever soils and irrigation methods are unfit and pruning is careless, particularly on older trees. Moreover, a poor bud union can weaken the roots and make them more susceptible to rot. P. citrophthora and less frequently P. nicotiana var. parasitica are causal agents of this rot. As a rule Sour orange has a good resistance to the disease as well as Ichang lemon, Alemow, Troyer and Carrizio citranges, many selections of Trifoliate orange,etc.

1.3 Phytophthora gummosis

In the second half of the past century the disease destroyed virtually all citrus plantings in the Mediterranean region. New plantings were then made using the resistant Sour orange rootstock.

The Phytophthora species involved in the disease in the Mediterranean region are P. citrophthora and P. nicotianae var. parasitica. In Italy has been reported also P. nicotianae var. nicotianae (12). However, very often only one species is isolated from trunk-crown lesions according to the prevalent temperatures.

1.4 Brown Rot

Lemon and other citrus fruits sometimes become infected by the same fungi that cause Phytophthora gummosis, as well as by other species of the same genus. The species P. cactorum, P. citricola, P. citrophtora, P. siringae an P. hibernalis have been recorded in Sicily (7,3). In Turkey has been also recorded P. hibernalis (16) and in Morocco P. syringae (18). P. nicotiana var. parasitica, P. hibernalis and P. citrophthora have been isolated in Spain, but only the last species seems important cusal agent of brown rot, mainly on oranges (17).

Serious outbreaks usually develop only in years when the autumn and winter are very wet. The soil conditions are also important. For instance, soil cultivation in autumn and winter, perfomed sometimes in lemon orchards to increase fruit size, and nontillage carried out by weedkillers, are conducive to heavy attaks by these soil-inhabiting pathogens.

1.5 Mal secco
a
This is/vascular disease caused by the fungus Phoma tracheiphila (Petri) Kanc,et Ghik. At present it is undoubtely the most seriuos disease of Italian citriculture, occurring in nearly all lemon and bergamot horchards. Other species of Citrus and of the related genera Poncirus, Severinia and Fortunella are also affected, although whith less severity. The commonly used Sour orange rootstock is susceptible as well as Rough lemon, Volkamer lemon and Alemow. With the exception of Spain, Morocco and possible Egypt, all countries bordering the Mediterranean Sea to some extent are affected by the disease, which also occurs on the east coast of the Black Sea.

As a rule mal secco infections occur in winter months through unhea-
led/in the leaves and branches. Infections via roots are also possible and
sometime very frequent. They are responsible for the form of the disease
more specifically known as "mal nero", characterized by a blackish discolouration of deep wood in the basal part of the trunk (4).

The Mediterranean lemon industry, because of its outstanding economic
importance, is particularly plagued by the disease. Severe wilting or
death of the trees and reduced application of nitrogen, as Italian growers
do in order to slow disease evolution, reduce yelds to as little as less
than half the potential ones. Moreover, the value of the trees killed has
to be added to this figure (15).

1.6 Gum diseases

The gum disease associated with Botryosphaeria ribis, the imperfect
state of wich is Dothiorella ribis, is the most common gummosis on lemon
in the Mediterranean basin. Fusoid gum pockets form on the trunk or large
limbs and gum oozes from them mainly in spring and summer. Gum production
stops as the lesion become older and soon after the bark splits and dries.

The gummosis due to Phomopsis citri, imperfect state of Diaporthe citri, is very similar to that caused by B. ribis. The disease has recently
been reported in Italy (8) where it seems rather serious because trees
still young and species other than lemon, as sweet orange, are affected
(9). In addition, in the case of this gummosis the canopy of the affected
trees shows a widespread decline, since the lesions are more frequently
localized on trunk near the grafting point.

Another lemon gummosis due to Hendersonula toruloidea has been recently reported in Israel (1) and in Lebanon (10). Moreover, a gum disease
due to Diplodia natalensis is considered widespread in the Mediterranean
basin (11).

1.7 Anthracnose

The disease is serious only when the cultural and sanitary conditions
of the trees are dificient. Anyway, anthracnose shows a different appearan
ce and severity in relation to the infected organs, as smal twigs, leaves,
fruits or fruit peduncles. The disease is incited by protracted rainy weather, persistent high humidity in the air, high planting density, and care
less pruning. Damages have been recently reported in Italy, Israel, Yugoslavia and, mainly, in Spain, where in some areas the disease seems a problem. The causal agent of anthracnose is Colletotrichum gloeosporioides,
ubiquitous and polyphagous fungus, weak pathogen.

1.8 Blast and Black pit

The disease, caused by yhe bacterium Pseudomonas syringae pv. syringae, is present in almost all the citrus growing countries of the Mediterranean region, because here both moisture and the right temperature condition for infection (not exceeding 20°C) occur simultaneously (13).Always
present in endemic form it becames cyclically epidemic, when cold and

rainy winters follow one another. There are two types of disease manifestation; blast (leaf and twig lesions) and black pit (fruit lesions).

Infections leading to development of blast occur during autumn and winter, following mechanical injuries, especially by wind. The bacteria usuall invade the wings of leaf petioles of leaves that are less than one year old (6). Black pit occurs principally on lemon fruits, in Italy mainly on the last picking of the winter crop, on which as many as 50% of fruits may be affected. The infections leading to development of black pit, are promoted by mechanical injuries, particulary those caused by hailstorms. Even light hailstorms that cause no visible damage to the rind can promote infection (15).

REFERENCES

1. ACHILEA, O. and SZTEJNBERG, A. (1975). A newly discovered gummosis disease on lemon trees in Israel caused by the fungus Hendersonula toruloidea. Phytoparasitica 3(1), 80 - 81.
2. CHAPOT, F. and DELUCCHI, V. L. (1964). Maladies, troubles et ravageurs des agrumes au Maroc. Institut National de la Recherche Agronomique, Rabat 339 pp.
3. CUTULI, G. (1975). Osservazioni di campo e di laboratorio sul marciume bruno dei frutti di agrumi da Phytophthora hibernalis. Ann. Ist. Sperim. Agrumicoltura, Acireale 5 (1972), 99 - 105.
4. CUTULI, G. (1975). Il "mal nero": una particolare forma di "mal seco" / Phoma (Deuterophoma) tracheiphila (Petri) Kanc. et Ghik. / osservata su specie diverse di agrumi. Ann. Ist. Sperim. Agrumicoltura, Acireale 5(1972), 281 - 290.
5. CUTULI, G. (1981). Negli agrumeti siciliani si diffonde il marciume radicale da Armillaria mellea. L'informatore Agrario, Verona 37, 18539 - 18541.
6. DE CICCO, V., LUISI, N. and SALERNO, M. (1978). Epidemiology and control of citrus blast. Proc. Int. Soc. Citriculture 204 - 206.
7. FAVALORO, M. and SAMMARCO, G. (1973). Ricerche sul marciume del colletto e radicale degli agrumi. Specie di Phytophthora presenti negli agrumeti della Sicilia orientale. Phytopath. medit. 12, 105 - 107.
8. GRASSO, S. (1982). Un deperimento di piante di limone associato a Phomopsis citri. Riv. Pat. Veg., S. IV, 18, 65 - 70.
9. GRASSO, S. (1983). Funghi associati a cancri gommosi di agrumi in Sicilia. Inf.tore Fitopatol. 33 (12), 43 - 46.
10. HARTMAN, G. and NIENHAUS, F. (1974). Gummosis agents Phytophthora citrophthora (Smith et Smith) Leonian and Hendersonula toruloidea Natrass on Citrus limon in Lebanon. I. Identification and symptomatology. Z. Pflkrankh. Pflschutz. 81 (5), 269 - 286.
11. KLOTZ, L. J. (1978). Fungal, Bacterial and Non - parasitic Diseases and Injuries originating in the Seedbed, Nursery, and Orchard, 1 - 66. In W. Reuther, E. C. Calavan, and G. E. Carman (ed.). The Citrus Industry, vol. 4. Berkeley, University of California Press.
12. MAGNANO di SAN LIO, G. and DAVINO, M. (1977). Un isolato di Phitophthora nicotianae var. nicotianae patogeno su piante di agrumi. Riv. Patol. veg. 13, 85 - 93.

13. REICHERT, I. (1964). A northern bacterial invader in citrus plantings. A pathogeographical study. Ann. Inst. Phytopath. Benaki, N.S. 6, 146 - 155.

14. SALERNO, M. and CUTULI, G. (1977). Control of Citrus mal secco in Italy today. Proc. Int. Soc. Citriculture 1001 - 1003.

15. SALERNO, M. and CUTULI, G. (1981). The management of fungal and bacterial diseases of citrus in Italy. Proc. Int. Soc. Citriculture 360 - 362.

16. SALIH, H. (1974). A Phytophthora species new for Turkey determined in citrus orchards in Adana, J. Turk. Phytopath. 3(3), 113 - 115.

17. TUSET BARRACHINA, J.J. (1977). Contribucion al conocimiento del género Phytophthora De Bary en Espana. An. Inst. nac. Invest. Agrarias, Proteccion veg. 7, 11 - 106, CRIDA (Levante), INIA, Spain.

18. VANDERWEYEN, A. (1966). Phytophthora syringae sur agrumes au Maroc. Al Awamia, 18, 31 - 34.

Epidemiological aspects of Mal secco disease of lemons

C.C.Thanassoulopoulos
Regional Direction of Agriculture, Patras, Greece

Summary

Observations made for several years in three lemon groves in Patras and Poros areas of Peloponnessus in Greece, showed that Mal secco disease has a rapid and increasing intensity when frosts are preceded, while mild winters reduce disease intensity. Significant negative correlation was found in disease intensity versus yield, reducing yield more than 50% when disease index is over 2.5 in a scale 0-5. With the exeption of cv Lapithos which showed resistance to the disease but has not good lemon quality, all other cvs tested, Adamopoulou, Karystini, Maglini and Santa Tereza, showed various degrees of sensitivity. Rootstocks, Troyer, Volkameriana and Sour Orange had not found to have any particular influence on disease development; Troyer seems to be more sensitive than other two rootstocks. The combination Maglini on Troyer showed higher disease intensity and development.

1. Introduction

Limited epidemiological aspects of Mal secco disease (Phoma tracheiphila(Petri)Kantschaveli and Ghikachvili) are known, particularly under natural conditions of infection (2). The aspects referred up to now, summarized by Cutuli et al(2), mainly examine infection period, primary sources of inoculum, pathogen survival and dissemination, hosts and influence of meteorological agents to fungus development. However, there was not found in literature a long term work concerning disease development on several lemon cvs and on several rootstocks. Frost influence also in disease development, under natural infection, has not been adequately studied. In Greece there are only some preliminary epidemiological observations published (6), so the knowledge on this subject should be considered nule.[1]

In the present work are some data presented, collected for a period of 14 years, concerning disease development on five lemon cvs and three rootstocks, as well as some correlation of frost injuries to disease outbreaks.

2. Materials and Methods

Observations were made in lemon orchards planted in two Peloponnessus regions, Patras with an average of 5.8 frosty

Table I. Mal secco (<u>Phoma tracheiphila</u>) development on lemon trees cv Maglini on Sour Orange rootstock for 14 years, in Patras area.

	Years					
	1973	1974	1975	1976	1977	1978
Number of trees with[1]:						
Symptoms	14	64	90			
Red discoloration		29	90			
Positive isolations		24	87	90		
Dead trees			4	2	8	10
Disease index[2]	0.19	0.88	1.50	2.00	2.57	3.16
Days of frost[4]	3	3	3	7	5	3

	Years					
	1979	1980	1981	1982	1983	1984	1985
Dead trees (continued)	--[3]	--	4	0	3	--	1
Disease index	--	--	2.08	2.21	1.79	--	2.40
Days of frost	6	5	6	3	9	0	2

1. The total number of trees was 90 until 1979, thereafter beeing only 23; rest trees survived Mal secco were losted by frost happened in 1979. 2. Mean of all survived trees of each year. 3. Serious frost with damaging results on trees. 4. Days with temperature below $0^{\circ}C$, from October of the previous year to March of the year indicated on the top.

Table II. Number of diseased lemon trees by Mal secco (<u>Phoma tracheiphila</u>) the second and seventh years after planting.

Cultivars	Second year				Seventh year			
	T[1]	V	SO	Total	T	V	SO	Total
Adamopoulou	3[2]	2	3	8	10	10	10	30
Karystini	2	4	3	9	10	10	10	30
Maglini	7	3	6	16	10	10	10	30
Santa Tereza	0	6	4	10	10	10	10	30
Total	12	15	16	43	40	40	40	120

1. T:Troyer as rootstock, V:Volkameriana, SO:Sour Orange.
2. In each combination 10 trees.

days per year, and Poros with milder winters. Three groves
were planted as follows: In the first, in the area of Patras,
90 lemon trees cv Maglini grafted on Sour Orange were planted
in 1971. In the second grove, in Poros, they planted in 1978,
120 trees, 30 of each cv, Adamopoulou, Karystini, Maglini, and
Santa Tereza, grafted on rootstocks Troyer, Volkameriana and
Sour Orange, so in each combination there were 10 trees; other
15 trees of cv Lapithos on Volkameriana only were planted also
in the same orchard. Finally in the third grove 45 lemon trees,
15 from each cv Adamopoulou, Maglini and Santa Tereza, grafted
on Volkameriana, were planted in Patras area in 1981.

The trees were left to be infected under completely natural conditions. The only treatment given it was a pruning
of all diseased trees during summer months, rejecting all twigs
and branches, dead or with red discoloration.

Observations were usually taken once a year, in May, u-
sing a disease index 0-5, for evaluation of external symptoms,
developped as follows: 0: apparently healthy trees; 1: few de-
foliated twigs, even some tips dead, leaves of the tree green-
off colour, slight yellowing of the veins; 2: a lot of dead
twigs or greater shoots, but in no case covering more than
half of the whole foliage; 3: half of the tree foliage dead;
4: more than half of the tree foliage dead; 5: lemon tree dead
(in several cases rootstock was alive giving a new diseased
growth).

In case that there was doubt for the presence of the disease, particularly in the disease index 1, it was trying to
find the characteristic red discoloration with longitudinal
sections. In the first orchard isolations were made from all
trees, while in others only eratically for certification of
the disease presence. Isolations were made on Potato Dextrose
Agar and the fungus when it was grown on it was easily identified.

During winter of 1979 a serious frost with a temperature
$-3.2^{\circ}C$ for several hours killed or destroyed several of the
trees in the no 1 grove, so regrowth two years later showed
that only 23 trees were appropriate for further observations.
In the same field, yield was counted until 1978, the year before the frost. Correlation between disease index and yield
was calculated getting the mean disease index of the years
1973 to 1978 and the mean yield of the years 1975, first year
of fruiting, 1978.

3. Results

The results reveal that early diseased trees appeared the
first or second year after planting, also all trees diseased
by Mal secco within 4th and 7th years since their planting
(Tables 1 and 2). It was also found that within five first
years of their life all the trees presented the characteristic
red discoloration of the xylem and the fungus was easily isolated from all of them (Table 1). Disease development, as e-
valuated by disease index and dead trees, was gradually increased as the trees were aging (Table 1, 4 and 5). From tables 1 and 4 is clearly evident that outbreak of the disease
happened the years 1978 (Table 1) and 1985 (Table 1 and 5)

251

Table III. Number of dead lemon trees by Mal secco (<u>Phoma</u> <u>tracheiphila</u>) each year, in Poros area.

Cultivars	Dead trees each year in each cultivar					Total number of dead trees in 1985 on rootstock-cultivar combination		
	1981	1982	1983	1984	1985	T[1]	V	SO
Adamopoulou	2[2]	0	3	--	12	7[3]	5	5
Karystini	1	2	1	--	7	6	2	3
Maglini	0	2	2	--	13	6	7	4
Santa Tereza	0	1	2	--	8	5	4	2
Total	3	5	8	--	40	24	18	14
Days of frost[4]	0	2	10	0	1			

1. T:Troyer, V:Volkameriana, SO:Sour Orange.
2. Each figure represent dead trees in each year; total number of dead trees are the sum of all figures; 30 trees in each cv.
3. In each combination 10 trees.
4. Days with temperature below 0°C, from October of the previous year to March of the year indicated on the top.

Table IV. Disease index of lemon trees attacked by Mal secco (<u>Phoma</u> <u>tracheiphila</u>) grafted on rootstoks Troyer, Volkameriana and Sour Orange, in Poros area.

Cultivars	Years							The 7th year on rootstock		
	1979	1980	1981	1982	1983	1984	1985	T[4]	V	SO
Adamopoulou[1]	0.26[2]	0.46	0.98	0.33	1.55	--[3]	3.50	3.8	3.0	3.7
Karystini	0.45	0.96	0.50	0.66	0.96	--	3.10	3.7	2.9	2.7
Maglini	0.66	1.23	0.87	0.95	1.78	--	3.83	4.1	3.9	3.5
S. Tereza	0.50	0.56	0.43	0.33	1.13	--	2.83	3.7	2.8	2.1
Lapithos	0[5]	0	0	0	0	--	0		0	
Mean total								3.8	3.2	3.0
Days of frost[6]	2	1	0	2	10	0	1			
Rain in mm	542	532	540	648	282	497	488			

1. Average disease index (0-5) of 30 trees, independent of the rootstocks. 2. Mean index of all survived trees each year.
3. Observations not received. 4. Mean index of 10 trees or less if there were dead; T:Troyer, V:Volkameriana, SO:Sour Orange. 5. Only 14 trees grafted only on Volkameriana. 6. Days with temperature below 0°C and total rain in mm from October of the previous year to March of the year indicated on the top.

after winters 1976-1977 and 1977-1978 and 1982-1983 with serious (temperature below -3°C) and prolonged (more than six hours) frosts. Dead trees by Mal secco were also sharply increased the year with serious frost and thereafter (Table 1 and 3). Precipitation level does not seem to have any influence on disease development (Table 4).

From tables 2, 3 and 4 was evident that differences among lemon cultivars or rootstock-cultivar combination were practically not existed. Disease incidence, disease development and dead trees were approximately on the same level, indicating only some increased sensitivity in the combination of all cultivars with Troyer rootstock. Maglini also showed an increasing sensitivity compared to the other three cvs, as it is indicated by disease incidence (Table 3) and disease index (Table 4). Only the Cyprian cv Lapithos did not show any symptom of Mal secco disease the same period and under the same weather conditions.

Correlation between mean yield of four years versus mean disease index of five years in 75 trees, showed a correlation coefficient r=-0.48, which was significant in 0.1% level of significance.

4. Discussion

The results of this work indicate that the disease has a rapid and increasing intensity when frosts are preceded. Precipitations does not seem to have any particular role in disease development, at least in Greece, in which winters are not dry or with excessive rains. Goidanich and Ruggieri (3) considered that frosts predisposed the trees in more serious disease development. Scaramuzzi et al (4) in experimental work certified that low temperatures, 5°C, increased disease intensity. Solel and Oren (5) also referred that there is unusual disease intensity in lemon orchards when during the preceding winter there were exceptionally favorable conditions, but did not explain what exactly they meant, and probably they did not discuss about frosts, as in this country frosts are rather unusual. From the results of the present work there are clear evidences that disease was seriously outbreaked after frosts 1977, 1978 and 1983, while frost of 1979 was so serious in Patras area that trees succumbed from frost itself. It was also clearly evident that disease intensity was decreased when preceding winter was rather mild with none or very few chilly days and the temperature in frosty days was not lower than -1°C, as this was apparent during 1981, mainly in Poros, and 1982 in both regions.

The highly significant negative correlation observed among yield and disease index, indicate that disease influence on yield is beyond any question. The results of four only years do not permit to fix a formula for probable calculation of expecting damages in given disease index, but it is probable that disease intensity with an index more than 2.5 will reduce yield more than 50%, as it was observed from the results of this work. Akteke (1) in Turkey observed 12.3% in one year only experiment, but this low difference is logical if the conditions for disease development were not so favourable.

Table V. Disease index, number of diseased and dead lemon trees grafted on Volkameriana attacked by Mal secco (<u>Phoma tracheiphila</u>), in Patras area.

Cultivars	Number of diseased trees				Dead trees 1985	Disease index (0-5)			
	1982	1983	1984	1985	1985	1982	1983	1984	1985
Adamopoulou	4[1]	2	-	6	1[2]	0.28[3]	0.14	--	0.88
Maglini	8	5	-	10	5	0.71	1.65	--	2.45
S. Tereza	6	5	-	10	1	0.58	0.71	--	1.33
Days of frost						3[4]	9	0	2

1. From each cultivar were planted 15 trees.
2. Dead trees of each year were not replaced.
3. Mean index of all survived trees.
4. Days with temperature below 0°C, from October of the previous year to March of the year indicated

The results concerning on cultivars, rootstocks or their combinations, indicate that practically there are not differences. Neither of the cultivars nor of the rootstocks showed any tolerence; there are only various degrees of sensitivity. In this respect cv Maglini is the more sensitive while all cv scions were more sensitive on Troyer rootstock with some higher sensitivity of Maglini-Troyer combination. It was considered useless to examine if there was statistical differences. Even if it was so, eg between cultivars conserning disease index, practically is meaningless, as a disease index over 2 is quite significant. The cv Lapithos seems to be highly resistant to Mal secco, at least on Volkameriana rootstock, and this is an interest point of view giving promices for better results from a further selection within this cv, mainly for horticultural characteristics, concerning fruit sizes and quality.

REFERENCES

1. AKTEKE, A.Ş. and KARAKA, J. (1977). Studies on the survey and biology of Mal secco disease (<u>Phoma tracheiphila</u> (Petri) Kanciaveli et Ghikascvilli) of lemon trees. J. Turkish Phytopath. 6:91-102.
2. CUTULI, G., LAVIOLA, C., PERROTA, G., SALERNO, M. and SPINA, P. (1984). Le Mal secco des agrumes. Federazione Nazionale dei Dottori in Scienze Agrarie e Forestali, Roma, p. 131.
3. GOIDANICH, G., e RUGGIERI, G. (1949). Effetti del freddo e "mal secco" negli agrumeti siciliani. Ann. Sperim. Agraria N.S. 3:391-397.
4. SCARAMUZZI, G., SALERNO, M. e CATARA, A. (1964). Ricerche sul "Mal secco" degli agrumi (<u>Deuterophoma tracheiphila</u> Petri). Riv. Pat. Veget. S. 3, V. 4:319-327.
5. SOLEL, Z. and OREN, Y. (1975). Outbreak of Mal secco disease in Israel on normally tolerant citrus cultivars. Plant Dis. Reptr. 59:945-946.

6. THANASSOULOPOULOS, C.C. (1983). Introduction and development of integrated managment system in the control of Mal secco disease of lemon (Phoma tracheiphila). 9th Inter-balkanic Conference on Plant Protection, Nov. 1983, Athens, Greece, p.10.

Observations on the localization of *Phoma tracheiphila* in canopy infections

N.Butera, L.Cupperi & W.V.Zucker
AID SpA Research Center, Catania, Italy

A.Catara & S.Grasso
Plant Pathology Institute, University of Catania, Italy

Summary

Phoma tracheiphila, the causal agent of the citrus disease, mal secco, was investigated for its presence in   sampled branches of severely infected trees at two sites in Sicily. We find the fungal pathogen at high probability in natural wounds and at leaf abscission zones. Hail clearly creates wounds through which fungal invasion can occur. We do not find a strong effect of branch orientation to the pathogen presence. Infection sites appear to be at random. We demonstrated a strong correlation between the visual presence of Phoma tracheiphila and outgrowth of the fungus in culture, from samples taken at random from the same visually examined specimens. The relationship of these findings to chemical control is discussed.

INTRODUCTION

The knowledge of penetration sites of Phoma tracheiphila (Petri) Kanc. e Ghik, causal agent of mal secco in citrus, is still incomplete and fragmentary (1). Petri (2) hypothesized that the fungus could penetrate through the stomata, but numerous reports and the results of a recent scanning electron microscope study (3), would appear to exclude this possibility.

Experiments have shown that infections originate in wounds. Field observations confirm this since, under natural conditions, very serious symptoms of disease can be seen after hailstorms and strong winds (4,5,6,7). However, there is still a lack of experimental evidence even though the acquisition of this data is unanimously acknowledged to be of fundamental importance for a rational control program (1,6,8).

Accordingly, it is useful to consider studying those topics which have practical applications, viz,, the extent of infections that can be due to hail lesions. We have studied this at two different, separated orchards of Sicily, located in traditional lemon growing areas. The data collected refers only to the frequency and localization of infections in the young branches of these orchards.

SAMPLINGS AND RESULTS

Investigations concerned lemon orchards located in the areas of Acireale (near Catania), Contrada "Tocco", and Siracusa, Contrada "Carrozzieri".

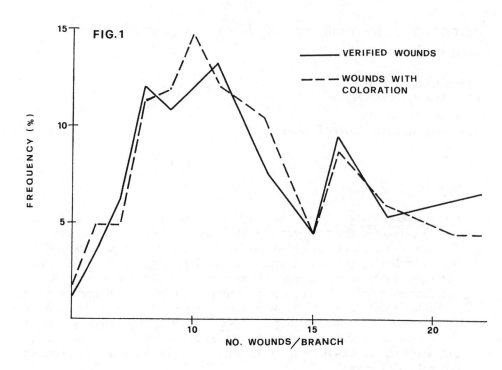

FIG. 1

FREQUENCY (%)

VERIFIED WOUNDS

WOUNDS WITH COLORATION

NO. WOUNDS/BRANCH

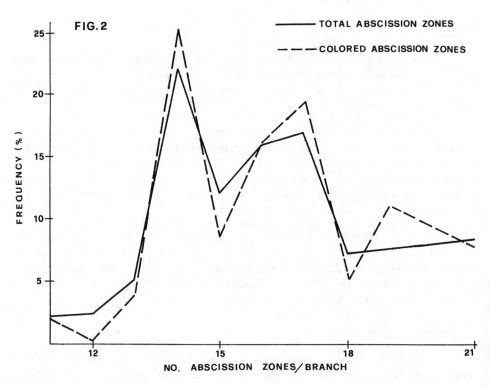

FIG. 2

FREQUENCY (%)

TOTAL ABSCISSION ZONES

COLORED ABSCISSION ZONES

NO. ABSCISSION ZONES/BRANCH

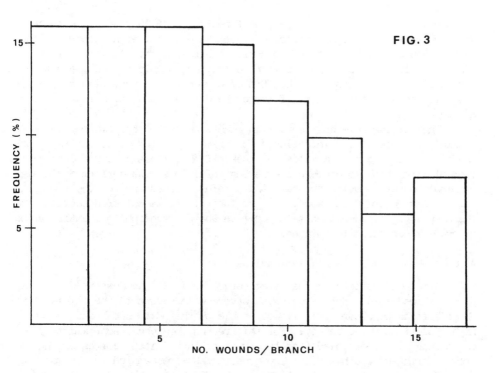

FIG. 3

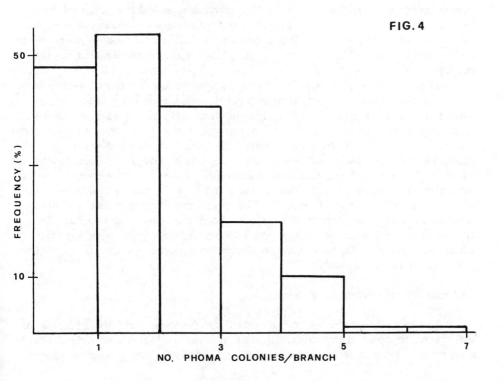

FIG. 4

259

At Tocco, the trees were of widely different ages, 3-30 years, due to progressive infections of mal secco. It sustained a violent hailstorm on the 14th of February 1983 that caused a severe defoliation. At Carrozzieri, trees were 15 years old, in excellent condition, highly productive but which had received a minor hail on the 18th of November 1983, that resulted in serious infections in three areas of the orchard. Many plants, here and there, showed modest disease symtoms.

The surveys were carried out on plants chosen at random from those not showing evidence of old infections, for example, drying up of entire branches and/or severe pruning. Samples were taken from various canopy sections (1-2 year old branches) to enumerate existing infections. We performed visual examinations by removing the bark and carefully observing for the typical salmon-pink color in the wood. In the laboratory, we cultured sections of the samplings in order to obtain positive Phoma tracheiphila identification using well described techniques.

A) Contrada Tocco - Visual Examination

After the hailstorm, the first survey (25/2/84) demonstrated that all plants showed some defoliation, with severe wounds down to the cambial layer. These effects were particularly noticeable on the north side, and in the higher sections of the canopy. For this reason, branches (approximately 50 cm in length) were sampled from two trees at three sites, corresponding to crown, north and south sides. Two other trees were sampled in the same way on 24/3/84. More than half (54.5%) of the 772 branches visually examined showed symptoms of mal secco (table 1) with average values of 29.8% in February and 91% in March. Negligible differences were found between the two trees (data not shown). The frequency of infected branches sampled in the canopy varied, with highest values in the north, most evident in the February survey.

From two trees, groups of 32 branches were sampled from the more exposed portions of the canopy (top and north side) from each tree, and examined in the laboratory to study infection density at wound sites and at leaf abscission zones. For this examination, the apical parts of 6-12 month old stems (about 30 cm in length) were utilized. We counted the number of wounds $>$ 5 mm, the number of leaf abscission zones and, after bark removal, the cases in which the typical coloration was observed. The percentage of estimated infections in the branches was the same at both wound sites and at leaf abscission zones (table 2). However, the average number of leaf abscission zones per branch was greater than the number of wounds. The frequency distribution of wounds and leaf abscissions are very similar (figures 1 and 2). We have also observed the typical coloration in wounds of $<$ 5 mm (data not shown).

Contrado Tocco - Laboratory Tests

In the laboratory, we investigated the frequency and localization of infections in sampled branches. Pieces of stem having either a leaf abscission or a wound and showing the typical coloration, were placed on carrot

TABLE 1

Frequency of symptoms of mal secco in branches of lemon trees sampled in various positions of the canopy at two different times (Contrada Tocco - Acireale).

Infected Branches

%(N)

POSITION	25/2/84	24/3/84
North	38.3 (99)	96.2 (105)
Crown	29.7 (195)	96.1 (102)
South	24.7 (166)	80.9 (105)
TOTAL	29.8 (460)	91.1 (312)

TABLE 2

Average number of abscission zones and wounds with and without symptoms of mal secco (wood coloration) found in terminal branches of lemon trees (Contrada Tocco - Acireale, 25/2/84).

Branches Examined

	Total	With Symptoms	%
	N (S.E.)	N (S.E.)	
ABSCISSION ZONES	16 (.41)	10 (.65)	62.5
WOUNDS (> .5 cm)	11 (.76)	7 (.49)	63.6
	t = 5.94	t = 4.14	
	P < .01	P < .01	

261

TABLE 3

Percentage of infected branches and frequency of isolates of Phoma tracheiphila from the wood below abscission zones and wounds of lemon branches (Contrada Tocco- Acireale, 25/2/84).

	Infected Branches % (S.E.,N)	Colonies/Branch Number (S.E.,N)
ABSCISSION ZONES	75 (.33,69)	2.18 (.15,151)
WOUNDS (> .5 cm)	57 (.32,42)	.81 (.15,34)

F = 14.7, P < .01

TABLE 4

Percentage of infected branches and number of colonies of Phoma tracheiphila isolated from deep wounds and abscission zones sampled at various positions in the canopy of two naturally infected lemon trees (Contrada Tocco - Acireale).

POSITION	Infected Branches [*] %		Colonies/Branch Number (S.E.,N)	
	Abscission Zones	Wounds	Abscission Zones	Wounds
Crown	66	56	3.1(.22,53)	2.7(.25,43)
North	50	41	2.5(.29,43)	2.1(.21,31)
South	51	36	2.4(.29,39)	1.6(.30,25)

[*] The differences in the three positions were not statistically significant (ANOVA).

agar and incubated at 20°C for 8 days. Pieces showing P. tracheiphila colonies were then counted (table 3). These results show percentages of
infection slightly different from what was found by visual examination
(compare with table 2). It can be mentioned that the number of pieces showing
Phoma infections was slightly more than 2 per branch which is 40% of the
pieces put on carrot agar.

In another test, the influence of branch orientation on infections
was investigated. From two trees, groups of 25 branches were sampled from
three different sites, at the crown, north and south. For each stem,
pieces of wood were sampled at 5 leaf abscission zones, and also at 5
wounds; each was placed on carrot agar and incubated as described (table 4).
We obtained P. tracheiphila colonies in 50% of all the samples taken, but
abscission zones were more infected than wound sites, with the crown
appearing to be the most susceptible to infection.

Another test was carried out to investigate the relationship between
the number of wounds and the number of P. tracheiphila infection sites.
Branches of 10 cm were sampled, and after counting the wounds greater than
5mm, the branches were cut at random into 1 cm pieces, and these pieces
were placed on carrot agar and incubated (figures 3 and 4).

B) Contrada Corrozzieri - Visual Examination and Laboratory Tests

Due to the generally lower infections in this orchard, we decided to
carry out a survey using a larger sample size, i.e., with more trees and
more branches.

In the first test (table 5, part A) performed between the 11th and 20th
of May 1984, ca 50 cm long branches were randomly sampled from the crown of
18 trees. We selected 5 abscission zones per branch for analysis. In the
second test (table 5, part B) performed on the 31st of May 1984, we made
separate samplings of branches from the crown and the sides. We scored all
branches for signs of mal secco infection and segregated out the apparently
healthy branches. From these apparently unaffected branches, we then sampled
10 pieces of wood, chosen at random, below each wound site or abscission zone
with the aid of a dissection microscope. These were placed on carrot agar and
incubated in order to determine if they were infected with P. tracheiphila.

From random branch sampling in the crown of 18 trees, 10.3% showed
visual signs of mal secco. Among the apparently healthy branches, in fact
3.8% were infected with P. tracheiphila. The level of infection in these
branches was quite low; on the average, a branch had a single infection only
9.6% of the time (table 5, part A).

In the second test, we randomly sampled at 20 different places on each
apparently healthy branch (10 each, wounds and abscission zones; table 5B). In
this case, the number of infected branches was almost doubled (6.5%). The
reason for this difference is not known. No significant differences were
observed between branches sampled at the crown and laterally. On average,
only 1.24% of all the samples taken were infected, 0.58% from wounds and
1.61% from leaf abscission zones. In this test, visual examination of the
branches showed a relatively low number of wounds, equal to only half of
the leaf abscission zones (data not shown).

TABLE 5

Mal secco infections found in branches of lemon trees sampled at Contrada Carrozzieri - Siracusa.

Branches Examined	Number of Trees	Branches With Wood Coloration % (N)	Branches Minus Coloration but Infected % (N)	Colonies of Phoma tracheiphila % (N)	/Branch (N)
A) From different positions in the canopy selected at random	18	10.3 (624)	3.8 (260)	1.93(1292)	.09(260)
B) From only one part of the canopy					
......CROWN	7	11.4 (140)	6.4 (124)	.97(1240)	.09(124)
......SIDES	7	12.4 (140)	6.5 (123)	1.50(1230)	.15(123)

DISCUSSION

Our results show that Contrada Tocco had some very infected trees, obviously due to the large number of wounds that occurred in the autumn-winter period. Also important is the higher quantity of inoculum that probably existed there. From our data, we would conclude that leaf absciss ions are of great importance in providing P. tracheiphila penetration sites. Our data is consistent on this point in both orchards. Also, it should be pointed out that the number of abscission zones per branch varies from 11 to 22 (average 16) in 50 cm of branch. Abscission zones may be the primary means of P. tracheiphila infection.

The role of hail in predisposing trees to P. tracheiphila penetration is demonstrated by the large diffusion of mal secco in   Contrada Tocco. Highest densities of infected branches were found in the north and in the crown, areas less protected from hail effects. In Contrada Carrozzieri, where hail damage was considerably less, branch infections were more evenly distributed.

Our work leads to the following practical applications in the control of the mal secco disease:

a) Hail is extremely important in the infection process due to the wounds and defoliation that it causes.

b) Increasing infections due to hail calls for an unusual degree of care in planning chemical and surgical control of the disease. It must be considered in order to cope with the difficulties that are encountered in pruning the crowns, removal of infected branches, and for fungicide applications.

c) Leaf abscission zones and wounds are the major routes for P. tracheiphila penetration. They are numerous and always present, due to the natural leaf fall or to leaf loss due to special events (wind, water, nutritional inbalances, etc.). More research is certainly needed on this aspect.

Successful chemical control of mal secco can be obtained by regular, frequent treatments with fungicides in the autumn when natural leaf fall occurs, and by timely applications within 48 hours after any event which can cause wounds and lacerations.

ACKNOWLEDGEMENTS

The authors gratefully acknowledge the owners of the orchards, Professor Giovanni Continella (Contrada Tocco) and Mr. Francesco Nicotra (Contrada Carrozzieri) for their kind hospitality and collaboration during the surveys.

REFERENCES

1. Cutuli, G. e C; Laviola. 1977. Attuali conoscenze sulla epidemiologia del "mal secco" degli agrumi. Ann. Ist. Sper. Agrumicol. 9-10, 93-102.

2a. Petri, L. 1930. Ulteriori ricerche sulla morfologia, biologia, e parassitismo della Deuterphoma tracheiphila. Boll. Staz. Pat. Veg. 10, 191-221.

2b. Petri, L. 1930. I risultati di alcune ricerche sperimentali sopra il "mal secco" degli agrumi. Boll. Staz. Pat. Veg. 10, 353-359.

3. Zucker, W.V. e A. Catara. 1985. La penetrazione di Phoma tracheiphila attraverso ferite visualizzata attraverso il microscopo elettronico a scansione. Inf.tore Fitopatol. (in corso di stampa).

4. Gassner, G. 1940. Untersuchungen über des "mal secco" oder "kurutan" der limonbaume. Phytopath. Z. 13, 1-90.

5. Petri, L. 1941. Sul creduto originale della "Phoma limonis" Thumen. Boll. Staz. Pat. Veg. 21, 157-160.

6. Ruggieri, G. 1948. Fattori che condizionano e contribuiscono allo sviluppo del "mal secco" degli agrumi e metodi di lotta contro il medesimo. Ann. Sper. Agr. N.S.; 2, 255-305.

7. Cutuli, G. e O. Li Destri Nicosia. 1976. L'influenza della grandine sulle infezioni di "mal secco". Terra e Vita 17; 8, 60-61.

8. Cutuli, G., C. Laviola, G. Perrotta, M. Salerno e P. Spina. 1984. Il mal secco degli agrumi. Seminario internazionale di studio organizzato nell'ambito del programma di ricerche Agrimed. 131 pp.

Variability in *Phoma tracheiphila*

G.Magnano di San Lio & G.Perrotta
Plant Pathology Institute, University of Catania, Italy

Summary

Non chromogenic isolates of P. tracheiphila (Petri) Kanc. et Ghik., closely resembling "DP race" previously described by Baldacci have been found in nature in a location near Palermo (Sicily). Pathogenicity of non chromogenic strain did not differ significatively from that one of chromogenic isolates.

Variation in cultural charateristics and pathogenicity among strains of Phoma tracheiphila (Petri) Kanc. et Ghik., the "mal secco" disease fungus was first reported by PETRI (6,7). Twenty years later BALDACCI (1) suggested that two variant types of the fungus exist in nature: chromogenic strains ("DRP race") which in colture on artificial media produce red pigmentation and non chromogenic strains ("DP race") which do not produce red pigments. A third variant type ("R race") originates by sectoring from cultures of chromogenic strains grown on artificial media. According to BALDACCI (1) pycnidia of chromogenic strains are smaller than those ones of non chromogenic strains while an essential feature of red-albino variants is the irreversible loss of capacity to differentiate pycnidia. SCRIVANI (9), by root inoculating sour orange seedlings, ascertained that a DP isolate supplied by Baldacci and his coworkers (3,1) was less aggressive than DRP strains.

Afterwards variation in pathogenicity, expressed as different degree of virulence, among isolates of P. tracheiphila has been recognized by many AA. (2,5,8,11) but cultural characteristic of strains have not been always defined. In most cases chromogenic isolates have been tested (2,8). Following SALERNO and PERROTTA definition of races given by BALDACCI (1) should be reexamined because also chromogenic strains produced large pycnidia (7). Anyhow pycnidium size does not seem a suitable criterion for distinguishing chromogenic and non chromogenic variant types since it is extremely variable within the same isolate (PERROTTA, unpublished data).

Recently we examined 600 isolates of P. tracheiphila obtained from "mal secco" infected lemon twigs, collected in Sicily, from several localities near Palermo, Catania, Messina, and Syracuse. In order to be characterized isolates were grown on the following substrates: potato-dextrose agar (PDA, approx pH 5.6), carrot agar (CA, approx pH 5.5), Czapek Dox agar (CzA, approx pH 6.8), V_8–$CaCO_3$ agar (approx pH 6.9), and coconut agar (CoA, approx pH 5.9).

Mass isolates obtained from infected twigs appeared quite uniform with respect to their cultural characters when grown on PDA, CA, and V_8 –$CaCO_3$ in the dark.

Most isolates (chromogenic isolates) exposed to either daylight or U.V. produced red pigmentation also on above substrates while few isolates (non chromogenic isolates), recovered from a lemon tree near Palermo, did not. Colonies of chromogenic isolates turn red on CoA and CzA, also in the dark, while non chromogenic isolates on these substrates produced yellow pigmentation.

Sectors spontaneously differentiated from colonies of some chromogenic isolates grown on PDA, CA, and CzA. Mass transfers from variant sectors gave colonies similar for all respects to "R race" described by Baldacci (1). Sectoring was noted also in single–phyaloconidic progenies (*) of the variant producing isolates. There was no end to the appearence of sectors from some isolates which showed the tendency to sector as far as the sixth single phyaloconidic generation. Single phyaloconidic isolations from variant red–albino sectors gave colonies with cultural characters intermediate to those of "race R" and race DRP sensu BALDACCI (1).

Some chromogenic isolates generated uniform single–phyaloconidic progenies while other ones gave rise to several culturally distinct types that varied in their capability in producing red pigmentation on CzA. Some of these variants mantained their cultural characteristics in subsequent progenies. Cultural characteristics of chromogenic strains often changed with time culture. However there were no remarkable differences in colony morphology among chromogenic strains when plated on CoA.

Non chromogenic variants were never obtained in progenies of chromogenic strains.

Non chromogenic strains isolated from naturally infected lemon twigs appeared stable in their cultural characteristics even when they were repeatedly subcultured. Moreover monophyaloconidic progenies of these isolates were extremely uniform independently of the culture medium used and gave parent-like colonies.

To compare the pathogenicity of above variants inoculation trials were conducted using seedlings of sour orange (Citrus aurantium L.) and troyer citrange (C. sinensis Osbeck x Poncirus trifoliata (L.) Raf.) as test plants. The inoculum was in the form of a water suspension of

---

(*) Since both chromogenic and nonchromogenic isolates lost capability of producing pycnidia on artificial culture media after repeated transferring, for studying variation six successive single phyaloconidic series, consisting of 30 to 60 cultures each, were made of each isolate.

phyaloconidia (*). It was applied by inserting a bistoury blade bearing droplets of conidial suspension on the lower stem of seedlings. Four parent strains (two chromogenic strains, one non chromogenic strain, and one red- albino variant strain) and four (one for each parent isolate) progeny isolates were tested. The amount of symptom expression and the rate of symptoms development among these isolates were not significatively different. Moreover troyer citrange seedlings appeared as susceptible as sour orange to all isolates. Cultural characteristics of colonies reisolated from diseased seedlings two months after inoculation were similar to those of isolates inoculated into the seedlings.

Although the term "races" used by BALDACCI (1) for indicating variants of P. tracheiphila would be avoided since it is not accepted by the international code of botanical nomenclature (11), it appears unquestionable on present evidence that the fungus exist in nature as distinct entities. At least two of them have been characterized here: a chromogenic variant with an high phenotypic variability and a non chromogenic variant. Interestingly chromogenic strains are widely distributed in all parts of Sicily while non chromogenic strains were found in a restricted area near Palermo, the same where BALDACCI and his coworkers recognized DP race (1,3).

Although pathogenicity tests do not support any correlation between cultural characteristic and pathogenicity the geographical specialization of non chromogenic strains could have epidemiological implications. Because performance of strains in nature could depend not only upon virulence but also on other characteristics, such as competitive saprophitic ability and host specialization, it appears of obvious interest to know if variants of P. tracheiphila, mainly those ones occuring in nature, differ in other biological characteristics beside pathogenicity.

Another open question is the origin of above variants. It has been suggested that the non-chromogenic variant could have arisen from the chromogenic one by mutation(1) however neither convincing explanation nor experimental evidence have been given. On the other hand the results of single-monophyalidic isolations are not consistent with the hypothesis that variability showed by P. tracheiphila in culture on artificial media is due to the "dual phenomenon" described by HANSEN (4). With respect to the character "capability in producing red pigmentation" neither chromogenic strains occurring in nature ("DPR race" sensu BALDACCI) nor red-albino variant ("R race" sensu) could be considered an heterotype and an homotype respectively. Single-phyaloconidic series of chromogenic strains infact do not give rise to non-chromogenic variants while red-albino variants separate from parental chromogenic strains only by sectoring. In addition progenies of red-albino variants are still provided with brown aerial hyphae and show indefinitely the tendency to sector.

(*) Three concentrations (10^3, 10^4, and 10^5 phyaloconidia/ml) per isolate and five seedlings of each citrus variety per concentration were used.

REFERENCES

1. BALDACCI, E. (1950). Caratteri colturali delle razze di Bakerophoma tracheiphila. Ricerca scient., vol. 18, 5, 6.
2. DE CICCO, V., and LUISI, N. (1977). Variabilità della patogenicità e di alcuni caratteri colturali in isolati di Phoma tracheiphila provenienti da aree ed ospiti diversi del Mediterraneo. Annali Fac. Sci. agr. Univ. Bari, vol. 29, 565-573.
3. GAROFALO, F. (1948). Sul ceppo non cromogeno di "Bakerophoma tracheiphila" isolato dagli agrumi della Sicilia. Naturalista sicil., ser. III, 2.
4. HANSEN, H.N. (1938). The dual phenomenon in imperfect fungi. Phytopathology, vol. 30, 442-455.
5. LUISI, N., DE CICCO, V., CUTULI, G., and SALERNO, M. (1979). Studio della patogenicità del fungo del mal secco degli agrumi. Atti II Seminario di studio sul miglioramento genetico del limone. Giovinazzo (Bari), 57-67.
6. PETRI, L. (1930). Lo stato attuale delle ricerche sul "mal del secco" dei limoni. Boll. Staz. Patol. veg., vol. 10, 63-107.
7. PETRI, L. (1930a). Ulteriori ricerche sulla morfologia, biologia e parassitismo della Deuterophoma tracheiphila. Boll. Staz. Patol. veg., vol. 10, 191-221.
8. SALERNO, M., and PERROTTA, G. (1966). Ricerche sul "mal secco" degli agrumi (Deuterophoma tracheiphila Petri). V. Virulenza e caratteri colturali del fungo in Sicilia. Riv. Pat. veg., Ser. 4, vol. 2, 303-312.
9. SCRIVANI, P. (1954). Patogenesi, riproduzione sperimentale del malsecco da Deuterophoma tracheiphila Petri e ricerche sulla formazione di metaboliti tossici in coltura. Phytopath. Z., vol. 22, 83-108.
10. STAFLEU, F.A., VOSS, E.G., BURDET, H.M., CHALONER, W.G., DEMOULIN V., HIEPKO, P., McNEILL, J., MEIKLE, R.D., NICOLSON, D.H., ROLLINS, R.C., SILVA, P.C., and GEUTER, W. (1983). International code of botanical nomenclature. Utrecht, the Netherlands.
11. SURICO, G., DE CICCO, V., and IACOBELLIS, N.S. (1981). Osservazioni sulla patogenicità di Phoma tracheiphila (Petri) Kanc. et Ghik. in relazione alla produzione in vitro di metaboliti fitotossici. Phytopath. medit., vol. 20, 17-22.

Observations in the scanning electron microscope on the foliar penetration of *Phoma tracheiphila*

W.V.Zucker
AID SpA Research Center, Catania, Italy

A.Catara
Plant Pathology Institute, University of Catania, Italy

Summary

Leaves from a susceptible lemon vartiety were purposely wounded and infected with the fungus Phoma tracheiphila, the causal agent of the "mal
secco" disease. Samplings at the wound sites were examined under the
scanning electron microscope which revealed an abundant micelial growth
surrounding and inside the wounds. Hyphae passed around and over stomatal openings but in no case, even when the hyphae were very concentrated, was there any hyphal penetration of stomatal openings.

INTRODUCTION

The foliar penetration of Phoma tracheiphila Kanc. et Ghik, the causal
agent of mal secco disease in citrus, has been studied ever since the first
appearance of the fungal disease. Petri (1,2) assumed that the fungus could
also penetrate through stomatal openings and carried out extensive research
to determine how. "Would it be necessary -he wondered- for germinating
spores on the leaf surfaces and the resulting micelia to penetrate into the
leaves through the stomatal openings? Or, driven by the wind, could the
spores be passively transported through those openings and reach the hypo-
stomatic cells? In the first case, preventative treatments with fungicides
would be quite effective, in the latter case they would be ineffective.
Therefore, it is of the greatest importance to answer this question".

In spite of repeated trials, this hypothesis has never been verified.
Nevertheless, the fungal penetration through stomata has still been considered possible. Recent experiments, even if not specifically directed
toward this issue, have repeatedly shown that Phoma tracheiphila penetrates
through wounds, while healthy leaves do not show symtoms of the disease (3).
Because of this evidence, Petri's hypothesis is unsupported; however, more
definitive evidence is highly desirable. Scanning electron microscope
studies have been considered as the most effective and direct approach to
investigate this aspect. In this report, we present preliminary results
on the mode of penetration of Phoma tracheiphila into citrus plants.

MATERIALS AND METHODS

Inoculations were made on leaves of three-year-old white Femminello
Zagara lemon plants (Citrus limon F.) grafted on Volkamer lemon (C.volk-

ameriana Pasq.). All plants were grown in 25 liter pots in the greenhouse.
A picnoconidial suspension (2 x 10^6 conidia/ml) from a highly pathogenic
isolate was sprayed on some leaves previously perforated with an entomological neeedle to create small wounds. Plants were then kept covered with
a transparent, polyethylene bag for two days, and maintained in the greenhouse at 18-30°C, relative humidity 50-95%, with natural illumination.
After three days, some leaves were removed for observations under the
scanning electron microscope (SEM). Samplings of leaves at wound sites were
removed and metallized with gold, according to the currently used technique
All sections were examined under a 15-25 KV Stereoscan 100 electron microscope (Cambridge Instruments Ltd.).

RESULTS

In the 30 day observation period, the inoculated plants kept in the
greenhouse showed typical mal secco symptoms of leaf chlorosis, starting
from the area surrounding the wounds, very similar to natural infections of
P. tracheiphila. No symptoms were observed in the healthy leaf sections.
Samplings examined under the SEM three days after inoculation showed abundant micelial hyphae on both sides which appeared slightly squashed due to
the preparation technique used. They were much more numerous surrounding
the wounds and appeared very concentrated inside them, where they had penetrated the damaged cells (figures 1-5). On the lower leaf surfaces, micelial
hyphae could be observed moving along the surface, but in no case was there
any hyphal penetration of stomata. Hyphae passed around and over the stoma
in spite of their large openings (figure 4 and 5), and even when the hyphae
were very concentrated. Hyphae had a diameter of 1.5 to 2μ, while the
stomatal openings measured from 3-7μ.

DISCUSSION

The observations made in the SEM show that the micelial hyphae of
P. tracheiphila are considerably smaller than the stomatal openings and
could easily penetrate them (2). Even though stoma appeared fully open
in the preparations, and hyphae were numerous on the leaf surface, they were
never observed entering the stomata. Taking into account that samplings
were examined three days after inoculation and that plant growth conditions
certainly provided normal ambient for openings of stomata, we can conclude
that P. tracheiphila does not utilize this mode of penetration in the infection process. While it cannot be excluded that some elements of the
fungus (conidia, hyphae, etc.) could accidentally penetrate through the
stomata, such an event would have no practical meaning.
Petri (2) tried to observe, without success, the penetration of micelial tubes into stomata. Solel (4), examining the origin of the disease
in seedlings exposed to natural infections, observed that they always start
from leaf wounds and lacerations. No one who has investigated the mal secc
disease by leaf-wound inoculation, has ever reported observing infection ex
cept in the purposely wounded sites (5,6,7) or seen elements of the fungus
inside the stomata (8).

Results obtained with the SEM provide clear evidence that the presence of micélial hyphae is particularly concentrated near the wound sites where they probably find nutrients for their growth. Such mode of penetration is of great importance since plant leaves are very easily subject to wounds. In this way, the pathogen may pass from leaves into stems, particularly in the period from October to May (9).

These findings greatly contribute to the understanding of the biology and the parasitism of P. tracheiphilia, and once again confirm the importance of the chemical control of the disease, by timely treatments, after any wound causing events to the plant. In a future communication, we will report quantitative observations, and will discuss in a probabilistic manner, the possibility that conidia and hyphae can enter the plant through stomata.

REFERENCES

1. Petri, L. 1930. Ulteriori ricerche sulla morfologia, biologia e parassitismo del Deuterophoma tracheiphila. Boll. Staz. Pat. Veg. 10, 191÷221.
2. Petri, L. 1930. I risultati di alcune ricerche sperimentali sopra il "mal secco" degli agrumi. Boll. Staz. Pat. Veg., 10, 353-359.
3. Cutuli, G., C. Laviola, G. Perrotta, M. Salerno e P. Spina. 1984. Il malsecco degli agrumi. Seminario internazionale di studio organizzato nell'ambito del programma di ricerche Agrimed, 131 pp.
4. Solel, Z. 1976. Epidemiology of mal secco disease of lemons. Phytopathol. Z., 85, 90-92.
5. Salerno, M. e G. Cartia. 1965. Ricerche sul "mal secco" degli agrumi (Deuterophoma tracheiphila Petri). III. Prove "in vitro" e "in vivo" sull'efficacia di alcuni anticrittogamici. Riv. Pat. Veg., Ser. 4; 1, 71-82.
6. Salerno, M. e A. Catara. 1967. Ricerche sul "mal secco" degli agrumi (Deuterophoma tracheiphila Petri). VI. Indagini sulla riproduzione sperimentale della malattia. Riv. Pat. Veg., Ser. 4; 3, 89-97.
7. Luisi, N., V. De Cicco, G; Cutuli e M. Salerno. 1977. Ricerche su un metodo di studio della patogenicità del "mal secco" degli agrumi. Ann. Ist. Sper. Agrumic., 9-10, 167-173.
8. Bassi, M., G. di San Lio Magnano e G. Perrotta. 1980. Morphological observations on the host-parasite relations in sour orange leaves infected with Phoma tracheiphila. Phytopath. Z., 98, 320-330.
9. Somma, V., G. Scarito and C. Laviola. 1981. Research on the epidemiology of "mal secco" disease of citrus: preliminary results on the role of leaf infection. Proc. Int. Soc. Citriculture, pp 353-354.

FIGURE 1. 16 days after spraying lemon leaves with a picnoconidial suspension of Phoma tracheiphila, chlorotic regions
are observed near the site of the experimentally induced
wounds.

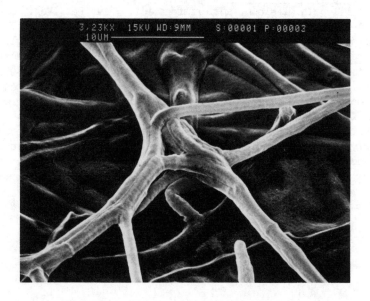

FIGURE 2. Hyphae of Phoma tracheiphila on a lemon leaf
three days after inoculation.

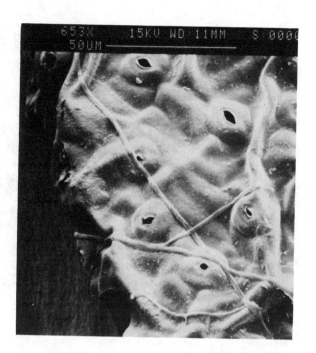

FIGURE 3. Lower surface of a lemon leaf 3 days after spraying with a suspension of picnoconidia of Phoma tracheiphila. The hyphae pass over and near stomatal openings without penetrating them.

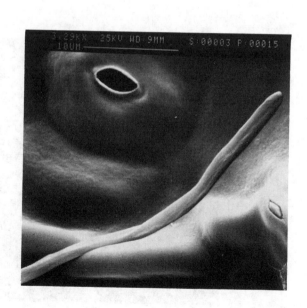

FIGURE 4. A hypha passing near a stomatal opening. The difference in diameters (hypha vs stomatal opening) can be clearly seen.

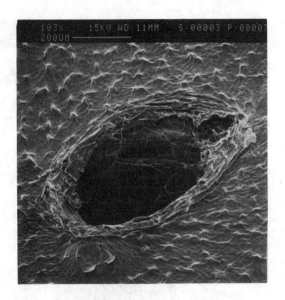

FIGURE 5.

FIGURES 5,6, AND 7. Penetration of Phoma tracheiphila through a wound. The hyphae are more abundant near the wounds where they occur most frequently on the inside edges.

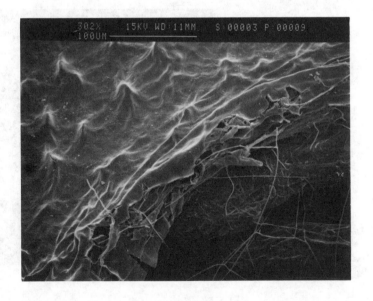

FIGURE 6.

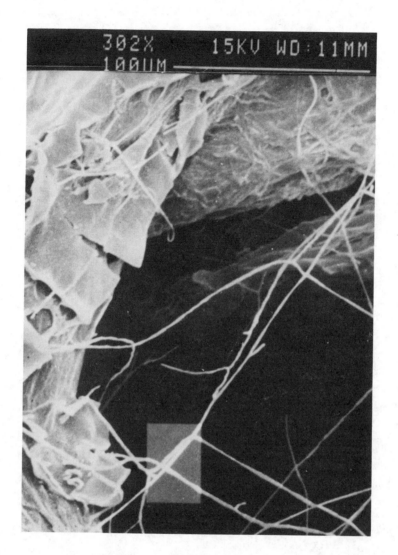

FIGURE 7.

277

Phytophthora blight – A destructive disease of ornamental citrus*

G.Magnano di San Lio, R.Tuttobene & A.M.Pennisi
Plant Pathology Institute, University of Catania, Italy

Summary

Phytophthora blight is a disease that severely damages container grown ornamental citrus in Eastern Sicily. The following varieties of ornamental citrus resulted affected in Sicily: calamondin, kumquats, bergamot, narrow leaf and bouquet bitter oranges, chinotto, lemon, and Eustis limequats. The disease is caused by the fungus Phytophthora citrophthora (Sm. & Sm.) Leon. Conditions favouring the disease are described and control measures are suggested.

1.1 Introduction

Potted citrus are only a component of the total industry production of ornamental plants in Sicily, however for some nurseries they are one of the principal crops.

The most commonly container-grown ornamental citrus are calamondin (Citrus mitis Blanco) and kumquats (Fortunella margarita (Lour.) Swing. and F. japonica (Thunb.) Swing.). These species may be easily propagated as rooted cuttings, but nurserymen prefer to bud them on sour orange (C. aurantium L.) or volkamer lemon (C. volkameriana Pasq.) rootstocks.

Phytophthora blight (1), that seldom causes economic important damages in citrus orchards is a serious threat to the commercial production of ornamental citrus plants. In recent years severe outbreaks of the disease have been noticed in nurseries of coastal area of Sicily. Misdentifications of the disease made by unexperienced technichians and lack of adequate control measures may account in part for so devastating effects of the disease.

Owing to considerable economic significance of Phytophthora blight, since 1982 we started a research program with the following objectives: 1) to define symptoms of the disease; 2) to ascertain the distribution and severity of the disease in commercial nurseries of Sicily; 3) to determine the species of Phytophthora involved; 4) to devise more effective control measures.

The work has taken the form of laboratory and greenhouse research combined with surveies and field trials in commercial nurseries. It was carried out in Province of Catania, where most of ornamental citrus cultivations are located.

* This work has been financially supported by "Cassa per il Mezzogiorno". Research project 11/28.

279

1.2 Disease symptoms

Tipically in a production nursery the first indications of the disease are dead shoots that must be pruned away before affected plants can be marketed. The dry, dead shoots are usually observed two to several weeks after a rain. Earlier stages of the disease, however, can be detected by careful inspection soon after favourable conditions for infection occur.

The first sign of blight is the appearance on leaves of large brown lesions with a water-soaked irregular outer margin. In raining weather the lesions enlarge and turn black, while, if fine weather follows, lesions usually become dark brown. Lesions starting at the leaf tip cause tip burn. Affected leaves drop or dry up in place.

Other symptoms of the disease are blight of blossoms and fruit brown rot.

Serious damages are caused by infections originating on shoots and twigs. Shoots at first appear slighty wilted with brown discoloration of the green stem and leaf petioles. Shriveling and browing progress down the stem until the entire shoot is gradually wilted.

The most evident symptom on twigs and woody stems is a profuse gumming on the surface of bark. Removal of the bark reveals gum impregnation of the cambial surface and of the outer wood-layers. Gum reaction extends far away beyond the infected area. Phytophthora gummosis can usually be distinguished from other forms of infections indeed by the severity of the gumming.

Attacks of aerial parts of the plants are frequently associated to brown rot gummosis infections on the lower stem of rootstock.

In young budlings, that are very susceptible to blight, affected scion damps off while rootsock may remain alive. On more than one year old saplings the injurious effects of the disease may act slowly over many weeks. Single branches are usually affected as a result of blight. In humid conditions infections may result in a serious reduction of tree-canopy and even death of the entire plant. Anyway badly affected saplings are unmarketable.

1.3 The causal agent

By utilizing selective agar-media and baiting techniques (7), Phytophthora citrophthora (Sm. & Sm.) Leon., P. nicotianae van Breda de Haan var. nicotianae Waterh., and P. citricola Sawada have been isolated from rhizosphere soil of potted citrus affected by blight. Both P. citrophthora and P. nicotianae have been consistently isolated from the infected roots and the lower stem, but only the former species has been recovered from aerial parts of blight affected plants.

Water suspension of zoospores of P. citrophthora were sprayed on healthy plants of calamondin and kumquats which were then kept humidified in the greenhouse under plastic canopies for 48 hs. Typical symptoms of blight were reproduced within six to fifteen days and cultures of the fungus were reisolated from the affected tissues. The evidence indicates that

P. citrophthora is the causal agent of blight in Sicily.

However, it should be investigated why only P. citrophthora is responsible of blight in Sicily, notwithstanding also P. nicotianae and P. citricola were pathogenic to calamondin and kumquat in pathogenicity tests. Distribution of species could depend, in part, upon temperature; thus in Japan, where rains are frequent in summer, P. nicotianae, which grows better at temperatures above 25°C, is the species that most frequently causes blight in citrus nurseries (3).

1.4 Varietal susceptibility

In Japan lemons, navel oranges, kumquats and early maturing satsuma mandarins are considered susceptible to Phytophthora blight, while median to late maturing satsuma mandarins are considered fairly tolerant (3).

Our field observations led to conclusion that apparently there are no differences in susceptibility to Phytophthora blight of various ornamental citrus varieties, although they vary greatly in their susceptibility to root and foot rot.

Among pot-grown ornamental citrus, beside calamondin and kumquats, bergamot (C. bergamia Risso), narrow leaf and bouquet bitter oranges (C. aurantium L.), myrtle leaf orange (chinotto), lemon (C. limon (L.) Burm.) and Eustis limequats (F. japonica (Thunb.) Swing. x C. aurantifolia (Christm.) Swing.) resulted affected by blight in Sicily.

1.5 Epidemiology

Soil borne inoculum is by far the most important source of P. citrophthora. In most ornamental citrus nurseries inoculum density of P. citrophthora in soil of containers, as determined by soil-dilution plating techniques in conjunction with BNPRAH selective medium (5), was several hundred propagules per gram of soil. Propagules of the fungus have also been recovered from irrigation water when uncovered basins were the source. Secondary inoculum is provided by established infections on aboveground parts of the plant, because the pathogen can produce sporangia externally on the plant, given favourable conditions, i.e. high humidity. Inoculum is dispersed by water splashing. Furthermore it can be wind-blown in water droplets. The most favourable conditions for infection in outdoor blocks of plants occur during rainy weather when temperatures are mild, usually in late winter or autumn. However rains may favour infection also in colder months. Practically, in Sicily, infections may occur all year round.

Effect of irrigation on disease development depends largely on the macroclimate. When this is extremely dry, the effect of irrigation practise is minimal. In production nurseries, although most irrigation is by overhead sprinklers, closely simulating rainfall, outbreaks of blight have not as yet been noted in Sicily during the warm season from June through August, notwithstanding diseased plants may be present also in this period. We can only speculate that the longer moisture periods occurring

Table I - Incidence of Phytophthora blight in overhead-sprinckler irrigated citrus nurseries. Damage was rated by utilizing a disease severity scale based on visual assesment of disease incidence on stem, twigs, leaves, flowers, and fruits. Plants were grouped in five classes: 0= No disease; 1= Trace, 1-25% blight or no cankers; 2= Light, 26-50% blight or cankers on twigs; 3= Moderate, 51-75% blight or cankers on lateral branches; 4= Severe, 75% blight or cankers on stem; 5= Dead plant. Blight includes stem, leaves, and flowers.

Location	Date	Remarks	Rootstock	Citrus varieties	Age of bundlings (years)	No. of plants examined	% of diseased plants rated in a disease severity scale						
							0	1	2	3	4	5	TOTAL
Fiumefreddo (Catania)	December 1982	Under greenhouse. Copper fungicide sprays applied at weekly intervals from September to November	Sour orange	Kumquat	1	329	48	4	8	8	3	29	52
			Sour orange	Calamondin	1	747	31	13	22	17	10	7	69
			Sour orange	Chinotto	1	334	19	1	5	7	3	65	81
Giarre (Catania)	January 1985	Outdoor. No chemical treatments	Volkamer lemon	Calamondin	3	280	28	7	10	20	15	20	72

during rainy weather are required for the infection process to take place, whereas drying conditions after sprinkling are unfavourable.

According to KNORR (1973), epidemics of fruit brown rot, a disease caused by the same species of Phytophthora, develop only after fruits have been continously wetted with dew or rain for apptoximately 18 hs. In Sicily the occurrence of blight is practically determined by the presence and duration of wetting.

Under greenhouse frequent irrigations during daytime may prolonge the wet period provided by dew at night thus favouring the disease. Moreover, splashing of inoculum is an important drawback of sprinkler irrigation systems. Devastating effects of blight following overhead sprinkling in late summer were observed on ornamental citrus grown under vynil canopy (tab. I).

In ornamental nurseries infested soil is the source, but also the most important reserve of inoculum. Infested soil, collected in an ornamental citrus nursery and mantained in pot, at home temperature for over one year without host plants, yelded P. citricola colonies when plated in selective medium, or caused infections on lemon fruits, notwithstanding low residual soil moisture (5%) after so prolonged storage. In another experiment a water suspension of soil naturally infested with P. citrophthora was mantained at 4°C. Although periodic plating showed a gradual decline in the number of Phytophthora colonies on the isolation medium, inoculum density kept high also after six months (Fig. 1). TSAO (7) observed that P. nicotianae could withstand in the soil several years without loss viability, provided the soil was incubated in moist conditions. These selected examples give a good idea of how long inoculum of Phytophthora can survive and persist in the soil. Nonmycelial propagules (chlamydospores and oospores) are believed to be responsible for this persistance. The pathogen can also survive indefinitely as mycelium or chlamydospores in infected roots of the host.

1.6 Control

Owing to its destructive effects Phytophthora blight requires adequate control in ornamental citrus cultivations.

The most obvious control strategy is to esclude primary inoculum from pots by utilizing either sterilized or virgin soil. Other uncontaminated substrates, alone or mixed with sterilized soil, could be employed. These preventive measures, however meet, in the practice, several difficulties:
1) some materials (as perlite, peat, ecc.) are too expensive;
2) most of nurserymen have not appropriate equipment for soil sterilization;
3) primary inoculum usually is carried with soil adhering to roots of rootstock-seedlings supplied in open-field nurseries;
4) ornamental citrus nurseries in Sicily are located in intensive citrus areas. Therefore, reinfestation of sterilized soil from neighbouring citrus cultivations, through irrigation water. Moreover the likelihood that soils supplied in these areas are contamined is very high.

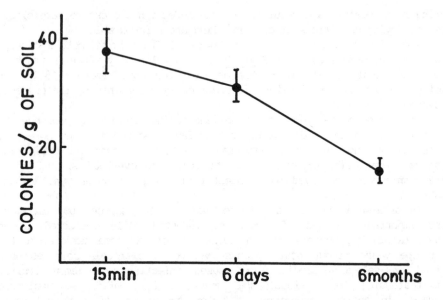

Fig. 1 – Inoculum density of Phytophthora citrophthora as determined by the soil dilution plate method, at several time intervals in a naturally infested soil suspended in water (1:10 w/v) and stored at 4°C.

Current control approaches are based on reduction of secondary inoculum and protection from infections. Secondary inoculum usually is reduced by removing infected shoots. Foliage and twigs are protected from infections by fungicide sprays applied at regular intervals. The most commonly used by nurserymen are copper fungicides that are active also against black pit caused by Pseudomonas syringae van Hall. In severe disease situations however protectant fungicides did not give satisfactory control. Conceivable poor fungicide performance could depend upon timing of sprayes and leaching of fungicides either by rains or overhead sprinkler irrigation water. Furthermore protectant fungicides have no curative activity on established infections. Lastly, foliar applications of protectant fungicides have little or no effect on soilborne inoculum. Accumulation of visible residue and phytotoxicity may be additional drawbacks of repeated copper fungicides applications.

Researches on citrus blight control now in progress include:

1) The use of narrow-spectrum systemic fungicides, as acilalanines and ethil phosphites, in conjunction with residual protectant fungicides (phtalimides, copper compounds, dithiocarbammates). Such combinations show sinergistic effects that allow an extension of the standard spray intervals. Moreover systemic fungicides could exert both protective and curative activity, and should protect also rapidly growing and highly susceptible new shoots. Their systemic properties also offer additional flexibility in spray coverage. A continous and uniform fungicide film

over the plant, for example, should be not necessary for protection, because these chemicals are redistributed throughout the plant organs (6).

Owing to these characteristics systemic fungicides appear well adapted, in many respects, for integrated control strategies of Phytophthora blight. In our view, however, their effectiveness in controlling citrus diseases caused by P. citrophthora has not yet been sufficiently proved.

2) The proper timing of fungicide application. Both protectant and systemic fungicides perform better as preinfection treatments. Furthermore foliar spray applications should be scheduled according to irrigations in order to avoid leaching. Fungicides with long leaving residual activity, such as captafol and copper compounds, could be opportunely applied though overhead sprinkler irrigation systems.

3) The chemical control of soilborne inoculum by soil treatments.

4) The use of mulching in order to avoid splashing of soilborne inoculum. Similarly primary inoculum coming from neighbourhoods may be excluded from uncontaminated soils by taking pots on benches.

5) The management of cultural practices, i.e. irrigation. Reduction in the number and frequency of overhead sprinkler irrigations, expecially those at the end of summer, lower disease in absence of rainfall. Thrickle irrigation systems could be suggested because minimize inoculum dispersal. Benefical effects of irrigation systems are more evident under greenhouse. In greenhouse cultivations with an high incidence of blight, epidemic progress was consistently delayed when irrigation shifted from sprinkler to trickle systems (unpublished data).

Beside irrigation, all environmental modifications, as row spacing, canopy density and ventilation, that 'shorten the wetness period, may influence epidemic progress.

These measures, alone, or combined, could improve Phytophthora blight control strategies. Anyhow the earliest possible detection of the disease is of paramount importance for successful control. We found that, by monitoring inoculum density (ID) of Phytophthora in the soil of pots, epidemics of Phytophthora blight may be forecasted before first symptoms of disease have appeared. Indicatively, if values of ID, determined by the standard soil dilution plate method (4), are above three propagules per gram of soil the likelihood of disease to occur is high, whenever environmental conditions are favourable. Because infections of P. citrophthora occur in a wide range of temperatures, it can be assumed, as discussed earlier, that in our climatic conditions, outbreaks of Phytophthora blight will depend upon duration of foliage wetness, provided that inoculum in the soil is high.

REFERENCES

1. KLOTZ, L.J. (1978). Fungal, bacterial, and nonparasitic diseases and injuries originating in the seedbed, nursery, and orchard. pp. 2-66. In: The Citrus Industry (W. Reuther, E.C. Calavan, and G.E. Carman eds.). University of California, Berkeley, California, vol. 4, 362 pp.

2. KNORR, L.C. (1973). Citrus diseases and disorders. The University Presses of Florida, Gainesville, 163 pp.
3. KURAMOTO, T. (1981). Phytophthora diseases. pp. 19–20. In: Citrus diseases in Japan (T. Miyakawa and A. Yamaguchi eds.). Japan Plant Protection Association, Tokio, 64 pp.
4. MAGNANO DI SAN LIO, G., G. PERROTTA, and R. TUTTOBENE (1983). Indagine sui funghi patogeni presenti nel terreno di semenzai di arancio amaro in Sicilia. Inf.tore fitopatol., vol. 33, 49–55.
5. MASAGO, H., M. YOSHIKAWA, M. FUKADA, and N. NAKANISHI (1977). Selective inhibition of Pythium spp. on a medium for direct isolation of Phytophthora spp. from soils and plants. Phytopathology, vol. 67, 425–428.
6. SCHWINN, F.J. (1983). New developments in chemical control of Phytophthora. pp. 327–334. In: Phytophthora: its biology, taxonomy, ecology and pathology (D.C. Erwin, S. Bartnicki-Garcia, and P.H. Tsao eds.). The American Phytopathological Society, St. Paul, Minnesota, 392 pp.
7. TSAO, P.H. (1969). Studies on the saprophytic behaviour of Phytophthora parasitica in soil. Proc. 1st Int. Citrus Symp., vol. 3, 1221–1230.

Post-harvest rots of citrus fruit in Greece

A.Chitzanidis

'Benaki' Phytopathological Institute, Kiphissia, Greece

Summary

The major-post harvest rots of citrus fruits in Greece are caused by
Penicillium digitatum, P. italicum and Phytophthora spp.. Penicillium
species invade the fruit through wounds. Great efforts are made to
minimize skin injuries on the fruit during harvesting and processing.
Post-harvest treatments with thiabendazole, methylthiophanate, sodium
orthophenylphenate and biphenyl are used in packinghouses. To be
effective, fungicides must be applied promptly after harvest. Resistant
strains to fungicides in use were detected but are still in low
percentage in packinghouses and orchards. Sanitation measures are
essential to reduce the inoculum of Penicillium spp. Low temperature
during storage and transit retards the development of decay.
Phytophthora citophthora and P. syringae are the main species causing
brown rot in Greece. The disease is effectively controlled by one spray
in the orchard in autumn with a copper fungicide. Pruning up the
trees to avoid infections is not practical. Fruit with incepient
infections of P. citrophthora may develop decay and be discarded at
inspection if they are degreened for more than two days. Metalaxyl as
a post harvest dip delayed and reduced the decay. Its effect was
better if applied at 45° C.

The major post-harvest rots of citrus fruits in Greece are caused by
Penicillium digitatum (green mold) and P. italicum (blue mold) as well as
by species of Phytophthora.

Penicillium species invade the fruit through wounds. Poor handling
practices during harvesting and subsequent processing are the main causes
of wound induction on fruit. An important factor influencing the development of Penicillium rot is temperature. The optimum temperature for both
species is around 24° C. At this temperature the incubation period of the
disease lasts three days. At temperatures above or below 24° C the development of the disease is slowed down. Rain is also a factor that favors the
disease. Thus fruit picked during wet weather is more prone to decay than
fruit picked under dry conditions.

The most effective method to control Penicillium decay consists in
careful handling of the fruit during harvesting, packing and marketing, in
order to avoid skin injuries. To help prevent injuries, pickers should
wear gloves and use special clippers, fruit should be placed carefully in
picking boxes and loading and unloading should be done with care. Also
fruit should not be picked in rainy weather. Cooperatives of citrus growers
and Agricultural Services at the beggining of the harvest season give clear
instructions to the growers on the right handling of the fruit. However, due
to the high cost of labor, the lack of experienced pickers and often the
demand pressure of the market the application of these procedures is not
feasible. For this reason the use of fungicides after harvest is unavoidable.

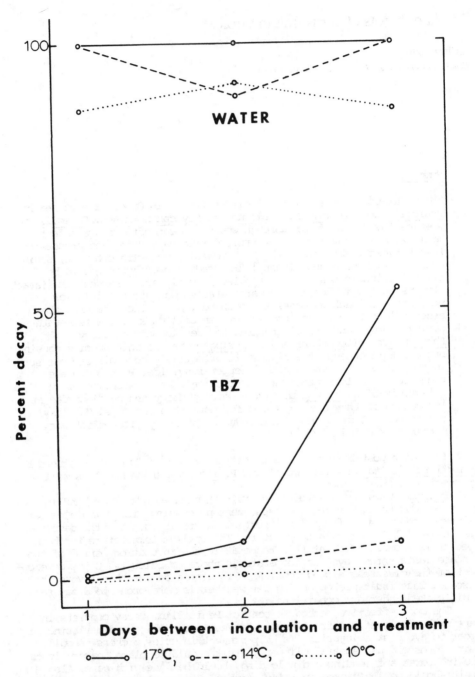

Figure 1. Effect of time elapsed between inoculation and TBZ treatment upon decay of oranges at different temperatures.

Fungicides are effective when applied promptly after infection (Fig.1). As most infections take place during harvesting, fruit should be treated soon after harvest. Severe Penicillium decay has been observed in Greece in lots of fruits that had been kept in storage or had been degreened before treatment (1).

The fungicides used for post harvest treatment of citrus fruits in Greece are : the benzimidazoles, maily thiabendazole (TBZ) and to a lesser extent methylthiophanate sodium orthophenylphenate (SOPP) and diphenyl. TBZ is incorporated in the water wax emulsion and some times is added in the bath. Methylthiophanate is added in the bath. SOPP is applied in the bath or as a foam wash and biphenyl is impregnated in the wrapping papers.

In countries where these fungicides have been used for several years, resistant strains to one or more of them have been observed. In order to find out if resistant strains had been selected in Greece, surveys were carried out in orchards and packinghouses. Low percentage of TBZ-resistant strains were found even in areas where packinghouses were not present and where benzimidazole fungicides have never been used (Table I). In the packinghouses the incidence of TBZ-resistant strains was very low, even in those where the fungicide had been used for several years (Table II). The percentage of TBZ-resistant strains was found to be highest in culled fruit piles and in packed fruit kept in storage (1). SOPP-resistant strains were also detected but at very low numbers. Most of the SOPP-resistant strains were less pathogenic to fruit than the susceptible ones (2).The fact that up to now there was no problem from resistant strains in Greece, although they exist in nature, can be attributed to the practice that the fruit is not drenched in a fungicide before processing,as it is done in other countries, and also that the fruit is not stored after treatment. Thus the selected resistant strains do not accumulate in packinghouses.

Sanitation measures are essential to reduce the inoculum of Penicillium spp. in packinghouses. In many packinghouses the importance of sanitation has been realized; the processing area is kept clean and all decayed fruit is removed frequently. In some packinghouses the packing areas are being disinfested with driol.

Low temperature during transit and storage delays the development and spread of Penicillium and other decays. Greek citrus fruits are not stored after packing. Low temperatures are kept in carriages of exported fruit. The temperature is selected according to the species and variety of the fruit.

Brown rot of citrus caused by Phytophthora spp. is a common disease in all the citrus growing regions of Greece. It is particularly severe in areas with heavy soils and orchards where the ground is bare due to cultivation or the use of herbicides. Zoospores of Phytophthora spp. are splashed by rain onto the fruit hanging in the lower skirts of the trees. In wet years the amount of infected fruit may reach 50% of the fruit hanging up to 70 cm from the ground. However, brown rot is most serious as a post harvest disease because symptoms often develop after picking and inspection and there are no effective post harvest treatments against Phytophthora spp.

In a survey carried out over five years several species of Phytophthora were found to cause brown rot of citrus fruit in Greece, but P. citrophthora and P. syringae are the most common ones. P. citrophthora infects the fruit early in the season, while P. syringae prevails after December, when the temperatures in the orchard are very low (3).

Phytophthora brown rot is best controled by spraying the skirts of the trees in autumn with a fungicide, thereby preventing infection of the fruit before harvest. Several fungicides were tested in the field. Copper

Table I. Penicillium isolates from decayed fruits in citrus orchards

Area	Penicillium digitatum isolates		Penicillium italicum isolates	
	Total	TBZ resistant	Total	TBZ resistant
Packinghouses present	205	0	305	3
Packinghouses not present	262	0	231	1

Table II. Number of Penicillium colonies in 15 plates exposed for 15″ in commercial packinghouse that used TBZ for various periods

Packinghouse	Number of seasons of TBZ use	Penicillium digitatum colonies		Penicillium italicum colonies	
		Total	TBZ resistant	Total	TBZ resistant
A	9	0	0	1	0
B	9	0	0	40	0
C	1	5	0	61	0
D	5	0	0	2	0
E	9	0	0	6	0
F	2	51	0	331	1
G	3	116	0	2	0
H	5	88	0	1	0
I	9	111	0	0	0
J	3	45	0	0	0

Table III. Duration of effectiveness of one fungicide spray application in the orchard against brown rot of oranges. Fruits were inoculated with P. citrophthora four months after treatment (mean of 40 fruits)

Treatment and dosage		Number of spots/fruit
Copper oxychloride	0.150% Cu	0.18 a*
Copper hydroxile	o.085% Cu	0.05 a
Control		22.90 b

* Mean separation within column by Duncan's multiple range test 5% level.

Table IV. Effect of one fungicide spray application on the soil under the canopy of orange trees on the control of brown rot of oranges (mean of 12 trees)

Treatment and dosage (g.a.i/m^2 of soil)	Insfected fruits (%)*
Metalaxyl 0.15	30.5 bcd[+]
Folpet 0.60	42.1 ab
Mancozeb 1.12	38.3 abc
Copper oxychloride 0.75	42.4 ab
Metalaxyl 0.15 + folpet 0.60	29.4 cd
Metalaxyl 0.15 + mancozeb 1.12	20.7 d
Metalaxyl 0.15 + copper oxychloride 0.75	26.6 cd
Water	45.9 a

* Up to 70 cm from the ground

[+] Mean seperation within column by Duncan's multiple range test 5% level

fungicides gave the most satisfactory results (3). A single spray with a copper fungicide can protect the fruit for four months at least (Table III). However, copper fungicides have the disadvantage that treated fruit are stained. This is a problem for fruits that are not washed during the packing process like tangerines and also for fruits that are not processed like the ones destined for the local market. Captan compared to the copper fungicides does not stain the fruit but more than one spray is necessary to protect them.

It is generally recommended that when fungicides are applied in citrus orchards the soil under the canopy of the trees should also be sprayed, in order to suppress the inoculum of Phytophthora species. Harris (5) mentions that P. syringae in apple orchard soils was suppressed by spraying the soil with fungicides. In a field experiment the soil under the canopy of orange trees was sprayed in late October with metalaxyl, folpet, copper oxychloride and combinations of metalaxyl with the other three fungicides. Percent Phytophthora infestation of hanging fruits was significantly lower on plots treated with metalaxyl and its combinations with the other fungicides than on untreated controls. Copper oxychloride was less effective than the other fungicides (Table IV).

Besides chemical control, pruning up the trees is recommended so that none of the fruits hangs within 70 cm from the ground. The method is not entirely practical because a large amount of easy to pick fruit is produced near the ground.

Greek fruits picked before the end of November need degreening from 1-5 days and subsequently are kept for one more day to cool in the packing-house before processing and packing. The main species causing brown rot at this period is P. citrophthora. If fruits with incipient infections of P. citrophthora are degreened for two or more days, the temperature in the degreening room being very close to the optimum for the development of this species, most of the infected fruit show well developed symptoms at the end of the process and can be discarded at inspection. If, however, fruits need only one day of degreening, fruits do not develop symptoms before inspection. For this reason, after heavy rains that might incite infections, the picking of fruits should be postponed for more than three days. Thus after degreening infected fruits will show welldeveloped symptoms and can be discarded (3).

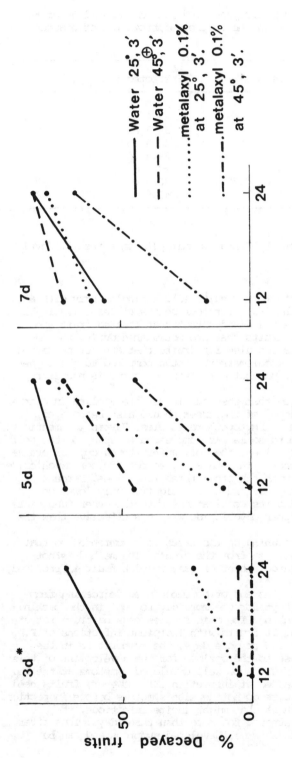

Days after inoculation

Water 25°, 3'
Water 45°, 3' ⊕
.... metalaxyl 0.1%
           at 25°, 3'.
—··—··— metalaxyl 0.1%
           at 45°, 3'.

* Days between inoculation and treatment.
⊕ Fruits were previously dipped for 3' at 35°.

Figure 2. Effect of timing of post harvest dips on oranges inoculated with P. syringae. Fruits were incubated at 12°.

Research was carried out in order to find treatments that could be applied in the packinghouses against brown rot. Dipping of fruits with incipient infections in metalaxyl after harvest delayed the development of brown rot and generally resulted in a smaller percentage of fruits with symptoms. This effect is more pronounced in fruit treated soon after infection and kept at temperatures at which the fungi grow slowly (3). It must be noted that metalaxyl is not registered for citrus fruit. In California it was recommended to submerge citrus fruit after harvest in water at 48° C for 2′- 4″. This treatment might eradicate incipient infections of P. citrophthora (6). Fruit treated at this temperature may show oleocellosis and becomes also susceptible to attacks by Penicillium spp. (4). P. syringae is more susceptible to high temperatures than P. citrophthora . Therefore an attempt was made to eradicate the fungus at lower temperatures. Fruits with incipient infections of P. syringae were dipped for three minutes in water at 45° C. At this temperature the development of symptoms was suppressed if the fruits were treated within three days from inoculation. The effect of the temperature improved when metalaxyl was added in the water (Fig.2). Up to now it doesn't seem that an effective method to control brown rot after harvest is available.

In conclusion,a combination of several measures is needed to control post-harvest rots of citrus fruits in Greece : (a) application of one spray with a copper fungicide in autumn; (b) careful handling of the fruit during harvesting and processing; (c) delay of picking after rains if fruits are to be degreened; (d) use of a fungicide drench if the processing in the packinghouse is postponed; (e) use of two chemically unrelated fungicides after harvest; (f) removal of culled fruit piles from packinghouses and orchards; (g) avoiding storage of fruit in the packinghouse; (h) Application of sanitation measures in packinghouses and equipment; (i) keeping proper temperature in storage and transport of fruit.

REFERENCES

1. CHITZANIDIS, ANNA (1983). Resistant strains of Penicillium digitatum and Penicillium italicum to thiabendazole in Greece. IVth International Congress of Plant Pathology, Melbourne Australia. Abstracts of Papers : 108.
2. CHITZANIDIS, ANNA and LASCARIS, D. (1980). Detection of fungicide-resistant strains of Penicillium digitatum and Penicillium italicum in Greece. IInd National Phytopathological Conference, Athens. Abstracts of Papers, In Phytopath. medit. 19 : 67
3. CHITZANIDIS, ANNA and LASCARIS, D. (1981). Research of Phytophthora rots of oranges in Greece. Proc. Int. Soc. Citriculture, Tokyo, Japan : 796-799.
4. HARDING, P.R. and SAVAGE, D.C. (1964). Investigations of possible correlation of hot-water washing with excessive storage decay in coastal California lemon packing houses. Plant Disease Reptr., 48 : 808-810.
5. HARRIS, D.C. (1979). The suppression of Phytophthora syringae in orchard soil by furalaxyl as a means of controlling fruit rot of apple and pear. Ann. appl. Biol. 19 : 331-336.
6. KLOTZ, L.J., DE WOLFE, T.A. and MACRILL, J.R. (1965). Control of brown rot of citrus in orchard and packing house. Calif. Citrograph 50 : 456, 474-476.

Virus and virus-like diseases of citrus, plant quarantine regulation, clonal and sanitary programs in the Mediterranean area and criteria for handling virus-free clones

A.Catara & M.Davino
Plant Pathology Institute, University of Catania, Italy

F.Russo & G.Terranova
Citrus Experimental Institute, Acireale, Italy

Summary

After a review of virus and virus like diseases reported in the mediterranean countries the need of quarantine regulations inside and outside the mediterranean area and of clonal and sanitary programs under standard procedures is discussed. Criteria for growing healthy material and for maintenance of genetic and sanitary purity and fast multiplication procedures are also described.

1. Introduction

Among the problems concerning the integrated pest control of citrus the diseases caused by virus and virus like agents are very important because healthy material is an essential condition to achieve good results. Furthermore, expecially with woody plants, infections may become a focus for pathogen dispersal (through propagative material, pruning, grafts or regrafts, etc.) and cause large losses (10).

Citrus diseases caused by virus and virus like agents are numerous (4,5). All of them are transmitted through graft, some are transmitted by insect vectors (tristeza, stubborn, greening, vein enation) or are strongly suspected to be insect-borne (i.e. blight, ringspot), only few are mechanically (i.e. exocortis, cachexia-xyloporosis) or seed transmitted (psorosis A).

The continous exchange of nursery material among mediterranean countries, the hazard of introduction of infectious agents from abroad, the lack of adequate sanitary program and the use of topworking to change the variety, arise some problems which should be analyzed.

This review concerns with the diseases caused by virus and virus like agents reported in the mediterranean area, the problems of plant quarantine, the need of clonal and sanitary programs. The criteria for growing healthy material and for their maintenance in genetic and sanitary purity and fast multiplication are also reported.

2. Citrus virus diseases in the mediterranean area

Citrus diseases caused by virus, viroids, or procaryotes have been reported in the mediterranean area since 50 years ago when researchers from the States travelling in the mediterranean countries discovered some of them.

In Italy the first described disease was impietratura but its infectious nature was shown only many years later. The economic importance of the virus diseases was evaluated only at the beginning of fifthees.

The availability of suitable and rapid indexing (17) tests have allowed an increase of knowledges on the distribution of these diseases in many mediterranean countries. Fiftheen have been already reported: four are present everywhere, eleven only in some countries (1). Many others known in other citrus area are still unknown in the circummediterranean countries.

a. Among the first group are: psorosis A, concave gum-blind pocket, exocortis, cachexia-xyloporosis.

Psorosis A. Young leaf symptoms have been reported quite frequently but scaly bark and decline of affected trees have been described only in few cases and are more frequent in the varieties of the "navel" group. Since the local varieties of sweet orange appear tolerant the disease has no economic importance unless the affected (but symptomless) plants are regrafted to navel varieties.

Concave gum-blind pocket. Concavities of the trunk or branches and gum impregnation of xylem are quite common, and cause reduction of yield and decline of mandarins, tangelos, and navel oranges. Once again the choice of varieties for the regraft has to be considered carefully.

Exocortis. Its wide distribution does not cause problems in old citrus groves grafted on tolerant (mainly sour orange) rootstocks but it is a limitation for the introduction of some rootstocks (i.e. citranges) which would allow better agronomic results. Graft (or regraft) of plants on susceptible rootstocks with untested budsticks is risky.

Cachexia-xyloporosis. It is largely diffused in many old lines of citrus. Same species and varieties used as scion (mandarin, tangelo, satsuma, etc.) or rootstock (alemow) are susceptible and their use in the nursery or in the regrafts must be carefully evaluated.

b. Diseases belonging to the second group are: tristeza, stubborn, cristacortis, impietratura, infectious variegation-crinkly leaf, ringspot, vein enation-woody galls, gummy bark, gum pocket, kumquat incompatibility.

Tristeza. Decline and death of trees grafted on sour orange have been widely reported in Istrael and Spain. Reduction in growth have been reported in Italy, Jordan and Morocco. Stem pitting and seedling yellows have not been reported in the mediterranean area.

In Spain the use of sour orange as rootstock has been discontinued by law, whereas in Israel an eradication program is in progress since ten years. In Italy, a wide survey for eradication of trees, has been started after the detection of infected trees in Calabria area.

The widespread use of susceptible sour orange is the major hazard for the citriculture of the mediterranean area.

Stubborn. The disease has been reported to cause losses only in Morocco whereas in other areas the pathogen has not been found or has been isolated only after a growth at high temperature. The climatic conditions of the mediterranean area seem not suitable for the development of the disease, but the presence of the vector makes high the menace of a diffussion.

Cristacortis. This disease has not been reported in Cyprus, Jordan and Yugoslavia. The effect on yield and growth has not been shown.

Impietratura. The disease has not been reported in Egypt, Jordan and Yugoslavia. It causes heavy losses and poor quality of fruits of sweet orange and grapefruit. Sweet orange varieties of local origin or introduced from abroad are equally susceptible. Since the disease is detectable only through the symptoms on the fruits a careful inspection of trees during the ripening time is suggested to avoid affected mother trees.

Infectious variegation-crinkly leaf. It is widely diffused (probably in all the countries even if not reported) mainly in lemon groves (6). Heavy losses and poor quality of fruits have been sometimes reported in orchards where lemons were regrafted to mandarin, clementine or sweet orange.

Vein enation-woody galls. The disease has been reported only in Spain and that is surprising since many citrus varieties from that country have been distributed in the mediterranean area.

Ringspot. Rings on leaves and fruits have been reported in some citrus species and varieties but the economic importance has not been evaluated. The isolate found in mediterranean countries appear different from those present in Florida and Texas.

Gummy bark. Described in Egypt, Syria and Turkey probably has not been investigated in other countries.

Gum pocket. Suspicious symptoms of the disease have been reported in Italy but further investigation are needed also to evaluate the economic importance of the disease.

Kumquat incompatibility. A new graft transmissible disease inducing incompatibility between Nagami kumquat and Troyer citrange has been found recently (13). Altough incompatibility has been observed on purposely inoculated sweet orange grafted on Troyer citrange the risk that the disease could eventually affect other commercial varieties cannot be excluded.

In conclusion, the need of large surveys and deeper investigations in many countries is evident from the above discussion. Infact, due to the climatic conditions, to the history of citriculture in this area and to the large exchange of plants and budsticks, it is likely that with few exceptions the virus problems in the different countries are quite similar.

A third group of virus and virus like diseases of citrus have not been reported in the mediterranean area. Among them are many destructive

declines well known in the eastern countries of Asia (leaf mottle), South
Africa (greening), South America (declinio, declinamiento, etc.) and
Florida (blight, young tree decline, etc.). Other unreported diseases are:
citrange stunt-tatter leaf, citrus mosaic, leaf curl, leprosis, moultiple
sprouting, yellow mottle and yellow vein (5).

3. Plant quarantine

The regulation 1977/93 of the Council of EEC and later revisions
establish the phytosanitary surveys and plant quarantine for the countries
in EEC. Each country has been invited to uniform its laws to those
suggestions. The phylosophy was to make easier the exchanges inside the
community and to avoid the introductions from abroad, but some gaps
appeared and must be filled. In particular, no mention is reported for
what concerns virus and virus like diseases of citrus so that countries
not citrus producer allow the introduction of citrus plant material from
other continents and there from the spreading in EEC countries is much
easier. It is clear that in such a situation it is meaningless for a
country to be restrictive if the neighbouring countries open the door to
pathogens and parasites.

Therefore, it would be useful an international cooperation to offer
the best possible garantee in safe garding citrus (and other plants)
health in the mediterranean area. That can be achieved through: a)
embargoes; b) inspection at the port of entry; c) inspection at the port
of dispatch; d) field inspection during the growing season (preclearance);
e) controlled entry; f) postentry quarantine stations (9).

a) Embargoes has some serious economic implications. Furthermore, are
more likely to be imposed against the more developed countries where the
pests or diseases are recorded as a result of more advanced scientific
investigations.

b) Inspection at the port of entry has certain limitation due to the
fact that it is not possible to carefully examine all of consignement and
it is rather difficult to detect citrus virus (or other) diseases in the
introduced material.

c) Inspection at point of origin are more faisible and, specially if
applied at growing season may be useful for the detection of viruses and
other diseases.

d) Controlled entry is useful for pest control but is not reliable
for other diseases unless combined with the following points.

e) Postentry and intermediate quarantine stations are probably the
best approach to the problem but there is a serious shortage at an international level. A good program has been organized for citrus in Spain
(12). The general procedure is to permit entry of only small lots.

f) Public awareness and information dissemination. A wide information
of people involved in plant quarantine, nurserymen, citrus growers

traveler is probably the best approach to have a general cooperation in a quarantine program.

4. Clonal and sanitary selection

After what has been shown it is hopefull that the citrus producer countries develop clonal and sanitary programs according to standard procedures (17). Up to now similar programs are in progress in a large basis only in Spain (11) and Israel (2).

In Italy, a scheme for the genetic and sanitary improvement of citrus has been developed since 1974 on a voluntary basis. Such a program (8), concerns the main citrus species (sweet orange, clementine, mandarin, tangelo and grapefruit) and the main virus and virus like diseases (tristeza, exocortis, psorosis, infectious variegation-crinkly leaf, cachexia-xyloporosis, impietratura and cristacortis). The mother trees have been selected in citrus orchards (old lines carrying only mild strain of exocortis), or among nucellar clones indexed for viruses, and the propagative material has been distributed to the citrus growers by the Istituto Sperimentale di Agrumicoltura (Acireale). Since some years many laboratories carry out the shoot tip grafting (STG) technique to obtain virus free propagative material but a complete indexing of such material is still lacking. No budwood has been yet released to the growers.

In Corse large efforts have been developed to obtain good selections of clementine, certified for genetic and sanitary aspects, which have been distributed in many countries of the Mediterranean area. A STG program has been started since 1980.

In Greece, no certification program is operating though some laboratories are involved with citrus virus disease researches and STG technique.

We believe that if a cooperative efforts will be made adequate programs could be operating in short terms in all the mediterranean countries with large benefits for the citriculture of the entire area.

5. Handling, preservation in purity and rapid multiplication of virus-free clones

When the virus-free clones are obtained, the next steps to do are: 1) the maintenance in healthy condition and 2) the rapid multiplication to meet the need of virus-free plants.

To obtain these two aims the combined action of the genetist, plant pathologist and horticulturist is needed, as, of course, during the first step aimed to the selection and sanitation of the citrus clones.

5.1. Genetic and horticultural aspects

The bud mutations. From the genetic and horticultural point of wiew it is important to be sure that the virus-free plant is true to type. Extensive surveys (20,21) have shown that up to 25% of trees in an orchard were variants, most of them of inferior quality or worthless, regardless whether the orchard was of oranges, lemons or grapefruits. Bud mutations express themselves in a variety of ways: some exhibit obvious differences in tree size or growth habit; leaf size, shape or clorophyll content; fruit size, shape, rind texture or color (3,7,15,20,21,22). Others, however, exhibit subtle and less obvious differences and are often difficult to detect. Such mutants or bud variants which possess undesiderable characteristics are usually called "off type".

Several "Washington navel" orange variants have an acid content in the fruit greater than normal. Such type of variants originating from a normal Frost nucellar "Washington navel" have been not detected until substantial propagations have been made (22).

This case and many others, not mentioned here for the sake of brevity, illustrate the difficulty of detecting the subtle and incospicious "variant" or "off-type". To avoid such hazard some rules must be followed in the selection and handling of the budwood donor trees. These are (15):

- Be sure that the trees from which buds are cut are true to type for the variety. Never use trees that have not fruited as primary bud sources.

- Be sure that the bud-source trees are productive, according with records or careful observations (at least 3 years).

- Inspect periodically, in different seasons of the year, bud source trees carefully for any abnormal growth (trunk, branches, foliage) or fruiting.

- Remove any branch, or the whole tree, that appear "off-type" to prevent inadvertently taking buds.

- Do not cut heavely the trees. Heavy cutting promotes excessive vegetative growth and while this type of growth produces strong buds the risk that some of these vigorous shoots my be mutants is increased.

- Avoid to cut budwoods from a branch that appears excessively vigorous compared with the rest of the tree. Again excessive vegetative vigor my be associated with a non-productive or other type of variant.

- Cut budwoods evenly all around a tree. If an undetected mutant limb does occur the percentage of variant buds will be kept low.

- Cut budwoods from behind fruit when possible.

- Be sure that the bud cutter understands the importance of a careful selection and the need to avoid variants.

The use of seedling trees to produce seeds of the rootstock variety is unwise because a seedling variant can appear very frequently. This variant can grown and produce fruits tue to type or very similar to the rootstock variety but the off-spring can have different physiological characteristics as in the case of the "Troyer" and "Carrizo" citrange.

Nursery increase blocks. In some situation, as in the case of a new selected clone, the demand for budwood is so great that cannot be satisfied without heavy cutting. In this situation it is preferable to take the budwoods from what are known as nursery increase blocks. These include trees propagated from budwoods taken from a selected and virus-free mother tree already in production at least for tree year (8-10 years of age). Cutting of budwoods from nursery increase blocks is limited to a period not exceeding eighteen months from the budding. In such way the budsticks available the next year is decuplicated.

The use of nursery increase blocks as a budwoods source does not follow the suggestion of cutting budwoods only from bearing trees. However, if correct procedures are followed, it results in no greater risk percentage-wise of propagating variant trees, than does taking buds only from primary bud source trees. These procedures (14) are as follows:
- Cut budsticks uniformly to include all sectors of the donor tree, so that those taken from the possible undetected variant limb will be at a low percentage.
- Insert all the buds in seedlings of the nursery increase block.
- Cut the buds from the nursery increase block uniformly over the entire block, regardless of the needed number of buds, so the percentage of undetected variants remains at the original.

As mentioned before, the complete exclusion of all "variants" or "off-type" in citrus propagation is virtually impossible. The widespread repropagation of indesiderable types can be avoided, however, if the propagator or the nursery man follows careful and critical bud selection.

5.2. Phytopathological aspects

Sanitary measures. To maintain virus-free the trees of the selected clones many precautions must be taken in the foundation blocks, in the budwood mother blocks and in the nursery increase blocks and during all the multiplication steps. Two trees of the selected clones are maintained in insect proof glass or screenhouse, to avoid infections by insect vectors, to serve as repository.

To prevent diseases transmitted by insect vectors the trees are raised in screenhouse or insect proof glasshouse or in an area where the vectors are not present. All the tools (knives, pruning tools etc.) must be sterilized with sodium or calcium hypocloride to avoid virus and virus like agents transmitted mechanically.

Although is known that the Phytophthora spp. and the citrus parasite nematodes are not vectors of the virus diseases of citrus they must be eliminated from the soil where the virus-free trees have to be raised. Soil fumigation is fective for elimanating nematodes, Phytophthora spp. and also weeds seed. Rigid precautions must be taken to prevent reintroduction of Phytophthora spp. in the fumiganted plots: the block is completely fenced to prevent the entrance of wild animals that can carry soil infested with Phytophthora spp. and the gate is kept locked when no work is in progress. Workmen entering the block must change into clean shoes kept in lockers at the entrance and treat their shoes with Bordeaux powder or other fungicides.

If the irrigation water does not come from deep, clean wells but from reservoirs or canals, it must be treated with cupper sulfate (20 ppm) or with active clhorine (1 ppm).

301

Seeds are immerged in a hot water bath at 52°C for 10 minutes and treated with T.M.T.D. to damping- off fungi (19).

Handling and multiplication of virus free (Foundation) stocks. As soon as a virus-free clone has been obtained the first step is the propagation on different rootstocks beside the sour orange. If we deal with orange, mandarin and grapefruit the following rootstocks can be used: "Troyer" and "Carrizo" citrange, trifoliate orange, "Swingle" citrumelo and eventually "Cleopatra" mandarin; if we deal with lemon, alemow or volkameriana lemon can ben employed. A large number of trees of each clone, according with the need for the variety, are raised in the field in the so called foundation block. This block shall be located far away of the commercial citrus orchards particulary if in the area is present a virus disease that is spread by insect vectors.

The trees in the foundation block are planted at least at the distance of m 6 x 6 and between a clone and the next the space must be more (m 15); this to avoid roots grafting. In the foundation block the trees are carefully examined up to 8-10 years from the planting before the cutting of boodwoods for propagation is started.

In the same time virus free clones are being grown in orchards in different areas for comparative studies of fruit, tree characters and horticultural performance. In the foundation block the trees are carefully examined in different periods of the year for the growth characteristics, for the yield and for internal and external fruit characters; if an "off-type" limb appears is soon removed. As previously demostrated, only after the productivity and the quality of the yieldhave been observed for at least three consecutive years the first budwoods can be cutted and released to nurserymen and growers for the production of nursery increase blocks and for the istitution of budwood mother blocks or for the production of nursery trees.

The seed mother block is realized following the same rules.

To increase the quantity of propagation material produced in the mother trees block or in the sursery increase block the budwoods can be cutted in the winter time (December-February) when the trees are in the rest stage. These budsticks, dept in polythene bags, are stocked in a contitionated room at the temperature of 14-15C. During the stockage the budsticks are periodically checked to remove the decayed ones. So doing can be utilized for grafting or budding also the extreme part of the shooth already woody that otherwise, when the budwoods are cutted in the spring, must be discarded because is already with the new soft flush of growth and useless (16,18).

6. Concluding remarks

It appears from the above discussion that a considerable effort must be produced to overcome the need of more informations on virus and virus like diseases of citrus in the mediterranean area, specially in some countries. A more realistic picture of the situation will be fundamental for adequate plant quarantine measures to regulate the exchange of citrus plant (or vegetative parts) among the countries of the mediterranean area and the importation from abroad. The hazard connected with the introduction of destructive diseases from abroad the mediterranean countries must be emphasized in the actual regulation asking also the cooperation of not citrus producer countries.

Clonal and sanitary programs will help to reduce the infectious diseases and to improve yield and fruit quality.

To reach that goals a combined effort of genetists, horticulturists and plant pathologists is needed in each country and a better exchange of scientists and informations. The importance of these points can be easily understood if we consider that in a long term program only through such a cooperation it will be possible to keep high the citriculture of the mediterranean area.

REFERENCES

1. BOVE, J.M. and VOGEL, R. (1981). Geographical distribution of citrus virus and virus-like diseases. In: Description and illustration of virus and virus-like diseases of citrus: A collection of colour slides. Secto-Irfa, Paris.
2. CALAVAN, E.C., MATHER, S.M. and McEACHERN, E.H. (1978). Registration, certification, and indexing of citrus trees. In: The Citrus Industry, 4: 185-222, Rev. edition.
3. CAMERON, J.W. and FROST, H.B. (1968). Genetics, Breeding and nucellar embriony. In: The Citrus Industry, 1: 951-952. Rev. edition.
4. CATARA, A. (1984). Virus, viroidi e micoplasmi degli agrumi in Italia. Inf.tore Fitopatol., 34 (3): 15-35.
5. CATARA, A. and DAVINO, M. (1984). Malattie degli agrumi da virus e agenti virus simili non presenti in Italia. Inf.tore Fitopatol., 34 (11): 9-21.
6. DAVINO, M. and LA ROSA, Rosa (1983). Indagini sulla diffusione del virus della variegatura infettiva degli agrumi. Atti Giornate Fitopatologiche, 3: 257-264.
7. IWAMASA, M., NITO, N., YAMAGUCHI, S., KURIYAMA, T., EHARA, T. and NAKAMUTA, T. (1981). Occurence of very early mutants from the Wase early reapening Satsuma. Proc. Int. Soc. Citriculture 1981, 1: 99.
8. MARTELLI, G.P. and RUSSO, F. (1983). La certificazione della vite e degli agrumi in Italia: attualità e prospettive. Inf.tore Fitopatol., 33 (2): 83-88.
9. MATHYS, G. and BAKER, Elisabeth A. (1980). An appraisal of the effectiveness of quarantines. Ann. Rev. Phytopathol., 18: 85-101.
10. MATTHEWS, R.E.F. (1970). Economic importance and control, p. 587-617. In: Plant Virology, Academic Press, New York and London.
11. NAVARRO, L., BALLASTER, J.F., JUAREZ, J., PINA, J.A., ARREQUI, J.M. and BONO, R. (1981). Development of a program for desease-free Citrus budwood in Spain. Proc. Int. Soc. Citriculture, 1: 70-73.
12. NAVARRO, L., JUAREZ, J., PINA, J.A. and BALLESTER, J.F. (1984). The citrus quarantine station in Spain, p. 365-370. In: Proc. 9th Conf. IOCV, Univ. California, Riverside.

13. NAVARRO, L., PINA, J.A., BALLESTER; J.F., MORENO, P. and CAMBRA, M. (1984). A new graft transmissible disease found in Nagami Kumquat, p. 234–240. In: Proc. 9th Conf. IOCV, Univ. California, Riverside.
14. NEWCOMB, D.A. (1970). A program for cutting buds from nursery increase blocks Yerbk. Calif. Citrus nurs. Soc., 9: 54–55.
15. PLATT, R.G. (1973). Importance of avoiding undesiderable off tipes and mutants in Citrus propagation. I° Congreso Mundial de Citricultura – Murcia – Valencia Espana, 3: 549–555.
16. PUGLISI, A. and RUSSO, F. (1978–79). La conservazione prolungata delle marze di agrumi in ambiente condizionato. Ann. Ist. Sperim. Agrum. Acireale, 11–12: 81–100.
17. ROISTACHER, C.N. (1976). Detection of citrus viruses by graft transmission: A review, p. 175–184. In: Proc. 7th Conf. IOCV, Univ. California, Riverside.
18. RUSSO, F. (1977). Scelta del materiale di propagazione in Agrumicoltura. Frutticoltura, 39 (5): 53–56.
19. RUSSO, F. and LO GIUDICE, V. (1980). Tecniche vivaistiche in Agrumicoltura. Cassa per il Mezzogiorno, Progetto speciale n. 11, 86 pp.
20. RUSSO, F., STARRANTINO, A. and SPINA, P. (1981). La selezione clonale dello arancio e del mandarino: ricerche e risultati ottenuti presso l'Istituto Sperimentale per l'Agrumicoltura di Acireale. Riv. Ortoflorofrutticoltura It., 65: 1–10.
21. SHAMEL, A.D. (1948). Bud variation and bud selection. In: The Citrus Industry, 1: 915–952.
22. SOOST, R.K., CAMERON, J.W., BITTERS, W.P. and PLATT, R.G. (1961). Citrus bud variation – old and new. Calif. Citrog., 46: 176, 188, 190–193.

Preliminary results of citrus tristeza virus transmission by aphids in Italy

M.Davino
Plant Pathology Institute, University of Catania, Italy

I.Patti
Institute of Agricultural Entomology, University of Catania, Italy

Summary

Transmission tests of four isolates of CTV, two found in Calabria (CTV–GB and CTV–SW), one from Japan (CTV–J), and one from Florida (CTV–T4) by aphids was attempted inside a cabinet under a greenhouse. Aphis citricola was able to transmit CTV–T4 to Mexican lime seedlings in three tests, whereas Aphis gossypii, Toxoptera aurantii, Myzus persicae and Aphis fabae failed in all the trials carried out.
Colonies of Aphis craccivora, A. fabae, A. gossypii, Aulacortum solani, Macrosiphum euphorbiae and M. persicae collected from young shoots of infected trees in Calabria were also assayed by ELISA method and by feeding on Mexican lime seedlings. Both tests were negative.

1. Introduction

Many aphid species can transmit semipersistently different isolates of citrus tristeza virus (CTV): Toxoptera citricidus (Kirk.) (11), Aphis citricola van der Goot (12) (sub A. spiraecola Patch.), A. craccivora Koch (18), sub A. medicaginis Anct.), A. gossypii Glover (8), Dactynotus jaceae L. (18), Myzus persicae (Sulzer) (19), T. aurantii (B.d.F.) (13). The first species is undoubtedly the most effective in the transmission of CTV.

Recently has been also ascertained (5) by ELISA test that A. jacobeae Schrank, A. nerii B.d.F., Hyalopterus pruni (Geoffr.) and Macrosiphum rosae (L.) are able to acquire the virus. But, among these nonvector aphid species, only A. nerii has been reported as living on citrus.

In recent years in several citrus–growing areas of the world (3, 4, 17) a change has been observed in transmissionof CTV vectored by A. gossypii. About thirty years ago (9) estimated that in California 5,600 aphids were necessary to have a single infection in the field. In laboratory, with more than seven hundred of transmission tests the same authors reported a very low rate of infection (not exceding 6% of transmission) using a number of aphids per test between 21÷221. Similar low transmission were obtained in Florida by Norman and Grant (13). In more recent studies, carried out in California, Roistacher (14) reported ratio transmission of 100% with A. gossypii for many isolates of seedling yellows–tristeza (SY). In another recent paper Roistacher et al. (16)

305

reported that a small population of A. gossypii can readily transmit some isolates of SY and in one case even a single aphid was able to transmit it.

Therefore the transmission ratio of CTV changes with different isolates, the vectors' efficiency, the different donator or acceptor plant and the environmental conditions (15).

Most aphid vectors of CTV are present in the Italian citrus groves. Only three of the ten citrus aphid species present in Italy have a considerable economic importance; they are: A. citricola, A. gossypii and T. aurantii (1). Such aphid species and A. craccivora have not been able to transmit an isolate of CTV found on sticks of Satsuma mandarin (Citrus unshiu Marc.) imported from Japan (6). The experience of the other countries however suggests that transmissible mutants may arise (2, 17). In view of this possibility, introduced propagation material should be carefully indexed and a continous survey made of the declining citrus trees for CTV. In addition it should be clear the role of the aphid species present in Italy, and this paper reports preliminary results of aphids transmission tests.

2. Materials and Methods

Virus isolates: Four isolates of CTV were used in this study. They included the recently isolates of CTV discovered near Monasterace (Reggio Calabria) (7) coded CTV-GB, CTV-SW, besides a severe isolate described earlier on Satsuma mandarin (6) coded CTV-J, and an isolate already described in Florida (10) and coded CTV-T4.

All the isolates were increased in seedling or small trees of Mexican lime (C. aurantifolia (Christm.) Swingle) and sweet orange (C. sinensis Osbeck), about one or two years old. All isolates produced distintive symptoms on Mexican lime.

Plants: Field samples were collected from 30 years old of Golden Buckeye sweet orange and Satsuma Mandarin type Wase, naturally infected, and growing in Monasterace. For greenhouse tests we used Mexican lime seedlings and small trees of Madame vinous sweet orange, propagated by seeds or by grafting on sour orange (C. aurantium L.) or volkamer lemon (C. volkameriana Pasq.). All these plants were grown in a steam-sterilized potting medium and were fertilized periodically to keep a good growth.

During the test period all the plants were protected by nets, to avoid infestation with aphids or other insects.

3. Experimental procedures

Two different procedures have been used: 1. Aphid colonies collected in the field on Satsuma mandarin and Golden Buckeye sweet orange, tested positive for CTV infection, were assayed for the presence of the virus by ELISA test and feeding on Mexican lime seedlings. The following aphid species were test: A. craccivora, A. fabae Scop., A. gossypii, Aulacortum solani (Kalt.), Macrosiphum euphorbiae (Thomas) and Myzus persicae. The colonies, collected on young shoots, were put in a plastic bag refrigerated, and then transferred to the laboratory. After 24 hrs a sample of 100÷120 specimens (adult apterous and 4th instar nymphs), with the help of a brush, were transferred to a pot and homogenized in a phosphate buffered saline solution, pH 7.2, additioned with polyvinyl-pyrrolidone (1%) and tested by ELISA method. Other groups of the same colonies were transferred to Mexican lime seedlings and the aphids allowed the citrus plants to colonize them for 24 hrs under a cage.

2. In other experiments the following aphid species were collected in field: A. citricola, A. gossypii, M. persicae, T. aurantii and A. fabae. The first four aphid species were collected in different citrus trees near Catania, while A. fabae were collected on broad bean crops, growing in the same area. The young shoots of citrus trees, infested with one aphid species, were placed on young leaves of donor Mexican lime or Madame vinous sweet orange inoculated by one isolate of CTV. After a 24 hrs acquisition period, aphids were removed. The young leaves of the acquisition host that were covered with feeding aphids were cut and attached to the young leaves of the acceptor host (Mexican lime). The specimen number of A. citricola ranged from 15-140, the A. gossypii from 15-115 the M. persicae from 20-108, the T. aurantium from 75-125 and the A. fabae from 30-118 for each test. After an inoculation feeding of 24 hrs, plants were sprayed with an aphicide (pirimicarb). Mexican lime seedling were inspected every 7 days for CTV symptoms and tested by ELISA (7), during one year, every month for CTV infections. All transmission tests were carried out in May, June, July, September and October in an isolated greenhouse, where temperature ranged from 22-32°C (in some cases 38C were reached) and relative humidity from 30% to 100%.

4. Results and discussion

The different aphid species collected from citrus plants carrying CTV resulted negative when tested by ELISA method and were not able to transmit the virus to Mexican lime seedlings. In the transmission trials carried out with aphids collected from CTV free citrus and non citrus hosts only A. citricola was able to transmit CTV-T4 isolate. Table 1 shows the results of transmission rates for the four CTV isolates and the five species of aphids tested. The percentage of aphids found alive after the feeding period in our greenhouse were rather low.

These studies indicate that a very small population of A. citricola could give a significant threat to the citrus groves in Italy. It is also convenient to note that the results may have been mostly negative owing to the conditions in which the trials were carried out.

In our positive transmission only a few specimens were able to transmit CTV-T4 isolate. Other researches have transmitted CTV with a small population of A. gossypii (16).

In conclusion CTV establishes a heavy menace for the italian citrus groves and for other Mediterranean areas, since:

1) T. citricidus can be accidentally introduced into the Mediterranean basin.

307

Table I - Results of transmission trials of five aphid species and four CTV isolates.

Aphids	N. infected plants/ N. inoculated with CTV isolates				
	CTV-T4	CTV-GB	CTV-SW	CTV-J	TOTAL
A. citricola	3/23	0/10	0/12	0/23	3/68
A. fabae	0/ 9	-	-	0/1	0/10
A. gossypii	0/ 9	-	-	-	0/ 9
M. persicae	0/ 6	-	-	-	0/ 6
T. aurantii	0/19	0/5	0/8	-	0/32
TOTAL	3/66	0/15	0/20	0/24	

2) The actual populations of citrus aphids present in Italy may differentiate new clones more effective in the transmission of CTV.
3) From the isolates of CTV present in Italy may arise mutants transmissible by our aphid populations.
4) More virulent CTV isolates including the seedling yellows component, easily transmissible from different aphid species, can be introduced accidentally in Italy.

REFERENCES

1. BARGABALLO, S. and PATTI, I. (1983). Citrus aphids and their entomophagous in Italy. In: Aphid antagonists, Balkema, Rotterdam, 116-119.
2. BAR-JOSEPH, M. (1978). Cross protection incompleteness: a possible cause for natural spread of citrus tristeza virus after a prolonged lag period in Israel. Phytopathology, vol. 68: 1110-1111.
3. BAR-JOSEPH, M. and LOEBENSTEIN, G. (1973). Effects of strain, source plant, and temperature on the transmissibility of citrus tristeza virus by melon aphid. Phytopathology, vol. 63: 716-720.
4. BAR-JOSEPH, M., ROISTACHER, C.N. and GARNSEY, S.M. (1982). The epidemiology and control of citrus tristeza disease. In: Plant Virus Disease Epidemiology, Blackwell Scientific Publications, Oxford, 368 pp.
5. CAMBRA, M., HERMOSO, A., MORENO, P. and NAVARRO, L. (1982). Use of enzyme-linked immunosorbent assay (ELISA) for detection of citrus tristeza virus (CTV) in different aphid species. In: Proc. Int. Soc. Citriculture, 1981, vol. 1: 444-448.

6. CARTIA, G., BARBAGALLO, S. and CATARA, A. (1980). Lack of spread of citrus tristeza virus by aphids in Sicily, p. 88–90. In: Proc. 8th Conf. IOCV, Univ. California, Riverside.

7. DAVINO, M., CATARA, A., RUSSO, F., TERRANOVA, G. and CARBONE, G. (1984). A survey for citrus tristeza virus in Italy by the use of enzyme–linked immunosorbent assay, p. 66–69. In: Proc. 9th Conf. IOCV, Univ. California, Riverside.

8. DICKSON R.C., FLOCK, R.A. and JOHNSON, M.M. (1951). Insect transmission of citrus quick–decline virus. J. econ. Ent., vol. 44: 172–176.

9. DICKSON, R.C., JOHNSON, M.M., FLOCK, R.A. and LAIRD, E.F. Jr. (1956). Flying aphid populations in southern California citrus groves and their relation to the transmission of the tristeza virus. Phytopathology, vol. 46: 204–210.

10. GONSALVES, D., PURCIFULL, D.E. and GARNSEY, S.M. (1978). Purification and serology of citrus tristeza virus. Phytopathology, vol. 68: 553–559.

11. MENEGHINI, M. (1946). Sôbre a natureza e transmissibilidade de doenza "tristeza" dos citros. O Biologico, vol. 12: 285–287.

12. NORMAN, P.A. and GRANT, T.J. (1954). Preliminary studies on aphid transmission of tristeza virus in Florida. The Citrus Industry, vol. 35: 10–12.

13. NORMAN, P.A. and GRANT, T.J. (1956). Transmission of tristeza virus by aphids in Florida. Proc. Fla. Hort. Soc., vol. 69: 38–42.

14. ROISTACHER, C.N. (1982). A blueprint for disaster. III. The destructive potential for seedling yellows. Citrograph., vol. 67: 48–53.

15. ROISTACHER, C.N. and BAR-JOSEPH, M. (1984). Transmission of tristeza and seedling yellows–tristeza by Aphis gossypii from sweet orange, grapefruit and lemon to Mexican lime, grapefruit and lemon, p. 9–18. In: Proc. 9th Conf. IOCV, Univ. California, Riverside.

16. ROISTACHER, C.N., BAR-JOSEPH, M. and GUMPF, D.J. (1984). Transmission of tristeza and seedling yellows tristeza virus by small populations of Aphis gossypii. Plant Disease, vol. 68: 494–496.

17. ROISTACHER, C.N., NAUER E.M., KISHABA, A. and CALAVAN, E.C. (1980). Transmission of citrus tristeza virus by Aphis gossypii reflecting changes in virus transmissibility in California, p. 76–82. In: Proc. 8th Conf. IOCV, Univ. California, Riverside.

18. VARMA, P.M., RAO, D.G. and CAPOOR, S.P. (1965). Transmission of tristeza virus by Aphis craccivora (Koch) and Dactynotus jaceae (L.). Ind. J. Ent., vol. 27: 67–71.

19. VARMA P.M., RAO, D.G. and VASUDEVA, R.S. (1960). Additional vectors of tristeza disease of citrus in India. Curr. Sci., vol. 29: 359.

Biological and electrophoretic characterization of some isolates of tristeza virus complex found in Italy

M.Davino & R.La Rosa

Plant Pathology Institute, University of Catania, Italy

Summary

Two isolates of Citrus tristeza virus recently discovered in Calabria (CTV-GB and CTV-SW) and one from Japan (CTV-J) were inoculated under safe conditions into indicator plants.

CTV-J induced a remarkable reduction in growth on grapefruit and inclusion bodies in all the inoculated species. CTV-GB and CTV-SW gave a moderate stunting and a mild pitting on the inner face of bark of sour orange rootstock as showed by the donor infected trees in the field.

Polyacrylamide gel electrophoresis of dsRNA extracted from CTV-infected tissues revealed a similar pattern for the three isolates and one from Florida (CTV-T4). None of them showed the specific band associated to seedling yellow component.

1. Introduction

Tristeza is a citrus disease widespread in almost all citrus areas of the world including Italy. It is caused by a virus which induces three main sindromes: "tristeza", "stem pitting" and "seedling yellows" referable to different components (7, 8, 9).

The name tristeza indicates the syndrome which is responsible for the incompatibility of sweet orange (Citrus sinensis Osb.) and mandarin (C. reticulata Blanco) grafted to sour orange (C. aurantium L.). It causes the necrosis of the phloem sieve cells just below the graft line and the decline of the plant which die.

Stem pitting is easily found in Mexican lime (C. aurantifolia (Christm.) Swing) and grapefruit (C. paradisi Macf.); it also affects other citrus scions and rootstocks such as Rangpur lime (C. limonia Osb.), rough lemon (C. jambhiri Lush.) and citrange (P. trifoliata Raf x C. sinensis). It may be accompanied by a slow decline of the plants.

Seedling yellows (SY) is the most destructive syndrome of the three ones and it is associated with a characteristic dsRNA electrophoretic pattern (4). Its transmissibility by aphids is high and the destructiveness is impressive (10). Therefore, it is very important to characterize the isolates present in an area in order to evaluate their transmissibility and their potential effect on the yield. That can be done by inoculation on test plants (5, 6) and by polyacrylamide gel electrophoresis of dsRNA (4).

The recent discovery of CTV infections in some italian citrus groves suggested to characterize these isolate by indexing on biological indicators and by definition of the electrophoretic pattern, that is the object of this report.

2. Material and Methods

Virus isolates. Three CTV isolates have been tested: two of them (CTV-GB and CTV-SW) were found on sweet orange Golden Buckeye and Satsuma mandarin (C. unshiu Marc.) Wase type (2) and the third (CTV-J) on budsticks introduced by chance from Japan (1). A fourth isolate from Florida was used (CTV-T4) in some studies.

Citrus hosts. All the isolates were inoculated into seedlings of sour orange, Femminello S. Teresa lemon (C. lemon (L) Burm.) and Mexican lime as well as into Madam Vinous sweet orange and Marsh seedless grapefruit grafted on sour orange. Plants were cultivated in containers on a medium sterilized by moist temperature and fertilized to assure right nourishing levels; adequate precautions were used in order to prevent mechanical contamination (11).

Ten young plants per each isolate were inoculated by bark patch (two inocula per plant) and as many, not inoculated, were used as control. During the testing period the plants were kept in a thermoconditioned room (T25-26C and U.R. 60-80%) and protected with thin-mesh nets. The plants and the nets were periodically sprayed with an aphicide (Pirimicarb). The plants were examined every ten days and after six months they were pulled out to ascertain the collar diameter of the plants, the height, the weight of roots and canopy, any symptom on wood and bark.

Inclusion observations. Observation of inclusion in petioles of inoculated plants was performed by light microscopy on sections stained with azure A as previous done (3).

Extraction and PAGE of dsRNA. The dsRNA was isolated from bark peeled from 10-20 cm long twigs, collected from plants grown in greenhouse. Extraction was performed by TSE buffer containing 3% sodium dodecyl sulfate (SDS) and 0.5% 2-mercaptoethanol, 0.5 mg of bentonite and 0.7 ml of 2XTSE buffer-saturated phenol. After mixing 0.7 ml of chloroform and isopropyl alcohol (24:1) was added. It was mixed well and allowed to stand for 20 minutes at 4°C, with occasional mixing.

After centrifugation (5,000 g for 15 min.) the acqueous phase was mixed with 50 mg of cellulose powder (Whatman CF-11), 95% ethanol was added and the mixture was shaken for 15-20 minutes and centrifugated at 5,000 g for 5 minutes at 4C. This cellulose was washed three times or more with 16.5% ethanol in TSE buffer and once with TSE buffer.

The cellulose powder was then resuspended with 500 ul/g of TSE buffer and the supernatant was recovered. dsRNA were precipated by sodium acetate pH 5.5 and 95% ethanol after storage at -20°C and recovered after centrifugation for 15 min at 5,000 g.

The pellet was resuspended in 2.5% bromophenol blue tracer dye in 12%

sucrose solution in electrophoresis buffer (90 mMTris, 90 mM borate, 3mM EDTA, pH 8,3).

The samples (10 ul) were electrophoresed in a 5% polyacrilamide gel slab buffered in 1:10 TBE for 4 hours at 100V and the slab was stained in ethidium bromide (100 ng/ml) for 10 minutes in electrophoresis buffer or, alternatively, with silver stain (Bio Rad, Richmond, California).

3. Results

Regardless of CTV isolate the lime seedlings showed the characteristic vein clearing of young leaves after nearly one month and later on vein corking and stem pitting. Those inoculated with CTV-J began to decline after four months. Some time after also the sour orange seedlings, inoculated with CTV-J and CTV-GB isolates, showed some phylloptosis followed by twig wilting beginning from the top. Like symptoms appeared on lemon seedlings some time later. The grapefruit seedlings which had been inoculated with CTV-J and CTV-GB isolates showed a remarkable decline and appeared to be much smaller than those inoculated with CTV-SW and the control plants. Sweet oranges inoculated with CTV-J appeared smaller than those inoculated with CTV-SW and CTV-GB.

CTV-J reduced significantly all the examined parameters in the indicator plants (Table 1). CTV-GB induced a reduction in growth and caused very small pegs of the wood connected with pits of the inner bark of sour orange rootstock. CTV-SW isolate showed a similar behaviour but a smaller difference.

None of three tested isolates induced yellowing of the leaves.

Only CTV-J induced inclusions in the petioles of indicator plants.

The electrophoretic pattern of dsRNA of citrus extracts was similar to that reported for CTV by Dodds et al. (4) whitout any difference among the three isolates. No dsRNAs band specifically associated with SY component was found.

4. Conclusions

The indexing on indicator plants pointed out that among the different CTV isolates found in Italy during the latest years the most virulent one is that introduced by chance from Japan.

The other two isolates reduced – especially CTV-GB – the test plant growth. The same behaviour was observed on the field affected trees; in fact, the plants of Golden Buckeye showed small pegs of the wood connected with a pitting of the bark cambial face on sour orange while these symptoms were less remarkable on Satsuma type Wase mandarin plants (2) even if grafted on sour orange. These results let us refer the tested isolates to the tristeza component 'sensu lato".

Since no yellowing has been observed on leaves of indicator plants we may assume that seedling yellows component was not present in the CTV isolates found in Italy. This assumption is also supported by the electrophoretic pattern, which actually is the only physical method to discriminate CTV isolates (4).

313

Tab. 1 – Effects of some CTV isolates on various citrus species and cultivars inoculated in greenhouse.

Species and CTV isolate	Collar (cm)	Height (cm)	Weight (g)	
			roots	canopy
Sour orange				
CTV-J	0,71 Bc	47,20 Cc	3,90 Dd	12,06 Cd
CTV-SW	0,80 Bb	66,80 ABab	8,62 ABab	24,62 ABab
CTV-GB	0,74 Bbc	53,16 Cc	5,82 Cc	14,36 Cc
Control plant	0,90 Aa	68,20 Aa	9,00 Aa	26,20 Aa
Madam Vinous sweet orange				
CTV-J	0,84 Bb	53,10 Cd	10,34 Ab	13,34 Dd
CTV-SW	0,98 ABab	71,80 ABab	12,62 Aab	30,44 ABab
CTV-GB	0,99 ABab	61,50 BCc	13,00 Aab	20,94 Cc
Control plant	1,10 Aa	73,20 Aa	14,20 Aa	35,20 Aa
Femminello S. Teresa lemon				
CTV-J	0,79 Dd	51,70 Cc	9,36 Cc	13,58 Cd
CTV-SW	1,01 Bb	82,80 ABab	13,84 ABab	31,66 ABab
CTV-GB	0,95 BCbc	63,20 Cc	7,46 Cc	22,04 Cc
Control plant	1,19 Aa	86,20 Aa	15,80 Aa	35,20 Aa
Marsh seedless grapefruit				
CTV-J	0,92 Bc	48,30 Cd	6,80 Dd	11,46 Dd
CTV-SW	0,95 Bbc	62,60 ABab	14,36 ACbc	35,76 ABab
CTV-GB	1,04 Bb	57,40 BCbc	15,82 ABab	25,28 Cc
Control plant	1,18 Aa	68,20 Aa	18,20 Aa	42,00 Aa

The reported values are the average per ten repetitions. They were elabo-
rated with analysis of variance and compared with Duncan test. The same
lettered values are not statistically different. Capital letters: P=0,01.
Small letters: P=0,05.

REFERENCES

1. CARTIA, G., BARBAGALLO, S., and CATARA, A. (1980) – Lack of spread of
   citrus tristeza virus by aphids in Sicily, p. 88–90. In: Proc. 8th
   Conf. I.O.C.V., Univ. California Riverside.
2. DAVINO, M., CATARA, A., RUSSO, F., TERRANOVA, G., and CARBONE, G.
   (1984). A survey for citrus tristeza virus in Italy by the use of
   enzyme-linked immunosorbent assay, p. 66–69. In: Proc. 9th Conf. IOCV,
   Univ. California, Riverside.

3. DAVINO, M., PAVONE, P. and LA ROSA, R. (1985). Osservazioni sulla tecnica delle inclusioni cellulari nella diagnosi della tristezza degli agrumi. Riv. Pat. Veg., S. IV (in press).
4. DODDS, J.A., TAMAKI, S.J., and ROISTACHER, C.N. (1984). Indexing of citrus tristeza virus double-stranded RNA in field trees. p. 327-329. In: Proc. of 9th Conf. IOCV, Univ. California, Riverside.
5. FRASER, L. (1957). The relation of seedling yellows to tristeza. p. 57-62. In: Citrus virus diseases, Univ. California, Riverside.
6. Mc CLEAN, A.P.D. (1974) - The tristeza virus complex, p. 59-66. In: Proc. 6th Conf. I.O.C.V., Univ. California Divison Agricoltural Sciences.
7. ROISTACHER, C.N. (1981) - Blueprint for disaster. Part. one: The history of seedling yellows disease. Citrograph., 67, 46.
8. ROISTACHER, C.N. (1981) - Blueprint for disaster. Part. two: Recent changes in trasmissibility of seedling yellows. Citrograph., 67, 28-32.
9. ROISTACHER, C.N. (1982) - Blueprint fo disaster. Part. three: The destructive potential for seedling yellows. Citrograph., 67, 3, 48-53.
10. ROISTACHER, C.N., BAR-JOSEPH, M., and GUMPF, D.J. (1984) - Transmission of tristeza and seedling yellows tristeza virus by small populations of Aphis gossypii. Plant. Disease, 68, 494-496.
11. ROISTACHER, C.N., NAUER E.M., and VAGNER, L.W. (1980) - Transmissibility of cachexia, dweet mottle, psorosis, tatterleaf and infectious variegation viruses on knife blades and its prevention, p. 225-229. In: Proc. 8th Conf. I.O.C.V., Univ., California, Riverside.
12. SCHNEIDER, H. (1959). The anatomy of tristeza-virus-infected citrus, p. 73-84. In: Proc. Conf. Citrus Virus Disease, 1957, Univ. California, Div. Agr. Sci.

Symptoms of xyloporosis disease on 'Mapo' tangelo

G.Terranova & A.Caruso
Citrus Experimental Institute, Acireale, Italy

Summary

Field observations of declining trees of "Mapo" tangelo, hybrid of "Avana" mandarin (Citrus deliciosa Tenore) and "Duncan" grapefruit (Citrus paradisi Macf.) topworked on "Comune" clementine having as rootstock the sour orange, allow the description of the symptomatology shown by this hybrid when infected with xyloporosis viroid.

Both from the biological assays and from the symtomatology, it was established that the trees are infected with exocortis, xyloporosis, concave gum and cristacortis.

The symptoms of cristacortis and of concave gum are typical of these diseases.

The exocortis does not show any symptom in the infected trees as a result of the behaviour of the parents of "Mapo" tangelo regarding this viroid.

The xyloporosis in the intermediate stock of "Comune" clementine shows the typical symptomatology with pits on the bark correspondent to pegs on the wood, whereas on "Mapo" tangelo it causes a particular discolouration of the wood, enlarging more than in other species.

1.1 INTRODUCTION

"Mapo" tangelo hybrids of "Avana" mandarin (Citrus deliciosa Tenore) and of "Duncan" grapefruit (Citrus paradisi Macf.) were obtained in 1950 at the Citrus Experimental Institute of Acireale and released for cultivation from 1972 (2).

For its good characteristics such as earliness, juiciness and taste it has interested the growers and is so esteemed that at present its cultivation reaches already 300 ha, comprehensive of new plantings and regraftings.

Unfortunately the regraftings, especially at first, were done by growers who did not know the susceptibility of this new hybrid to viruses, viroids and mycoplasmas.

At the same time little importance was given to the presence of these agents on the tree on which topworking was done.

Thus, on a certain number of declining trees of "Mapo" tangelo showing a particular symptomatology, observations and biological assays were done with the aim to evidentiate the ethiology.

1

Figure 1 - "Mapo" tangelo tree showing yellowing and defoliation.

Figure 2 - "Mapo" tangelo tree showing same symptoms as in the first
picture, but in a more advanced stage.

The trees resulting were originally formed by the sour orange as
rootstock and by "Comune" mandarin as scion. When 15 years old the trees
were topworked and grafted with budsticks of "Mapo" tangelo.

In some trees an interstock of clementine was left, obtaining in this
way a three-membered plant.

From the regrafting operation to the time when the observations were
made a period of 4 years passed.

1.2 DESCRIPTION OF SYMPTOMS

The greater part of the trees show yellowing of foliage, defoliation
and wilting which at first is limited to the twigs (Fig. 1), but
progressively extends to branches of ever increasing diameter (Fig. 2) with
the whole branch and sometimes all the canopy involved above the bud-union
point.

The branches and twigs up to a certain diameter show numerous cracks
with the margins of the bark wilted and raised (Fig. 3).

Sometimes on the branches of a certain diameter can be observed
vertical depressions more or less sunken and wide. By removing the bark in
correspondence to these vertical depressions one can observe pitting of the

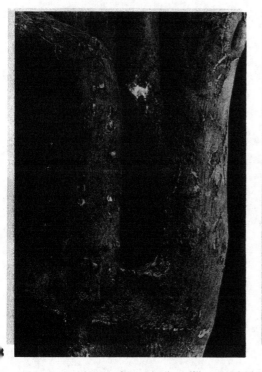

Figure 3 – Branches of "Mapo" tangelo showing cracks with the margins of
the bark raised.

Figure 4 – Branch of "Mapo" tangelo showing the vertical depressions
typical of cristacortis.

wood and pegging in the cambial surface of the bark (Fig. 4), typical
symptoms of cristacortis (5). By removing the bark at the bud-union point,
prominent areas are observed on the cambial surface of the bark,
alternating with sunken ones.

The same type of alteration can be observed on the wood that presents
underlying veinings of a deep chestnut colour turning towards brown and
attenuating after little time of exposition to the light (Figs. 5 - 6).

The picture extends to the affected branch until 20-30 cm from the bud
union point and generally does not attain the whole circumference of it.

By effecting transversal cutting of the branches, also if distant from
the bud-union point, gum impregnations interesting some rings of wood can
be observed (Fig. 7).

These are typical symptoms of concave gum, described by Vogel and Bové
(6) encountered on oranges of the "Navel" group, and by Terranova on
"Brasiliano" orange (3).

The bark and the wood of the sour orange under the bud-union look
perfectly normal without deformations or particular colourings.

In the cases in which, when operating the topworking a piece of
clementine trunk was left, it can be observed that this shows the typical

319

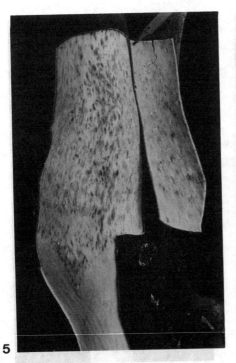

5

Figures 5 and 6 - Bud-union area between "Mapo" tangelo and clementine. When the bark is removed, it is possible to see some little prominent and depressed areas. On the wood of tangelo are present veinings of a deep chestnut colour that are not present on wood of clementine.

symptoms of xyloporosis; i.e. gum impregnations, pits of the bark and little pegs on the wood (Fig. 8) (1,4).

1.3 <u>INDEXING</u>

Some of these trees were indexed with the following indicator plants: "Hamlin" orange, "Etrog 60.13" citron, "Messican" lime, "Parson special" mandarin, and "Orlando" tangelo.

From one year after the innoculation, the citron reacted positively, on which appeared the epinasty, and the "Hamlin" orange that showed the typical "oak leaf" pattern on young leaves, whereas no symptom appeared on other indicator plants.

Regarding the "Parson special" mandarin and the "Orlando" tangelo, the fact that they did not show symptoms could be ascribed to the brief interval from the innoculation.

However, the clementime on which the "Orlando" tangelo is grafted shows the typical symptoms of xyloporosis. Thus, the trees are infected with this disease.

The same is true for cristacortis.

 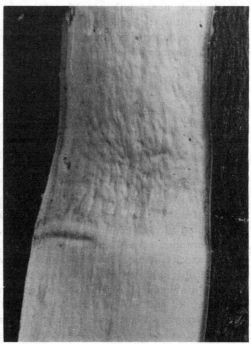

Figure 7 - Transversal cutting of a branch of "Mapo" tangelo showing gum impregnation in some rings of wood, typical symptoms of concave gum.

Figure 8 - Trunk of clementine with typical symptoms of xyloporosis.

1.4 CONCLUSIONS

From the observations made in the field and from the results of the indicator plants, it was possible to ascertain that the trees are infected with exocortis, xyloporosis, concave gum and cristacortis.

Among the various types of symptomatology encountered on the trees, it is believed to distinguish the cracks on the branches and the gumming as caused by the virus of concave gum and the vertical depressions and the pitting of the wood in the twigs as caused by the virus of cristacortis.

No specific symptom is thought to be ascribed to exocortis, also considering the behaviour towards this viroid of the parents of "Mapo" tangelo, i.e. the mandarin and the grapefruit (7).

The xyloporosis shows a typical symptomatology on the intermediate stock clementine, whereas on the tangelo, besides to the pits and the pegs on the bud-union line, causes a particular discolouration of the wood and the interested area extends on the branch more than in other species.

These peculiarities were not encountered in other species and it is likely that they are due to higher sensibility and to a particular reaction of the tangelo.

A last practical consideration is to recommend to the growers that they exercise the greatest prudence in topworking, particularly when species and cultivars are employed whose behaviour, as in the case of "Mapo" tangelo, towards virus and virus-like diseases is not yet well known.

REFERENCES

1. ROISTACHER, C.N. (1983). Cachexia disease: virus or viroid. Citrograph. Vol. 68; 111-113.
2. RUSSO, F., STARRANTINO, A. and REFORGIATO RECUPERO, G. (1977). New promising mandarin and mandarin hybrids. Proc. Int. Sic. Citriculture. Vol.2; 597-601.
3. TERRANOVA,G.(1974-75).La concave gum sull'aranci cv "Brasiliano". Annali dell'Istituto Sperimentale per l'Agrumicoltura. Vol.VII-VIII; 95-100.
4. TRIBULATO, E., CARTIA, G., CATARA, A. and CONTINELLA, G. (1980). Performance of a Clementine mandarin with cachexia-xyloporosis on eleven rootstocks. Proc. 8th Conf.IOCV, Univ. California, Riverside, 232-238.
5. VOGEL,R. (1975). Cristacortis. Descrip. & illustr. of virus and virus-like diseases of citrus: a collection of colour slides. IFAC, Paris, 4.
6. VOGEL, R. and BOVE', J.M. (1976). Effect of various concave gum isolates on mandarin and sweet orange trees: absence of correlation between reduction of growth and severity of symptom expression. Proc. 7th Conf. IOCV, Univ.California, Riverside; 119-124.
7. WEATHERS,L.G.(1975). Exocortis. Descrip. & illustr. of virus and virus-like disease of citrus: a collection of colour slides. IFAC,Paris,27.

Indexing of citrus for exocortis by bidirectional polyacrylamide gel electrophoresis analysis

G.Albanese, R.La Rosa & G.Polizzi

Plant Pathology Institute, University of Catania, Italy

Summary

A bidirectional PAGE technique has been applied to diagnose citrus exocortis viroid (CEV) infections in different species and varieties of citrus growing in greenhouse. Detection of different strains isolates was accomplished from leaves of citron, sweet orange, sour orange, lemon, mandarin and volkameriana lemon. Even samples of 1.2 g were suitable for the diagnosis. Both refrigerated or frozen samples behaved satisfactorily. Young succulent leaves gave better results, but CEV recovery was also possible from mature ones. The described procedure allows to shorten to less than one month the indexing for CEV.

1. Introduction

Citrus exocortis viroid (CEV) is usually detected by graft inoculation of tissues onto Etrog citron (5,10).Though that is a great advance over the test on trifoliate orange (4) and Rangupur lime (11) this method of indexing still has some drawbacks. Therefore attempts have been made to develop alternative methods (1,2,3,6,7,8). Among them polyacrylamide gel electrophoresis (PAGE) procedures (1,2) are very interesting because they have been shown reliable to detect other viroids also in symptomless plants.

In a recent paper, we described a bidirectional PAGE technique on relatively crude nucleic acid extracts of citrus trees (2), which gave very promising results (3). Further experiments, carried out to investigate the possibility to apply the technique for routine diagnosis of CEV infections, are here reported.

2. Materials and Methods

Plant materials. Different donor plants were used (table 1): budlings of Etrog citron, lemon, sweet orange and mandarin, seedlings of sour orange and volkameriana lemon. Inoculation was carried out with different strains of CEV by slash or bark patch or by propagation of infected buds. The plants were indexed for exocortis on Etrog citron S-1 (10) and for other virus and virus like agents on standard test plants (9). Some of them were also affected by psorosis,crinkly leaf-infectious variegation, concave gum, cristacortis, cachexia-xyloporosis, impietratura. Uninoculated (healthy) plants were assayed as checks. Samples of greenhouse (22-34C) or field growing plants were collected at various times from spring to summer. Ten grams were routinely tested but 1 to 30 grams were used in some tests. In a separate assay leaves of different age were compared. Leaves collected from single flushes of sour orange seedlings were grouped as expanding, succulent partly expanded,fully expanded (mature) and tested. In few attempts old hardened leaves were also processed.

Electrophoresis. Samples either freshly collected or stored at -20 C, were frozen in liquid nitrogen and ground to a powder with mortar and pestle in the presence of 4 vol (v:w) of 0.2 M Tris, 0.1 M NaCl, 10 mM ethylenediaminotetraacetate (disodium salt) (EDTA), 2% (w:v) polyvinyl pyrrolidone (PVP), 1% (w:v) sodium dodecyl sulphate (SDS), pH 8.5 - 9.5. The slurry was homogenized with 1 vol of water-satured phenol containing 0.1% (w:v) 8 - hydroxyquinoline and 1 vol of a 25:1 mixture of chloroform : isoamyl alcohol and processed as previously reported (2). The Li-soluble fractions were recovered and electrophoresed in 5% polyacrylamide gel slabs buffered in 90 mM Tris, 90 mM borate, 3 mM EDTA, pH 8.3. The zone of the gel containing the CEV RNA was reelectrophoresed in the upwards direction under denaturing conditions (8 M urea, 225 V, 50 C) (2).Each sample was electrophoresed at least twice. Chrysanthenum stunt viroid (CSV) and coconut cadang cadang viroid (CCCV) were used as markers.

Stains. Three different stains were used to develop the slabs after electrophoresis: o-toluidine blue(0.1 %) in water,silver stain (Biorad Richmond California), ethidium bromide (0.5 ug/ml) in buffer electrophoresis . After staining in ethidium bromide the slabs were examined on a UV light box and photographed on 52 Polaroid film with UV and orange filters.

3. Results

In the first direction (non denaturing) gel, the only visible bands (where viroids are expected to migrate) were those of CCCV and CSV only, whereas citrus extracts showed a high background, which impaired the visualization of CEV band. After the second run viroid bands were also shown in citrus extracts, including those of asymptomatic plants (Fig.1). The mobility of CEV was always very close to CSV and slower than CCCV. Repeated assays showed that in the first direction viroids migrated between 5.5 and 8.5 cm. from the loading line.
All the samples of CEV infected citron and almost all other donor plants gave a viroid band. Etrog citron, lemon, sour orange and volkameriana lemon allowed the detection of CEV even in relative small samples (table 1).

In the assays with leaves of different age CEV was always detected in expanding leaves,succulent partly expanded leaves

324

TAB. 1 - RESULTS OF PAGE TESTS APPLIED TO LEAF EXTRACTS
         OF DIFFERENT DONOR PLANTS GROWN IN GREENHOUSE

DONOR PLANT	TYPE OF INOCULATION (x)	SAMPLE WEIGHT	No POSITIVE SAMPLE (y)
CITRON ETROG	s	1-30	15/15
	b	3-18	13/13
	h	6-18	0/14
SWEET ORANGE			
Madam vinous	b	2.5-30	5/10
Vaniglia apireno	bp	8.7	0/1
Sanguinello moscato	bp	10	0/2
Ovale	b	8	0/1
Golden buckeye	b	8	1/1
SOUR ORANGE	s	2	1/1
	b	4-30	8/8
	h	4-30	0/5
MANDARIN			
Parson special	b	10	1/2
Tardivo di Ciaculli	bp	10	0/1
LEMON			
Femminello(many clones)	bp	4-10	5/6
Interdonato	bp	10	2/2
VOLKAMERIANA LEMON	b	2-10	4/5
	h	10	0/2

x s= slash; bp= bud propagation; b= bark; h= healthy
y Results were considered positive if coincident with
   citron Etrog.

and mature leaves. Occasional extraction from old hardened
leaves gave very poor results showing that this tissue is less
suitable.
After the first run staining with o- toluidine blue did not
revealed any viroid band in the extracts, though CSV and CCCV
markers were easily detected. Silver stain developed slabs
showed many bands: CCCV and CSV were easily detected whereas
CEV was very faint. Ethidium bromide allowed to detected CCCV
and CSV but less consistent results were obtained with CEV.
After the second direction, the best detection of CEV was
again obtained with silver stain and ethidium bromide.

Positive results were also obtained after sample storage for
four weeks in the freezer. Longer intervals were not tested.

Extracts from field growing plants gave a dark background and
rather poor results as in previous assays (2,3).

325

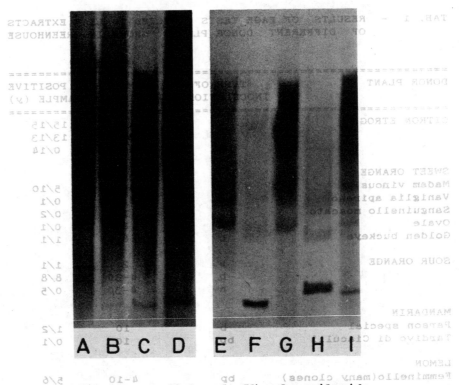

Fig. 1. Slabs of 5% polyacrilamide
gel loaded with crude extracts of
citrus leaves, electrophoresed in
upwards direction under denaturating
conditions and stained with silver
stain. Extracts from CEV infected sour
orange (B), lemon (C), sweet orange
(D), citron (F) and mandarin (I) show
a band with a mobility close to coconut
cadang cadang viroid marker (H). A,E,G
healthy leaves of citron.

4. Discussion

The described PAGE procedure allowed a reliable detection of
CEV from samples as small as 1 g and from different species. It
let to process many samples at once in a short time. A two
person team is able to process 10-20 samples in 2-3 days.
Since the detection by such a procedure depends on
isolation and visualitation of the viroid RNA and not on
biological symptoms, it is possible to detect mild CEV
isolates which are difficult to detect by biological indexing.
In fact, these isolates may take several months to induce
symptoms which sometimes appear difficult to identify. Results
show also that the bidirectional PAGE technique we used

is more sensitive than the conventional PAGE (12) used for CEV (1) though somewhat slower. Therefore this bidirectional procedure provides faster detection than indexing based on test plants. It may be applied to new leaves of citron plants shortly after the conventional inoculation of bark from candidate trees or directly to new flushes of bud propagation obtained in warm greenhouses. That allows to shorten to less than one month the indexing for CEV and to increase the sensitivity. Samples may be processed fresh or stored frozen. Moreover, the method appears particularly suitable for special purposes as a quarantine programs, indexing of small plants obtained by shoot tip grafting, separation of CEV strains from other citrus viruses. Further studies are in progress to improve the extraction procedure in order to make reliable the test also from field taken samples.

Acknowledgments

The excellent technical assistance of Miss Fausta Domina is acknowledged.

REFERENCES

1. BAKSH N., LEE R.F. and GARSNEY S.M. (1983). Detection of citrus exocortis viroid by polyacrylamide gel electrophoresis,p. 343-352. Proc. 9th Conf. IOCV, IOCV Univ. California, Riverside.

2. BOCCARDO G., LA ROSA ROSA and CATARA A.(1983).Detection of citrus exocortis viroid by polyacrilamide gel electrophoresis of glasshouse citrus nucleic acid extracts, p.357-361. Proc.9th Conf. IOCV, IOCV Univ. California, Riverside.

3. BOCCARDO G., ALBANESE G., LA ROSA R. and CATARA A. (1984). " PAGE " detection of exocortis viroid in citrus, p. 382. Proc. 6th Congr. Med. Phytopath. Union.

4. CALAVAN E.C. (1968). Exocortis, p.28-34. Indexing procedures for 15 virus diseases of citrus trees. U.S. Dep.Agric. Handb. 333.

5. CALAVAN E.C., FROLICH E.F., CARPENTER J.B.,ROISTACHER C.N. and CHRISTIANSEN D.W. (1965). Rapid detection of exocortis in citrus. California Agriculture 19: 8-10.

6. CATARA A. and CARTIA G.(1971).Possibilita' d'impiego di un metodo cromatografico per la diagnosi delle infezioni di "exocortite" degli agrumi in materiale siciliano. Riv. Pat. Veg., S. IV, 7: 173 -180.

7. CHILDS J.F.L. (1968). Color test for exocortis, p. 83-86. Indexing procedures for 15 virus diseases of citrus trees.U.S.DEP. Agric. Hand. 333.

8. FUDL-ALLAH A.E.-S.A., SIMS J.J. and CALAVAN E.C. (1974). Indexing of exocortis virus-infected citron by using thin-layer chromatography. Plant Dis. Reptr. 58: 82-85.

9. ROISTACHER C.N. (1976). Detection of citrus viruses by graft trasmission: a review, p. 175-184. Proc. 7th Conf. IOCV, IOCV Univ. California, Riverside.

10. ROISTACHER C.N., CALAVAN E.C., BLUE R.L., NAVARRO L. and R. GONZALES (1977). A new more sensitive citron indicator for detection of mild isolates of citrus exocortis viroid (CEV). Plant Dis. Reptr. 61: 135-139.

11. SALIBE A.A. and MOREIRA S. (1965). Strains of exocortis virus, p. 108-112. Proc. 3rd Conf. IOCV, Univ. Florida Press, Gainesville.

12. SCHUMACHER J.J., RANDLES J.W. and REISNER D. (1983). Viroid and virusoid detection : an electrophoretic technique with the sensitivity of molecular hibridization. Anal. Biochem. 135: 288-295.

Susceptibility of Italian cultivars of lemon to membranosis and influence of some mineral elements*

P.Bertolini, M.Maccaferri & G.C.Pratella
CRIOF, Department for the Protection and Improvement of Agricultural Produce, University of Bologna, Italy

E.Di Martino & G.Lanza
Citrus Experimental Institute, Acireale, Italy

Summary

Lemons of the Femminello comune, Femminello Continella, Femminello zagara bianca, Interdonato, Lisbon and Monachello varieties grown in Sicily were stored at 12°C for 60 days. The Femminello Continella and Monachello varieties showed some resistance to membranosis, where as the other varieties proved highly susceptible. The disease spread most rapidly in the first 30 days of storage. An analysis of the mineral composition of the membranes showed no direct relationship between it and susceptibility to the disease. The chemical and organoleptic properties of the juice were not affected by the progress of the disorder.

INTRODUCTION

Like other species of plant, particularly those grown in tropical and sub-tropical climates, lemons in cold storage are subject to membranosis a physiological disorder caused by chilling injury. The disorder exists in most lemon-growing areas. It manifests itself at temperatures below 15°C, hence this species is not suitable for cold storage at low temperatures (1, 3, 4, 6, 8, 11, and 14). The symptoms are irregularly-shaped blotches, initially appearing as dots and then gradually spreading over the carpellar membranes. They are reddish-brown at first and later turn dark brown and - in the worst cases - affect most of the membrane (3, 4, 6, 8 and 14).

Like other chilling injury of citrus it is favoured by temperatures of between 2°C and 5°C and is slowed by temperatures of more than 10°C or in the region of 0°C. In most cases, there was a lower incidence of the discoloration in fruit stored at 0°C for one or two weeks and subsequently at 10°C or 15°C, than in fruit stored continuously at 10°C to 15°C (4). Waxing tends to reduce the injury (4). The part played by carbon dioxide is not clear: when it is present in large quantities it curbs

_____

*Research conducted as part of the Ministry of Agriculture and Forestry project "Development and improvement of industrial fruit-growing, early fruit growing and citrus fruit growing", Publication no. 87.

329

Table 1: (1982) Incidence of membranosis before and during cold storage at 12°C (% of lemons affected).

No. of days in cold storage Varieties	0	30	60
Femminello comune	2.0 aB[1]	21.6 aA	24.2 aA
Interdonato	0.0 aC	5.7 bB	29.8 bA
Monachello	0.0 aA	0.0 cA	0.0 cA

Table 2: (1983) Incidence of membranosis before and during cold storage at 12°C (% of lemons affected).

No. of days in cold storage Varieties	0	30	60
Femminello comune	0 aC[1]	11.7 aB	16.4 aA
Monachello	0 aA	0 bA	0 bA

Table 3: (1984) Incidence of membranosis before and during cold storage at 12°C (% of lemons affected).

No. of days in cold storage Varieties	0	30	60
Femminello comune	4.6 aB[1]	22.9 aA	25.9 bA
Femminello Continella[2]	1.5 aB	6.2 cA	6.7 dA
Femminello zagara bianca	1.5 aC	25.6 aB	33.3 aA
Lisbon	0.0 aC	12.0 bB	18.7 cA
Monachello[2]	0.0 aB	1.6 dAB	3.1 dA

(1) Data followed by same letters indicate not significant differences within columns (lower-case letters) and within lines (upper case letters) by Duncan's multiple range test, 5% level.

(2) Minute dots of membranosis, negligible from the commercial standpoint.

330

injury (4), whereas limited quantities of carbon dioxide and oxygen do not appear to have any effect (1).

Research on lemons grown in Sicily has shown that susceptibility varies with the microclimates of the different areas and throughout the harvesting season. Winter harvest (January and February) proved to by far the most susceptible (11).

The aim of this study was to determine the susceptibility of lemon varieties and selections grown in Sicily to membranosis and its relationship with mineral composition.

MATERIALS AND METHODS

The lemons used for the experiment were grown in Acireale (Catania) by the "Istituto Sperimentale per l'Agrumicoltura", employing the routine cultivation techniques adopted in the area. The investigation covered a three-year period. It was initially limited to three varieties (Tables 1, 2 and 3) and considerably expanded in the last year.

In the first year the lemons were picked on 15 February 1982, in the second year on 13 February 1983 and in the third year on 13 February 1984. The fruits were completely yellow in colour and were selected, calibrated and sampled at the CRIOF in Bologna, to form samples each comprising 60 lemons, 60-65 mm in diameter. They were then treated by dipping in a 0.1% Imazalil solution and put into cold storage at 12°C with r.h. of 87-90% within a week of being picked. Inspections to determine the incidence of membranosis were carried out at the time the lemons were picked and after 30 and 60 days in cold storage. Five samples, each comprising 60 lemons, were examined for the disease, after having been cut in half and squeezed so as to leave only the carpellar membranes in position. Furthermore, in order to determine the effect of membranosis on the qualitative properties of the juice, the juice yield, the soluble solids (ss), the acidity and the pH were determined.

Before each variety was put into cold storage, a sample of 20 lemons was selected, some of the membranes were removed with a knife, and an atomic absorption spectrophotometer was used to determine their content of magnesium, potassium and calcium.

RESULTS

The maximum increase of the disorder on the cultivars tested occurred in the first 30 days of cold storage (Tables 1, 2 and 3). After that, it either stabilized or increased slightly. This finding coincides with the results of previous research we have undertaken (11).

It should be noted that, in the case of some varieties, and particularly Femminello in 1982 and 1984 (Tables 1 and 3), some of the lemons were already affected by membranosis before they were put into cold storage. Although only a short time elapsed between picking and cold storage (approximately a week), the early onset of the disease might be due to non conditioned lorry transport. The incidence and intensity of the symptoms in the varieties studied in 1984 differed; the varieties could be listed in order of their severity: Femminello zagara bianca, Femminello comune, and Lisbon. In the case of the Monachello and Femminello Continella varieties, the figure for the percentage of lemons affected by the discoloration does not reflect the actual picture concerning the symptoms. Whereas the other varieties had numerous discoloured carpellar membranes covering up to a third of their surface, only a few dots of discoloration (1-2 mm in diameter) were to be found on the Monachello and Femminello Continella, and these were negligible from a commercial standpoint.

As stated in previous reports (11), the intensity of the discoloration varies from one year to another, particularly in the case of Femminello comune.

The Interdonato variety was omitted after the first year because it did not keep well (symptoms included softening of the fruit, discoloration of the membranes and the carpellar column, and microbiological disorders).

Many disorders caused by chilling are known to be influenced by the mineral composition (9, 10, 12, 13 and 14). We therefore wished to establish whether the differing degrees of susceptibility of the varieties to membranosis could be related to the mineral composition of the membranes, and particularly the ratio of magnesium + potassium to calcium. Although the analyses which were made (Table 4) highlighted substantial differences among the varieties, no conclusions can be drawn in this connection.

Like other researchers (2), we too noted an increase in the juice yield during cold storage, this being particularly marked in the case of varieties belonging to the Femminello group (Table 5). However, in the case of the varieties affected by membranosis, we did not notice any significant difference in the properties of the juice which was detectable by chemical analysis (ss, acidity, pH) or organoleptic tests. Research carried out in Israel, albeit on different varieties, did however bring to light differences in the quality of the juice of lemons affected by membranosis (5).

Table 4 - 1984: Mineral composition of the membranes (ppm of dry weight)

Minerals Varieties	Mg	K	Ca	Mg + K/Ca
Femminello comune	7.3	55.4	53.3	1.2
Femminello Continella	5.8	55.6	51.2	1.2
Femminello zagara bianca	7.1	67.1	36.5	2.0
Lisbon	5.6	45.1	44.2	1.1
Monachello	7.4	76.5	43.8	1.9

Table 5 - 1984: Analysis of the juice before and during cold storage at 12°C.

No. of days in cold storage	0				30				60			
Varieties	Juice yield	S.S.	Acid.	pH	Juice yield	S.S.	Acid.	pH	Juice yield	S.S.	Acid.	pH
Femminello comune	29.5	7.6	6.2	2.3	32.5	7.7	6.4	2.3	35.8	7.7	6.4	2.3
Femminello Continella	37.8	7.1	5.9	2.3	42.7	7.0	6.3	2.3	46.0	7.2	6.3	2.2
Femminello zagara b.	29.1	7.7	5.9	2.4	33.2	7.7	6.3	2.3	37.6	7.8	6.4	2.2
Lisbon	31.8	8.8	6.8	2.3	32.7	8.8	–	–	35.7	9.1	6.9	2.3
Monachello	25.6	7.2	6.0	2.3	26.2	7.3	6.1	2.3	31.1	7.3	6.2	2.3

CONCLUSIONS

The research showed that there is a varying degree of susceptibility to membranosis in varieties of lemon grown in Italy. For instance,
the Femminello Continella and Monachello varieties have been proved to
have some resistance to the disease. Further research is needed, however,
since, as established by our previous research (11), the weather conditions in the individual years and the microclimates found in the areas
in which lemons are grown in Sicily can have a significant impact on
the occurrence of the disorder.

Although the influence of some mineral elements, and particularly
calcium, magnesium and potassium, on the development of chilling injury
(9, 10 and 12) has been established in the case of other plant species,
research carried out to-date on citrus fruits, including this project,
has not produced conclusive results (11 and 13). Contrary to the findings
of other researchers, we did not detect significant changes in the chemical and organoleptic properties of lemons affected by membranosis.

BIBLIOGRAPHY

1. ARTES CALERO, F., GUILLEN, M.C. and ESCRICHE, A. (1981). Physiological disorders in the storage of lemon fruits. Proc. Int. Soc. Citriculture, Vol. 2; 768-772.
2. BARTHOLOMEW, E.T. (1957). The lemon fruit. Its composition, physiology and products, University of California Press.
3. BERTOLINI, P. (1972). La Membranosi del limone, Notiziario del CRIOF,
   Vol. 3, No. 3; 5-6.
4. BROOKS, C. and McCOLLOCH, L.P. (1937). Some effects of storage conditions on certain diseases of lemons. J. of Agricultural Res., Vol.
   55, No. 11; 795-809.

5. COHEN, E., SHUALI, M. & SHALOM, Y. (1983). The use of intermittent warming to prevent development of chilling injury in lemons in cold storage. Inst. for Tech. and Storage of Agr. Products. Special pub. No. 216; 69.

6. GOIDANICH, G. and PUPILLO, M. (1951). La Membranosi dei frutti di limone. Rivista della Ortofrutticoltura Italiana. Vol. XXXV, No. 3-4; 1-6.

7. GRIERSON, W. (1981). Physiological disorders of citrus fruits. Proc. Int. Soc. Citriculture. Vol. 2; 764-767.

8. FAWCETT, H.S. (1936). Citrus diseases and their control. McGraw-Hill Inc.

9. LYONS, J.M. (1973). Chilling injury in plants. Ann. Rev. Plant Physiol., 24; 445-466.

10. PERRING, M.A. (1968). The mineral composition of apples: the relationship between fruit composition and storage disorders. J. Sci. Food Agric.19; 186-191.

11. PRATELLA, G.C. and BERTOLINI, P. (1976). Indagine sulla eziologia della Membranosi nel limone Femminello siciliano frigoconservato. Notiziario del CRIOF, vol. VI, No. 2; 9-18.

12. PRATELLA, G.C., BERTOLINI, P., TONINI, G. and MACCAFERRI, M. (1979). The effect of calcium and growth regulators on the incidence of soft scald during refrigerated storage of apples and pears. XV Int. Congress of Refrigeration, Venezia, vol. III; 439-444.

13. SLUTZKY, B., GONZALES-ABREU, A. and BERDAM, I. (1981). Chilling injury related to mineral composition of grapefruit and limes during cold storage. Proc. Int. Soc. Citriculture. Vol. 2; 770-782.

14. SMOOT, J.J., HOUCK, L.G. and JOHNSON, H.B. (1971). Market diseases of citrus and other subtropical fruits. U.S.D.A. Handbook No. 398.

PART II. INTEGRATED PEST MANAGEMENT IN CITRUS-GROVES

Session 4
Means of control

Chairman: R.Prota

Address of the session chairman

R.Prota
Institute of Agricultural Entomology, University of Sassari, Italy

Although there have been frequent alterations to the complexity of the citrus-fruit ecosystem caused by unsuitable human interactions, it is beginning to show signs of a positive recovery, especially when the systems are defended by means and technologies which can be integrated.

For the correct application of the means of control which will be examined by the various lecturers, one must have a knowledge of the basic elements of the key species to be combatted, their dynamics and an estimate of the infestation and damage caused together, with other principles which are no less important.

Many of these elements are sufficiently known in research laboratories but not in agricultural practice, where they have not been welcomed because of both technical and economic difficulties.

Many operators want methods which are simpler and less expensive and also want adequate assistance services with specific professional capacities also for carrying them out and technical and organizational structures which can develop programmes over broad areas.

In this meeting one can give some answers.

Entomophagous insects in the strategy for control of citrus pests

A.Jimenez
National Institute of Agricultural Research, Madrid, Spain

Summary

In the last years, several exotic entomophagous species of citrus pests have been artificially introduced in various Mediterranean coun tries. Some of them have adapted better to weather conditions than indigenous parasites.
Introduced parasitoids reduces the populations of several important citrus pests-mainly coccids and aleurodids-to insignificant densities during large periods of year.
An increase of our knowledge on the use of entomophagous organisms – must continue if we want to achieve the best phytosanitary protection of citrus trees.

1. Introduction

Citrus protection presents some peculiar characteristics. They are mainly attacked by insects that for a long span of their lives are protec- ted by their own excretions as it happens with scales and aleurodids and although the adult stage of many species is fixed the plant, larvae and nymphs are easily propagated by the wind.

Another characteristic of these insects is their polyphagia, they can find their food in many different host plants, mainly ornamentals that usually are out of sanitary control. These ornamental plants often growing in the gardens of the urban areas are the focus that will contaminate neigh bouring groves.

A lot of species are partenogenetic. They need not the cooperation of the male. The eggs laid by female are fertile. This sort of reproduction – increases the possibilities of survival of the insect.

These reasons make clear the difficulties in citrus pest control. Due to the presence of untreated "hot spots", the chemical sprays do not solve the problem and only reduce more or less the level of pest populations.

It is thus understood the interest in the use of entomophagous because the "hot spots" untreated with insecticides could be reached by entomopha- gous insects.

In the last few years several papers provide a large amount of infor- mation about the success in biological control of citrus pest in various parts of the world.

We will consider shortly some aspects of biological control strategy by means of indigenous and exotic entomophagous insects.

2. Potential fecundity and predation

The first point to consider will be the potential fecundity or average progeny of a female ovipositing on a preferential host that is kept under optimal conditions.

High fecundity is estimated like an important attribute of an effective entomophagous. The female of Rodolia cardinalis (Muls.) can lay about
800 eggs in its life (25). However the number of eggs laid by a female does
not determine the parasitoid efficiency since other aspects must be considered like number of generations an a year, duration of preimaginal develop
ment, parasitoid feeding on host stages, etc. Both larvae and adults of
Rodolia cardinalis (Muls.) feed on Icerya purchasi (Mask.) eggs and larvae.
On the other hand, egg predation is more advantageous than larvae or
adult predation, with the same amount of food ingested a larger number of
hosts are destroyed. An excellent attribute is a higher number of generations of the entomophagous respect with the host. Encarsia lahorensis (How.)
can develop up to 6 generations compared with 2-3 of its host, Dialeurodes
citri (Ashm.) (28).

3. Host finding

The entomophagous efficiency would be limited by its ability to find
hosts. In a lot of species, the parasitoid emerges near its host but in
other they must look for their food.
Odour emited by the host can be perceived by the parasite. Some entomo
phagous are attracted by the sexpheromone of the host. Aphytis melinus
(DeBach) and Aphytis coheni (DeBach) respond to the sexpheromone of Califor
nia red scale female (26).
Ohgophagous parasites have a primary host and several secondary ones,
in these, the parasitism is lower than in the main host. Lisiphlebus gomesi
(Quilis) achieve its highest efficiency on Toxoptera aurantii (Boy.) from
March to July, then in November is recorded upon Aphis fabae (Scop) (23).
The host shape influences the parasite oviposition; i.e. the female
of Metaphycus lounsburyi (Howard) lays its eggs into the adult female of
Sarssetia oleae (Oliv.) The convexity of the host body seems to stimulate
the oviposition.When the coccid presents flat body, the parasite escapes
after having examined it (19).
The dispersion of entomophagous insects depends on a large number of
factors. Some of them are related to the population fluctuations of host
and entomophagous, to the potential fecundity and to the parasite life-cicle
Some aspects of the Cales noaki (How.) dispersion have been pointed out; the
aphelinid sets up a primary focus in the initial point of release, then
strengthens and later establishes a new secondary focus not far from the
release point (16). The emigration of the parasite happens when the fly
population decreases. Free of the parasite, the aleyrodid expands apain.
When it has reached a certain level, the parasites acts again (13).

4. Synchrony host-entomophagous

An adequate temporal synchrony betwen host and entomophagous ensures
a better efficiency in the biological control. This allows the auxiliar
insect to have enough food when it emerges. In the Spanish East citrus
growing area,Icerya purchasi (Mask.) is recorded from March to the end of
November, while Rodolia cardinalis (Muls.) appears from May to October.
The delay of the coccinelid is compensated with the faster development of
the auxiliar insect when the temperature tends to go up (11).
Amitus hesperidum (Silv.) adults emerge when the mejority of the citrus
black fly nymphs of the following generation are in their early instar.
It is known that Amitus hesperidum (Silv.) prefers the 1st and 2nd instars
for oviposition (12).
Sometimes the host-entomophagous temporal synchrony is not achieved
due to the short longevity of the parasitoid.The female dies before ovipo-

342

sition if during the few days of its life it does not find a host. On the other hand, the spatial synchrony is reduced at low host densities; the female needs more time to meet an adequate host.

5. Environmental factors

Entomophagous insects, mainly parasites, are relatively sensitive to environmental factors. Climatic conditions have an active role in the establishment of entomophagous. Temperature and humidity condition the auxiliar insects conservation. Larvae are usually more sensitive than adults. It is quite possible that hot dry summers have been the motive of several failures to introduce Cryptolaemus montrouzieri (Muls) in dry areas of the Spanish South East. Young larvae died from the contact with the not soil (14).

Low Winter temperatures cause high mortality in Aphytis lepidosaphes (Comp.) on the California red scale. Aphytis lingnanensis (Comp.) is more effective in the coast areas, while A. melinus (DeBach) is more active in inner valleys (11). In the citrus black fly control in Mexico by means of several entomophagous it was observed that Amitus hesperidum (Silv.) was affected by long hot periods while Prospaltella opulenta (Silv.) tolerates warm seasons and P. clypealis(Silv.) shows better efficiency in wet environments.

Males and females of Aphytis lingnanensis (Comp.) are sterilized by the cold because the sperm stored in its spermateka becomes inactive and the females do not repeat the mating (11).

6. Coexistence and competition among entomophagous

Several entomophagous insects can accumulate their predator or parasitic action if they act on differente stages of the phytophagous. Amitus spiniferus (Breth) destroys 1st instar larvae of Aleurothrixus floccosus (Mask), Cales noaki (How.) parasites mainly 2nd and 3rd instar larvae. An early attack of the aphelinid modifies the composition of the larval population in their different instars, reducing the number of 2nd and 3rd instar larvae.The a larger proportion of 1st instar larvae would facilitate the action of Amitus.

The degree of effectiveness of the entomophagous as to the host density has been registered in citrus blackfly. Its parasite is more effective at high host densities and is responsible for sharp reduction in high initial host populations. In opposition Prospaltella opulenta (Silv.) appears more active at low host densities and can maintain population of A. woglumii (Ashby.) below economically demaging levels. Both parasites coexist and exert mortality on low host populations (9).

An exotic parasite can replace a native one. It has been pointed out that Aphytis chrisomphali (Mercet) has been replaced by A. melinus (DeBach) introduced against Aonidiella aurantii (Mask) (11). The endoparasites Comperiella bifasciata (How) and Prospaltella perniciosi (Tower) would complete the aphelinid action.

Together Prospaltella inquirenda (Silv.) and Aphytis hispanicus (Merc.) are able to control Parlatoria pergandei (Comst.). This scale is very affected by the environment.

On the contrary the entomophagous contend when they take possesion of homologous ecological niches. So it has happened with Metaphyens helvolus (Comp.) and Metaphycus flavus (How.) both parasites of Saissetia oleae (Oliv.) (3). However Metaphycus lounsburyi (How.) developing on 3rd instar larvae and adults coexists with M. Helvolus (Comp.) whose female lays its eggs on 2nd and young 3rd instar larvae (10).

7. Introduction of entomophagous

As it is known, the appearance of a pest in a country is not usually accompanied simultaneaously by the appearance of its natural enemies. Then a fast progress of the insect can take place if the conditions are advantageous to it, i.e, optimal temperature and humidity, abundance of host plants etc.The phytophagous free of its natural enemies multiplies causing a great deal of harm.

An aim of biologal control is the importation and establishment of auxiliar insects in areas where a phytophagous has been accidently introduced so as to reach a stable equilibrium as it happens in their countries of origin.

The literature on citrus protection reports a lot of examples about successes in biological control. We can not forget the case of Rodolia cardinalis shen almost a century ago Riley and koebele introduced that coccinelid to California from Australia for control of Icerya purchasi (Mask).

In the last few years several exotic entomophagous insects have been introduced in various Mediterranean countries. Those colinizations affect a larger number of citrus pests, mainly coccids and aleyrodids.

Several species of Aphytis were imported so as to control Diaspididae scales (1, 4, 5, 8, 21, 27). In the seventies biological control of Saisse tia oleae (Oliv.) was intensified by means of introduction of several species of Metaphycus (3, 7, 18, 20, 21). With that, the already existent useful fauna of predators and parasites, like Chilocorus bipustulatus (L.) and Scutellista cyanea (Motsh) was enriched.

Against citrus mealybug, the hopes are set in Cryptolaemus mountrouzieri (Muls), Leptomastix dactylopii (How) and Nephus reunioni (Furs). (2, 15, 21, 22).

The panic caused by the introduction of Aleurothrixus floccosus (Mask) in the Mediterranean area was contained with the establishment of Cales noaki (How.). A stable host-parasite equilibrium has been reached which is only broken by higher Summer temperatures that cause a large mortality — among the entomophagous (6, 17, 24).

8. Conclusion

Several entomophagous insects are important regulators of citrus pest populations and must be sufficiently known so as to be use in biological and integrated control.

More information about parasites and predators of citrus pests is required mainly concerning their behaviour, population dynamics and hos-entomophagous relationships.

For auxiliar species that have already been established the definition of a methodology so as to evaluate the populations would be useful. In this subject, methods of adult monitoring by means of olfactory and colour traps ought to be set. This knowledge would allow us to know the necessity of increasing the entomophagous populations in the groves.

To develop methods of artificial predator and parasite rearing would be worth because then we could arrange entomophagous to be released in the required places.

At present, secondary citrus pest can be controled with biological methods. As in other pests and outbreak can take place by the own dynamics of the citrus integral complex. A fast introduction of their entomophagous would then be the most accurete action. The first failures must not discourage us. An entomophagous can be no useful in any places but be very efficient in other.

Biological control tries to find the ways so as to use the natural resources more rationally so that a qualitative and quantitative increase of the crop is not achieved at the expense of Nature degradation.

REFERENCES

1. ABASSI, M., EUNERTE, G. (1974). Etude de l'efficatite et de l'aclimatation d'Aphytis melinus DeBach au Maroc. WPRS Bull. 1974, 3:159-168.
2. ALEXANDRAKIS, V.Z. (1984). Integrated control on Citrus Mealybug Planococcus citri Risso. CEE Programme on Integrated and Biological Control. Final Report 1979/1983.
3. ARGYRIOU, L.C. & DEBACH, P. (1968). The establishment of Metaphycus hevolus Compere (Hym. Encirtidae on Saissetia oleae Berh. (Hom. Coccidae) in olive groves in Greece. Entomophaga, 13,3:223-228.
4. ARGYRIOU, L.C. (1974). Data on the biological control of citrus scales in Greece. WPRS Bull.1974, 3:89-94.
5. BENASSY, C., BIANCHI, H., FRANCO, E. (1974). Note sur l'introduction en France d'Aphytis lepidosaphes COM. (Hymenop. Aphelinidae) parasite de la cochenille virgule des Citrus (Lepidosaphes beckii Newm) (Homop. Diaspididae). C.R. Acad. Agric. France, 60, 191-196.
6. BENASSY, C., ONILLON, J.C., BRUN, P. (1981). Integrated control against citrus pests. CEE. Programme on Integrated and Biological Control. Progress Repor 1979-81:67-68.
7. CARRERO, J.M., LIMON, F., PANIS, A. (1977). Note biologique sur quelques insectes entomophages vivan sur olivier et sur agrumes en Espagne. Fruits 32, 9:548-551.
8. CAZELLES, J.P., BERTIN, A., CULTRUT, G. (1974). Dix huit mois d'activite de l'insectarium de Mechra bel Ksir. WPRS Bull 1974,3:121-129.
9. CHERRY, R. (1980). Variations in population levels of citrus black fly Aleurocanthus woglumi (Hom.: Aleyrodidae) and parasites during an eradicative program in Florida. Entomophaga 25, 4:365-368.
10. CLAUSEN, S.P. (1956). Biological control of pests in the Continental United States. U.S. Dept. Agric. Tech. Bull nº. 1939.
11. DEBACH, P. (1977). Lucha biológica contra los enemigos de las plantas. Ed. Mundi-prensa, Madrid.
12. DOWEL, R.V. (1979).Synchrony and impact of Amitus hesperidum (Hym.: Platygasteridae) on its host, Aleurocanthus woglumi (Hym.: Aleyrodidae) in Southerm Florida. Entomophaga, 24,3:221-227.
13. GARRIDO, A., TARANCON., DEL BUSTO, T., MARTINEZ,M.C. (1976). Repartición y estudio poblacional deAleurothrixus floccosus Mask a nivel de árbol y equilibrio con su parásito Cales noaki How. An. INIA, Ser. Prot. Veg. 6:89-121.
14. GOMEZ CLEMENTE, F. (1951-52). Estado actual de la lucha biológica contra algunas cochinillas de los agrios (Pseudococcus citri y Pericerya purchasi). Bol. Pat. Veg. y Ent. Agric. XIX:19-35.
15. NUCIFORA, A. (1984). Integrated pest control in Lemon Groves in Sicilia Five years of Demonstrative Tests and Present Feasibilities of Transferring Results. CEE. Programme on Integrated and Biological Control Final Report 1979/83.
16. ONILLON, J.C., ONILLON, J. (1974). Contribution a l'etude de la dynamique des populations d'homopteres infeodes aux agrumes. III.2. Modalites de la dispersion de Cales noacki How. (Hymenopt, Aphelinidae), parasite d'Aleurothrixus floccosus Mask (Homopt, Aelurodidae) WPRS BUll 1974, 3: 51-66.
17. ONILLON, J.C. (1975). Contribution a l'etude de la dynamique des popula tions d'homopteres inféodés aux agrumes V.3. Evolution des populations d'A.floccosus Mask (Homopt. Aleurodidae pendant les trois années suivant l'introduction de Cales noacki How. (Hymenopt Aphelinidae). Fruits, 30, 4:237-245.

18. ONILLON, J.C., PANIS, A, BRUN,P. (1984). Summary of the Studies and Works carried out in the Framework of the Programme on Integrated Control in Citrus Fruit Groves against Aleyrodes and Lecaninae and Pseudo coccinae Scale-incects.CEE. Programme on Integrated and Biological Control Final Report 1979/83.
19. PANIS, A., MARRO, J.P. (1978). Variation du comportement chez Metaphy - cus lounsburgi. Entomophaga 23, 1:9-18.
20. PANIS; A. (1981). Bases d'utilisation des parasites de la cochenilla noire d l'olivier, Saissetia oleae (Oliv.) (Homoptera, Coccoidae, Cocci dae CEE. Reu. Group. Experts. Antibes 4-6 Nov. 162-168.
21. PELEKASSIS; C.D. (1974). Historical review of biological control of citrus scale insects in Greece WPRS Bull. 1974, 3:14-20.
22. PROTA; R. ORTU, S. DELRIO, G. (1983). CEC. Programme on Integrated and Biological Control Final Report. 1979-83:147-163.
23. QUILIS, M. (1983). Los Aphidiidae fósiles de Wittenheim (Haut-Rhin, Francia) Hym. Brac. Eos, 14:23-61.
24. SANTABALLA, E., BORRAS, C., COLOMER, P. (1980). Lucha contra la mosca blanca de los cítricos Aleurothrixus floccosus Mask Bol. Ser. Plagas 6:109-118.
25. SILVESTRI, F. (1939). Comprendio di Entomologia Applicata. Portici.
26. STERNLICHT, M. (1973). Parasitic wasps attracted by the sex pheromone of their coccid host. Entomophaga, 18,4:339-342.
27. TUNCYUREK, M. & ONCUER, C. (1974). Studies on Aphelinid parasites and their hosts, citrus deaspine scale insects, in citrus orchards in the Aegean Region. WPRS Bull 1974,3:95-108.
28. VIGGIANI, G. MAAZONE, P. (1978). Morfologia, biologia e utilizzazione di Prospaltella lahorensis How. (Hym. Aphelinidae), parassita esotico introdotto in Italia per la lotta biologica al Dialeurodes citri (Ashm) Bull. Lab. Entomol. Agrar. F. Silvestri Portici 35:99-161.

Use of entomophagous insects to replace one of the chemical treatments for *Planococcus citri* Risso (Homoptera, Coccoidea, Pseudococcidae) in citrus-groves

V.Z.Alexandrakis

Subtropical Plants and Olive Trees Institute, Chania, Greece

Summary

The predators Cryptolaemus montrouzieri L and Nephus réunioni FÜRS were used in a grapefruit grove, with a view to one of the two applications of chemicals usually used in Crete to control Planococcus citri RISSO by biological control techniques. Substitution of the second, June, application by predators yielded satisfactory results in that the density of the P. citri population never exceeded the economic infestation threshold. By contrast, high densities were observed when the first, May, application was replaced by predators, thus showing the need for this application. Application of chemicals in both May and June kept citrus mealy bug populations under control, whereas the use of biological means alone resulted in high Pl. citri populations throughout the period.

1.1 INTRODUCTION

Among the major citrus pests, the citrus mealy bug, Planococcus citri Risso (Homoptera, Coccoidea, Pseudococcidae) is the most harmful in Crete (Alexandrakis, 1983), as well in a number of citrus-growing countries in the Mediterranean basin such as Spain, France (Panis, 1968), Italy (Viggiani, 1975; Zinna, 1959) and Israel (Rosen, 1969).

It is the cause of major damage in Crete, particularly in navel oranges. Minor and very minor infestations respectively have been noted on lemon and mandarin trees.

Over recent years, the citrus mealy bug has caused major damage in the recently established grapefruit groves of Crete.

Control techniques for citrus mealy bug in Crete generally depend on the use of insecticides to kill off young, first-generation insects. Spraying against Pl. citri is generally done twice in Crete, first at the end of May and then a month later.

As part of a programme for integrated control in citrus groves, various trials of biological or integrated control of Pl. citri have been carried out in Crete using the predators Cryptolaemus montrouzieri L and Nephus réunioni FÜRS (Alexandrakis, 1983), as well as in other citrus-growing countries of the Mediterranean basin such as France (Panis, 1978, 1981), Italy (Zinna, 1960; Viggiani, 1975), Israel (Rosen, 1969) and Spain, whereby use was made either of the predators described above or the

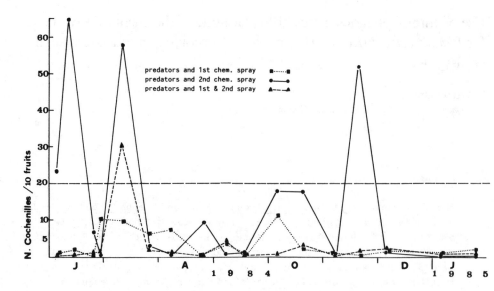

Fig. 1: Changes in the population of <u>Planococcus</u> <u>citri</u> (including eggs) on grapefruit after spraying with pesticides and releasing predators.

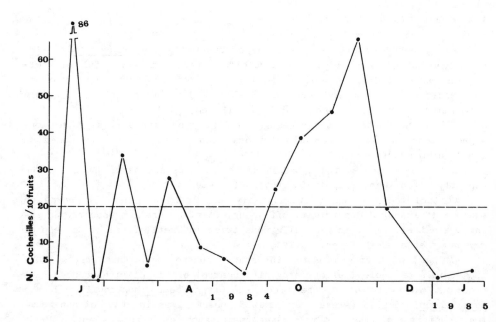

Fig. 2: Changes in the population of <u>Planococcus</u> <u>citri</u> (including eggs) in plots receiving biological control measures alone.

Ensyrtides parasites Leptomastix dactylopii How and Anagyrus pseudococci (Girault).

In spite of biological control trials, citrus growers in Crete still rely on applications of chemicals at the periods described above to protect their groves. In this experiment we tried replacing one of these two applications by the use of predators.

1.2 EQUIPMENT AND METHODOLOGY

Trials were carried out in a grapefruit grove at Armeni, 20 km west of Chania, at an altitude of 30 m on trees of the Marsh-Frost variety. The following four treatment programmes were carried out:

   a) biological control and a chemical spray in May
   b) biological control and a chemical spray in June
   c) both the earlier and later chemical sprays
   d) biological control alone

Biological control involved the release of 18 and 9 adult C. montrouzieri and N. réunioni predators respectively. In those plots receiving the first treatment programme, release began 15 days after the chemical spray, while the predators were released into the plots receiving the second treatment programme from 20 May, except for a break at the end of June and the beginning of July when there were still chemical residues on the trees. In those plots in which biological control alone was used, the total number of predators mentioned above was released over the period between 20 May and the end of November 1984.

In the plots receiving both chemical sprays, the usual double applications were administered on 4 and 25 June. The pesticide employed was morfotox (68% mecarbam) at a concentration of 100 cc of active ingredient per 100 litres of water. Sampling started when the trees were bearing fruit after the spring blossoming. We examined 10 fruit in each plot, selected at random from those at head height, every fortnight until January 1985. In view of the significant loss of eggs, two separate density calculations were made: with and without mealy bug eggs. Finally, the figures were transformed into logarithms (X+1) for the calculations, and the Duncan test (Dalianis, 1972) for comparison of means was applied.

1.3 RESULTS

It appears that biological control alone, using the predators C. montrouzieri and N. réunioni, is not capable of keeping citrus mealy bug populations at a negligible level. If all the living stages of the bug are calculated (Fig. 1), all four summer generations of the mealy bug exceeded the economic threshold of two insects per fruit (unpublished results). The density remained at the level of 6.6 live insects per fruit throughout the maximum of the last generation during the month of November. Similar results were recorded in the plots receiving predators and the second chemical spray, the density over the same period being 5.2 insects per fruit. It was only during the third summer generation that the density

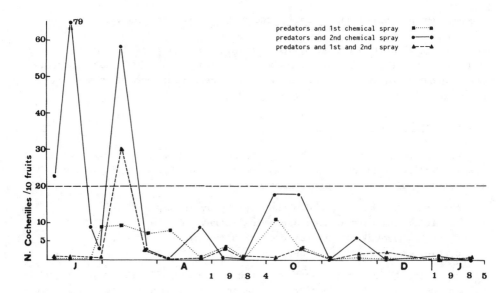

Fig. 3: Changes in the population of _Planococcus citri_ (excluding eggs) after treatment with insecticides and predators.

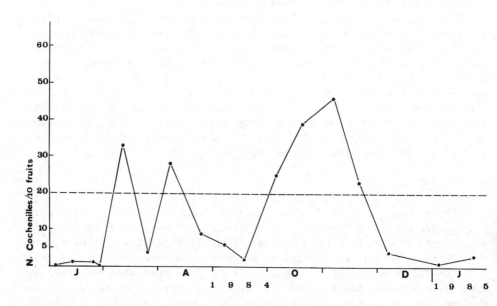

Fig. 4: Changes in the population of _Planococcus citri_ (excluding eggs) after employment of biological control alone.

remained below two live insects per fruit. The plots receiving the first chemical spray and then the predators, and those with both chemical sprays, had low populations throughout the experimental period. The absence of reference plots makes it impossible to show the influence of predators alone (Fig. 2) on Pl. citri populations. If the mealy bug eggs are excluded (Fig. 3), the populations show the same trends. A low density is always recorded on the plots with the first spray and predators, while similar results are achieved in those plots with two chemical sprays. In the latter cases the population densities exceeded the threshold of two live insects per fruit only during the second summer generation. This threshold was considerably exceeded in the plots receiving the second chemical spray and the predators, with the exception of the density recorded during the last generation of old insects, which is much lower than the mealy bug density in plots receiving only biological control measures (Fig. 4). The density of live insects in the latter case was 4.6 insects per fruit, while those plots receiving the second chemical spray and predators had an average of 1.8 insects per fruit. In the two other cases, the Pl. citri density did not exceed 1.1 old live insects per fruit over the same period.

An analysis of the 15 dates between July and the end of 1984 on which checks were made showed very significant differences between the four control methods tested (Table 1). In particular, the study of all the stages of the insect's life cycle, including eggs, provided the following results.

The plots with biological control alone had population densities 8.3 times higher than those with two chemical sprays, and 7.2 times higher than those receiving the first chemical spray and predators. No significant difference was found between plots receiving biological control measures and those with the second chemical spray and the predators. The latter plots had populations 6.5 times higher than the ones with two chemical sprays and

Table 1. Live Planococcus citri / 10 fruit as a function of the control methods used (n = 15)

	including mealy bug eggs	without mealy bug eggs
First spray and predators	3.40 (a)	3.27 (ac)
Second spray and predators	19.13 (b)	15.13 (ab)
Both chemical sprays	2.93 (a)	2.93 (c)
Predators alone	24.40 (b)	14.73 (b)
F	7.01**(1)	4.93**(1)

(1) F0.05 for 3 and 42 d.1 = 2.83
F0.01 for 3 and 42 " = 4.29

351

5.6 times higher than those with the first chemical spray and predators. There was no significant difference in pest density between plots with two chemical sprays and those with the first spray followed by predators.

Similar results emerge from an analysis ignoring the mealy bug eggs. Highly significant differences again appear, with the plots receiving biological control alone having 5 and 4.5 times more mealy bugs than the plots with two chemical sprays and those with the first spray followed by predators respectively. The plots receiving only the second chemical spray showed densities 5 times higher than the plots with both chemical sprays.

1.4 CONCLUSIONS

Control measures for Pl. citri in Cretan citrus groves generally comprise the application of two chemical sprays, the first being at the end of May or the beginning of June, and the second a month later. Experimental work carried out in 1982 and 1983 (Alexandrakis, 1983) has shown that the predators C. montrouzieri and N. réunioni, used as a means of biological control, are not in themselves capable of keeping down mealy bug populations, in spite of the inroads that they do make. These trials also showed that the application of the first chemical spray alone was not sufficient to control the pest.

This trial, carried out on grapefruit trees in 1984, has shown the need for the first chemical spray. It appears that the second spray could be replaced by use of the predators described above. By contrast, the use of entomophagous species throughout the trial as a replacement for the first chemical spray did not yield satisfactory results. In this trial, the absence of reference plots did not affect the comparisons between the various control methods tested, because the density of mealy bug populations in the plots exposed to biological control were always much higher than the economic infestation threshold.

This trial again showed that predators are ineffective during the autumn.

Finally, the two chemical treatments achieved satisfactory control of mealy bug populations.

REFERENCES

1. ALEXANDRAKIS, V. (1983). Integrated control on citrus Mealybug Planococcus citri Risso. Rapport final contract 0715 C.E.E., Brusselles 1983; 15.
2. ANONYME. (1970). Instrucciones para la aplicacion de colonias de "Cryptolaemus Montrouzieri" en los focos de COTONET ("Pseudococcus Citri"). Suc de vives Mora-Valencia, 1970; 1.
3. DALIANIS, K.(1972). Ebauche et Analyse des Expériences. Athènes. p.586.
4. PANIS, A. (1968). Les Cochenilles farineuses des Agrumes. Agrumes Presse, January 1968; 4.
5. PANIS, A. (1978). Modalités des auxiliaires contre les Cochenilles Farineuses et Lecanines. B.T.J. 332-333, 1978, L4-AGRO-436; 1-4.

6. PANIS, A. (1981). Note sur quelques insectes auxiliaires régulateurs des populations de Pseudococcidae et de Cocciade (Homoptera, Coccoidea) des agrumes en Provence orientale. Fruits, vol. 39, n° 1; 49-52.

7. ROSEN, D. (1969). The parasites of coccids Aphids and Aleurodids on Citrus in Israel: Some Zoogeographical considerations. Isr. Journ. of Entom. Vol. IV; 45-53.

8. VIGGIANI, G. (1975). Possibilità di lotta biologica contro alcuni insetti degli Agrumi (Planococcus citri RISSO e Dioleurodes citri-Ashm.-). Boll. Lab. Entom. Agr. Portici, Vol. XXXII-1975; 1-10.

9. ZINNA, G. Richerch sugli insetti entomophagi. I Specializzazione negli Encyrtidae: Studio morfologico, etologico del Leptomastix dactylopii Howard. Boll. Lab. Entom. Agr. Portici, Vol. XVIII, 1-150.

10. ZINNA, G. (1960). Esperimenti di lotta biologica contro il cotonello degli agrumi (Pseudococcus citri (RISSO) nell'Isola di Procida mediante l'impiego di due parassiti esotici, Pauridia peregrina Timb. e Leptomastix dactylopii How. Boll. Lab. Entom. Agri. Portici. Vol. XVIII; 257-284.

The role of scale size in the biological control of California red scale

R.F.Luck

Department of Entomology, University of California, Riverside, USA

Summary

The competitive displacement of Aphytis lingnanensis Compere by A. melinus DeBach was studied in Southern California in an attempt to explain why biological control works on California red scale, Aonidiella aurantii (Mask.), in southern California but is ineffective against this scale in the San Joaquin Valley. Our results indicated that A. melinus displaced A. lingnanensis because the former used a smaller scale size than the latter for the production of female progeny. Climatic effects on the age structure of California red scale created bottlenecks in the availability of suitable scale stages and hence a resource shortage ensued. Climate effects on the scale's age structure in the San Joaquin Valley created periods during which suitable scale stages for Aphytis were scarce or absent. These lengthy periods of resource scarcity alternating with periods of resource abundance prevented Aphytis from economically controlling the scale.

Biological control involves three general practices: classical (introduction of exotic natural enemies), augmentative (release of insectary reared natural enemies) and manipulative biological control (managing endemic natural enemies to enhance their effectiveness). Augmentative biological control supplements endemic natural enemies by initiating a natural enemy population at the beginning of each season, by increasing the densities of natural enemies at critical periods during the season or by inundating prey or host populations to reduce their densities immediately. Manipulative biological control enhances natural enemy effectiveness by choosing compatible control strategies and minimizing or eliminating factors that interfere with the natural enemy, e.g., insecticide drift, dust or ants.

Classical biological control, when successful, has proven to be inexpensive, effective and, most importantly, permanent (4). Approximately 34% of the classical biological control projects (for all crops) have resulted in at least partial economic control of the targeted pest (4,11,12,16). Although less widely practiced, both augmentative and manipulative biological control have been employed with success (e.g. 4,10,11,13,15,17,18,21, 23). But not all biological control projects have succeeded. Some introduced natural enemies fail to become established, those that establish fail to reduce a pest's abundance to economically acceptable densities, or the release of insectary reared natural enemies or attempts at manipulating natural enemies yields inconsistent results. Yet if we are to implement integrated pest management, we must increase both our effectiveness in the use of natural enemies and our ability to forecast the results of such use. Unfortunately, we do not yet understand the processes or attributes of a natural enemy/pest interaction well enough to forecast the outcome of a biological control effort consistently. In an attempt to improve our understanding, we have chosen to investigate, in detail, the elements that have led both to the effective and the ineffective biological control of California red scale, Aonidiella aurantii (Mask.), in California.

I have two objectives with my studies of California red scale, its host plants and its natural enemies: a specific objective that seeks to improve the biological control of the scale, especially in the San Joaquin Valley, and a general objective that seeks to understand how biological control works in southern California and what constrains its success in the San Joaquin Valley. This latter objective assumes that understanding the reasons for success and failure and the associated processes and mechanisms in a specific case will improve our ability to predict the outcome of this specific natural enemy-pest interaction (i.e., will the natural enemy complex provide acceptable economic control of the scale population in this grove this year?) and of such interactions in general (i.e., the success or failure of a biological control project). This assumption is also a hypothesis I am testing. I also include within this interaction insect predator-prey, parasitoid-host interactions in arthropod communities of non-economic interest (i.e., those involving arthropods in more or less natural systems). Thus, biological control of California red scale serves as both a specific example with which to test hypotheses about biological control and a model system with which to develop general hypotheses for application to other biological control projects and to population dynamic studies of noneconomic arthropods.

To understand why I have chosen this research strategy, one needs to know something about the history of the biological control effort against California red scale. A biological control project was initiated against this pest soon after its introduction into California in the latter half of the 19th Century. It was not until 1947, however, that an effective natural enemy, Aphytis lingnanensis Compere, was first introduced. It was not the first Aphytis present on red scale in California, however, as A. chrysomphali (Mercet) had apparently been introduced sometime before 1900 (24). Aphytis lingnanensis replaced A. chrysomphali presumably because the former was better adapted to the climate of California's citrus districts (8). However, A. lingnanensis was unable to achieve economic scale control in the more interior citrus districts, hence additional natural enemies were sought. One of these, A. melinus DeBach, was established in the interior in 1956-57 and quickly spread throughout the area, displacing A. lingnanensis in the process. It provided economic scale control throughout much of southern California. As of the late 1970's the geographic distribution of these three Aphytis species was reported as (1) A. melinus in the interior citrus areas, (2) A. lingnanensis along the coast and inland approximately 10 km, and (3) A. chrysomphali as a relic in a few coastal enclaves (24).

DeBach and his coworkers monitored the establishment and spread of these introduced Aphytis species by sampling red scale populations at a large number of locations throughout southern California (5). It was during this monitoring program that DeBach noticed A. melinus displacing A. lingnanensis and, in some groves, this displacement occurred even when third stage female scale, the putative host stage for the two Aphytis species, was abundant. DeBach interpreted these observations as indicating that hosts for the parasitoids were not limiting, hence competition for scarce hosts could not explain the displacement (6,9,24). Laboratory experiments designed to test this hypothesis, however, contradicted the field observation: A. lingnanensis displaced A. melinus in the laboratory in contrast to the displacement that occurred in the field (9). DeBach provided no explanation for the contradiction and went on to suggest that

356

since A. melinus was both a better searcher and better adapted to southern California's climate than A. lingnanensis, it replaced A. lingnanensis. No experimental data are provided to support these conclusions.

DeBach's hypothesis also presents a theoretical dilemma for biological control. Biological control views successful control as arising through the ability of natural enemies to limit the density of their host or prey via intra- and interspecific competition among the entomophages for hosts or prey (7,14). Since Aphytis melinus is documented to be an effective biological control agent of California red scale, DeBach's hypothesis seemed to undermine the theoretical foundations of biological control.

Thus DeBach's hypothesis raised two important questions: (1) how did A. melinus replace A. lingnanensis in southern California, and (2) how was A. melinus able to control California red scale? I reasoned that if we wanted to understand biological control of the scale we had to understand the mechanism by which A. melinus displaced A. lingnanensis. The two questions were really the same. I suspected that it was important to first test DeBach's hypothesis that A. melinus and A. lingnanensis utilized the same host resource (scale stage and size) (6,9,24).

Haggai Podoler and I tested this hypothesis by conducting a series of behavioral experiments in which we observed the host selection behavior of A. melinus and A. lingnanensis on California red scale. We found that four phases characterized the ovipositional behavior of these two species: (1) drumming and turning, (2) drilling, (3) probing, and (4) ovipositing (Fig. 1) (20). The drumming and turning phase commenced when the female parasitoid initially contacted the scale and mounted it, centering herself on the scale cover. She then walked forward, drumming the scale cover with her antennae. Upon contacting the scale-cover's margin she backed up, recentered herself on the scale cover, rotated right or left and repeated the drumming behavior. She continued these behaviors 10 to 15 times, at which point she initiated drilling by walking half way off the scale cover, engaging the ovipositor by lowering the tip of her abdomen and exerting backward pressure on the ovipositor. When the ovipositor penetrated the scale cover, the parasitoid briefly hesitated before inserting the ovipositor into the scale body.

Insertion of the ovipositor into the scale body marked the probing phase, evinced as extensive movements of the ovipositor within the scale's body. Oviposition followed removal of the ovipositor from the body, although the ovipositor remained beneath the scale cover. The egg was either inserted between the scale body and the scale cover or the scale body and substrate.

If a second egg was laid, the behavioral sequence, depicted in Figure 1, was repeated. Similarly, if a third egg was laid the behavioral sequence was again repeated. However, once the parasitoid left the scale and resumed her search, she rejected the scale as a suitable host if she recontacted it. She also rejected a scale previously parasitized by a conspecific or a congeneric. Such rejection was manifested during the drumming and turning phase as only one or two drums and turns before the parasitoid abandoned the scale.

These observations also revealed a behavior that indicated the deposition of an egg. If a second "vibration" occurred (Fig. 1), we were 98% certain that an egg had been deposited. The identification of this behavior allowed us to determine the number and gender of eggs laid on different scale sizes (19). We introduced single, mated female A. melinus or A. lingnanensis into a cage attached to a lemon containing 30-40 adult virgin California red scale of different sizes. After 72 hours we measured the scale size (length x width of the scale's body) and transferred the parasitoid larva to an oleander scale to provide the larva with an adequate

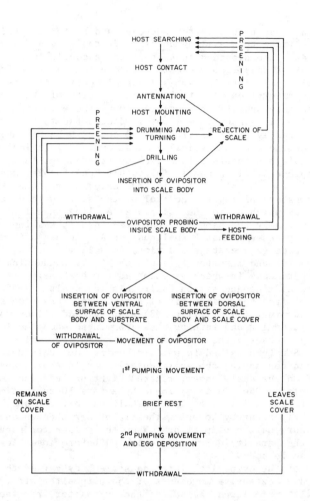

Figure 1. The sequences of behaviors exhibited by Aphytis melinus
and A. lingnanensis during host selection and acceptance.

food supply to complete its development. We transferred the larva to
exclude differential mortality of female larva as a possible explanation
for sons arising from smaller scales than daughters. In many parasitoid
species female larvae are thought to require more food than male larvae to
complete their development.

Figures 2 and 3 show the results of these experiments for A. melinus
and A. lingnanensis, respectively. The distribution of scale sizes chosen
for parasitization by both A. melinus and A. lingnanensis did not differ
from those offered (Fig. 2e vs. 2f, 3e vs. 3f). With A. melinus, only six
scales received a single male egg (Fig. 2d). Most single eggs were females
(Fig. 2c). In every case in which A. melinus laid two eggs, one was a male
and the other was a female (Fig. 2b). Also, in all three cases in which A.
melinus laid three eggs, one was a male and two were females (Fig. 2a).
The minimum scale size on which A. melinus laid a female egg was 0.39 mm^2
and on which it laid two eggs was 0.55 mm^2. Abdelrahman (1) has shown that

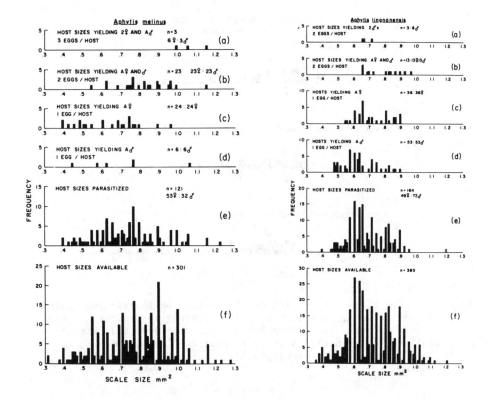

Figures 2 and 3. (a-f) The size distributions of third-stage California red scales on which Aphytis melinus (Fig. 2) and A. lingnanensis (Fig. 3) laid the indicated numbers and sexes of eggs. (e,f.) The size distributions of scales chosen by and offered to A. melinus (Fig. 2) and A. lingnanensis (Fig. 3) for parasitization.

parasitized second stage female scales yield mostly male A. melinus. In contrast to A. melinus, A. lingnanensis laid mostly single male eggs on third instar scales (Fig. 3d). The minimum scale size on which this species laid a single female egg was 0.55 mm^2 (Fig. 3c) and on which it laid two eggs, one male and one female, was 0.65 mm^2 (Fig. 3b). Thus our research shows that A. melinus lays female and multiple eggs (one male and one female) on smaller scale than A. lingnanensis. Subsequent research (22) has confirmed this pattern.

We hypothesize that the ability of A. melinus to use smaller scales than A. lingnanensis to produce female offspring permitted A. melinus to displace A. lingnanensis from California's citrus groves. As California red scale increases in size with maturation, it enters the size classes used by A. melinus for female progeny production before it does so for A. lingnanensis. When A. melinus parasitizes a scale it paralyzes it; hence further scale growth (and hence increase in size) ceases. Moreover, both A. lingnanensis and A. melinus avoid ovipositing on hosts previously parasitized by the other species. Thus, the ability of A. melinus to use smaller California red scale substantially reduced the scale density avail-

359

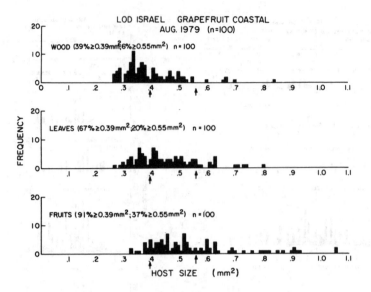

Figure 4. The size distribution of third-stage California red scale
present on wood, leaves, and fruits within grapefruit trees planted
at Lod, Israel. The arrows beneath the abscissas indicate the
smallest scale insect on which A. melinus (left arrow) and A.
lingnanensis (right arrow) produced progeny. n = No. of scales
counted.

able to A. lingnanensis for female progeny production and thus, potentially, created a resource shortage for this parasitoid species.
      The question that now remains is: what is the pattern of resource
availability (scales of suitable age and size) in the field? Our research
to date has identified two answers to this question. First, by measuring
the body size of adult virgin red scale in the field we have found that
this scale stage is largest on the fruit, smallest on the wood and of
intermediate size on the leaves (Fig. 4) (3,19). We have also found that
in the San Joaquin Valley adult virgin red scale are smaller on the wood
than they are at other locations where we have measured them (Fig. 5) (19).
Since the fruit are harvested, the scale they support are a transient
resource for Aphytis because it is absent for one to several months during
the year (2). Also, when compared to scale on fruit, less of the scale on
leaves and wood are of sufficient size for the production of Aphytis
females. However, more of them are of sufficient size for the production
of female A. melinus progeny than of female A. lingnanensis progeny. Thus
we suspect that the supply of suitable scale stages are periodically in
short supply for both A. melinus and A. lingnanensis, especially following
harvest and it is this period of short supply that favors A. melinus.
      We have also found that the age structure of California red scale
varies seasonally as a consequence of climate. Few crawlers are produced
during winter. In the San Joaquin Valley the more severe climatic conditions, in comparison to those of coastal southern California, result in
the scale population in the Valley being dominated by one or two age
classes (adult virgin and gravid female stage scales) at the end of winter.

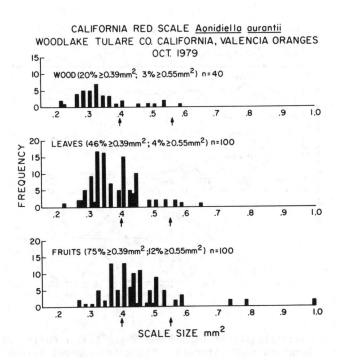

CALIFORNIA RED SCALE Aonidiella aurantii
WOODLAKE TULARE CO. CALIFORNIA, VALENCIA ORANGES
OCT. 1979

WOOD (20%≥0.39mm², 3%≥0.55mm²) n=40

LEAVES (46%≥0.39mm²; 4%≥0.55mm²) n=100

FRUITS (75%≥0.39mm²;12%≥0.55mm²) n=100

FREQUENCY

SCALE SIZE mm²

Figure 5. The size distribution of third-stage California red scale
present on wood, leaves, and fruits within navel orange trees at
Woodlake, California.

Reproduction by these stages in late spring leads to a large pulse of
crawlers which settle and mature during May and June as a wave of scale
that sequentially passes through each succeeding scale stage from first
instar to gravid female. This pattern leads to periods in which second and
third instar scales (host stages for A. melinus) are unavailable for
substantial periods during the late spring. Thus, A. melinus is faced with
alternating periods of scale abundance and scarcity of such magnitude that
it is unable to economically control California red scale in the San
Joaquin Valley (Fig. 6). In southern California such periods of scale
scarcity seldom occur. When they do they are more often the result of chemical intervention as was the case at Fillmore in 1982 (Fig. 6). Thus, it
is the combined requirement by A. melinus and A. lingnanensis for scales of
a particular quality for female progeny production (i.e., unparasitized
scale of a particular stage and size coupled with as yet other unidentified
qualities) coupled with the availability of such scale in the field that
appears to govern both the competitive displacement of A. lingnanensis by
A. melinus and the biological control of California red scale by A. melinus
in southern California (the coastal and intermediate coastal areas). It
also explains the absence of such control in the San Joaquin Valley.
Consequently, of paramount importance in evaluating biological control is
an understanding of the resource requirements of a potential biological
control agent, especially for female progeny production. The only means of
determining these requirements is through a combination of detailed behavioral experiments in the laboratory and the field.

361

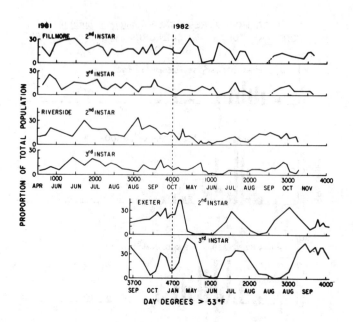

Figure 6. The availability of second and third instar California red scale
at three location in California: Fillmore (coastal inland southern
California), Riverside (interior coastal southern California) and
Exeter (San Joaquin Valley).

REFERENCES

1. ABDELRAHMAN, I. (1974). Studies in the ovipositional behavior and control of sex in Aphytis melinus DeBach, a parasite of California red
   scale, Aonidiella aurantii (Mask.). Aust. J. Zool. 22: 231-247.
2. CARROLL, D. and LUCK, R. F. (1984) Within-tree distribution of
   California red scale, Aonidiella aurantii (Maskell) (Homoptera:
   Diaspididae), and its parasitoid Comperiella bifasciata Howard
   (Hymenoptera: Encyrtidae) on orange trees in the San Joaquin Valley.
   Environ. Entomol. 13: 179-183.
3. EBELING, W. (1959). Subtropical Fruit Pests. Univ. Calif. Div. Agric.
   Sci., Berkeley. 436 pp.
4. DeBACH, P. (1964). Biological Control of Insect Pests and Weeds.
   Chapman and Hall, Ltd., London. 844 pp.
5. DeBACH, P. (1965). Some biological and ecological phenomena associated
   with colonizing entomophagous insects. Pages 287-303 in H. G. Baker
   and G. L. Stebbins, editors. The genetics of colonizing species.
   Academic Press, New York.
6. DeBACH, P. (1966). The competitive displacement and coexistence principles. Ann. Rev. Entomol. 11: 183-212.
7. DeBACH, P. (1974). Biological Control by Natural Enemies. Cambridge
   Univ. Press, New York. 323 pp.
8. DeBACH, P. and SISOJEVIC, P. (1960). Some effects of temperature and
   competition on the distribution and relative abundance of Aphytis
   lingnanensis and A. chrysomphali (Hymenoptera: Aphelinidae).
   Ecology 41: 153-160.

9. DeBACH, P. and SUNDBY, R. A. (1963). Competitive displacement between ecological homologues. Hilgardia 34: 105-166.
10. DeBACH, P., ROSEN, D. and KENNETT, C. E. (1971). Biological control of coccids by introduced natural enemies. In Huffaker, C. B. (ed.), Biological control. Plenum Press, N.Y., pp. 165-194.
11. HALL, R. W. and EHLER, L. E. (1979). Rate of establishment of natural enemies in classical biological control. Bull. Ent. Soc. Amer. 25: 280-282.
12. HALL, R. W., EHLER, L. E. and BISABRI-ERSHADI, B. (1980). Rate of success in classical biological control. Bull. Entomol. Soc. Amer. 26: 111-114.
13. HASSEN, S. A., LANGENBRUCH, G. A. and NEUFFER, G. (1978). Der Einfluss des Wirtes in der Massenzucht auf die Qualität des Eiparasiten Trichogramma evanescens bei der Bekämpfung des Maiszünslers Ostrinia nubialis. Entomophaga 23: 321-329.
14. HUFFAKER, C. B. and LAING, J. E. (1972). "Competitive displacement" without a shortage of resources? Res. Pop. Ecol. 14: 1-17.
15. HUSSEY, N. W. and BRAVENBORE, L. (1971). Control of pests in glass-house culture by the introduction of natural enemies. In C. B. Huffaker (ed.), Biological Control. Plenum Press, New York, pp. 195-216.
16. LAING, J. E. and HAMAI, J. (1976). Biological control of insect pests and weeds by imported parasites, predators and pathogens In C. B. Huffaker and P. S. Messenger (eds.), Theory and Practice of Biological Control. Academic Press, New York, pp. 685-743.
17. LENTEREN, J. C. VAN and WOETS, J. (1977). Development and establish-ment of biological control of some glasshouse pests in the Nether-lands. In F. F. Smith and R. E. Webb (eds.), Pest Management in Protected Culture Crops. USDA AS ARD-NE-85, pp. 81-87.
18. LENTEREN, J. C. VAN, PAMALCEPS, P. M. J. and WOETS, J. (1980). Inte-grated control of vegetable pests in galsshouses. P. Gruys and A. K. Minks (eds.), Integrated Control of Pests in the Netherlands. Pudoc. Wageningen, pp. 109-118.
19. LUCK, R. F., and PODOLER, H. (1985). The Competitive exclusion of Aphytis lingnanensis by A. melinus: potential role of host size. Ecology. In press.
20. LUCK, R. F., PODOLER, H. and KFIR, R. (1982). Host selection and egg allocation behaviour by Aphytis melinus and A. lingnanensis: A comparison of two facultatively gregarious parasitoids. Ecol. Entomol. 7: 397-408.
21. NATIONAL ACADEMY OF SCIENCES. (1975). Pest control: an assessment of present and alternative technologies. Contemporary pest control prac-tices and prospects: report of the executive committee. Vol. I, 506 pp.
22. OPP, S. B. and LUCK, R. F. (Submitted). Effects of host size on selected fitness components of Aphytis melinus DeBach and A. lingnan-ensis Compere (Hymenoptera: Aphelinidae). Ecol. Entomol.
23. RIDGEWAY, R. L. and VINSON, S. B. (1976). Biological Control by Aug-mentation of Natural Enemies. Plenum Press, New York. 480 pp.
24. ROSEN, D. and DeBACH, P. (1979). Species of Aphytis of the World (Hymenoptera: Aphelinidae). W. Junk, Den Hague, 801 pp.

Distribution on the Côte d'Azur of *Encarsia lahorensis* (How.), a parasite introduced against the citrus white-fly, *Dialeurodes citri* (Ashm.)

J.C.Malausa & E.Franco
INRA, Zoological & Biological Control Station, Laboratory 'E.Biliotti', Valbonne, France

J.C.Onillon
Zoological Station, CNRA, Versailles, France

Summary

After Encarsia lahorensis was released in Antibes in 1976, a systematic survey of citrus trees in 1984 showed that the parasite was
present in almost all citrus-growing areas on the Côte d'Azur.

As part of the integrated control of Dialeurodes citri, Encarsia
lahorensis (How.), a specific parasite originally found in India, was
introduced into France by Onillon (Onillon and Brun, 1983). This parasite
has also been successfully introduced into several other countries in the
Mediterranean region - Italy, Greece, and possibly Israel (Viggiani, 1980).

The D. citri larvae, infested by the parasite E. lahorensis, were
brought in from California in November 1975. From July 1976, adult E.
lahorensis were released among the citrus trees in the Station's experimental orchard and in an orchard with about 100 clementine trees on the
Antibes Cape. Some 80 female and 50 male E. lahorensis were released
during July 1976. For several years, E. lahorensis did not seem to be
having much effect, and the percentage of D. citri affected by the
parasites remained very low. Although the parasite appeared to have become
established, it did not seem to be spreading from the point of release.

When D. citri was sampled in 1982, E. lahorensis was found in the
citrus orchard at Valbonne, 11 km from the release point. An assessment of
the rate of infestation in D. citri larvae in this orchard in 1984 showed
that it remained fairly low. No more than 2% of first generation whitefly
larvae, and only about 15% of the second and partial third generation
larvae were affected. This overall infestation was measured at the end of
each whitefly generation on the shoots of the first rising of the sap.

The fact that the parasite was found this far from the point of release led to a systematic survey of the citrus trees on the Côte d'Azur and
in the neighbouring hills. This survey in 1984 showed that E. lahorensis
was present throughout the citrus-growing areas of the Alpes-Maritimes
region up to an altitude of about 300 m (Vence, Grasse). Distribution runs
from Mandelieu to Menton, the latter being 40 km from the point of release
(Fig. 1). The Italian Riviera was not surveyed.

The rate of infestation varies greatly from one spot to another, and
it seems that it is inversely proportional to the density of D. citri. In
several places the few larvae of D. citri found were infested, whereas

365

Fig. 1: Distribution of <u>Encarsia lahorensis</u> in the Alpes-Maritimes
region in 1984.

where there were high densities of the pest it was sometimes difficult to
detect the presence of <u>E. lahorensis</u>. These observations indicate that the
parasite is well able to find the host, even if it is present only in low
density, but on the other hand it seems unable to control growth in
populations of whitefly, and this explains the low rate of infestation
observed in orchards with a high density of <u>D. citri</u>. The fact that the
whitefly forms only two complete generations per year in our climate, and
that there is a lack of continuity in the presence of receptive stages in
<u>E. lahorensis</u>' host, is definitely a handicap in establishing the parasite,
which cannot respond to rapid increases in the number of whitefly.

The study of distribution was not carried out every year and therefore
only reflects the situation about 10 years after release. This does not
simplify the task of comparison with the distribution observed in the other

countries where E. lahorensis has been introduced. The discovery of the parasite in Menton means that it has advanced on average 4 km per year, which is important and compares with the spread observed in Italy by Viggiane and Mazzone (1978) of 3 to 4 km in 1 to 2 years. A similarly rapid spread was also noted in Florida, whilst in California the spread remained extremely limited because of the specific climatic conditions (Ru and Sailer, 1979).

Efforts to acclimatize E. lahorensis began in 1984 in Corsica, where the species is not yet present. The releases took place in the citrus fruit station's experimental orchard in San Giuliano from pieces of twig and young citrus plants in pots infested with whitefly larvae affected by parasites and taken from our nurseries; only a few adults were released directly. This release will allow us to study the spatio-temporal development of E. lahorensis in the conditions found in Corsica.

REFERENCES

1. ONILLON, J.C. and BRUN, P. (1983). Integrated control against Citrus pests. C.E.C. programme on integrated and biological control. Progress report 1979/1981, C.E.C. publications 1983; 75-78.
2. RU Nguyen and SAILER, R.I. (1979). Colonization of a citrus whitefly parasite, Prospaltella lahorensis, in Gainesville, Florida. Fla. Entomol. 62 (1); 59-65.
3. VIGGIANI, G. (1980). Progress toward the integrated control of citrus pests in Italy. Proceedings Inter. Symposium IOBC/WPRS. 1980; 293-296.
4. VIGGIANI, G. and MAZZONE, P. (1978). Morfolgia, biologia e utilizzazione di Prospaltella lahorensis HOW. (Hym. Aphelinidae), parassita esotico introdotto in Italia per la lotta biologica al Dialeurodes citri (Ashm.). Boll.Lab.Ent.Agr.Portici. 35; 99-160.

Some considerations on the role of mites in the biological control of the citrus ecosystem

A.Nucifora & V.Vacante
Institute of Agricultural Entomology, University of Catania, Italy

Summary

Reference is made to the role which the mite predators of phytophagous mites and insects have in the biodynamics of the citrus–fruit ecosystem. In Sicily the finding so far of 28 species of useful mites gives a reason for the very good results obtained so far with the application of integrated control in the lemon orchards of the island.

1.1. Introduction

We have always been interested in the link between mites and the quality of production of agrarian plants and have direct experience of certain aspects of their role in the ambit of agricultural ecosystems; we know the importance which the game of antagonistic symbiosis plays in crops defend with integrated control systems, to try to encourage in them, where possible, the action of phytophagous species.

Today we know that in citrus agroecosystems, for example, Tetranychus urticae Koch and Panonychus citri (Mc Gregor) may be well contained by the action of Phytoseiulus persimilis Athias–Henriot and Amblyseius stipulatus A.H. respectively, as well as by the complementary action of other phytoseiids, many of which live on the crops to pray on the former, when Man does not intervene to alter the situation negatively by spreading tons of acaricides and insecticides. One would not need these if one trusted in the action of the mite predators; they by themselves could keep the infestation of phytophages to non–damaging levels. If their actions were researched, protected and not obstructed by the irrational spraying of phytochemicals, as commonly occurs at the level of phytosanitary defence, it would ensure permanent results which would be very interesting in the defence of crops against the attack of phytophagous mites.We are sure of this not because we have heard it said, but because of our considerable experience in this sector, both in the area of the defence of citrus fruits in the open field and in that of market–garden crops and flowers in green houses.

1.2. State of the situation and new knowledge acquired

The study of the citrus–fruit ecosystem under the profile of acarofauna and its interrelationships with other Arthropods, with fungi, viruses and other animal and vegetable species awaits further and more detailed knowledge.

Among modern acarologists one may count in Italy specialists of value in specific sectors, but an overall study of agricultural ecosystems and of the species which characterize them has not yet been or is only in the initial stages.

We have recently begun, in collaboration with the Universities of Palermo and Bari, an enquiry into the acarofauna of Mediterranean agricultural crops which, limited to the citrus fruit ecosystem, has led up to now to the determination (Vacante and Nucifora, 1985) of 67 species, 28 of which are predators of mites and insects, 12 are phytophagous in habit, while the others are micophagous or coprophagous; 11 of these species were new to Italian acarofauna. For most of them there is little ethobiological knowledge.

We have some information but are still looking for answers on the phytophagy of some species or the predation which others perform or the parasitic activity of which some are capable, living permanently or temporarily on the bodies of vertebrates or invertebrates (see the example of Acarapis woodi(Rennie)and of Varroa jacobsoni Oudemans as the expense of bees). The case mentioned above as an example shows us how some species can indirectly have a negative relationship with agricultural ecosystems. To understandt this it is sufficient to think of what would happen to our agrarian crops if the pollinating activity of Apis mellifica L. were limited in any way.

But where do the mites which are useful or damaging in the economy of the citrus ecosystem live? They live on the ground and they live on the parts of the plant which are below ground. The soil is one of the most important components of the agricultural ecosystem and the role which the mites have with their given possibility of "taking up the most meagre organic waste to feed themselves and their great powers of multiplications by which in a few days one female, together with her offspring, may breed colonies of thousands and thousands of individuals, so that their job in the economy of nature is performed rapidily, in a way that neither chemical decomposition not the competition of larger forms ..... could do ..... The activity of mites in the continuous modification of organic waste and of decomposing substances is very important" (Berlese, 1925). One should therefore always fear the disruptive effects which may be induced in the dynamism of this transformation by the administration of phytopharmaceuticals directly on the ground or by herbicides or simply by falling, when the plants are irrigated. Little is known about these phenomena.

Other mites live on the foliage or on the trunk. Of these those phytophages which live at the expense of the plant are small in number when compared to the number of known species, but are those which are the most studied and which play a role of decisive importance in the ecosystem for the crops and their production. One need only think of what can be done by Aculops pelekassi(Keifer)or the even more feared Phyllocoptruta oleivora(Ashmead), which is not yet present with us, or Aceria sheldoni (Ewing) on citrus fruits, not to speak of the already-mentioned P. citri and T. urticae.

Various disorders at the plant physiological level in citrus fruits and other plants are attributed to the Brevipalpus of the Tenuipalpidae family. It seems that the species of this genus inject a persistent toxin into the plant, when they feed on it. An alteration in the orange called "Leprosis" (Vergani, 1945) with ring marks and gummy phenomena on leaves, fruits and buds, is attributed to the action of Brevipalpus californicus (Banks).

The presence of these phytophages leads to interventions by Man, who wants to halt or limit their deleterious effects; because of this disturbing factors are introduced into the agroecosystem, such as the use of acaricide phytopharmaceuticals,as we have mentioned above. The use of these and other products in the environment leaves its traces, disturbing in a more or less severe way the relationship between the entomoacarofaunula of the plants and with the manifest or concealed phytotoxicity phenomena which they induce.

Thus, neither the broad spectrum insecticides not the acaricides but the use of biological agents solve the problem rationally. These are the very important fac tors which we must use to reduce the populations of phytophagous mites, containing their development and regulating their cycles. They are pathogenetic factors, they may be viruses diseases, but they are mainly mites, predators insects and spiders. There are more than 65 recorded predators of Panonychus ulmi (Koch) (Jeppson et al., 1975), but usually the number of species which we can set directly against one or another species is fairly limited. Thus P. persimilis works against T. urticae and it is this, little helped by other phytoseiids, which has the job of keeping the effects of T. urticae within acceptable limits. Another phytoseiid, A. stipulatus, works against P. citri and ensures that this phytophage is practically absent in citrus fruits, and in lemons in particular. In the pilot lemon orchards of integrated control which were set up in Sicily in 1979-1983 with the financial aid of the C.E. C. and the Italian MAF, the infestation of this and some other phytophagous species was kept permanently below the economic threshold levels by the action of Phytoseiidae, Ascidae, Stigmaeidae and Cheyletidae. There are 19 predator acaro phage species which can be used as acarophages in Sicily in citrus orchards where integrated control techniques are used. All of them are useful in one way or another. Other mites of the Hemisarcoptidae, Tydeidae and Cunaxidae families form part however of the group of auxiliaries which are useful for our crops. Thus we have found Hemisarcoptes malus (Shimer) in this same habitat preying in all its sta ges on eggs, intermediate forms and adults of various species of Diaspididae. Mites easily find protection under the shield of the host and may reach, according to the evidence of Ewing and Webster (1912) control levels of 50% and more of the diaspidids preyed upon. In Sicily on the lemon tree we have seen them preying at the expense of Aspidiotus nerii Bouche and of Mitylococcus beckii(Newan)with attack levels up to 40%.

Of the Tydeidae, of which 11 species have been identified up to now in our citrus orchards (Vacante and Nucifora, l.c.), Orthotydeus californicus (Banks) would seem, according to Baker (1965), to prey upon Aceria sheldoni (Ewing); thus also Tydeus ferulus (Baker) may live on eriophyds and scale eggs. The other 9 species, 4 of which are new in the Italian acarofauna, would seem to be micophagous or detritivorous or a few may be phytophagous; there is much still to be studied on the ethology of this group of mites. Finally the Cunaxidae with the two species Cunaxa capreoleus (Berlese) and Cunaxa setirostris (Hermann) are predators, and have been found several times in the Mediterranean area, but little or nothing is known of their ethobiology yet.

But this does not complete the story and neither is everything left to mites when we talk about the restriction of phytophagous species in agricultural ecosystems and in citrus crops in particular. Predator insects are at work every day to alter the influence and to modify this role. This is known when one looks at the action of Stethorus punctillum Weise against tetranychids and other similar species, that of Staphylinidae and mainly that of Chrysopidae, and of various general of emet terids, tisanopterids and Cecidomiidae. Concerning this last family we are pleased to able to cite the case of Therodiplosis persicae Kieffer, which lives in the larva state preying on the T. urticae. Vacante (1981) found it on citrus fruits in the open field and has studied its behavious on market-garden fruits and flowers in greenhou ses. In cucumber greenhouses, treated with integrated control techniques, without use of acaricides or insecticides, up to 150 live forms of the diptera (larvae and pupae) per leaf were seen in April-May. In a rose bed, 100% of the middle-lower leaves of which were attacked by T. urticae in mid-January, the cecidomid multiplied from January to the end of March in such a way that the prey-predator ratio fell from 234 to 6, reducing the infestation and altering it by a factor of 40 in only 70 days. These facts clearly show the determining role of antagonistic biocenosis and the weight which biological factors have in regulating it; these same factors also show the quantity of damage which Man can cause, often acting in unconsidered ways and thus breaking the equilibria present in agricultural ecosystems.

2. Conclusions

The examination of the results obtained from our research on the acarofauna of the citrus ecosystem in Italy gives a motivated explanation of the causes which, in citrus orchards treated with integrated control criteria, have led to the disappearance of infestations of some phytophagous mites, which are however of primary importance in citrus orchards treated with other control systems. In citrus orchards and in particular in lemon orchards where integrated control is practiced, supported by just white mineral oil (one or two treatments at most per year) (Nucifora, 1983), the acarophages have a wide range of action and develop almost undisturbed, interacting together and ensuring stable conditions of biological equilibrium.

The role of this useful acarofauna on mites and damaging insects, as well as that of other valuable auxiliaries, is encured by this control system, which has been widely tested in Sicily with complete success in the five years of demonstrative integrated control tests, as part of the C.E.C. project 0730 (Nucifora, l.c.) and has been further recognised in its application by several citrus growers in the lemon crop of the island in the last two years. A wider dissemination of the results with updating courses at the technical-informative level is to be recommended.

REFERENCES

1. BAKER, E.W. (1965). A Review of the Genera of the Family Tydeidae (Acarina). Advances in Acarology, Cornell University Press, Ithaca, New York, 2: 95-133.
2. BERLESE, A.(1925). Gli insetti. S.E.I., Milano, pp. 992.
3. EWING, H.E. and WEBSTER R.L. (1912). Mites associated with the oyster-shell scale (Lepidosaphes ulmi Linne). Psyche, 19: 121-134.
4. JEPPSON, L.R., KEIFER, H.H. and BAKER, E.W. (1975). Mites Injurious to Economic Plants, University of California Press, Berkeley, pp. 614.
5. NUCIFORA, A. (1983). Integrated Pest Control in Lemon Groves in Sicily: Five Years of Demonstrative Tests and Present Feasibilities of Transferring Results. Final Report 1979/1983, C.E.C. Programme on Integrated and Biological Control: 129-146.
6. VACANTE, V. (1981). Notizie sulla presenza di Therodiplosis persicae Kieffer (Diptera, Cecidomiidae) in serra su piante orticole e floreali, attaccate da Tetranychus urticae Koch (Acarina, Tetranychidae). Tecnica Agricola, 5: 303-312.
7. VACANTE, V. and NUCIFORA, A. (1985). A First list of the mites in Citrus Orchards in Italy. Proceedings Integrated Pest Control in Citrus-Groves, Acireale, 26-29 March (in press).
8. VERGANI, A.R. (1945). Transmission y naturalez de la "Lepra explosiva" del Naranjo. Argentina Inst. Sanidad. Veg. 1 series A (3), pp. 10.

Promising results of biological control of citrus Mal secco

V.De Cicco, M.Paradies & A.Ippolito
Plant Pathology Institute, University of Bari, Italy

Summary

Fusarium oxysporum, *Verticillium dahliae* and a hypovirulent isolate of *Phoma tracheiphila* were used to induce resistance to Citrus mal secco in sour orange seedlings. The tests were performed in a growth chamber using 1-2 year old sour orange seedlings. The isolates were inoculated simultaneously , 15 days 1,2,4 months before the inoculation of the virulent isolate of *Phoma tracheiphila*. Only the hypovirulent isolate induced resistance to mal secco: protection was observed when one month elapsed between the first and second inoculation; and, the greater the time interval between the two inoculations, the greater the protection. The results obtained seem to suggest that the control of mal secco by inducing resistance with inoculations of a hypovirulent isolate of the pathogen is promising and deserves further research.

1.1 Introduction

Over these last years the use of biological control of plant diseases has become an interesting reality. In fact, the vast bibliography on this subject reports results that are not only satisfactory, but also sometimes economically competitive (3,6). Since the beginning of this century the study and the utilization of "premunity" has aroused great interest. Premunity consists of an increase of host resistance towards the pathogen, mainly induced by the inoculation of the plant with forms of the same pathogen or other microrganism, even if taxonomically distant.

The phenomenon of premunity has found interesting practical application against some viroses (2,7); while it is still far from being valid in the field of the fungal and bacterial diseases, although there are positive results for some leaf and vascular diseases (5).

There is not much research on Citrus mal secco, a vascular disease caused by *Phoma tracheiphila* (Petri) Kanc. et Ghik., except for a trial by Grasso and Tirrò (4) carried out by pre-inoculating an isolate of *Verticillium dahliae* Kleb. Moreover, to this subject are related the results obtained by Salerno et al.

(10,11), who preinoculated suor orange seedlings (Citrus aurantium L.) with some virus; these seedlings showed resistance to
the subsequent infections of mal secco, associated to modifications of the phenolic metabolism of the plant.

In view of the considerable difficulties and uncertainties
of the control of this disease (12), it seemed useful to under-
take research on biological control. The results obtained so
far, a part of which have been reported elsewhere (8), are stated briefly here.

1.2 Materials and methods

All tests were performed in growth-chamber artificially
lit for 12 h (Gro-lux Sylvania tubes, type F40T12/GRO), with a
light intensity of 6500 lux. One to two year-old sour orange
seedlings, grown on a mixture of soil (70%) and peat (30%) in
1 litre plastic containers were used.

The inoculum of P. tracheiphila was obtained according to
Salerno and Catara technique (9) with slight modification, that
is using 250 ml fluted shake flasks,each containing 60 ml carrot broth and a rotary shaker adjusted to 180 rpm. The inoculum
of F. oxysporum f. sp. lycopersici S. H. and V. dahliae was
prepared by suspending in sterile water the conidia obtained
from the conidified cultures of both fungi.

The fungi utilised were: one isolate of F. oxysporum f. sp.
lycopersici, two isolates of V. dahliae, respectively from tomato (Solanum lycopersici L.) and olive (Olea europea L.), one
hypovirulent (Pt52) and one virulent (Pt1) isolate of P. tracheiphila.

The inoculations of the non-pathogenic and hypovirulent
isolates were performed on opposite sides of the stem, about
3cm above soil level. The inoculation of the virulent isolate
(Pt1) was made 2cm above the previous ones. F. oxysporum, V.
dahliae and Pt52 isolates were inoculated simultaneously,15,30
60 and 120 days before inoculating the virulent isolate Pt1.
All the inoculations were performed by incising the stem of the
plant transversally with a blade on which a drop of conidial
suspension had been placed. The concentrations of the conidial
suspension were 8×10^6/ml for F. oxysporum and V. dahliae; and
10^6/ml for the two P. tracheiphila isolates. The controls were
made by placing a drop of sterile water on the blade.

The progression of the disease was assessed about every 15
days, using an empirical scale (13) which comprises 5 degrees
of disease intensity, from "0" (no disease symptoms) to "4"
(all, or almost all leaves wilted or fallen, with the drying up
of the plant starting from the top).

For the determination of the number of propagules in the
wood of the inoculated seedlings, the Buchenauer and Erwin technique (1) was followed, making slight modifications. Two por

374

Table I - Average intensity of leaf symptoms of Citrus "mal
secco" 30 and 60 days after the inoculation of a
virulent isolate (Ptl) of P. tracheiphila,on sour
orange seedlings preinoculated with V. dahliae(Vd).

Treatments	days after inoculation	
	30	60
Vd and Ptl after 30 days	1,44 B	4,00
water and Ptl " " "	2,67 AB	4,00
Vd and Ptl " 60 "	2,22 AB	4,00
water and Ptl " " "	4,00 A	4,00

Means in column not followed by the same letter are significantly different (P=0,01) according to Duncan's multiple range test. The data are the average of 10 replicates .

tions were taken from the stem of the seedling respectively
just above the inoculation site and 20-25 cm above it. After
removing the bark, a piece weighing 0,5 g was homogenized in
50 ml of sterile water and then diluted 10 times. 2.5ml of it
were mixed in Petri dishes with 12.5ml of potato dextrose agar
(55-60°C) containing 30 ppm of streptomycin. The colonies in
each Petri dish were counted after a 4-6 day incubation period
at 22°C.

1.3 Results and discussion

The isolates used as "inducers" were previously grown in
Petri dishes together with the virulent isolate of P. tracheiphila used as "challenger" to evidence some possible antagonism.
The results showed no antagonism.
Preliminary trials were performed to identify the best interval of time between the two inoculations for achieving protection. The inducer isolates were inoculated simultaneously,
15, 30, 60 and 120 days before the inoculation of the virulent
isolate. In these preliminary trials, F. oxysporum did not induce any resistance to the disease, V. dahliae slightly delayed
the appearance and evolution of disease simptoms, and the hypo-
virulen isolate Pt52 induced resistance only in the plants
which had been inoculated at least 30 days before the inoculation of the virulent isolate.
On the basis of these preliminary results, other trials
were performed using V. dahliae and Pt52 as "inducers" and time
intervals that seemed to provide better results.
V. dahliae again induced only a slight delay (with no

Table II - Average intensity of leaf simptoms of Citrus "mal secco" 30 and 60 days after the inoculation of a virulent isolate (Ptl), on sour orange seedlings pre-inoculated (30, 60 and 120 days before) with a hypovirulent isolate (Pt52) of P. tracheiphila.

Treatments	1st trial		2nd trial	
	30 days	60 days	30 days	60 days
Pt52 and Ptl after 30 days	0,10 C	1,91 B	0,27 B	1,34 B
water and Ptl " " "	1,60 B	4,00 A	0,53 B	4,00 A
Pt52 and Ptl " 60 "	0,50 C	1,61 B	0,26 B	1,27 B
water and Ptl " " "	0,30 C	4,00 A	2,87 A	3,80 A
Pt52 and Ptl " 120 "	0,00 C	0,00 C	0,00 B	0,61 B
water and Ptl " " "	3,10 A	4,00 A	0,47 B	4,00 A

Means in column not followed by the same letter are significantly different (P=0,01) according to Duncan's multiple range test. The data are the average of 10 replicates for the first trial and 15 for the second one.

Table III - Results of isolations and propagule evaluations
at inoculation site (a) and subapical part of
the stem (b) of sour orange seedlings inoculated
45 days before.

Treatments	Isolations		Propagules	
	(a)	(b)	(a)	(b)
Pt52	+	+	+	-
Pt1	+	+	+++	++
Pt52 and Pt1 after 120 days	+	-	+	-

The + signe indicates a positive result and - sign , a negative result. With regard to the propagules more than one
+ sign indicates a proportionally higher number of the same.

statistical significance) in the evolution of the symptom of
the disease (Table I), confirming the results achieved by Grasso and Tirrò (4). Sixty days after the inoculation of Pt1, all
the plants showed symptoms with the highest value on the scale.
The Pt52 isolate always induced resistance when the Pt1 inoculation followed 1, 2 and 4 months later. The greater the time
interval between the two inoculation, the greater the protection (Table II).
    Some sour orange seedlings of the last trial, those with
120 days between the first and second inoculation, were transferred to the field for further observations. For comparison,
seedlings inoculated only with the hypovirulent isolate Pt52
were also transferred. Nine months after the transplantation,
that is about one year after inoculation of the virulent isolate, both the seedlings inoculated with Pt52 and those with double inoculation did not show any leaf symptom of mal secco.
However, some plants showed chlorosis in the old leaves and
sections of the woody cylinder above the inoculation sites revealed the typical orange colour, which was more evident in
seedlings with double inoculation.
    During the trials isolations were performed and the number
of propagules was determined. When F. oxysporum and V. dahliae
were inoculated singly, the isolations showed positive results
only at the inoculation site. On the contrary, the Pt52 isolations were positive also in the apical part of the plants, as
in the plants inoculated only with Pt1 or with double inoculation. The propagules were determined only in plants inoculated
with Pt52, with Pt1 and with both isolates but the results were
not always uniform. Table III shows the results of isolations
and propagule valuation 45 days after the inoculation.

The results of these trials seem to indicate the validity of preinoculation with a hypovirulent isolate of P. tracheiphi-la to induce resistance against citrus mal secco. The same cannot be said about the results obtained with preinoculation of F. oxysporum and V. dahliae although the latter showed a small delay of the symptoms.

In view of the fact that most of the plants pre-inoculated with the hypovirulent isolate (in particular the seedlings where the two inoculations were 4 months apart) did not show leaf symptoms one year after inoculation of the virulent isolate, it may be said that the control of mal secco by inducing resistance with inoculations of a hypovirulent isolate is promising and deserves further research.

REFERENCES

1. BUCHENAUER, H. and ERWIN, D.C. (1972).Control of Verticillium wilt of cotton by spraying with acid solutions of Benomyl, Methyl 2-benzimidazole carbamate, and Thiabendazole. Phytopath. Z., 75, 124-139.
2. COSTA, A. S. and MUELLER, G. W. (1980). Tristeza control by cross protection. Plant dis. Rptr, 64, 538-541.
3. DARPOUX, H. (1960). Biological interference with epidemics. In Plant Pathology, Horsfall J.G. and Dimond A. E., ed. Academic press, N. Y. 521-565.
4. GRASSO, S. and TIRRO', A. (1982). Primi risultati sull'effetto della preinoculazione di Verticillium dahliae in piante di arancio amaro inoculate con Phoma tracheiphila.Tec. Agric., 34, 179-186.
5. MATTA, A. (1979). Resistenza biologicamente indotta verso malattie batteriche e fungine. Inf.tore fitopatol.,8,17-29.
6. MATTA , A.(1982). Prospettive di lotta biologica contro le malattie fungine e batteriche delle piante. Annali Accademia di Agricoltura, Torino, 124,23.
7. MIGLIORI, A. et al.(1972). The use of immunization against tobacco mosaic virus in glasshouse tomato crops. Pepinieristes, Horticulteurs, Maraichers No 132, 15-19.
8. PARADIES, M., DE CICCO, V. and SALERNO, M. (1985). Prove di lotta biologica contro il mal secco degli agrumi a mezzo di un ceppo ipovirulento del patogeno. Atti convegno su "la lotta biologica". Torino 16 febbraio 1985. In corso di stampa.
9. SALERNO, M. and CATARA, A. (1967). Ricerche sul mal secco degli Agrumi (Deuterophoma tracheiphila Petri).VI. Indagini sulla riproduzione sperimentale della malattia. Riv. Pat. veg., 3, 89-97.
10. SALERNO, M., SOMMA, V. and EVOLA, C. (1970). Influenza di alcune virosi sul decorso del mal secco degli Agrumi e pri-

mi risultati relativi al contenuto fenolico delle tesi a
confronto. Phytopath. medit., 9, 22-28.

11. SALERNO, M., EVOLA, C. and SOMMA, V. (1971). Modificazione
    del metabolismo fenolico e mal secco degli Agrumi in semen-
    zali di Arancio amaro con precedenti infezioni da virus.
    Phytopath. medit., 10, 195-201.
12. CUTULI, C. and SALERNO, M.(1981). On the epidemiological
    meaning of phialospores in Phoma tracheiphila (Petri) Kanc.
    and Ghik. Proc. 5th Congr. Mediter. Phytopath. Union, Patras
    Greece, 72-73.
13. SCARAMUZZI, G., SALERNO, M. and CATARA, A. (1964). Ricerche
    sul mal secco degli Agrumi (Deuterophoma tracheiphila Petri).
    II.Influenza delle basse temperature sul decorso della ma+
    lattia. Riv. Pat. veg., 4, 319-327.

Biotechnical pest control methods and agro-ecosystem

E.F.Boller

Swiss Federal Research Station for Arboriculture, Viticulture & Horticulture, Wädenswil

Summary

Integrated Pest Management has changed the former plant protection
strategies in the sense that the individual pest species or disease is
no longer considered in an isolated manner when suitable control
measures are planned and applied. The pest or disease as well as undesirable weeds are components of an agro-ecosystem that interact with
each other in various ways. Hence, control measures have to be evaluated not only with respect to their effectiveness against the target
species but also with respect to their impact on other components of
the system. Biotechnical methods are of increased importance in modern
IPM programs because they can exert a complementary role to biotic
regulating factors acting within the agro-ecosystem. Biotechnical
methods available or under investigation are reviewed in the context
of citrus groves.

1. Agro-ecosystems and their importance for the design of pest control programs

Many plant protection specialists working with perennial crops have
recognized with increasing concern that research and development is mostly
carried out within the established scientific disciplines (such as entomology, phytopathology, weed control etc.) but that the application phase
of pest control is of interdisciplinary nature - a serious problem that
has to be solved by the grower. There are strong tendencies to-day to make
the interdisciplinary synthesis already at the level of research and
development in order to implement balanced integrated pest control or
integrated production programs at the farm level.

It has also been recognized that entomologists, pathologists, herbologists, physiologists, and agronomists specialized on plant nutrition or
cultural practices need a common language and common terms of reference.
One tool that has proven its usefulness in interdisciplinary research is
the phenomenon of the agro-ecosystem, such as citrus groves, vineyards,
apple orchards, where the interaction of the components can be identified,
verified by experimentation and taken into consideration in the design of
plant protection or production strategies. An example of a generalized and
simplified model of a perennial agro-ecosystem is shown in Fig. 1. It has
been developed in 1983 for instruction purposes and is used in Switzerland
in the training of students in agronomy at the university or college level.

The system parameters (such as location, soil type, cultivar, clones,
pruning system, nutrient level) provide the overall characteristics and
possibilities of the system that shows in its center the crop species or

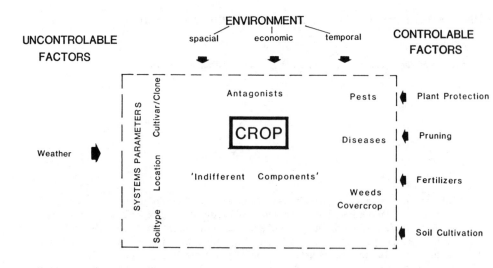

Fig.1: Components and possible structuration of an agro-ecosystem

crop mixtures. System parameters also influence heavily the 'furniture' of the system, i.e. the pest species complex, diseases, weeds, antagonists, and the large group of 'indifferent' components. The rôle of the 'indifferent' components of a perennial system is often unknown but there is increasing evidence that some of the faunistic or botanical components are important for the internal (or ecological) stabilization of the system (e.g. flowering plants as hosts for parasites and predators of pest species).

Many interactions between pest complex, weeds and diseases can be identified theoretically and verified by proper experimentation (e.g. insect damages favouring development of fungal diseases; insects acting as disease vectors; weeds and herbicide application causing spider-mite migrations from the green cover plants to the fruit trees; disease control by fungicides destroying important insect antagonists such as predators and parasitoids; insecticides stimulating spider-mite population build-up and eliminating at the same time the antagonists; excessive nitrogen input via fertilizers stimulating both fungal diseases and sucking pest species requiring again new pesticide application).

These few examples show that in fact we do not deal with bilateral interactions between two components of the system but in most cases with a network of interactions. One pest control measure selected on the basis of its efficiency against the target species alone can hence cause considerable unexpected side-effects within the agro-ecosystem and increase the problems instead of solving them.

We should also consider briefly the factors that act from the outside on the ecosystem and influence the events and developments taking place. Beyond control by man are weather conditions but by choosing the proper site for the orchards, groves and vineyards will take care of predictable incidences caused by weather such as frequent hail, late frosts, unsufficient or too much precipitation. Factors listed on the right side of the

382

diagram such as plant protection measures, pruning, fertilizers, soil cultivation and similar activities of agricultural practices and management are of evident impact on the components of the system. Often neglected is the aspect of the environment (geographic and economic) that is surrounding and severely influencing our given agro-ecosystem. In the horizontal dimension we can recognize neglected neighbouring fields as potential sources for reinfestation by diseases and arthropod pests whereas hedges, shrubs, forests or extensively managed pasture land might provide ecological reservoirs of antagonists moving into our citrus grove from the outside. In the vertical dimension we are more and more faced with the problem of immissions that have already caused severe problems in our forests as sensitive perennial bio-indicators.

After this look at the agro-ecosystem we can conclude that scientific meetings devoted to the discussion or planning of pest control strategies should be organized jointly for entomologists, plant pathologists, herbologists and agronomists. Because of increasing specialization at the research level we should be able to understand the problems occurring in the other disciplines and inform our professional colleagues about our own requirements and constraints. It is therefore interesting to note that this particular meeting is attended by specialists from different disciplines and that this is hopefully the start of a tradition to discuss and solve plant protection and general crop production problems in a holistic manner.

What is now the place of biotechnical pest control methods in the context of the agro-ecosystem and what is their rôle in integrated pest control programs? If we accept the definitions of integrated pest control proposed by either FAO or IOBC as general guideline for our considerations then we come to the conclusion that we have to manage our agro-ecosystem in such a way that plant protection measures and cultural manipulations have a minimal disruptive effect on the delicate ecological web within the given crop system. When we look at the inventories of pest species established in various citrus growing regions we realize that biotic factors (i.e. predators and parasites) are of paramount importance in citrus pest management. The range of beneficial organisms acting upon citrus pests such as scale insects, aphids or spider-mites is impressive. Hence IPM programs will try to eliminate or strongly reduce pest control measures that are detrimental to these antagonists, i.e. lack selectivity. If we accept the protection of natural or introduced antagonists as corner-stone of the pest management in citrus it follows that unselective and ecologically disturbing influences upon the agro-ecosystem have to be replaced. Biotechnical methods of pest control can make a significant contribution for the ecological stabilization of a system by supporting or supplementing the activity of antagonists. However, it must be pointed out, that biotechnical methods per se are not ecologically safer or more selective than other means of pest control if they are not evaluated critically and selected on the basis of their intended use.

When we look for those components of an integrated pest management program that can help us to stabilize the agro-ecosystem and to reduce the use of disruptive chemical pesticides we can summarize as follows: Selection of cultivars or clones that have been improved with respect to resistance against major pests and diseases; protection, stimulation or introduction of natural enemies wherever possible; application of selective biotechnical methods where biotic factors are not available (e.g. against Ceratitis capitata) or not sufficiently effective; reduce all negative factors that can aggravate pest problems (e.g. excessive nitrogen input).

Fig. 2 : Biotechnical Methods of Pest Control (* with high selectivity)

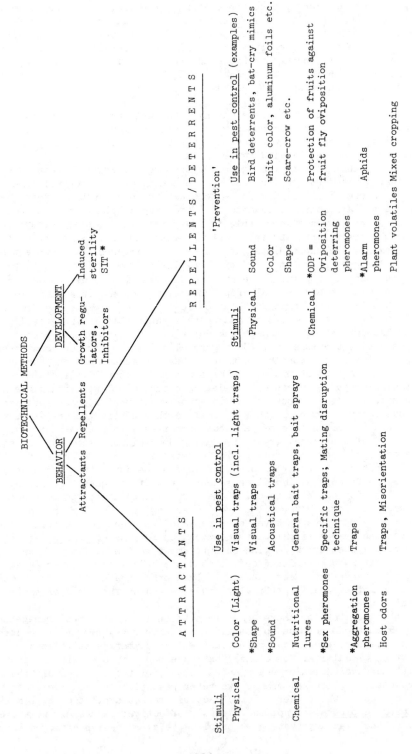

BIOTECHNICAL METHODS

BEHAVIOR — Attractants, Repellents

DEVELOPMENT — Growth regulators, Inhibitors ; Induced sterility SIT *

ATTRACTANTS

Stimuli		Use in pest control
Physical	Color (Light)	Visual traps (incl. light traps)
	*Shape	Visual traps
	*Sound	Acoustical traps
Chemical	Nutritional lures	General bait traps, bait sprays
	*Sex pheromones	Specific traps; Mating disruption technique
	*Aggregation pheromones	Traps
	Host odors	Traps, Misorientation

REPELLENTS / DETERRENTS

'Prevention'

Stimuli		Use in pest control (examples)
Physical	Sound	Bird deterrents, bat-cry mimics
	Color	white color, aluminum foils etc.
	Shape	Scare-crow etc.
Chemical	*ODP = Oviposition deterring pheromones	Protection of fruits against fruit fly oviposition
	*Alarm pheromones	Aphids
	Plant volatiles	Mixed cropping

2. Biotechnical Methods: Definition and Evaluation

The term 'biotechnical control' has been proposed by Franz for the first time in 1964 at the occasion of a FAO symposium on forest pests and hence used by an increasing number of European entomologists with changeing interpretation. I take the liberty to refer here to the definition proposed at the CEC/IOBC Symposium on Fruit Flies of Economic Importance held 1982 in Athens (1): "Biotechnical methods of pest control utilize chemical or physical stimuli or agents that influence the behavior or the development of the target species.

They include the use of attractants, repellents (deterrents), growth regulators and inhibitors including artificially induced sterility in pest species.

The advantage of the biotechnical methods is their high specificity and high compatibility with the environment as they do in general not create undesirable residue problems .....".

If we examine the list of possible biotechnical methods (Fig. 2) critically in view of their applicability to citrus problems and selectivity we arrive at the following conclusions:

Most efforts with respect to the development of biotechnical control methods have been undertaken with respect to Ceratitis capitata, the key pest in most citrus groves, that is difficult to be pushed below economic threshold levels by antagonists. Application of mating disruption techniques with sex pheromone or application of insect growth regulators with acceptable levels of selectivity could theoretically be considered as potential solutions in Prays citri (Lepidoptera), but biotechnical methods so far have to my knowledge not been investigated in other citrus pests as intensively as in Ceratitis capitata.

In the first place we have to mention the successful application of the Sterile-Insect-Technique (SIT) - per definitionem a biotechnical method - against the medfly in Mexico where eradication of this pest was achieved in 1982. Since genetic control will be discussed by other participants of this meeting I want to reduce my comments on SIT to the statement that this type of control is highly specific and hence without disruption of the ecosystem. However, we have to take note of possible side-effects caused by the sterile sting problem (sterile females can cause injuries on fruits with their ovipositors and hence possibly cause disease problems).

Whereas presently no other biotechnical method is in sight that influences the development of C. capitata we observe on the other hand certain research and development activities in the field of behavioral manipulation. As shown in Fig. 2 behavior can be influenced either by using attracting or repelling/deterring stimuli. Prominent in the first category are the attempts to develop powerful traps (super-traps) that could be used to attract and eliminate a maximum number of fruit flies - preferably females - before they have reached sexual maturity.

Characteristics of traps developed for detection, monitoring or control purposes have been discussed at an earlier meeting (2). Traps used for mass-application and direct population control require a high degree of selectivity in order to avoid major negative side-effects. Yellow sticky boards can be used for spot-sampling of medfly populations in ecological studies or can be used for monitoring purposes. However, their severe impact on many beneficial insects attracted by yellow colors make mass-application of such visual traps for control purposes a highly dangerous tool in IPM programs (1,3,4,5). Similar considerations have to be made with respect to general olfactory stimuli on food lure basis. Substances genera-

385

ting ammonia, such as ammonia salts or protein hydrolysates, are particularly attractive for fruit fly females but also to a vast array of potential beneficials that also get caught in these traps. Selective medfly attractants, such as trimedlure, are attracting only males and can therefore not be considered solutions for an efficient and selective population control in C. capitata. As Delrio (5) pointed out the male annihilation method seems not to offer successful possibilities. Unfortunately, no efficient specific female attractant is available to-day, despite the fact that the sex pheromone of C. capitata has been identified and tried under field conditions with unsatisfactory results (5). Present work in progress deals with the improvement of medfly trap efficiency and selectivity by testing different shapes, colors and olfactory stimuli in combined super-traps. These investigations are coordinated by the IOBC Working Group on Fruit Flies of Economic Importance.

Another area of research and potential application is the use of oviposition deterring substances that could protect the fruits from ovipositing fruit fly females (6,7). Reports about the deposition of a marking pheromone by medfly females after oviposition need re-examination and are also topics of research activities of the above mentioned IOBC working group. At the present stage of knowledge we can say that oviposition deterring pheromones (ODP) might exist in Ceratitis capitata but that the situation is not as clear-cut as in the cherry fruit fly, Rhagoletis cerasi, where cherry trees sprayed with crude extracts of ODP provided protection against fruit fly attack (8). Smaller field experiments with substances deposited by the medfly females after oviposition are planned in the near future in order to investigate the behavior of the fruit fly pest in nature when exposed to treated fruit. Marking pheromones or artificial oviposition deterrents found in laboratory and field screening tests could become an interesting biotechnical method of fruit protection against this key pest if the principles are the same as those observed in Rhagoletis species. It could also become a selective method that could be combined with other control procedures (such as SIT, super-traps, ripening retardants of the peel of citrus fruits (9)).

3. Prospectives

We anticipate that the present research activities on biotechnical control methods focussing on Ceratitis capitata as key pest in citrus groves will continue or expand. Not only was the straightforward collaboration between IOBC/WPRS and CEC stimulating this type of research in the past few years but we foresee an expansion of these investigations at the international level. A world-wide survey conducted by IOBC/WPRS among fruit fly specialists in fall 1984 revealed a high interest in further collaborative research programs on biotechnical methods for fruit flies, notably on attractants, deterrents and selective trapping systems. In view of the imminent establishment of an international fruit fly group under the umbrella of Global IOBC there is strong reason to assume that biotechnical technologists for fruit flies, and especially the medfly, will receive increased attention.

REFERENCES

1. BOLLER, E.F. (1983). Biotechnical methods for the management of fruit fly populations. Pages 342-352 in: Fruit Flies of Economic Importance

(R. Cavalloro, ed.). Proc. of CEC/IOBC Int. Symp. Athens, November 1982. Balkema, Rotterdam 642 pp.

2. BOLLER, E.F. (1980). The potential use of traps and sterile insects against Ceratitis capitata in citrus groves. Pages 73-79 in:Standardization of biotechnical methods of integrated pest control in citrus orchards (R. Cavalloro & R. Prota, eds.). Proc. CEC meeting 1980 in San Giuliano and Siniscola, 206 pp.

3. NEUENSCHWANDER, P. (1982). Beneficial insects caught by yellow traps used in mass-trapping of the olive fly, Dacus oleae. Ent. exp. appl. 32: 286-296.

4. DELRIO, G. and PROTA, R. (1980). Comparazione tra trappole cromotropiche e chemiotropiche per Ceratitis capitata Wied. Pages 87-102 in Proc. CEC meeting on 'Standardization of biotechnical methods of integrated pest control in citrus orchards (R. Cavalloro and R. Prota, eds.), 206 pp.

5. DELRIO, G. (1985). Biotechnical methods for Ceratitis capitata Wied. Proc. Fruit Fly Symposium, Hamburg 1984. (in print)

6. PROKOPY, R.J., ZIEGLER, J.R. and WONG, T.T.Y. (1978). Deterrence of repeated oviposition by fruit marking pheromone in Ceratitis capitata. J. Chem. Ecol. 4: 55-63.

7. PROKOPY, R.J. (1981). Epideictic pheromones that influence spacing patterns of phytophagous insects. Pages 181-213 in:"Semiochemicals: Their role in pest control (Nordlund, D.A., Jones, R.L. and Lewis, W.J., eds.), John Wiley & Sons, 306 pp.

8. BOLLER, E.F. (1981). Oviposition-deterring pheromone of the European cherry fruit fly: Status of research and potential applications. Pages 457-462 in:"Management of insect pests with semiochemicals" (E.R. Mitchell, ed.). Plenum Press, 514 pp.

9. GREANY, P. (1978). Citrus chemicals knock out fruit fly. Agr. Res. 27: 3-4.

Trapping *Ceratitis capitata* in McPhail traps baited with amino acids and ammonium salts

G.J.Tsiropoulos & G.Zervas

Biology Department, 'Demokritos' Nuclear Research Center, Aghia Paraskevi, Greece

Summary

The synergistic effect of single amino acids, in increasing Ceratitis capitata catches, when added to McPhail traps, baited with ammonium sulfate, was studied. More females were cought in traps containing the ammonium salt and one of the following amino acids:Isoleucine, valine, cystine, leucine, glutamic acid, tryptophane, lysine, alanine,arginine. Isoleucine trapped equal numbers of the two sexes, while, valine, tryptophane, aspartic acid, arginine, isoleucine, and glutamic acid trapped more males. The synergistic capacity of amino acids to ammonia salts, it is also discussed.

1. Introduction

The Mediterranean fruit fly, Ceratitis capitata (Wiedemann) (Medfly), is a serious pest of over 250 species of fruits, nuts and vegetables. The most potent lure for medflies is the commercially available trimedlure, a specific blend of the tert-butyl esters of 4- and 5- chloro-2-methylcyclo-hexanecarboxylic acids (1), which, however, attracts almost exclusively the males (2) (3).

Protein hydrolysates and ammonium salts, on the other hand, are used to attract and bait both sexes of several tephritid fly species (2). Commercial hydrolysates have been found to contain amino acids, either in free or bound form (4), some of which, in turn, have been reported as feeding stimulants for Ceratitis capitata (5). Ammonium salts alone in McPhail traps have a limited attractancy (3) probably due to increased gaseous ammonia emitted by them.

The scope of present preliminary investigation was to study the probable synergistic capacity of single amino acids in increasing the attractiveness of McPhail traps, baited with ammonium sulfate, to Ceratitis capitata flies and especially of the females.

2. Materials and Methods

Adult Ceratitis capitata flies, reared as larvae on a diet consisted of sucrose, brewer's yeast, alfalfa and straw meal, were provided sucrose and water, and held in plexiglass cages 40X20X20 cm. Temperature was 25±3 ^{o}C, relative humidity 60±5% and the natural photoperiod. Two hundred flies of each sex were released in a greenhouse measuring 10X7X5m, in which three pairs of McPhail traps, were hung at a height of about 3 m above ground. To avoid differences in trap color, due to the variation in the color of the various attractant solutions, all traps were painted outside with yellow color TEXOLAC No. 6 (Syntex Pry Ltd., Aspropyrgos, Greece), up to the liquid level. From each pair, one trap contained 200 ml of a 3% ammonium sulfate plus 2% of borax, and the other the same liquid, plus the tested amino acid. pH was adjusted at 8.5. The trapped flies were counted 3 days after release, followed by a second release and counting. To determine the flies´reaction to each amino acid, the ratio(r) of the flies trapped in the three traps containing the amino acid, over the flies trapped in the three traps containing the standard ammonium sulfate solution, was calculated. Thus, r>1 means a positive reaction, r<1 means a negative reaction, while r-values close to 1 denote a neutral situation.

3. Results and Discussion

Table I contains the 15 tested amino acids, ranked according to their r-value, for male and female flies, as well as the ratio between males and females.

Table I. Individual amino acids ranked according to Ceratitis capitata response to them as well as the ratio of males to females.

Males		Females		♂♂ : ♀♀	
amino acid	r-value	amino acid	r-value	ratio	r-value
Valine	1.84	Isoleucine	1.72	Aspartic acid	1.24
Tryptophane	1.40	Valine	1.43	Glycine	1.13
Aspartic acid	1.33	Cystine	1.39	Arginine	1.11
Arginine	1.25	Leucine	1.37	Isoleucine	1.00
Isoleucine	1.20	Glutamic acid	1.31	Leucine	0.94
Glutamic acid	1.20	Tryptophane	1.26	Serine	0.85
Tyrosine	1.10	Lysine	1.23	Valine	0.81
Lysine	1.06	Alanine	1.20	Tryptophane	0.65
Leucine	1.04	Arginine	1.20	Lysine	0.62
Cystine	0.97	Tyrosine	0.97	Phenylalanine	0.60
Glycine	0.96	Serine	0.95	Alanine	0.59
Phenylalanine	0.88	Aspartic acid	0.91	Tyrosine	0.58
Alanine	0.85	Phenylalanine	0.90	Methionine	0.43
Methionine	0.69	Glycine	0.84	Glutamic acid	0.38
Serine	0.62	Methionine	0.78	Cystine	0.35

The two sexes responded differently to the various amino acids. Males responded positively (r>1.20) to the amino acids: valine, tryptophane, aspartic acid, arginine, isoleucine, and glutamic acid. Females responded positively to the amino acids: isoleucine, valine, cystine, leucine, glutamic acid, tryptophane, lysine, alanine, arginine. As neutral could be considered the amino acids: tyrosine, lysine, leucine, cystine, and glycine, for males, and tyrosine, serine, aspartic acid, and phenylalanine for females. The remaining amino acids (r<0.90), could be considered as causing negative reactions to the flies. The amino acids which caused an increase in the number of the trapped flies, acted probably more as phagostimulants rather than as odor attractants, since amino acids have very low volatility, even at their isoelectic points (6). However, there is a posibility of exerting some attractancy by the products of their breakdown. In another rephritid, Dacus oleae, single amino acid water solutions have shown low-level attractancy (7) (8). Thus, it seems, the combination of ammonium sulfate and amino acid(s) act in two steps: first, the ammonia, which is a known powerful olfactory attractant, attracts the flies to the McPhail traps from a distance, and second, the amino acid(s), some of which are powerful phagostimulants (5), arrest the flies landing on the trap, when, by walking around the opening of the trap, develop a contact response, which finally brings them inside the trap, until they drown in the bait solution. The same process is probably taking place when the bait contains a protein hydrolysate. But, the advantage of the system: ammonium salt-amino acid(s), is in the capacity of the ammonia salt to produce higher amounts of gaseous ammonia, and to increase phago-stimulation by adding specific amino acid(s). This system, of course, allows the study of the phagostimulatory capacity of other substances, such as carbohydrates, aldehydes, etc.

Regarding the sex ratio of the trapped flies, it is worth noting that, out of the fifteen tested amino acids, eleven trapped more females than males, one equal numbers of the two sexes, and only three (aspartic acid, glycine, and arginine) trapped more males than females. This is an advantage of the system, since trimedlure traps attract almost exclusively males.

References

1. BEROZA, M. et al. (1961). Tert-Butyl and tert-pentyl esters of 6-me-thyl-3-cyclohexane-1-carboxylic acid as attractants for the Mediter-ranean fruit fly. J. Agric. Food Chem. 9:361.
2. CHAMBERS, D.L. (1977). Attractants for fruit fly survey and control, pp. 327-344. In Chemical Control of Insect Behavior: Theory and Application. John Wiley and Sons, Inc., New York.
3. DELRIO, G. and PROTA, R. (1980). Comparazione tra trappole cromotro-piche e chemiotropiche per Ceratitis capitata. pp. 87-102. In Standardization of biotechnical methods of integrated pest control in citrus orchards. Proceeding published by C.E.C. of an Expert's Meeting held in San Giuliano (Corsica) and Siniscola (Sardegna).
4. SIMPSON, R.J., NEUBERGER, M.R. and LIU, T.Y. (1976). Complete amino acid analysis of protein from a single hydrolysate. J. Biol. Chem. 251: 1936-1940.
5. GALUN, R., NITZAN, Y., BLONDHEIM, S. and GOTHILF, S. (1980). Respon-ses of the Mediterranean Fruit Fly Ceratitis capitata to Amino Acids. Proceedings Symp. Fruit Fly Problems, XVI Int. Congress of Entomology, Kyoto, Japan.
6. SVEC, H.J., and CLYDE, D.D. (1965). Vapor pressure of some a-amino acids. J. Chem. Eng. Data 10: 151-152.
7. ORPHANIDIS, P.S. and KALMOUKOS, P.E. (1968). Les acides amines et leurs sels en tant que facteurs de chimiotropisme positif des adultes du Dacus oleae (Gmel.). Meded. Fac. Landbowwet. Rijksuniv. Gent 33:937-943.
8. TSIROPOULOS, G.J. and HAGEN K.S. Unpublished data.

'In vitro' culture techniques to produce virus-free citrus plants*

A.Starrantino & A.Caruso

Citrus Experimental Institute, Acireale, Italy

Summary

Since 1949 the "Istituto Sperimentale per l'Agrumicoltura" in Acireale
has used selection from the field and the production and selection of
nucellar plants as a means of improving the sanitary status of citrus
trees.
In the last 15 years it has also used "in vitro" culture techniques
for nucellus, undeveloped ovules and shoot-tip grafting. This report
contains the results so far obtained using the latest "in vitro"
techniques.

1.1 INTRODUCTION

In the past, efforts to improve the sanitary status of citrus trees
were based solely on selection from the field and the production and
selection of nucellar plants.

By means of selection from the field, many interesting clones of the
different varieties were identified and kept under observation. Very few of
these proved to be free from virus infections and the majority were infected by exocortis and/or xiloporosis disease and, to a lesser extent, with
concave gum, blind pocket and other disorders. A clone of the "Clementine
Comune" and one of the "Tarocco Galici" have proved particularly
interesting and are already being distributed, although the latter is
infected with exocortis. It can be concluded from the foregoing that the
results achieved do not reflect the enormous amount of work which has been
done.

The other method of improving the sanitary status of citrus plants is
the production and selection of nucellar plants. It is common knowledge
that not only are nucellar plants almost always similar to the parent
plant, they are also usually free from virus infections. Thus, as a result
of a programme embarked upon in 1949 (13), many nucellar clones have been
obtained of the most common varieties of orange (Citrus sinensis L.
Osbeck), mandarin (C. deliciosa Tenore), grapefruit (C. paradisi Macf.) and
lemon (C. limon l. Burm.f.). Many of these clones have been distributed to

---

* Research conducted as part of the Ministry of Agriculture and Forestry
project "Development and improvment of industrial fruit-growing, early
fruit growing and citrus fruit growing", Publication n° 157.

citrus growers since 1965 and are used for almost all new commercial groves.

The drawbacks of this method are that some species are monoembryonic and do not produce nucellar embryos. Some of the most prized varieties are completely seedless or produce seeds only rarely. Furthermore, the nucellar plants have a very long juvenile phase, during which they are unproductive, very thorny and vigorous. This means that it is a long time, at least 10 to 15 years depending on the variety, before these plants can be used as mother plants and buds can be taken from them for commercial propagation. Furthermore, in the case of lemons, the nucellar clones have been used to only limited extent for commercial purposes, because they have proved to be more susceptible to "mal secco" infections than the old clones.

To overcome some of these difficulties, "in vitro" culture techniques have been used for the nucellus, undeveloped ovules and shoot-tip grafting.

This report is primarily concerned with the results obtained in the last 15 years at the Istituto Sperimentale per l'Agrumicoltura in Acireale, using these techniques.

1.2 DESCRIPTION

The first account of application of "in vitro" culture techniques to tissue to obtain citrus plants free from virus infection is given by Rangan et al (12) who devised a technique for isolating and culturing nucellus of monoembryonic species in order to induce nucellar embryo genesis. Many other authors report that they have used this technique successfully (4,11,15,17).

Plants of the clementine varieties "Comune", "Oroval", "Di Nules", "Monreal" and "Quattrostagioni" and the bergamot (C.bergamia Risso) varieties "Castagnaro" and "Fantastico" (species which are known to be monoembryonic (15,17), which we obtained using this technique proved to be different from the parent plants morphologically and to produce fruit with different organoleptic properties. From the information in our possession it appears that in California the plants obtained by "in vitro" culture of the nucellus of monoembryonic species (Murashige staff memo) are not similar to the parent plant. In Spain, approximately 20% of the plants obtained by Navarro et al (11) using this method had leaf abnormalities. These authors also report that the qualitative characteristics of the remaining, apparently normal, plants which had not yet been used in production had to be examined carefully before the plants could be used for commercial planting (11). In Japan, the "in vitro" culture of nucellus of various monoembryonic species by Iwamasa and Hirosue produced not nucellar embryos, but only zygotic embryos, which proliferated in a similar way to nucellar embryos. Kobayashi et al (7) have drawn attention to the role of the "primordial cell" in the formation of nucellar embryos which, in polyembryonic species, can be observed even before anthesis, whereas this is not the case in the monoembryonic species.

The same authors also report that the "in vitro" culture of the ovules of 23 monoembryonic varieties did not result in the production of nucellar embryos.

394

In the light of these reports and our own experience (16,18) we can state that the plants resulting from the "in vitro" culture of nucellus of monoembryonic species produced to date are not of nucellar origin, but are the result of the proliferation of zygotic embryos present but not identified in the nucellus at the time of its explantation for cultivation.

As to the completely seedless varieties, such as those belonging to the Navel group, Button and Bornman (2) and more recently other authors (1,8,11) report that they have obtained nucellar plants by cultivating unfertilized ovules taken from fruit of not more than 120 g days old. These authors thought that ovules taken from fruit weighing more than 120 days old would degenerate and not produce seedlings. From our research, it appears that these undeveloped ovules remain alive in the fruit even when it is in an advanced state of maturity and that when subjected to "in vitro" culture, they then produce nucellar embryos and subsequently seedlings (19,20).

With this method we have obtained nucellar plants from large numbers of "Navelina", "Navelate", "Thomson Navel", "Golden Buckeye" and "Skaggs Bonanza" varieties, which are free from virus infections and which have already started to produce. Of these a "Navelina" clone provisionally allocated the number 315, warrants special mention as being particularly interesting (21) for the quality of the fruit (Fig. 1), early bearing, the absence of thorns and its relatively limited vigour, compared with the nucellar plants of other varieties. It has accordingly been decided to start distributing it for commercial planting as of this year.

Of the other nucellar plants obtained using this method, some have been obtained from ovules imported under aseptic conditions from California, such as the "Robertson navel", "Atwood Early navel", "Workman navel", "Carter navel", "Dream navel", "Henniger navel" and "Newhall" varieties, and others based on recent valuable selected types of "Tarocco", which mature at different times (very early, early and late).

Since the shoots of the plants are known to be free from virus infections, it was decided to explant the tip of these shoots, comprising the cupule and the first two or three primordial leaves, this being not more than 0.3 mm in length, and to cultivate it "in vitro". This technique, which had been applied successfully to many herbaceous plants, had proved impossible to adopt for arboreous plants (and therefore citrus trees) because of difficulties in developing the shoots. This problem was resolved by Murashige et (9) al by micrografting the shoots onto seedling rootstocks plants cultivated "in vitro". This technique has since been applied and improved by a number of researchers (3,5,6,10,22).

The species to which we initially applied this technique were for the most part of the monoembryonic type, particularly bergamot and clementine (14). We have also worked on orange and lemon polyembryonic species (Fig. 2).

At the moment we have two bergamot clones, two "Oroval" and one "Nules" clementine clones and one "Favazzina" lemon clone, which biological tests have shown to be free from the main virus infections. A "Navelina" orange clone has been freed from psorosis and exocortis, although the test for impietratura has not yet been completed. Recently the health

improvement programme, involving shoot-tip grafting combined in some
instances with thermotherapy, has been stepped up and the scope expanded to
include a number of valuable early and late selected types of the "Tarocco"
orange, selected types of seedless lemons with sufficient resistance to
"mal secco", and also selected types of "Monreal" clementine with few
seeds, and an early clone of the "Comune clementine".

Large numbers of seedlings of these selected types have been obtained.
These are already in the process of cultivation (Fig. 3) and observation,
and are being subjected to biological tests.

REFERENCES

1. BITTERS, W.P., MURASHIGE, T., RANGAN, T.S. and NAUER, M. (1972).
    Investigations on establishing virus-free citrus plant through tissue
    culture. In Proc. 5th Conference Internat. Organization Citrus
    Virologists, University of Florida Press, Gainesville, p. 267-271.
2. BUTTON, J. and BORNMAN, C.H. (1971). Development of nucellar plants
    from unfertilized ovules of the "Washington Navel" orange through
    tissue culture. The Citrus Grower and Subtropical Fruit Journal, Sept.
    11-14.
3. CONTINELLA, G., BUSA', A. and VALENTI, C. (1983). Use of shoot-tip
    grafting (STC) technique in Italy I. World Congress International
    Association of Citrus Nurserymen. Valencia, Spain, 4-10 December 1983.
4. DEIDDA, P. (1973). Embrioni nucellari di clementine ottenuti in vitro.
    Riv. Ortoflorofruttic. Ital., 57:291-296.
5. DE LANGE, I.H. (1978). Shoot-tip grafting - a modified procedure. The
    Citrus and Subtropical Fruit Journal 539:13-15.
6. FODDAI, CORDA, A.P. and IDINI, G. (1980-81). Il risanamento degli
    agrumi dal viroide dell'exocortite mediante il microinnesto. Studi
    Sassaresi (S.III) 28:20-24.
7. KOBAYASHI, S., IKEDA, I and NAKATANI, N. (1978). Studies on nucellar
    embryogenesis in citrus. II. Formation of the primordium of the
    nucellar embryo in the ovule of the flower bud and its meristematic
    activity. J. Jap. Soc. Hortic. Sci., 48:179-185.
8. KOCHBA, J., SPIEGEL-ROY, P., SAFRAN, H. (1972). Adventive plants from
    ovules and nucelli in citrus. Planta, 106:156-162.
9. MURASHIGE, T., BITTERS, W.P., RANGAN T.S., NAUER M., ROISTACHER, C.N.
    and HOLLYDAY, P.H. (1972). A technique of shot-tip grafting in vitro
    for virus-free citrus. J. Amer. Soc. Hort. Sci., 100 (5):471-479.
10. NAVARRO, L., ROISTACHER, C.N. and MURASHIGE, T. (1975). Improvement of
    shoot-tip grafting in vitro for virus-free citrus. J. Amer. Soc. Hort.
    Sci., 100 (5):471-479.
11. NAVARRO, L. and JUAREZ J. (1977). Elimination of Citrus pathogen in
    Propagative budwood. II. In vitro propagation. Proc. Int. Soc.
    Citriculture 3:973-987.
12. RANGAN, T.S., MURASHIGE, T. and BITTERS, W.P. (1968). In vitro
    initiation of nucellar embryos in monoembryonic citrus. Hort. Science
    3:226-227.

13. RUSSO, F. and TORRISI, M. (1951). Il miglioramento delle nostre varietà di agrumi per selezione di forme derivate da embrioni nucellari. Annali Sperimentazione Agraria. Roma 5:5-12.

14. RUSSO. F. and STARRANTINO, A. (1973). Ricerche sulla tecnologia dei microinnesti nel quadro del miglioramento genetico-sanitario degli agrumi. Ann. Ist. Sper. Agrumicol. Acireale (CT) 6:209-222.

15. STARRANTINO, A. and RUSSO, F. (1972). Produzione "in vitro" di piantine dal tessuto nucellare delle specie monoembrioniche: clementine (Citrus reticulata Blanco), Bergamotto (Citrus bergamia Risso) e cedro (Citrus medica L.). Ann. Ist. Sper. Agrumicol., Acireale (CT) 5:67-79.

16. STARRANTINO, A. and TERRANOVA, G. (1976). Piante virus-esenti ottenute con la coltura "in vitro" di nucelle di Clementine '"Comune". Ann. Ist. Sper. Agrumicol., Acireale (CT) 9-10:201-208.

17. STARRANTINO, A., SPINA, P., RUSSO, F. (1978). Embriogenesi nucellare e sviluppo di piantine in vitro dalle nucelle di alcune specie di agrumi. Giornale Botanico Italiano, vol. 112, 1-2:41-52.

18. STARRANTINO, A., RUSSO, F. and SPINA, P. (1979). La micorpropagazione in Agrumicoltura. Atti Convegno su tecniche di colture "in vitro" per la propagazione su vasta scala delle specie ortoflorofrutticole. Pistoia 219-228.

19. STARRANTINO, A., RUSSO, F. (1980). Seedlings from undeveloped ovules of ripe fruit of polyembrionic citrus cultivars. Hort. Science. 15 (3):296-297.

20. STARRANTINO, A. and RUSSO, F. (1983). Reproduction of seedless orange cultivars from undeveloped ovules raised "in vitro". Acta Horticulturae, 131:253-258.

21. STARRANTINO, A. and RUSSO, F. (1985). Una nuova varietà di arancio precoce: la Navelina ISA 315. Frutticoltura (in stampa).

22. TUSA, N., DE PASQUALE, F. and RADOGNA, L. (1977). Research on the micro-grafting technology in citrus. Int. Soc. Citriculture 1:143-145.

Integrated pest control in citrus-groves: Problems posed by chemical methods

P.Brun

INRA, Agronomical Research Station, San Nicolao, France

Summary

Integrated pest control in citrus groves includes chemical methods. Decisions to apply chemicals involve various considerations, including the pest concerned, other pests present, the products to be used, the method of application, and the phenology of the plant.

1. INTRODUCTION

Over 70 different species of citrus pests are found in the citrus-growing countries of the Mediterranean basin. A score of these are capable of causing economically significant damage.

For several years, various control methods have been employed to limit the damage caused by the major pests or to check sudden infestations of secondary pests.

At the present time, there is increasingly widespread acceptance in a number of countries of the idea of integrated control in the citrus industry. It is, however, very rare to find a situation in which all pest-related problems in this sector can be solved without any use of chemicals. These therefore have a role to play alongside other methods, particularly biological control.

In the majority of countries, chemical means are used to combat a number of citrus pests. Certain rules must be observed if efficiency is to be improved and undesirable effects kept to a minimum.

2. CHEMICAL CONTROL IN CITRUS GROVES

In the operating budget for 1 hectare of citrus orchards at full production in Corsica, more than 12% of expenditure goes on plant protection, making it the third most important item after harvesting and pruning, and more expensive than fertilizer, herbicides and irrigation.

The percentage of expenditure devoted to chemical control may vary in the various countries of the Mediterranean basin, depending on the citrus species involved, the type of market, or the presence of particular pests requiring specific control measures.

Growers, however, always seek the cheapest possible plant protection methods in view of the spiralling cost of pesticides and of the labour needed to apply them.

Table I – The major citrus pests and the type of damage they cause

Pests	Parts of the plant attacked					Types of damage			
	Leaves	Fruit	Branches	Flowers	Roots	Sooty Mould	Lower yield	Fruit	Other
Scale insects:									
Diaspinae	+	+	+				+	+	
Lecaninae	+		+			+	+		
Pseudoc.	+	+				+	+	+	
Margar.	+		+			+	+		
Aphids	+		+	+		+	+		+
Whitefly	+					+	+		
Mites	+	+					+	+	
Lepidoptera	+	+		+			+	+	
Diptera	+	+					+	+	
Thrips	+	+		+				+	
Nematodes					+		+		
Leafhoppers		+						+	+
Ants	+		+						+ +
Grasshoppers	+								+ +
Slugs		+						+	

400

A grower's decision to use chemicals against one or more pests will be based on the following factors:

- the significance, in terms of intrinsic danger, he attributes to the kind of pest concerned;
- the type of damage caused by this pest and the speed with which it becomes evident in the orchard;
- the time of year in relation to the phenology of the trees;
- the type of pesticide used, the way it is applied and its effects on auxiliaries either naturally occurring or introduced.

2.1 Significance attributed to various pests

Pests are usually ranked according to the potential risk of loss or damage to the crop, i.e. the direct damage they can do to the fruit (Table 1). This direct damage may be caused by various species of scale insect (Diaspinae and Pseudococcinae), which attach themselves to the fruit or whose toxic saliva causes permanent discolouration, or by mites (spider mite, Panonychus citri Mc G.), leafhoppers (Empoasca sp.), Lepidoptera, thrips and the Mediterranean fruit fly.

The irreversible nature of damage to fruit explains why fruit growers attach such importance to it, since the alteration of the external appearance of the fruit means a reduction in its quality and hence its value. This in effect leads to crop losses, because this fruit will have to be rejected during sorting to comply with marketing standards or export legislation.

Where certain pests attack young fruit (in particular Diaspinae, Pseudococcinae and Lepidoptera), the relative importance attached to them depends on the extent of the direct crop losses which produce a drop in yield.

A less easily quantifiable form of damage occurs when insects disturb the various physiological functions of the tree by sucking out the sap or injecting toxic saliva into the plant tissue, which results in weakening of the tree, leaf-fall, drying out of the branches, reduced blossoming or a greater fall of young fruit after flowering.

In most cases, damage of this kind is done by very large pest populations. Moreover, such levels are rarely reached because the grower is given prior warning of the presence of pests in the orchard by the general appearance of the trees or, more often, by indirect damage in the form of sooty mould which is much more easily noticed.

Deposition of this sooty mould on the foliage helps to weaken the plant by interfering with respiration and photosynthesis. Sooty mould often indicates the presence of an insect producing honeydew, such as scale insects or white flies, the larval population of which often goes unnoticed on the plant.

When sooty mould damage appears on the leaves or fruits, pest populations are usually still quite small and not yet capable of affecting the physiological functions of the trees or the size of the crop.

The seriousness of this sooty mould deposit on the fruit can be reduced by post-harvest treatment to remove it, such as soaking and brushing at the packing plant.

One particular group of pests, Aphids (especially <u>Aphis citricola</u> P. and <u>Toxoptera aurantii</u> B. of F.), is associated with damage that is usually obvious and very quick to appear. Often this is regarded as far more serious than it really is - especially in the case of fully-grown trees. Decisions to spray aphids are usually based on the aesthetic appearance of the trees (deformed leaves), even though the sap-sucking and honey-dew production in fact have very little effect on the tree.

In practice, nothing is done to stem attacks by the citrus nematode <u>Tylenchulus semipenetrans</u> Cobb., because it is extremely difficult or even impossible to combat it with pesticides on adult trees. The risk of damage by this species is taken into account before the trees are planted and resistant rootstock is used.

It should be noted that the possible transmission of certain diseases by insects (aphids, leaf-hoppers) is rarely taken into account in the assessment of the indirect damage caused by pests.

2.2 Chemical treatment and citrus phenology

There are a number of stages at which tree phenology becomes a significant criterion in determining what, if any, chemical treatment is necessary.

During growth periods, attention is paid to the foliage and to those pests likely to hinder its proper development, such as aphids and mites, which deform and discolour the leaves respectively.

Pests capable of causing damage during the blossom period (aphids, thrips, Lepidoptera) often confront the grower with particular problems because spraying at this time is likely to cause flowers to fall and there-fore reduce the size of the next crop, and because he has to choose pesticides that will not harm bees or auxiliaries.

In the post-blossom period, when the young ovaries are falling, growers are highly reluctant to apply chemicals because they are afraid that spraying at this time will mean the fall of a higher proportion of young fruit. In the case of most citrus species, fruit begins to fall in mid-June, continuing to do so substantially until mid-July and then tailing off markedly until the beginning of August, at which point treatment can be recommenced without any fear of increasing the fall of young fruit, which now have an average diameter of 10 to 15 mm.

When one considers that the average proportion of young fruit that falls from citrus trees varies between 60% and 90%, the growers' fears are quite understandable.

From then on, during late summer and autumn, much closer attention is paid to the fruit on the tree and to any pest likely to attack it, with the grower's readiness to apply chemicals increasing as harvest approaches.

The various types of chemical treatment can be classified as follows, according to the phenology of the trees and the type of damage caused by the pests:

- preventive treatment carried out virtually every year at specific times depending on the presence and abundance of certain pests, such as summer spraying for scale insects;

402

- remedial treatment with the aim of checking a pest which has reached population levels considered to be too high and a potential risk to the proper functioning of the tree or to the next crop (e.g. aphids, whitefly, mites);

- remedial, preservative treatments during the two months before harvest, where the grower is no longer prepared to tolerate any pest attack that might reduce either the quantity or quality of the crop.

2.3 Choice of pesticide and spraying techniques

Once it has been decided to carry out chemical treatment, three new aspects of the problem have to be considered: the choice of active ingredient, application technique and the quantity of suspension to be sprayed.

The choice of active ingredient is very rapidly made if a specific pesticide is used (control of mites, aphids, Lepidoptera). The choice is more limited if pests show resistance to pesticides, particularly in the case of mites.

Where broad-spectrum pesticides are used, effectiveness of the active ingredient on the pest in question and its stage of development at the time of spraying are not the sole considerations. Account must also be taken of the impact this active ingredient will have on all the auxiliaries present in the orchard. The mode of action and, in particular, the systemic properties of the product will be examined for the control of certain pests or to avoid the pesticide being washed off in rainy periods or by over-tree sprinkler irrigation.

In practice, the grower sometimes looks for the broadest-spectrum pesticide available or even mixes a number of active ingredients with the aim of controlling several pests, even if only one of them has exceeded the threat threshold. His justification for this practice is that it makes "more economic" use of time and equipment. This use of chemical pest management is the most injurious to auxiliaries.

One specific case, i.e. treatment carried out at the end of winter after pruning, can be justified in a number of ways. On the one hand, it is the only time of year in which the quantity of foliage has been artificially reduced, by between 10 and 25% in the case of normal pruning (as with the clementine) or even by more than 40% where pruning is severe. In the case of most citrus species, the plant's habit and the quantity and density of its foliage make it very difficult to achieve correct penetration and spreading of the pesticide. Penetration is a great deal easier after pruning and it is much less difficult to achieve a proper spread of the suspension over the whole of the tree.

Secondly, treatment at this time of the year makes it possible to carry out a general 'cleaning' of the tree, either because attacks on the fruit by certain pests had been noticed during the previous harvest or because such treatment is regarded as a measure to reduce the damage done by insects or mites during the spring to summer period, the target being the generally small populations that have survived the winter.

Once the active ingredient has been chosen, the technique adopted to spread the pesticide again imposes a number of constraints that are rarely fully taken into account, yet have a major influence on the success or failure of chemical treatment. This is because, depending on the period of the year, the appearance of the trees and the pest concerned, a completely successful result will require the application of a greater or lesser quantity of suspension. The following are listed in increasing order of the quantities of suspension required:

- applications of very small quantities of liquid per hectare, particularly when combatting fruit flies, since only part of the surface is treated. Aerial spraying also falls into this category.

- treatments requiring a relatively small volume per hectare because the pest is restricted to parts of the tree that are easily reached by conventional spray techniques (combatting aphids on young leaves or localized pests on the flowers).

- intensive spraying of white flies or mites which are spread more widely through the foliage or located in places particularly difficult to reach, such as the under-side of the leaves.

- very intensive spraying to combat scale insects, made necessary by their presence throughout the tree or in favoured crannies that are difficult to reach. In this case, the use of active ingredients, such as white oils, which kill only on contact, demands that particular attention be paid to choosing exactly the right equipment to carry out the treatment.

3. CONCLUSION

The practical use of chemicals as part of integrated control in citrus groves requires the observation of a number of principles. Even though there are a number of levels at which advice on chemical control techniques can be offered to the grower, certain aspects of decision-making on whether and how treatment should be carried out remain the preserve of the grower himself. There is therefore a need to understand the reasoning involved, so as to avoid many failures and to help achieve proper integration of chemical control within the full range of pest control measures.

Trials of chemical control against *Aphis citricola* v.d. Goot on new grafts of Tarocco in the province of Catania

S.Inserra & G.Caltabiano
Institute of Agricultural Entomology, University of Catania, Italy

Summary

During 1984 chemical controls were made in a citrus orchard of cv. Tarocco lar
gely attacked by Aphis citricola v.d.G. in the province of Catania.
At the beginning of the tratment the infestation attacked 96% of the shoots and on
each of them there was an averange of 850 individuals counting adults and larvae.
The tested products were triazophos a 60 gr/hl and a 80 gl/hl, ethiofencarb a
69 gr/hl and heptenophos a 37,5 gr/hl.
The best results were obtained with the last two insecticides which destroied almost all the population of aphids and spared the useful predators.
The survived individuals were controlled by Coccinella 7-punctata L. and by Adalia bipunctata (L.) which were able to attack the infestation of aphids also on test.
On the contrary the use of triazophos was inadvisable.

1. Introduction

Aphids are a serious phytosanitary problem in particular years in new planted
citrus orchards, where their attacks interest young plants or in new grafted citrus orchards, where aphids colonize young grafts. Among 10 serveyed species on
citrus in Sicily (Patti, 1983) Aphis citricola v.d.G. causes the greatest difficulties
for its damages.

During 1984 in an orange-grove of cv. Tarocco grafted for a year in Motta S.
Anastasia (Catania), a very strong infestation of this species was noticed just on
young grafted shoots.

The attack of phytomizo, on treatment (30-5-1984) interested 96% of present
shoots; on each infested shoots there was an average of 850 individuals counting
adults and larvae.

Though there was a high presence of predators (Coccinellini, Sirfidi e Cecido
midi) whether adults or larvae, a chemical treatment was made all the same, becau
se the attack was strong and we coutd not expose the shoots to the infestation of
aphid further on. In fact, it was unknown the duration of its presence, in which mea
sure the predators found in the colonies would operate repressing it, and the effect
of aficidae we wanted to test on them. The comparison tests were the only means to
obtain our answers.

During our researchs no parasite was found that agrees with the result obtained in Sardinia by Ortu and Prota (1980). Who assert the population of A. citricola
is controlled by Sirfidi, Crisopidi and chiefly by Coccinellidi, and as a rule
with what was found by Tremblay and others (1980) in Southern Italy. They assert
that A. citricola is sparely controlled by indigenous Aphidiini.

TABLE I

Aphis citricola v.d.G. tests in comparison on Tarocco, in Motta S.Anastasia (Catania) and results of the treatment (30–5–1984).

Plots	Insetticides on dosage/hl in parenthesis	Average number of living individuals on 20 shoots (average of 4 replications)						
		Dates of samplings						
		31–5	1–6	2–6	4–6	7–6	11–6	14–6
1	triazophos (60)	115,25 Bb	78,75 b	113,5 b	170,5 b	181,25 b	252,75 a	291,25 b
2	triazophos (80)	53,0 Bb	60,0 b	71,5 b	90,5 b	142,25 b	357,75 a	632,0 a
3	ethiofencarb (69)	1,5 Bb	1,5 b	0,0 b	0,0 b	0,0 c	0,0 b	0,0 c
4	heptenophos (37,5)	3,0 Bb	3,0 b	2,0 b	0,0 b	0,0 c	0,0 b	0,0 c
5	test	4.172,0 Aa	3.700,0 a	3.075,0 a	2.100,0 a	443,0 a	34,25 b	0,0 c

– Values flanked by different capital letters are significant at P = 0,01

– Values flanked by different small letters are significant at P = 0,05

(Test of Duncan)

When we began our tests there was a very strong infestation and attack of a species (A. citricola) which according to Barbagallo (1966) is the most dangerous among the species infesting our citrus orchards, for the remarkable amount of shoots the infestation attack and because the attacked organs suffer irreversible alterations by the rolling up of leaves and strains of internodes, causing damages that other aphids living on citrus don't do.

2. Materials and methods

The test was made at the end of May (30–5–1984). On treatment, infestation attacked 96% of shoots on plants with an average of 850 individuals per shoot; each of them had an average of 94 leaves and a caulis about 18 cm long. There was the presence of a large population of predators on the colonies of aphid too, so it was possible to count an average of 5,3 predators on each examined shoot. Among them Chilocorus bipustulatus L. and Coccinella 7–punctata L. were the most frequent.

The examined plots had 96 plants in each plot we casually located and marked 20 shoots (4 replications x 5 plants x 1 shoot) on which drawings were made before and after the treatment. The other plants were utilized as separation among the plots and the replications they were treated too.

We tested the following 5 compared plots:

Plot 1 = triazophos   at 60 g/hl of water
Plot 2 = triazophos   at 80 g/hl of water
Plot 3 = ethiofencarb at 69 g/hl of water
Plot 4 = heptenophos at 36,5 g/hl of water
Plot 5 = test

The screening of the above insecticides aimed at proving the effect of two of them of new formulation (Hostaquick at 50% of triazophos and Hostation at 40% of triazophos) compared with a specific aficide of tested efficacy (Croneton at 46% of ethiofencarb).

Treatment was made by a knapsak – mist blower at normal concentration, it was important to wet the foliage without trickling with an average use of 0,5 lt insecticide mixture a tree.

The result of the treatment for the 5 compared plots was tested by means of drawings made every 24 hours for the first 3 samplings. The first of them was made on 31 May 1984. The following 4 samplings were made at intervals of 72 hours beginning the third day after the treatment.

In our periodical controls we closed the shoots into plastic sacks before removing them from the plants and then in laboratory counted the living samples of phitomize and its predators. In all samplings for each plot the shoots examined were 20 in all (1 shoot x 5 plants infested x 4 replications) taken casually.

Statistical elaborations were analyzed by the test of Duncan and the analysis of variance.

3. Results and discussion

Results are compared in table 1 and from their examination we note the effect aphicide of heptenophos in all similar to ethiofencarb, considered a selective specific aphicide (Melia and Blasco, 1980; Ortu and Prota, l.c.).

The selectivity- of the two insecticides was similar, in both it was restricted to Coleotteri, eliminating Sirfidi and Cecidomidi (Table 2). On the contrary the action of triazophos had a less effect on aphids and no selection; as a consequence of the non selectivity and   its partial action on aphids there was a new infestation on plots treated with this insecticide,absent on those treated with ethiofencarb and heptenophos (Graph. 1). For this reason we can assert that heptenophos as ethiofencarb may be considered specific aphicides with a partial selective action on Coc cinellini.

Our results agree with Dirimanov and others (1974) who assert ethiofencarb is not very toxic for Coccinella 7-punctata. Babrikova (1979) and (1980) finds that the above insecticide is not very toxic to Chrisopa carnea Steph. and Chrisopa formosa Brauer and ends saying that it can be used in the context of integrated control. On the contrary Grapel (1982) refers that in laboratory tests he has obtained mortality of 100% of Chrysopa carnea, Coccinella 7-punctata and Syrfus corollae L.

In our experiment the effect of treatment with ethiofencarb and heptenophos controlled almost all aphids with an insignificant presence of survived individuals, which were eliminated by the survived Coccinellini. We obtained the result 15 days before the predators on test.

4. Conclusion

The effect of predators is strong but tardive. In fact when they disinfest trees

407

TABLE II

Predators present in compared plots pre and post chemical treatment against Aphis citricola v.d.G.

Days from the date of treatment to control	Dates	Triazophos 60 gr/hl						Triazophos 80 gr/hl						Ethiofencarb 69 gr/hl						Heptenophos 37,5 gr/hl						Test					
		1		2		3	4	1		2		3	4	1		2		3	4	1		2		3	4	1		2		3	4
		A	B	A	B	B	B	A	B	A	B	B	B	A	B	A	B	B	B	A	B	A	B	B	B	A	B	A	B	B	B
—	30-5-84 pre-treatment	+	+	+	+	+	+	+	+	+	+	+	+	+	+	+	+	+	+	+	+	+	+	+	+	+	+	+	+	+	+
1	31-5-84	–	–	–	–	–	+	–	–	–	–	–	–	–	–	–	–	–	–	+	+	+	+	+	+	+	+	+	+	+	+
2	1-6-84	–	–	–	–	–	–	–	–	–	–	–	–	+	+	+	–	–	–	+	+	–	+	–	+	+	+	+	+	+	+
3	2-6-84	–	–	–	–	–	–	–	–	–	–	–	–	–	+	–	–	–	–	+	+	–	–	–	–	+	+	+	+	+	+
5	4-6-84	–	–	+	+	–	–	–	–	–	–	–	–	+	+	–	–	–	–	+	+	–	–	–	–	+	+	+	+	+	+
8	7-6-84	–	+	+	+	–	–	+	+	–	–	–	–	+	+	–	–	–	–	+	+	+	+	–	–	+	+	+	+	+	+
12	11-6-84	+	+	+	–	+	+	–	–	–	–	–	–	+	+	–	–	–	–	+	+	+	+	–	–	+	+	+	+	+	+
14	14-6-84	+	+	+	+	+	–	+	+	+	–	–	–	+	+	–	–	–	–	+	+	+	–	–	–	+	–	–	–	–	–

(1) = Coccinellini : Adalia bipunctata (L.).
(2) = Coccinellini : Coccinella 7-punctata L.
(3) = Sirfidi spp.
(4) = Cecidomidi spp.
(A) = Adults
(B) = Larvae
(+) = Presense
(–) = Absence

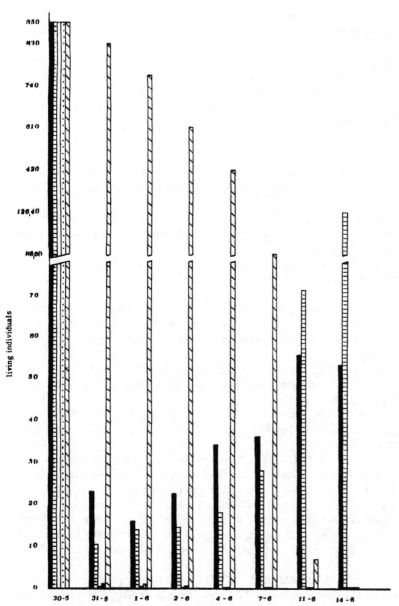

Graph. 1 – <u>Aphis citricola</u> v.d.G.. Average number of living individuals per shoorts in plots treated in different ways.

■ Triazophos at 60 g/hl   ⊡ Heptenophos at 37,5 g/hl

⊟ Triazophos at 80 g/hl   ⧄ Test

☐ Rthiofenearb at 69 g/hl

409

Aphids have already caused considerable damages.

Besides a tardy chemical treatment is unjustified in a common agricultural pratice.

It was useful only to show the aphicide effect of insecticides and their degree of selectivity in comparison with the useful fauna, present on treatment.

For this species, in fact, treatment is estimated when on orange tree the frequency of infested shoots is 10–15% (Ortu and Prota, l.c.) or 10% at most (Cavalloro and Prota, 1983).

REFERENCES

1. BABRIKOVA, T. (1979). The effect of pesticides on the individual stages of the common lacewing (Crysopa carnea Steph.). Rasteniev''dni Nauki, vol.16: 105–115.
2. BABRIKOVA, T. (1980). Studies on the effect of some pesticides on various stages of the lacewing – Crysopa formosa Br. Nauchni Trudove, Entomologiya, Mikrobiologiya, Fitopatologiya, vol. 25: 31–40.
3. BARBAGALLO, S. (1966). Contributo alla conoscenza degli afidi degli agrumi – Aphis spiraecola Patch. Boll.Lab.Ent.agr. "Filippo Silvestri" Portici, vol. 24: 49–83.
4. CAVALLORO, R. e PROTA, R. (1982). Lotta integrata in agrumicoltura: metodologia di campionamento e soglie di intervento per i principali fitofagi. Proceedings of the E.C. – Experts' Meeting, Siniscola – Muravera, 20–22 October 1982: 1–63.
5. DIRIMANOV, M., STEFANOV, D. and DIMITROV, A (1974). The effect of certain insecticides on the more important species of useful insects on tobacco. Rastitelna Zashchita, vol. 22: 18–20.
6. GRAPEL, H.(1982). Studies on the effects of some insecticides on natural enemies of aphids. Zeitschrift fur Pflanzenkrankheiten und Pflanzenschutz, vol. 17: 196–204.
7. MELIA, A. e BLASCO, J. (1980). Citrus aphids. The results of various tests of products to determine their effectiveness against different species. Servicio de Defensa contra Plagas e Inspeccion Fitopatologica, vol. 6: 67–73.
8. ORTU, S. e PROTA, R. (1980). Validità dei metodi di campionamento e delle relative soglie di intervento per il controllo dei principali fitofagi della arancicoltura. Atti Riunione CCE Corsica e Sardegna, 4–6 novembre 1980: 35–52.
9. PATTI, I. (1983). Gli Afidi degli Agrumi. Pubblicazione del C.N.R. 1983: 1–36.
10. TREMBLAY, E., BARBAGALLO, S. MICIELI DE BIASE, L., MONACO, R. e ORTU, S. (1980). Composizione dell'entomofauna parassitica vivente a carico degli Afidi degli Agrumi in Italia (Hymenoptera Ichneumonoidea, Homoptera, Aphidoidea). Boll.Lab.Ent.Agr. "Filippo Silvestri" Portici, vol.37: 209–216.

Planococcus citri (Risso) control in Sardinia*

S.Ortu

Institute of Agricultural Entomology, University of Sassari, Italy

Summary

Tests were carried out in 28 orange groves, each of 0.4 ha. The control
was programmed on the basis of male Planococcus citri (Risso) captures
by means of pheromone traps. A single chemical treatment effectuated
during the first ten days of September, coinciding with the second male
peak capture was sufficient, in the majority of cases, to keep the infestations within a harmless level. In January, at the beginning of the
harvest, treated fruit resulted almost entirely free of sooty mold and
directly marketable, while 37% of those untreated, were covered by sooty
mold and considerably depreciated.

1.1 Introduction

Planococcus citri (Risso) can be considered the most harmful pest for
citrus fruit cultivation in Sardinia (1). Serious scale infestions appear
in all fruit growing districts of the island, often determining vast economic losses above all due to sooty molds that disfigure fruit smeared with
honey–dew.

The application of a supervised control as the first form of rational
intervention and thus avoid massive use of ever more toxic insecticides,
permitted the limiting of chemical interventions to 1 - 2 (4). The contemporary examination of all the biocenose present in the crop permitted,
furthermore, a closer understanding of certain bioethological aspects essential for the reaching of integrated forms of control aimed at reducing
the scale population without disturbing the system as a whole.

The introduction of Nephus reunioni (Furs.) and Leptomastix dactylopii
(How.) and the greater diffusion of Cryptolaemus montrouzieri (Muls.) as
from 1980 strengthened the containment effect accomplished by other entomophagous present: Anagyrus pseudococci (Grlt.) and Leptomastidea abnormis
(Grlt.).

L. dactylopii activity although having reached at the moment of introduction in the experimental area a parasitization of above 90% (3) was con-

---

* Studies of the C.N.R. Working Group for the Integrated Control of
  Plant Pests: 258.

411

siderably reduced during the Winter season. Specific field controls showed
that the Encyrtid is not able to resist the island's adverse climate.
Coccinellids, on the contrary, including Nephus reunioni having a greater
ecological valency established colonies in the fruit growing areas where
they were introduced, showing themselves to be, however, less active than
L. dactylopii in the control of infestations.

From these initial observations it would seem possible that the efficiency of the biological control of the scale is connected to the constant
distribution of the quoted entomophages which can be assured only by the
presence of adequate breading structures. The lack of facilities adapted
for this purpose makes it necessary for us to direct our researches towards
alternate techniques which can keep the chemical intervention down to the
indispensable minimum. Of particular importance, for this purpose, is the
use of synthetic sexual pheromone which consents the accurate survey, even
by non-specialized personnel, of insect flight dynamics and therefore
greater selectivity of the chemical means employed (4).

As from the first experiments conducted on small areas, elements were
treated to extend experiments and obtain more complete information.

1.2 Materials and Method

Tests were conducted in a citrus grove of approximately 30 ha situated
in the countryside surrounding Sili (province of Oristano) on 28 groves of
adult orange trees of the following cultivars: Washington Navel, Ovale,
Tarocco, Belladonna, Valencia, three of which were chosen at random and kept
as untreated samples for comparison.

The survey of male Planococcus citri was carried out using cylinder
traps each one activated by 250 μg of synthetic sexual pheromone. The trap
consists of a 6cm dia. cylindrical tin, 8cm in height, with 6 (1.5 cm dia.)
entrance holes situated in the upper quarter. A third of the trap was filled
with a solution of Decis at 2% and the pheromone rubber capsule was hung at
the level of the entrance holes. 20 traps were used, baited every 15 days,
hung at eye-level in the south western quarter of the tree. The insect count
was carried out every day renewing the insecticidal liquid in the container.

The examination of fruit infestations was effectuated following the
sample methodology established by the C.C.E. Working Group " Integrated control in citrus orchards ". (Examination of 20% of trees, 10 fruit per plant
taken at random from around the foliage perimeter).

1.3 Results

During 1984 male captures of Planococcus citri began in May and continued up to the first few days of January (Fig. 1) confirming the presence
of three capture peaks in June-July, August-September and October-November,
showing the development in Sardinia of 3-4 generations per year.

The highest captures were noted towards the end of Summer. In this period the percentage of fruit infestation generally resulted considerably vari

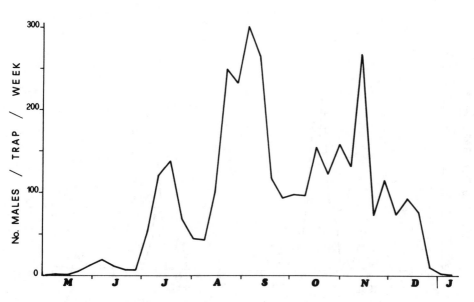

Fig. 1 Male captures of Planococcus citri (Risso)(Sili 1984: survey on 20 cylindrical traps baited with 250 μg synthetical pheromone).

Tab. I Planococcus citri (Risso) infestations on orange (Sili 1984: 28 groves each of 0.4 ha; estimated infestation on 20% of trees, 10 fruit per tree).

Date			Fruit infestation		
			< 15 %	15–30 %	> 30 %
7. 9.1984	Treated groves	(N°)	6	15	4
	Untreated groves	(N°)	2	0	1
28.12.1984	Treated groves	(N°)	23	2	0
	Untreated groves	(N°)	2	0	1

able from grove to grove with values oscillating between 7.7% and 52% demonstrating a considerable aggregation of the scale population.

The intervention threshold established in 15% of infested fruit was exceeded in 19 of the groves intended for insecticidal treatment while it was reached in only one of the untreated groves (Tab. I).

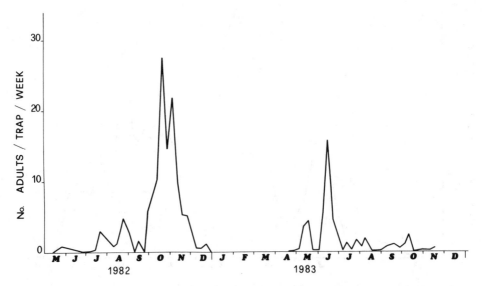

Fig. 2 Captures of <u>Anagyrus</u> <u>pseudococci</u> (Grlt.) and <u>Leptomastidea</u> <u>abnormis</u>
(Grlt.), principal natural enemies of <u>Planococcus</u> <u>citri</u> (Risso)(Sili
untreated grove: weekly count of 10 sticky yellow plexiglas traps).

The observations made on useful entomofauna in untreated groves revealed the higher presence of these entomophages during October-November (1982)
and June (1983)(Fig.2).

It appears evident, therefore, that in order to not disturb the biological action of these entomophages, repeated chemical treatments must be avoided during the Spring-Summer period especially if long persistance products
are used.

On the basis of these observations, treatments were carried out during
the first ten days of September immediately following the peak capture at
the end of August, corresponding to the highest presence of young scale
instars distributed on the fruit.

This moment appears to be, in fact, the most suitable for intervention
against the <u>Planococcus</u> population easily attackable also with low toxic insecticides (4).

Treatments were normally carried out with white oil to which was added,
only in exceptional cases of higher infested groves, a phosphorganic insecticide.

In all cases, copper oxy-cloride was added to the insecticide mixture
with the aim of preventing the formation of sooty molds.

In groves where the presence of <u>P. citri</u> on the fruit was lower than
the intervention threshold (less than 15 %) the successive infestation always remained within acceptable limits. In 19 groves in which the treatment
was carried out because the intervention threshold had been exceeded, only
2 (initially strongly attacked) presented a final infestation slightly
higher than the intervention threshold. In the untreated groves, on the con

Tab. II   Qualitative characteristics of production (Sili, January 1985: ob-
          servations of 20 % of trees, sampling 10 fruit per tree).

Groves	Healthy fruit	Fruit with sooty mold on	
	(%)	1/4 surface (%)	1/2 surface (%)
Treated	97.3	2.0	0.7
Untreated	62.8	0.2	37.0

trary, where the infestation percentage was 40.2 % a slight increase took
place, in the same period, and reached 42.5 % confirming the insufficient
action of the entomophages present including those of recent acclimatization.

In January, at the beginning of the harvest, the treated fruit resulted
almost entirely directly marketable, while those untreated, rather depreciated, were 37 % covered by sooty mold extending over a quarter of their sur
face (Tab. II).

1.4 Conclusions

The experiments confirmed the validity of the usage of pheromone traps
whether for the substantial reduction of the number of infestation observations (suggested by the sample methodology), thus limiting such control to
the period strictly corresponding to the second peak of male captures, or
whether for the more accurate programming of the timing of chemical intervention .

Chemical control is indispensable until it is possible to introduce con
stantly adequate quantities of Leptomastix dactylopii into vast areas, the
biological action of which, among other factors, has been shown to be as ef
ficient as that obtained by the application of chemical treatment (2,5). At
present it can be said that a single treatment can be sufficient, in most
cases for the control of P. citri; should this not be the case a further in
tervention 2-3 weeks after the first is recommended. Keeping in mind the aggregate distribution of the insect, the possibility to intervene exclusive
ly on the cultivation, localizing treatments to the infested plants, should
be considered, in all circumstances.

Pheromone traps consent, for the time being, to establish with preci-
sion, the optimal moment for the insecticidal intervention and it is possible, with more data on capture-infestation relations in the different
pedo-climatic situations, to identify a practical intervention threshold
based exclusively on the number of males attracted.

REFERENCES

1. DELRIO, G., ORTU, S., PROTA, R. (1979). Prospettive di lotta integrata nell'agrumicoltura sarda. Studi Sassaresi, sez. III, Ann. Fac. Agr. Sassari, 27: 205-232
2. MINEO, G., VIGGIANI, G. (1979). Su un esperimento di lotta integrata negli agrumeti della Sicilia. Boll. Lab. Agr. Portici, XXXIII: 219-231
3. ORTU, S., PROTA, R. (1980). Validità dei metodi di campionamento e delle relative soglie di intervento per il controllo dei principali fitofagi della arancicoltura. In: Standardizzazione di metodologie biotecniche nella lotta integrata in agrumicoltura. Cavalloro R. e Prota R., ed. Gallizzi, Sassari, 35-52
4. ORTU, S., DELRIO, G. (1983). Le trappole a feromoni nella programmazione della lotta al Planococcus citri (Risso) in Sardegna. Atti del XIII Congr. Naz. It. Ent., Sestriere - Torino, 235-258
5. VIGGIANI, G. (1977). La lotta guidata contro i fitofagi degli agrumi. Informatore Fitopatologico, nn. 6-7, XXVII: 39-43.

Infestation dynamics of *Ceroplastes sinensis* Del Guercio and *Panonychus citri* Mc Gregor in citrus undergoing differentiated control in Sicily

A.Nucifora, V.Vacante & S.Napoli*
Institute of Agricultural Entomology, University of Catania, Italy

Summary

The results are given of the comparative control tests using some insecticides, of their level of effectiveness against chinese wax scale (Ceroplastes sinensis Del Guercio) and of their direct and indirect effects on Panonychus citri (Mc Gregor). The present value of the intervention thresholds is considered, and it is suggested that the threshold based on the presence of females/10 cm of branch should be increased from 1 to 5.

1.1 Introduction

Infestation of Ceroplastes sinensis Del Guercio on citrus are not frequent, but in some areas they can be permanent, once established. This has been true for some years in a clementine orchard at 550 metres above sea level, in the area of Mascali (CT), Sicily, where we have performed the experiments reported in this report and where the species lives together with Saissetia oleae Bern.

It is a monovoltine species.

In the area investigated by us egg laying begins in the first half of July and continues throughout August. The neonates begin to hatch out at the end of July. They disperse out over the foliage and settle mainly along the median vein of the upper faces of the leaves. The maximum hatching rate of the eggs occurs in August and it ends in the first weeks of September.

The neanides transfer to the branches from the third age stage and overwinter there in this stage or in the young fertilized female stage. They reach their final somatic development in May – June.

During the winter there is considerable fall in population due to adverse environmental circumstances.

In the area investigated by us we found neanides still living on the leaves in December – January.

1.2 Intervention threshold and results of treatments

a) Material and methods

The main aim of our investigations was to discover the value of the interven-

* – S. Napoli has collaborated in the in-field and laboratory research, which is the subject of his graduate dissertation.
  V. Vacante dealt with the work connected with the mites infestation.
  A. Nucifora coordinated the research and supervised the writing up of the work.

tion threshold for this insect and to test what density/leaf its population reaches, starting from the specified levels of females per unit lenght of branch.

To do this some plants were chosen. In each of these a certain number of branches was marked and the number of females present per unit length (U.L. = 10 cm of branch) was counted. It was intended that these branches would be sampled at successive times, to enable us to detect periodically by laboratory examination for each branch the number of nymphs or new females present on the branches and/or on the wood, starting from the known infestation conditions.

Periodic samplings of unmarked branches were also planned to detect the general state of the infestation and its behaviour.

The observations were so performed every time as to detect not only the presence of the chinese wax scale, but also that of possible phytophagous mites and their variation as a function of the control actions applied in the various tests.

In the first year (1983) the dynamics of the Ceroplastes infestation in natural conditions were investigated. To do this 10 plants were chosen. On the 24th May 10 branches of each of the plants were marked and the number of females present on the wood of each of them was counted. These branches were later examined in the laboratory and the neanidal population on them was counted.

In the second year (1984) the neanide-killing effect of white mineral oil, alone or activated and of a pyrethroid, phenvalerate, and the collateral effect of these interventions on the development of the Panonychus citri (Mc Gregor) infestation were investigated. To do this 12 plants were chosen, experimenting upon 4 tests to compare with a simple randomized block with 3 repetitions of 1 plant each. This was done on the 24th September, 1984, when the hatching of the eggs could be considered over.

We worked with a shoulder atomizer, with commonly-used doses for high volumes. We made sure that all the foliage was covered by the jet, ensuring that the leaves, twigs and branches were thoroughly wet. In this way operating conditions were obtained,which were similar to those obtained using high volume, which is the most commonly used system in small and medium-sized farms.

The clementine orchard investigated has 12 year old plants which are about 2.5 m high, with branches which are rather intertwined because of insufficient pruning and thinning out, a fact which encourages the establishment of the coccids.

The first in-field counting operation was carried out in 1983 with the marking of 100 twigs. Later, on the 27th July, the egg-bearing females which had already been found on the branches, were recounted to confirm that they had all been found. There were not yet quantities of neonates which could be seen by the naked eye on the leaves.

The hatching of the eggs had yet begun.

The third counting was performed on the 7th October. This was performed in the laboratory, sampling from each plant 5 of the 10 marked twuigs. The operation was carried out using dissection binoculars to detect the nymphal population present on the 25 leaves, taken randomly from each of the branches sampled, and the population present on all the wood of the twigs examined. A second sampling of the branches (4th examination) was carried out on the 12th January, 1984: 2 of the remaining 5 marked twigs were removed from each plant and were examined in the laboratory in the same way as previously. The last sampling was carried out on the 30th May, sampling and observing the remaining 3 marked branches from each plant.

The overall data for the 100 branches examined are given in table 1.

Tab. 1 – Ceroplastes sinensis. Dynamic of infestation on marked leaves and branches, from May 1983 to May 1984.

Plant	Median length of examined branches	No.ovigerous females on branches (24 May 1983)		Median number of nymphs/leaf			Median number of nymphs and/or Jung ♀♀/10 cm of branch		
		Total	median/10 cm	6.10.83	12.1.84	30.5.84	6.10.83	12.1.84	30.5.84
1	412	2	1.38	0.10	0.1	—	—	0.9	0.28
2			5.40	4.92	0.3	—	—	1.90	1.28
3			0.65	0.16	—	—	—	0.07	0.41
4			0.46	0.56	—	—	—	0.01	0.07
5			0.15	0.24	0.1	—	—	0.01	0.005
6			0.17	0.08	—	—	—	0.01	0.003
7			0.15	0.08	—	—	—	0.01	0.006
8			0.48	0.08	0.2	—	—	0.06	0.06
9			1.05	1.64	0.1	—	—	0.09	1.10
10			0.50	—	0.2	—	—	0.01	0.60

Tab. 2 – Ceroplastes sinensis. Treatments effected on 24 September 1984 as soon as the intervention threshold has been overcome and their results.

Plots	Chemicals dose/hl	a.i. %	nymphs/leaf					Nymphs/10 cm of branch	
			t+1	t+22	t+30	t+70	t+120	t+70	t+120
1	Valoil (2.000)	White mineral oil (80%)	4.30	— A	—	0.01	—	—	—
2	Valoil (1.500) + Supracid (300)	White mineral oil (80%) + methidathion (20%)	4.90	— A	0.05	0.03	—	—	—
3	Sumicidin (30)	phenvalerate (11%)	4.75	1.55 B	1.15	0.35	0.30	0.90	1.50
Test	——	——	4.99	4.45 C	4.15	1.10	0.29	0.85	2.19

In 1984 the behaviour of the infestation was followed, counting the nymphs present on the leaves and the state of the egg-bearing females with periodic samplings in July and August, to established the best time for carrying out the treatment to be applied when the eggs are all hatched out. For each sampling 30 branches were removed and counted, one for each of the 30 plants under examination. The forms present on the wood and the leaves were counted in the laboratory. The state of infestation of the 30 plants was non-uniform and the average level of attack of the clementine orchard was below the intervention threshold levels. It was possible however to extrapolate 12 plants from the context, which formed a random block with 4 factors, with 3 repetitions each, in which the average infestation levels were 4.75, 4.90 and 4.30 respectively. The 4th factor with an attack level of 4.99 formed the control.

419

Tab. 3 – <u>Panonychus citri</u>. Dynamic of infestations in plots variously treated and in the test.

	Plots	Panonychus citri			Phytoseiids	
		% infested leaves	n° m.f./leaf	n° eggs/leaf	% leaves with predators	n° m.f./leaf
T–1	1. White mineral oil	6.30	0.12	0.65	3.55	0.06
	2. White mineral oil + Methidathion	5.95	0.13	0.51	2.90	0.07
	3. Phenvalerate	5.07	0.10	0.43	3.15	0.09
	Test	7.16	0.14	0.78	3.70	0.85
T+24	1. White mineral oil	0.07 A	—	0.01	—	—
	2. White mineral oil + Methidation	0.03 A	—	0.02	—	—
	3. Phenvalerate	1.58 B	0.03	0.47	—	—
	Test	3.67 B	0.06	0.38	1.27	0.01
T+70	1. White mineral oil	0.14 A	0.01	0.03	—	—
	2. White mineral oil + Methidation	0.10 A	0.01	0.05	—	—
	3. Phenvalerate	5.20 B	0.90	1.19	—	—
	Test	0.90 A	0.03	0.15	0.29	0.01
T+120	1. White mineral oil	— A	—	0.01	—	—
	2. White mineral oil + Methidation	— A	—	0.02	—	—
	3. Phenvalerate	38.50 B	1.05	0.98	—	—
	Test	0.26 A	0.01	0.05	0.05	0.03

The 3 factors were treated with white mineral oil at Kg 1.6 of a.i. (= active ingredient)/hl water, with white mineral oil at Kg 1.2 of a.i.   /hl + methidathion at 60 g/hl and with phenvalerate at 3.3 g/hl, as can be seen from table 2.

b) Results and discussion

As can be seen from the examination of the data of table 1, of the 10 plants examined in 1983 only 1 had a starting population which was definitely higher than that (1 female/10 cm of branch) indicated so far for other coccids as a probable intervention threshold (1). On this one there was, in fact, a mean level of 5.04 females/10 cm of branch. This density led in October to a mean presence of 4.92 neanides per leaf, higher than the threshold value, which for <u>Saissetia oleae</u> is 4 neani-

420

des leaf. On another 2 plants the starting population was 1 female/10 cm and the mean density of neanides in October was 1.21/leaf. On the remaining 7 plants there were starting values of 0.41 females/10 cm. On these in October a mean presence of 0.14 neanides per leaf was found.

If we wish to discuss these results critically it should first of all be said that the threshold value used so far at the level of females per 10 cm of branch for S. oleae seems low in the case of C. sinensis. The presence of 5 females/10 cm of branch with a resulting value of 4.9 neanides leaf indicates that the 2 values are directly correlated. This seems to be confirmed by the data from nearly all the cases examined.

As far as the adequacy of the value of 4 neanides/leaf as a threshold level is concerned, it must be checked by reference to the damage from blight which may affect the fruit at harvest time (November – December); it is not important that after the harvest and during the winter the populations are considerably reduced within the limits indicated in table 1. It was not however possible, in the case investigated by us, to evaluate the damage, because of the simultaneous presence on the plant of S. oleae, which made it impossible to distinguish when the blight on the fruits and leaves was due to it and when it was due to Ceroplastes. The data which we collected cannot give any information on this. They only give clear indications at the level of the numerical ratio between egg–bearing females and the nymphal populations deriving from them.

About the effect of the treatments on the C. sinensis populations which were just greater than the present threshold level, from the examination of the results obtained in 1984 and reported in table 2, one can see how the white mineral oil by itself was completely efficient. Factor 2 with active oil gave the same results as factor 1 and showed that adding an activating substance was useless.

The phenvalerate was however much less efficient; about 30% of the forms treated remained alive and it helped the development of Panonychus citri (Mc Gregor) which, was, however, contained by all the other 3 factors (tab. 3) compared. The lethal effect of white mineral oil on P. citri, already noted (2), (3) and (4) was fully confirmed, both when it was used by itself and when it was activated.

In the repetitions treated with phenvalerate, however, the infestation of this mite emerges from 30 days after the treatment has been carried out. After 70 days the attack level was at 5.2% of the leaves present and after 120 days 38.5%, although the density of mobile forms (m.f.) per leaf remained low.

1.3 Conclusions

The conclusions which can be drawn on the basis of the experiments carried out during the two years allow us to confirm that:

1. The use of white mineral oil by itself, at the dose of Kg 1,6 of a.i. allowed a disinfestation at the end of summer, in a fully satisfactory way, of the clementine trees infested with C. sinensis, with populations at a level scarcely higher than the threshold values, which for the moment is 4 nymphs on average per leaf.

2. The other threshold value, however, based on the number of egg–bearing females present in summer on unit lengths of branch, which for coccids is 1 female/10 cm of branch, must be increased, according to the results of our research, to the same level as the other. It is suggested that a threshold for C. sinensis of 4–5 egg–bearing females per 10 cm of branch should be considered.

421

3. The use of white mineral oil was directly efficient against the citrus red spider mite (P. citri) infestation on the citrus trees but at the same time it had a direct acaricide effect when dealing with the mobile forms of the predator mite. It does not however have a persistent action and so the useful populations reestablish more easily than when other insecticide or acaricide substances are used, as is seen in similar cases. The action of the former in the sample definitely brought the state of P. citri infestation to the same levels as those obtained in the plots where white mineral oil was used, either alone or activated.

4. The activation of white mineral oil with methidathion does not appear necessary or useful, as it made no difference to the results against C. sinensis.

5. The use of phenvalerate, which is considered by some people to be a reasonable acaricide, as well as a good insecticide, did not seem to produce good results against C. sinensis and encouraged the development of P. citri.

REFERENCES

1. BARBAGALLO, S. and NUCIFORA, A. (1981). Sampling methods and economic threshold for the control of lemon pests in Italy. In: Standardization of biotechnical methods of integrated pest control in citrus orchards. San Giuliano (F) - Siniscola (1) 4-6 nov. 1980. Eds R. Cavalloro and R. Prota, EUR 7342, EN-FR-IT, 27-34.
2. VACANTE, V., LONGO, S. and BENFATTO, D. (1980). Prove sperimentali di lotta contro Panonychus citri (Mc Gregor). Tecnica Agricola, 4: 223-228.
3. LANZA, G., CARUSO, A. and DI MARTINO, E. (1980). Prove invernali di lotta contro il Panonychus citri (Mc Gregor) in Calabria. Informatore Fitopatologico, 30 (6): 15-20.
4. NUCIFORA, A. (1983). Integrated Pest Control in Lemon Groves in Sicily: Five Years of Demonstrative Tests and Present Feasibilities of Transferring Results. In: C.E.C. Programme on Integrated and Biological control, Final Report 19+9/1983, Eds R. Cavalloro and A. Piavaux, EUR 8689, 129-146.

Influence of white mineral oil treatments on the population dynamics of some mites in a lemon orchard in Eastern Sicily

V.Vacante

Institute of Agricultural Entomology, University of Catania, Italy

Summary

The results are given of a field investigation on the effect of treatment with 2% white mineral oil on the main species of mite present in a lemon orchard in Eastern Sicily. The study confirms that white mineral oil gives a good control of Aceria sheldoni (Ewing); it appeared to be selective with respect to the phytoseiid Amblyseius stipulatus Athias-Henriot, which appreciably controls the populations of Tetranychus urticae Koch. The latter were adequately kept under control in the course of the year by A. stipulatus and by other acarophagi present in the field and which had been spared by the selective action of the white mineral oil. The tydeids and stigmaeids were instead weakened by the treatment, but over the year their populations were built up again.

1. Introduction

The use of white mineral oils to control the main phytophagi infesting citrus is of fundamental importance. Where the defense of crops against the attacks of Artropods is entrusted to the modern criteria of integrated control, white mineral oil is one of the few chemicals available today able to offer adequate guarantee from the toxicological and ecological points of view. The selectivity of this pesticide with respect to useful arthropodous fauna has been sufficiently established; repeated experiments in the field on lemons have shown that today in Sicily integrated control can be applied by using a combination of this product and the natural presence of entomophagi and acarophagi. No negative repercussions have been recorded on the overall balance of the citrus ecosystem when using it (Nucifora, 1983).

During 1984 an experiment was carried out to verify further in the field the influence of such treatments on the populations of the main species of mites present in a lemon orchard.

2. Materials and methods

The investigation took place in the province of Catania in a lemon orchard, cv. femminello, aged about 40 years.

On the 14th of April 1984 the crop was treated with 2% white mineral oil. The insecticide was mechanically sprayed at 20 atm. For each plant treated 5 liters of emulsion were used. A simple random block technique was adopted with plots of 20 trees replicated 4 times. Two plots were compared: the control plot and one treated with the white mineral oil. Random sampling of twigs (4 per plant from the 4 cardinal points) from the latest vegetation (length: 25-30 cm) was carried out before and after the treatment at a one-month interval. Leaves, branches and buds were exami-

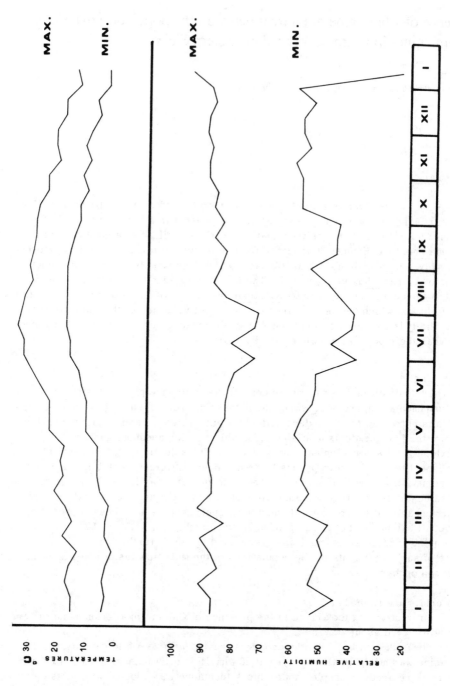

Fig. 1 – The temperature and relative humidity pattern with the maxima and minima at ten-day intervals recorded at the Citrus Fruit Experimental Institute in Acireale, near Catania.

ned stereomicroscopically the same day as they were sampled. On each twig all the leaves were thoroughly observed as well as the wood, and 4 buds were dissected (2 apical and 2 basal). All the mites found, except for those phytophagi already well known, were prepared and studied morphologically in order to identify them with cer tainty. The species found and dealt with were: the two-spotted spider mite (Tetrany-chus urticae Koch), the citrus red mite (Panonychus citri (Mc Gregor)), the citrus bud mite (Aceria sheldoni (Ewing)), the tydeids Tydeus formosa (Cooreman), T. fe-rulus (Baker), Orthotydeus californicus (Banks), O. foliorum (Schrank) and Prone-matus ubiquitus (Mc Gregor), the phytoseiid Amblyseius stipulatus Athias-Henriot and the stigmaeids Zetzellia mali (Ewing), Z. graeciana Gonzalez and Agistemus col-lyerae Gonzalez (1).

The temperature and relative humidity patterns reported in the graph of Fig.1 refer to data collected at the Experimental Citrus Fruit Institute in Acireale which is at a distance of several kilometers from the investigated field.

3. Results and discussion

The results of the investigation are broken down into individual species or sy-stematic group and are presented in summarized form in the respective graphs.

1) Aceria sheldoni. Fig. 2 proves the high efficiency of white mineral oil treatments in controlling this phytophagous mite. In the treated plot the eriophyid population was reduced and a month after treatment it was present on fewer than 15% of the buds examined; in the control plot, instead, the infestation rate was six times hi-gher. After 4 months the population of this mite in the treated plot had already cros-sed the threshold of 70% of infested buds (Vacante and Nucifora, 1984). That confir-ms once again, as experience has proved, that two treatments annually are often needed to control this mite, one in winter and the other in summer. Fig. 2 also shows how the eriophyid, while being active all the year, reproduce most intensely during the hottest months. In spring the reproduction rate is lower and the considerable ve getative energy of the plants leads to a drop in the percentage of buds infested due to the increase in the number of buds available for the mites; this phenomenon is absolutely not seen in summer and autumn, due to the fact the summer vegetation growth is more contained than the spring one and that the reproductive capacity of the mite is higher in summer. The coming of new shoots with new buds available for the mite in the summer and autumn months should have caused a slight decrease in the percentage of buds infested; it was, however, promptly counterbalanced by the increased number of mobile forms of mites moving about in the tops of the control plot trees. These promptly invaded the available sites and showed a considerably dynamic colonizing capacity. This phenomenon appears evident from the fluctuations in the number of mobile forms observed inside the buds. In the treated plot the in-crease of infestation was progressive reaching at c. 7 months from treatment ana-logous levels to those in the control plot. As regards damage to the produce, it was seen that c. 10% of the fruit (of the 800 lemons examined) in the control test bore the signs of infestation by this mite; of that fruit, however, only 3% circa could be considered deformed; the rest displayed slight alterations which did not compromi-se their marketability. In the treated plot damage of 5% was found (of the 800 lemons

(1) Various species have been found sporadically but in numbers such as not to need taking into consideration.

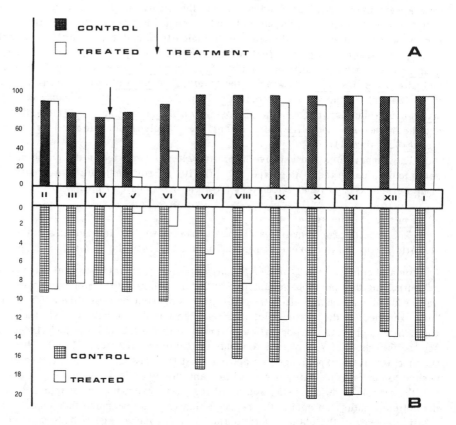

Fig. 2 – A: the percentage of buds infested by Aceria sheldoni (Ewing) in the control plot and in the treated plot with 2% white mineral oil; B: the average number of mobile forms of eriophyid per bud examined in the control plot and in the treated one.

examined), of which only 1.5% circa could be considered serious damage. The action of the native predators in controlling the citrus bud mite was rather modest.

2) Tetranychus urticae. Fig. 3 shows that in our environment the two-spotted spider mite is active also in winter on citrus. From this one can see that the use of white mineral oil offers no direct and immediate control of the mite in question, but partly reshapes the size of its populations, which in the treated plot were constantly lower than in the control plot. It should not, however, be excluded that other factors of a not-well-defined abiotic nature may have affected this phenomenon. An examination of the periodic pattern of the frequency of leaves infested and the density of mobile forms of this mite per leaf examined shows that spring – early summer and autumn are the times of greatest pullulation.

Temperatures of over 30°C are a limiting factor; optimal development seems to be reached with a temperature range of between a minimum of 10°C and a maxi-

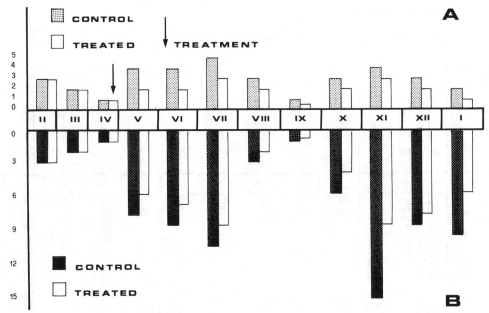

Fig. 3 - A: percentage of leaves infested with Tetranychus urticae Koch in the control plot and in the treated plot with 2% white mineral oil; B: the average number of mobile forms of tetranychid per leaf examined in the control plot and in the treated one.

mum of 25°C. In the case in question, the density and frequency of the tetranychid were constantly maintained within the limits of a wide tollerability, as is shown by the fruit infestation in the two comparative plots. In fact, in the control plot in January 1985 a percentage of fruit damaged of about 2% (out of 800 lemons examined) was found, whereas in the treated plot the fruit damaged was c. 1.5% (out of 800 le mons examined). Amoung the natural factors contributing to the containing of T.ur ticae mention should certainly be made of the phytoseiid mite A. stipulatus. This predator, although not having a specificity of action comparable to that of Phytoseiu lus persimilis Athias-Henriot, has made a decisive contribution to the redimensio ning of the tetranychid populations; together with other acarophagi (mites and insec ta), they seem to be the key to an explanation of infestation being contained in the terms described above. This also explains why for several years we have been re commending the exclusive use of white mineral oil in citrus pest control. The expe riments so far reported amply prove its usefulness.

3) Panonychus citri. The presence of this mite was revealed on two occasions only, in April and in November 1984, when rare eggs and some larvae were found on less than 1% of the total number of leaves examined. The discontinuousness of the finds and the insignificant presence of this mite can be explained by the fact that they we re certainly borne there accidentally from the already infested neighboring lemon orchards, and were swiftly suppressed by the numerous phytoseiids present on the crop.

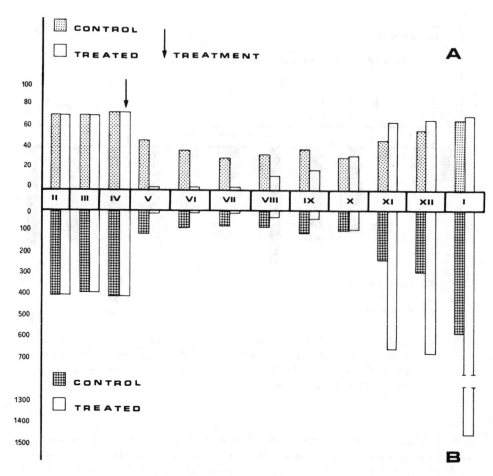

Fig. 4 – A: percentage of leaves with presence of tydeids in the control plot and in the treated plot with 2% white mineral oil; B: the average number of mobile forms of tydeids on 100 leaves examined from the control plot and from the treated one.

4) Tydeids. The evolution of the tydeid populations present in the field as the moment of treatment is shown in Fig. 4. These populations were formed 60% by T.formosa, 30% by O. californicus and 10% by O. foliorum. The use of white mineral oil was not found to be selective against them; in the treated plot the density and frequency of the mobile forms of the three mentioned species were fairly low one month after treatment, whereas in the control plot they remained considerably higher. A slow and continuous regression was then observed in both plots and in the summer period the three mentioned species were almost totally replaced by P. ubiquitus and T. ferulus. At the end of the summer T. formosa, O. californicus and O. foliorum slowly but progressively reappeared; in November the tydeids were more frequent and abundant in the treated plot than in the control one. That may depend on a concommitant infestation of Pseudococcidae whose development was enhanced by

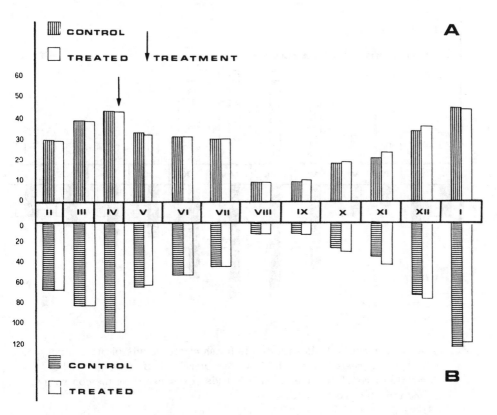

Fig. 5 – A: percentage of leaves with mobile forms of Amblyseius stipulatus
Athias–Henriot from the control plot and from the treated plot with
2% white mineral oil; B: average number of the phytoseiid on 100
leaves examined from the control plot and from the treated one.

the honeydew emitted and the relative sooty mold. Independently of the causes faci-
litating the rebuild–up of the tydeid populations mentioned above, it should be noted
that the use of white mineral oil, while proving to be not immediately selective
against these mites, has led to no side effects which might irreversibly prejudice
the equilibrium of the ecosystem. Among the biotic elements limiting the increase
of these populations, A. stipulatus is certainly one of the most efficient, frequently feeding on mobile forms of T. formosa and other tydeids. The popualtion dynamics of the predator and of certain species belonging to this important group of mites have been seen to coincide in a directly proportional way. That would confirm
the hypothesis that they can form a useful source of food for the phytoseiid. The
heterogeneous feeding habits of the predator means, however, that it is not necessarily dependent on the presence of the tydeids, in the sense that when they are
not available the predator can easily turn to other sources of food without any negative repercussions. A comparison of Figs 4 and 5 seems to confirm this.

5) Amblyseius stipulatus. The phytoseiids found in the course of the research were
various but the predominant species and more active one was A. stipulatus. Fig. 5

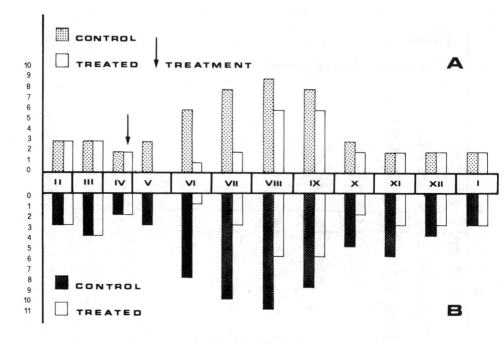

Fig. 6 – A: percentage of buds with mobile forms of stigmaeids in the control plot and in the treated plot with 2% white mineral oil; B: average number of mobile forms of stigmaeids on 100 buds examined from the control plot and from the treated one.

summarizes the frequency and density of the predator on the leaves from the plots under comparison. From the figure it can be seen how the use of white mineral oil had no negative effect on this predator. At a month after treatment no significant difference was to be observed between the two comparative plots. The higher densi ty and frequency of the predator were observed in the less hot and more humid months of the year; in summer its populations were less numerous. These findings show once again how the phytoseiid has a dynamic population analogous to those of some tydeids (T. formosa, O. californicus, O. foliorum). By comparing Figs 3 and 5 it can again be seen that as the frequency of the phytoseiid on the leaves increases, that of T. urticae decreases.

6) Stigmaeids. The Stigmaeids referred to in this paper were constantly found insi de buds infested with A. sheldoni. The most widespread species were Z. mali and Z. graeciana. At the time of treatment their populations were formed 66% of mobi le forms of Z. graeciana and 34% of Z. mali in mobile forms. The presence of A. collyerae was revealed in summer and autumn; it was at all times irrelevant. For this group of predator mites the action of white mineral oil was lethal. Fig. 6 shows, however, how their populations built up again over the year, although they never regained densities such as to lead one to think of their paying any large role in controlling the deformatory mite. Their population dynamics appeared rather to be independet of the density and availability of the prey in the field.

4. Conclusion

The conclusions that can be drawn from the field experiment outlined above lead us to claim that white mineral oil can be easily used in lemon cultivations without any negative repercussions for the main species of mites present in the treated environment. The use of such a substance moreover guarantees a satisfactory control of some species of phytophagous mites by acting directly on their eggs and mobile forms (A. sheldoni, P. citri); instead, it exerts no marked control on the eggs and mobile forms of T. urticae, but offers an effective and indirect control of this mite by safeguarding the useful species which in nature prey on its populations.

REFERENCES

1. NUCIFORA, A. (1983). Integrated Pest Control in Lemon Groves in Sicily: Five Years of Demonstrative Tests and Present Feasibilities of Transferning Results. Final Report 1979/1983, C.E.C. Programme on Integrated and Biological control: 129-146.
2. VACANTE, V. and NUCIFORA, A. (1984). L'acaro delle meraviglie (Aceria sheldoni (Ew.)): soglia d'intervento e trattamenti, 1° Contributo. Atti Giornate Fitopatologiche, 2: 525-533.

The influence of white mineral oil treatments on lemon trees: Three years of experiments*

E.Di Martino, G.Lanza & E.Di Martino Aleppo
Citrus Experimental Institute, Acireale, Italy

Summary

The authors report on a 3-year study of the influence which the spraying of lemon trees with white mineral oil at 2 and at 3% in summer and/or winter has on the fruit production of the trees.
The experiments showed that there were no measurable differences in the quality and quantity of fruit produced between the treated and untreated trees, apart from small variations in the average weight of the fruit produced by treated trees with respect to their controls.

Emulsified white oil can be used in the control of the many citrus phytophages, both mites and insects. As far as mites are concerned one only need remember the proved efficiency of white oil against Eriophyes sheldoni Ew., Tetranychus urticae Koch, and Phylloptruta oleivora (Ashmead) (and also, it seems, against the species with the same habits Aculops pelekassi K.) and against Panonychus citri McGreg. As far as insects are concerned it is almost needless to recall the positive results of control against scale insects, generally belonging to the Diaspini subfamily, and in general against all the other hemiptera families if the action is carried out on examples at the pre-image stage: this is also true for Aleurodids which have recently been included among the primary phytophages.
The repeated observations that after the distribution of oily products there were no repercussions on the ecosystem affecting the swarming of other harmful species and that the useful arthropodofauna soon reappeared in the citrus fruit groves has led us to review all the strategies for the use of those phytochemicals which were accused of damaging or at least of having a more or less great effect on the physiology of treated citrus fruits, especially oranges, and thus on production, both quantitative and qualitative (Sinclair, W.B. et al., 1941; Ebeling, W., 1950; Riehl, L.A. et al., 1956; Trammel, R. et al., 1966 and Dean, A.A. et al., 1967). Some experiments have thus been carried out at the Citrus Experimental Institute at Acireale on the "Taroc" orange (G. Lanza, E. Di Martino Aleppo, 1975) where it was seen that when there was considerable phytotoxicity, shown by

* Research conducted as part of the Ministry of Agriculture and Forestry project "Development and improvement of industrial fruit-growing, early fruit growing and citrus fruit growing", Publication n° 158.

433

the falling of leaves, there were considerable differences in that there were reductions in Vitamin C content and acidity while the other qualitative parameters remained the same (average weight, yield of juice and total solids).

Experiments were then carried out by the Acireale Citrus Experimental Institute (1981) in the summer on the most representative varieties of oranges grown in Italy, finding that even considerable leaf-drop did not have a great effect on the quantitative and qualitative parameters, apart from the already ascertained reduction of Vitamin C in the Taroc orange.

A further experiment which was carried out, on applications of oil on flowering in 1982 on the "Taroc" orange, did not find that white oil at 2% had any phytotoxic effect on closed flowers (Lanza, G. et al., 1982).

At the same time experiments were performed in other citrus fruit growing countries on the possible phytotoxic effects of oils. Veiverov, D. et al. (1979) have documented damage of orange peel caused by two commercial types of oil at various concentrations.

Beattie et al. (1980), in Australia, in control experiments against Aonidiella aurantii, did not find any harmful effects either on the fruit quality or on the phytotoxes. Finally, at the Conference of Citrologists in 1981, three papers were presented: L.A. Riehl refers to the characteristics, use, behaviour and methods of application with special reference to integrated control, with fairly satisfactory results in South California; Furness refers to the role of oil in Australia noting the declining use of phosphoorganics and other insecticides; and finally Ohkubo in Japan deals with oils in integrated control against Unaspis yanonensis (Kuw.) and Panonychus citri (McGreg.), key Japanese phytophages.

For our part, encouraged by previous experiments in which we ascertained the minimal phytotoxicity of the product used in winter together with its considerable efficiency against cocchinellae, we decided to carry out enquiries into the use of white oil on lemon trees.

As the lemon is cultivated in areas which are generally not subject to frosts it is normal crop-growing practice to spray with oils during the winter and after most of the fruit has been collected, to combat Eriophyes sheldoni Ew., Panonychus citri (McGreg.), Aonidiella aurantii (Mask.), Lepidosaphes beckii (Newm.), Aspidiotus hederae (Vall.), and also Dialeurodes citri (Ashm.). The treatment can also be used against the coccids Coccus hesperidum (L.), Saissetia oleae (Oliv.) and Ceroplastes rusci (L.), generally in concentrations of 2% of the commercial product, and sometimes 3% in cases of more severe infestations.

These same treatments can also be applied in the summer, especially where, as often happens in Sicily, forcing is practised for the production of summer lemons: the "verdelli".

We thus wanted to check the influence of single and repeated applications of oils at the two times of year when most "winter" and "summer" applications take place (see Fig. 1). We thus decided to carry out a series of experiments using white mineral oil at 2% and 3% in winter (February), in summer (August), and with a double winter and summer treatment. These applications which began in 1982 were carried out for one year only, for two years and for three consecutive years.

434

The same commercial product was used. It was kept in as good storage conditions as possible, the characteristics of which are shown elsewhere (Tab. 1).

A lemon orchard of "Common Femminello" (Citrus limon Burm.) was chosen. It was in the Taormina marshes and more than thirty years old on terrain of medium consistency; a motor pump with delivery nozzles was used for the sprayings with a manometer pressure of 20 atm., being careful to spray the plants until they were dripping. Five plants were sprayed for each trial. Each of them was repeated in the following sampling.

The phylotoxes assumed normally as an index of the phytotoxicity of the oil applications were monitored after every treatment for a month.

The productivity data were taken from the overall weight of the winter fruit collected from each plant. The data of phylotoxes and productivity refer to m^3 of foliage, to take account of the differences in size of the plants themselves.

The other parameters inherent in the fruit quality which were considered were: average weight, yield of juice, total solids, acidity and vitamin C. The data for the leaf-drop are given in the histogram (Fig. 2).

As general behaviour we can say that the greatest phylotoxes are generally found following the treatment at the end of winter, with an average of 30 to 64 leaves per m^3 of foliage, without noticing any significant difference between the trial with two oil concentrations and the appropriate untreated control.

In a preceding work written by one of us (Di Martino, 1977), however, again no significant difference was found between similar trials.

As far as the summer treatments are concerned, one can say that in general the phylotoxes are much lower; from 10 to 20 leaves per m^3 of foliage and again without significant difference between the trials and the control.

For the results of the double treatments, one can see from the data that there is considerable similarity with those of the summer treatment; the check of the fallen leaves only refers to after the last (summer) treatment.

The data of the second graph (Fig. 3) refer to the productions of the first, second and third year of the experiments.

From the data analysed statistically no significant differences were found between the trials at the two oil concentrations and the appropriate untreated controls.

The qualitative parameters are given in the appropriate tables (Tabs. II,III,IV,V,VI). The first refers to the average weight of the fruit; in these one cannot see any significant differences from the analysis of the covariance between the average weight of the fruits after one and two years of application. In the third year, however, there was such a difference, shown by the smaller size of the control. For the other qualitative data no differences were found between the untreated controls and the various trials.

In the light of these data, for the conditions in which we were working, we feel that we can conclude that the treatments had no influence on the phylotoxes which remained within normal values; in one of the cases which we cannot consider exceptional because it was statistically equal to its untreated control, there was a loss of about 65 leaves per cubic metre, which means that an average plant of 25 m^3 would have a loss of 1625 leaves in the critical month which we consider absolutely unconsistent, especially if we remember similar experiments on oranges (op. cit.) where there were 14,000 fallen leaves in two months.

The assertion of this is confirmed by the data for production where any type of treatment up to three consecutive years, even with two treatments every year, was found to have no effect.

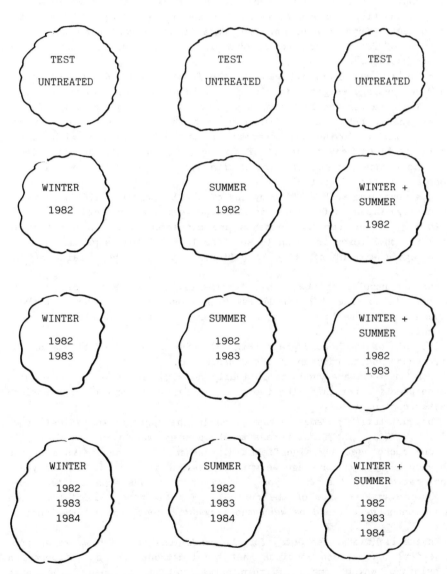

Fig. 1 - Combination of oil applications during three years.

Table I - Characteristics of the base oil used

Appearance: clear and limpid

Specific weight at 15/4 °C: 0.875 typical

Kinematic viscosity at 40 °C: cst 13.1 - 14.31

Kinematic viscosity at 100 °C: cst 2.92 typical

Viscosity index: 45 typical

Inflammability in closed vessel °C: minimum 150

Inflammability in open vessel °C: 165 typical

Yield value: maximum -39

Saybolt colour: minimum +20

Cloud test: 7 hours at 4.5°C, no clouding

Neutralisation number expressed in
mg of KOH per g: maximum 0.05

Maximum percentage weight of ashes: 0.01

Demulsification number, seconds: maximum 90

Nonsulphonated residue, percent volume: minimum 93

Distillation: initial point °C: minimum 290

        percentage by volume
        collected at 336 °C: 35/50

        95% collected at °C: maximum 390

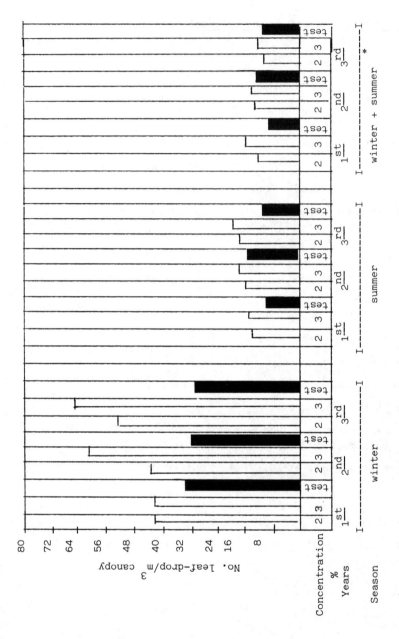

*) Leaf-drop only after summer treatment.

Fig. 2 – Leaf-drop during the first month after oil application

438

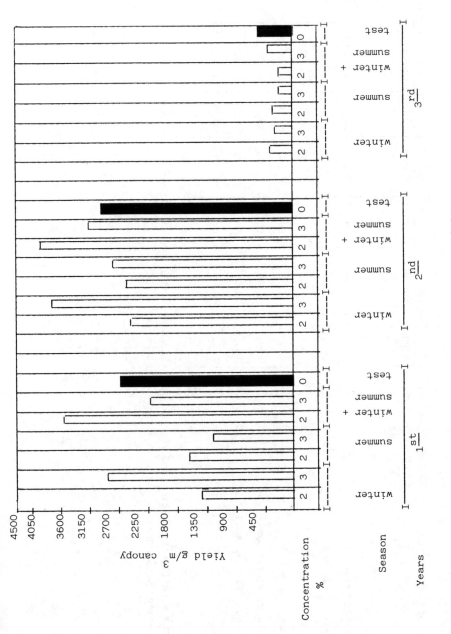

Fig. 3 - Influence of oil applications on yield

Table II - Analitical data: average weight of the fruit

Time of application	Concentration of mineral oil %	1st year real	1st year correct	2nd year real	2nd year correct	3rd year real	3rd year correct
february (winter)	2	95,70	94,46	98,30	96,30	98,10	A 90,93
february (winter)	3	105,30	96,97	100,60	97,76	82,82	A 92,00
august (summer)	2	96,10	94,58	101,60	97,83	89,28	AB 92,91
august (summer)	3	98,36	96,76	89,00	97,88	85,40	A 90,78
feb.+aug.(wint.+summ.)	2	109,56	97,05	85,90	97,53	94,03	A 90,34
feb.+aug.(wint.+summ.)	3	104,54	93,07	86,50	98,86	88,54	A 89,83
test	-	112,00	98,86	93,80	99,82	98,00 *	B 94,65

*) The significativity of third year is coming from depurated data by linear regression (relation between productions as kg and average weight as g)

Table III – Analitical data: juice percentage %

Time of application	Concentration of mineral oil %	1st year	2nd year	3rd year
february (winter)	2	47,32	41,13	34,05
february (winter)	3	44,44	37,70	35,31
august (summer)	2	42,71	42,64	35,14
august (summer)	3	41,50	41,40	31,04
feb.+aug. (wint.+summ.)	2	41,94	41,38	37,16
feb.+aug. (wint.+summ.)	3	39,96	43,36	35,65
test	–	42,49	41,80	30,33

441

Table IV - Analitical data: total soluble solids g% at 20°C

Time of application	Concentration of mineral oil %	1st year	2nd year	3rd year
february (winter)	2	8,81	9,18	9,85
february (winter)	3	9,00	9,43	9,60
august (summer)	2	8,70	8,99	9,20
august (summer)	3	8,98	9,13	9,40
feb.+aug. (wint.+summ.)	2	8,66	9,15	9,55
feb.+aug. (wint.+summ.)	3	9,24	9,08	9,65
test	-	8,79	9,34	9,33

Table V – Analitical data: acidity as citric acid anhidrous g%

Time of application	Concentration of mineral oil %	1st year	2nd year	3rd year
february (winter)	2	6,42	5,91	7,13
february (winter)	3	6,58	6,74	6,72
august (summer)	2	6,18	6,50	6,85
august (summer)	3	6,22	6,45	6,72
feb.+aug. (wint.+summ.)	2	6,22	6,37	7,10
feb.+aug. (wint.+summ.)	3	6,66	6,18	7,00
test	–	6,66	6,71	6,83

Table VI - Analitical data: vitamine C mg%

Time of application	Concentration of mineral oil %	1st year	2nd year	3rd year
february (winter)	2	67,23	64,77	56,67
february (winter)	3	67,93	68,99	60,90
august (summer)	2	66,17	66,53	57,73
august (summer)	3	64,94	70,05	59,49
feb.+aug. (wint.+summ.)	2	63,71	69,34	56,32
feb.+aug. (wint.+summ.)	3	65,12	73,22	58,08
test	-	64,06	63,71	59,14

444

REFERENCES

BEATTIE G.A.C. and RIPPON L.E.(1980) - Phytotoxicity and scalicide efficacy of citrus spray oils. Proceedings of the International Society of
Citriculture 1978.

DEAN H.A. and GLIFFORD E.H.(1967) - Responses of pineapple orange trees
to selected petroleum oil tractions. Journal of Economic Entomologoy 60
(6): 1668-1672.

DI MARTINO E.(1977) - Olio bianco su limone durante l'inverno contro
Mytilococcus beckii New. Informatore Fitopatologico, 6-7.

EBELING W.(1950) - Subtropical Entomology. Lithotype Process, S.Franci-
sco California.

FURNESS G.O.(1981) - The role of petroleum oil sprays in pest management
programs on citrus in Australia. Proceedings of the International Society of Citriculture 1981, 2: 607-611.

LANZA G. e DI MARTINO ALEPPO E.(1975) - Influenza di trattamenti anticoc
cidici sulla filloptosi e sulla maturazione dei frutti di arancio "Taroc
co". Tecnica Agricola 27(5).

LANZA G. e DI MARTINO ALEPPO E.(1981) - Aspetti collaterali di applicazioni oleose estive su cultivar di arancio a polpa bionda e pigmentata.
"La Difesa delle piante" (1): 25-38.

LANZA G. e DI MARTINO ALEPPO E.(1982) - Influenza delle applicazioni
oleose in fioritura sull'allegagione della cultivar di arancio "Tarocco".
Atti Giornate Fitopatologiche, 2° volume: 105-112.

OHKUBO N.(1981) - Role of petroleum oil sprays in on integrated pest management system of citrus crops in Japan. Proceedings of the International Society of Citriculture 1981, 2: 611-614.

RIEHL L.A.(1981) - Fundamental consideration and current development in
the production and use of petroleum oils. Proceedings of the International Society of Citriculture 1981, 2: 601-607.

RIEHL L.A., WEDDING R.T. and RODRIGUEZ J.L.(1956) - Effect of oil spray application timing on juice quality, yield, and size of Valencia oranges in a southern California orchard. Jour.of Econ.Ent., 49(3): 376-382.

SINCLAIR W.B., BARTHOLOMEW E.T. and EBELING W.(1941) - Comparative effects of oil spray and hydrocyanic acid fumigation on the composition of orange fruits. J.Econ.Entomol. 34(6): 821-9.

TRAMMEL K. and SIMANTON W.A.(1966) - Properties of spray oils in relation to effect on citrus trees in Florida. The Citrus Industry, 47(12): 5-11.

VEIEROV D., ERNER Y., SHOMER I. and AHAROUSON N.(1979) - Early evaluation of orange peel blotches caused by spray oils. Phytoparasitica,7(2): 79-88.

The action of white mineral oil against the fig wax scale (*Ceroplastes rusci* (L.)) on cvv Moro and Tarocco oranges and its influence on the organoleptic characteristics of the fruit*

C.Calabretta, S.Inserra, G.Ferlito** & F.Conti**
Institute of Agricultural Entomology, University of Catania, Italy

Summary

Tests carried out on cvv Moro and Tarocco oranges, using white mineral oil at 2 Kg/hl c.p. and to examine the influence of this on the fruit quality, lead the AA. to conclude that the oil does not have those phytotoxic properties erroneou sly attributed to it and that it offers an efficient disinfestation, allowing the bio logical equilibria to be rapidly reestablished after treatment.

1.1. Introduction

The use of white mineral oil on citrus is generally considered by technical experts and citrus farmers to be less effective than other methods and in particular than certain phosphoric esters (methyl–parathion, methidathion) and than carbaryl, which are widely used in scale insect control, either alone or more rarely in a mixture with 1% white mineral oil and with the addition of acaricides.

It is also a widely held opinion that at doses higher than 1% (Ebeling, 1950; Wedding et al., 1952) white mineral oil causes leaf drop, fruit drop, a decrease in fruit size, delayed and irregular pigmentation. In summer there is also the fear of other undefined phytotoxic actions due to the high temperatures, phenomena never verified by us or by other Authors (Liotta and Maniglia, 1974; Lanza and Di Martino Aleppo, 1975; Nucifora, 1981, 1983; Riehl et al., 1956) either on lemon or on orange.

In South Africa in 1980, Buitendag and Braukhorst, treated orange with 3% white mineral oil in full summer without observing damage either to the organoleptic properties inside the fruit or to the coloring of the skin. In the Lentini area there are cases of orange farms where white mineral oil has always been used every summer without any of the feared factors being verified. No infestation of Panonychus citri (Mc Gregor) has yet occurred on these farms whereas it has regularly been found in the neighboring farms which were treated with the synthetic products mentioned above.

These considerations led us to investigate the effects that withe mineral oil, used in high summer at doses of 1.6 Kg/hl, of active substance,can in reality have

*This work was subsidized by the Ministry of Agriculture under the "citrus cultivation" project.

**The research contribution in the field and in the laboratory of G. Ferlito and F. Conti to this study will be the subject of their respective degree theses.

447

Table I – <u>Ceroplastes rusci</u> L.. The results obtained from the experimental plot 25 days after the treatment of 10/8/1983 (Agnone, Conti farm).

Plot	Active substance and doses in g/hl in parentheses	Average mortality (%)
1	White mineral oil (1600)	88.5 A a
2	White mineral oil (800)	26.2 B c
3	White mineral oil (1600) + methyl–parathion (30) + zineb (160)	90.6 A a
4	White mineral oil (800) + methyl–parathion (30) + zineb (160)	77.9 A a
5	Methyl–parathion (60)	42.6 A b
6	Control	0.01 C d

Table II – <u>Ceroplstes rusci</u> L. cv Moro orange. Average size (= equatorial diam.), average number of fruit per plant and average unit weight at picking in the 6 comparative plots (Agnone, Conti farm, Jan. 1984).

Plot	Active substance and doses in g/hl in parentheses	Average equatorial diam.in cm	Average number of fruit per plant	Average unit weight in g
1	White mineral oil (1600)	6.82	406.5	95.65 A a
2	White mineral oil (800)	6.77	443.1	89.42 A a
3	White mineral oil (1600) + methyl–parathion (30) + zineb (160)	6.67	476.8	82.70 A a
4	White mineral oil (800) + methyl–parathion (30) + zineb (160)	6.85	363.9	100.44 A a
5	Methyl–parathion (60)	6.90	397.0	89.37 A a
6	Control	6.62	477.6	87.84 A a

– The values indicated with a capital letter are significant for $P = 0.01$.

– The values indicated with a small letter are significant for $F = 0.05$.

(The relative data have been covariance analyzed).

on the size and coloring of the fruit, as compared with the treatment usually applied in the area.

1.2. Experimental

a) Materials and Methods

The experiment was carried out during 1982–1983. In 1982 on a farm near Vittoria in the Ragusa area white mineral oil was testes against the fig wax scale (Ceroplastes rusci (L.)). In 1983 the experiment was extended also to a farm in the Syracusa area (at Agnone Bagni) on cv Moro oranges also infested by the fig wax scale.

In the Ragusa area (the Ferlito farm) white mineral oil was sprayed at 1.6% as compared with the mixture commonly used in that area of methyl-parathion (60 g) + dicofol (50 g) + tetradifon (15 g) + zineb (190.5 g). The two-plot comparative trials were carried out in two adjacent groves of cv Tarocco and Moro, each of approx 200 trees of equal health and planted in the same kind of ground which had been treated in the same way.

The insecticide treatment was given on the 10th of September 1982 with a turbosprayer applying approx 3.2 l of mixture per plant. The orange grove is about 10 years old. The ground is of medium consistency tending to calcareous; it had been treated in the way normal for that area. The comparative plots for each cv were set up using a random grid system with 4 replications of 10 plants each.

In the Syracuse area (the Conti farm) 6 plots were compared during 1983, as shown in Table I. The simple random grid system was used here with 6 replications of 1 plant each. The investigation was on cv Moro oranges, 8–9 years old.

On both farms the leaf drop, the growth of the fruit and the coloring of the skin and pulp were all followed up.

In the Syracuse area farm, treatment began on the 10th of August 1983 using a 15-atm pressure mechanical sprayer.

On both farms the irrigation plant had a 6 x 4 range and sprinkling irrigation.

b) Results and Discussion

On the Ferlito farm the treatment of the two comparitive plots, both the Moro and Tarocco, led to a satisfactory control of the infestation with a mortality rate of 80% of the forms present when using the oil, and of 73% when using the methyl-parathion. At the time of fruit picking the crop showed no trace of sooty mold. Not was any anomalous leaf drop observed in any of the plots.

On the Conti farm the data on the results of the control, of the charts of Fig. 1 and Table I, show that after 25 days of treatment there were satisfactory mortality rates, without any statistical difference between the plots sprayed with white mineral oil alone at 1600 g/hl and those with the activated product. The effect of the oil at half that dose was less good but satisfactory the same when used activated. There was statistically significant difference between the two wayes. Instead, the action of methyl-parathion used alone was insufficient, probably due to resistances which had been built up, seeing that this product had been used for some time in that area; it left 57.4% of the population present alive. All the plots showed significant differences with respect to the control plot.

In all the plots were white mineral oil at 1600 g/hl (= 2 Kg at 80%) was sprayed in summer, there were no kind of inconveniences found, nor any phenomena of qualitative, quantitative or commercial depreciation in the crops, as the results of the trials reported here show.

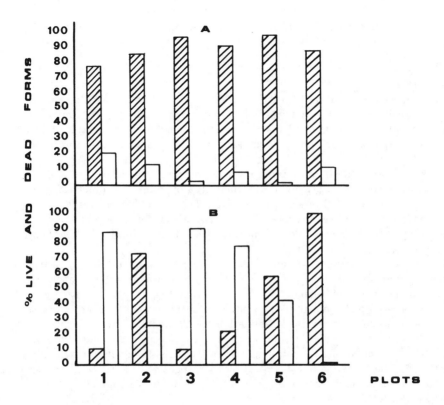

Key:

▨ live

☐ dead

Plot 1 = white mineral oil (1600 g)
  "   2 = white mineral oil (800 g)
  "   3 = white mineral oil (1600 g) + methyl-parathion (30 g) + zineb (160 g)
  "   4 = white mineral oil (800 g) + methyl-parathion (30 g) + zineb (160 g)
  "   5 = methyl-parathion (60 g)
  "   6 = control

Fig. 1 – Ceroplastes rusci L.. The state of vitality in the various experimental
plots before (A) and 25 days after (B) treatment (Agnone, Conti farm).

c) Influence on the quality, coloring and size of the fruit

I – cv Moro

The action of the various pesticides and of the doses used regarding leaf drop, fruit drop and the macroscopic organoleptic quality of the fruit has been investigated; the results are reported below.

On cv Moro in the Ferlito farm white mineral oil at 1600 g/hl somewhat delayed the coloring of the skin and pigmentation of the pulp compared with the control plot. The data were obtained in the two plots (the one treated with white mineral oil and the one with methyl-parathion) on 40 plants of the 100 treated in each plot. From the commercial point of view the slight delay in the turn of coloring had no repercussions on the picking time and thus for placing the produce on the national and foreign markets. The use of white mineral oil even caused a slight increase in the skin thickness and a finer grana membrane. The latter characteristic makes the fruit more marketable on the home market but less acceptable on foreign markets.

The most incisive factor, negative or positive, for the marketing of the fruit is the variation to the grana. As regards size and weight there was no statistically significant difference in the variously treated batches; that is true both for the Vittoria (Ferlito) farm and, as can be seen in Table II, for the Agnone (Conti) one. The varying of weight and size is to be attributed to the greater or lesser fruit load of each individual tree. That falsifies the commonly held belief according to which white mineral oil causes a decrease in size and overall decrease, in production.

As regards the coloring of the fruit, empirical estimates on the Conti farm by skilled commercial valuers did not reveal, at the sale of the fruit on the trees, any commercially appreciable differences in the six plots under comparison.

Concerning the effect of the oil treatment on fruit drop and leaf drop (Tables III and IV), the quantified investigation carried out on the Conti farm yielded no statistically significant differences between the control plot and the treated ones.

II – cv Tarocco

Instead, the reaction to treatment with oil in the cv Tarocco was different as regards the beginning of the turn of color and the skin coloring. In this plot they were earlier and more intense than in the other control plot, the one treated with methyl-parathion. In fact, the turn of color time was put forward by about a fortnight. The pulp, too, was slightly lighter in color. The grana, instead, displayed the same characteristics as was described for the cv Moro. The cost estimate made by skilled valuers for the produce treated with white mineral oil was a positive one. They were interested to know that the plot had undergone oil treatment, something which they, like the other people in the area, held to be partially phytotoxic and liable to cause a delay, rather than an anticipation in the turn of color time.

1.3. Conclusions

Certainly, the preliminary trials described here are on the whole indicative; they will be repeated over the coming years and studied in further depth.

For the moment on the basis of the results so far obtained, the following conclusions can be drawn:
1. The use of white mineral oil, alone and activated, has led to satisfactory results against the fig wax scale (C. rusci) but not complete ones (a maximum of 80% mor

451

Table III – Ceroplastes rusci L. cv Moro orange. Overall analysis of the fruit drop in the 6 comparative plots in the 12 days after treatment of 10/8/1983 (Agnone, Conti farm).

Plot	Active substance and doses in g/hl in parentheses	Average fruit drop (%)
1	White mineral oil (1600)	0.09 A a
2	White mineral oil (800)	0.07 A a
3	White mineral oil (1600) + methyl–parathion (30) + zineb (160)	0.08 A a
4	White mineral oil (800) + methyl–parathion (30) + zineb (160)	0.08 A a
5	Methyl–parathion (60)	0.06 A a
6	Control	0.08 A a

Table IV – Ceroplastes rusci L. cv Moro orange. Overall analysis of the leaf drop in the 6 comparative plots in the 12 days after the treatment of 10/8/1983 (Agnone, Conti farm).

Plot	Active substance and doses in g/hl in parentheses	Average leaf drop (%)
1	White mineral oil (1600)	109 A a
2	White mineral oil (800	98 A a
3	White mineral oil (1600) + methyl–parathion (30) + zineb (160)	125 A a
4	White mineral oil (800) + methyl–parathion (30) + zineb (160)	102 A a
5	Methyl–parathion (60)	104 A a
6	Control	66 A a

– The values indicated with a capital letter are significant for $P = 0.01$.

– The values indicated with a small letter are significant for $P = 0.05$.

(The relative data have been variance analyzed).

452

tality); there were no statistically significant differences between the two products;
2. there is no need to use activated oil; used alone at 1600 g/hl of a.s. it ensures
the maximum possible result in summer against this phytophagous;
3. none of the feared damage generally attributed to white mineral oil emerged from
the experiments; rather, there was an improvement in the putting forward of the
turn of color date and the pulp pigmentation in the cv Tarocco;
4. in the plots treated there was no anomalous leaf drop or fruit drop. All the products used yielded the same result in this respect.

REFERENCES

1. BUITENDAG, C.M. and BRONKHORST, G.S. (1980). Compressed-air spraying – a method for applying insecticides for pest management on citrus. Part 1 –
   Citrus and Subtropical Fruit Journal. N° 558, 4-5.
2. EBELING, W. (1950). Subtropical Entomology. Published by Lithotype Process
   Co, San Francisco, Calif., U.S.A.
3. LANZA, G. and DI MARTINO ALEPPO, E. (1975). Influenza dei trattamenti an-
   ticoccidici sulla filloptosi e sulla maturazione dei frutti di arancio "Tarocco".
   Tecn.agr. Anno XXVII, N° 5, 429-440.
4. LIOTTA, G. and MANIGLIA, G. (1974). Valutazione dell'efficacia dei fitofarmaci comunemente adoperati contro Aspidiotus hederae (Vall.) (Hom.Diaspididae)
   in Sicilia. Boll.Ist.Ent.Agr. e Oss.Fitopat.Palermo, Vol. IX, 207-212.
5. NUCIFORA, A. (1981). Pilot project for biological control in citrus culture.
   C.E.C. Programme on integrated and biological control. Progress report 1979/
   81, 79-88.
6. NUCIFORA, A. (1983). Integrated pest control in lemon groves in Sicily: five
   years of demonstrative tests and present feasibilities of transferring results.
   C.E.C. programme on integrated and biological control – final report 1979/83,
   129-146.
7. RIEHL, L.A. et al. (1956). Effect of oil sprays on the maturity of citrus fruits.
   Flo. St. Hort. Soc. Proc., 193-218.
8. WEDDING, R.T. et al. (1952). Effect of petroleum oil sprays on photosynthesis
   and respiration in citrus leaves. Plant Physiology, Vol. 27, N° 2, 269-278.

Effects of the distribution of chlorpyrifos on the lemon*

E.Di Martino & M.Romeo
Citrus Experimental Institute, Acireale, Italy

Summary

Trials were carried out during the winter months (February - March) spraying lemon trees with chlorpyrifos alone and mixed with white mineral oil.

I.1. INTRODUCTION

During the winter of 1983/4 a large number of fruit in a lemon orchard near Mascali were found to have unusual deformities. The epicarp and mesocarp were deformed to a wavy crest which became increasingly undulated with the maturity of the fruit (see Fig. 1). The crest ran around the fruit longitudinally from stalk to tip, about a centimeter high at the equator, and almost as wide. The phenomenon was widespread, and sufficiently frequent to rule out the possibility of either a chimera or a gemmaceous mutation. The fruit was thus unmarketable because of its appearance, and showed a tendency to rot, particularly by Penicillium spp., on account of the lesions which inevitably occurred in the carina which appeared all over the fruit. On-the-spot investigation revealed that the disease occurred only in those parts of the orchard where the trees had been sprayed; towards the end of the previous winter the affected trees had been treated with a mixture of chlorpyrifos and white mineral oil.

An attempt was then made to reproduce the phenomenon experimentally, in view of its potential importance and the high degree of efficacy of this treatment in control of Diaspinae, and Aonidiella aurantii (MASK.), in particular.

I.2 MATERIALS AND METHODS

A compact grove of eight-year old lemon trees (cv Femminello Siracusano) located near Syracuse was selected as a test ground. The mixes to be tested were 1.5% white oil plus 0.06% chlorpyrifos, to reproduce the spray used where the deformities were noted; 0.06% chlorpyrifos plus

---

* Research conducted as part of the Ministry of Agriculture and Forestry project "Development and improvement of industrial fruit-growing, early fruit growing and citrus fruit growing", Publication n° 159.

Fig. 1 - Typical injured lemon fruit

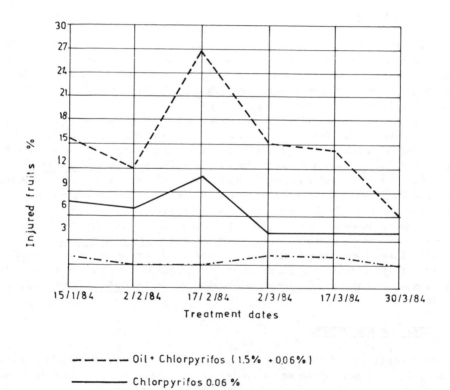

Fig. 2 - Percentage of injured lemon fruits for every date
of treatement and for used products.

456

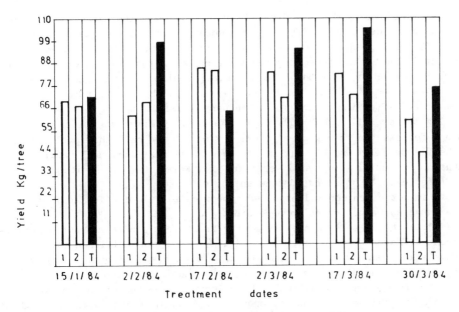

1 - Chlorpyrifos 0.06 % + mineral oil 1.5 %

2 - Chlorpyrifos 0.06% + wetting agent 0.05%

T - Test (untreated)

Fig. 3 - Lemon fruit production, related to dates of treatment.

anionic wetting agent at 0.05% to identify any effect of using the product without the adjunction of oil; and an untreated control. The first spraying was effected on January 15, 1984, followed by a further five at approximately fortnightly intervals, viz. on February 2, February 17, March 2, March 17 and March 30.

Blocks of trees were selected at random for each of the three groups treated on the same dates, each group being replicated four times. A suitable number of rows of untreated trees were left between test blocks and ignored in evaluation, to ensure that no cross-treatment occurred.

For each spraying a power pump operating at a pressure of 20 atmospheres was used to spray the lemon trees uniformly to the point at which the solution dripped from the leaves.

At the end of the season the crop was picked and, for each tree originally selected, the number of affected fruit counted and weighed.

Field observations began well before the harvest, but it was not possible to identify the deformities clearly until the fruit was approaching maturity.

I.3 RESULTS AND DISCUSSION

Data relating to the numbers of fruit affected were subjected to

457

statistical tests which indicated that, for each spraying, fruit were affected at dates differing significantly between the three groups (see Fig. 2). The plants treated with chlorpyrifos + oil bore consistently more affected fruit than those treated with chlorpyrifos alone, which in turn bore consistently more than the untreated control (where the damage rate never exceeded 1%). For the product + oil mix the damage rate rises from 16% to 24% after the spraying of February 17, falling off to 15% on March 2, 14% on March 17 and finally 6% for March 30. The curve for chlorpyrifos alone follows the same general line, but at barely half the level, peaking likewise on February 17, falling to around 4% next time and remaining at that level until the final spraying on March 30, at which point no difference could be detected between the samples.

The untreated control produced approximately 1% of affected fruit, but the curve shows no logical progression.

Quantity as well as quality of the fruit were monitored, but no significant differences were noted (see Fig. 3). Nor were differences noted in qualitative assessments of the fruit, i.e. average weight, juice yield, total solids, acidity, sugar/acidity, and vitamin C content.

The foregoing clearly shows the effect of the products and of the date of spraying on the incidence of crest formation in fruit, with the highest incidence arising from spraying between January 15 and February 17, and a steady falling-off in damage from its peak of 27% to a level of 6% and 4% with the March 30 spraying.

Greatest fruit damage resulted from spraying with chlorpyrifos + oil; less with chlorpyrifos alone and practically none in the untreated control. In the case of the control, it is likely that the "barrier" rows of trees were inadequate, and that spray was able to drift and reach some control trees, thus accounting for the presence of "crested" fruit in the control.

This lends support to the hypothesis that the product in question acts as a highly volatile regulator, causing the crests before the trees have come into fruit.

The product does not appear to have any other physiological effect on the plant, and the fruit are in no way affected by chlorpyrifos. The tests nevertheless demonstrate that the product should be used with circumspection, particularly during the months of February and March.

Effect of application methods on vertical distribution of 1,3 D in the soil

M.Basile, V.A.Melillo & F.Lamberti*
Institute of Agricultural Nematology, C.N.R., Bari, Italy

Summary

Studies were carried out to determine the effect of application methods on the vertical distribution of Telone II in the soil in relation to control Tylenchulus semipenetrans in established citrus groves. The results show that the best distribution of the chemical in the top 60 cm of a sandy-clay-loam soil is achieved by applying the emulsifiable formulation in 30 l/m^2 of irrigation water.

1.1 Introduction

The ban on the use of 1,2 dibromo-3-chloropropane has deprived farmers of an effective non-phytotoxic fumigant to control the citrus nematode, Tylenchulus semipenetrans Cobb, in established orchards.

It has been proposed that 1,3 dichloropropene (1,3 D), a fumigant extensively used at high dosages (1000-2000 l/ha) to control various nematodes in replanting fruit groves could be applied at non phytotoxic rates to reduce to non pathogenic levels nematode populations in vineyards and citrus groves. In fact Youngson et al. (6) have shown that this chemical can be used to control plant parasitic nematodes on grapevine in California, without causing adverse effects and Basile et al.(3) have investigated factors affecting its vertical distribution in different types of soil, showing an effect of soil structure, rate of application and time and depth of sampling.

To control plant parasitic nematodes in fruit and vine groves of the Mediterranean region it seems necessary to reach lethal concentrations of a toxicant to about 40 cm deep in the soil (2). Further information on the vertical distribution of 1,3 D is provided in this paper based on an experiment at Rutigliano, in the province of Bari, southern Italy.

1.2 Materials and Methods

An experiment was carried out in a sandy-clay-loam soil with physical characteristics as listed in Table I. Two adjacent fields each were divided into four blocks in which 4x4 m plots were distributed at random; plots were separated 0.5 m within the blocks and there was a 1 m separation between blocks. The soil was ploughed to a depth of 30 cm and rotavated.

---

* Technical assistance of Miss A.C. Basile is appreciated.

In the first field the chemical Telone II (92% of 1,3 D) was applied by injector gun on 3 September 1984, at the rates of 60, 90, 120, 180 and 240 l/ha with 30 cm separation between injection points. In the second field an emulsifiable formulation of the chemical Telone II SL was distributed in the irrigation water, at the rates of 60, 120 and 240 l/ha. The chemical was applied at each rate in 20 l of water per plot and after application of the chemical, the plots were overhead irrigated at the rate of 10, 20 and 30 l water/m^2 respectively, to carry the chemical deeper into the soil.

Each treatment was replicated four times.

Seven days after treatment a core of soil 50 cm in diameter was dug at the center of each plot and the sections between 0 and 30 and 30 and 60 cm were stored separately in glass jars. Aliquots of 250 g were then analysed by gas chromatography with an electron capture detector (1).

1.3 Results and Discussion

The concentrations of 1,3 D found in the soil samples are reported in Table II and show that the fumigant, seven days after its application by hand injector, was then almost evenly distributed in the two soil profiles checked, independently of the rates of application. As expected, the 1,3 D concentrations increased with the rate of application. The low concentrations detected in the soil treated with the highest doses of the chemical can be attributed to imperfect application or to sampling errors. However, these data indicate that if the treatment is properly performed it is not necessary to apply very high doses to reach lethal concentrations in the top 60 cm of the soil.

The trials with the emulsifiable formulation of 1,3 D have confirmed the above data with the additional that concentrations of the chemical increase as the volume of irrigation increases (Table III).

Comparing the application of 1,3 D by the two methods (injection and irrigation) (Table IV) it appears that generally the concentration of the chemical is higher when it is applied in the irrigation water, but its vertical distribution is differently affected by the two methods of application. In fact, when application is by injection the highest concentrations are always detected in the profile between 30 and 60 cm, but the reverse when application is in the irrigation water.

The first phenomenon can be explained by the fact that decomposition of 1,3 D is much slower in wet soils (5, 1). Conversely, the higher concentration of the chemical in the profile between 0 and 30 cm when applied in the irrigation water is probably due to the lower evaporation into the atmosphere from wet soils (4).

REFERENCES

1. BASILE, M. and LAMBERTI, F. (1978). Distribuzione verticale e persistenza dell'1,3 dicloropropene in tre tipi di terreno dell'Italia meridionale. Nematologia Mediterranea, 6: 135-145.
2. BASILE, M., ELIA, F. and LAMBERTI, F. (1980). Fattori che influenzano la distribuzione verticale dell'1,2 dibromo-3-cloropropano nel terreno. La Difesa delle Piante, 4: 241-248.
3. BASILE, M., LO GIUDICE, V., MELILLO, V.A. and LAMBERTI, F. (1984). Indagini sulla distribuzione verticale dell'1,3 D in 4 tipi di terreno dell'Italia meridionale. Atti Giornate Fitopatologiche, Sorrento, 26-29 marzo 1984, Vol. II: 113-119.

4. McKENRY, M.V. and THOMASON, I.J. (1974). 1,3 dichloropropene and 1,2 dibromoethane compounds: I. Movement and fate as affected by various conditions in several soils. Hilgardia, 42: 393-421.
5. WILLIAMS, I.H. (1968). Recovery of cis and trans dichloropropene residues from two types of soil and their detection and determination by electron capture gaschromatography. Journal Economic Entomology, 61: 1432-1435.
6. YOUNGSON, C.R., TURNER, G.O. and O'MELIA, F.C. (1981). Control of plant parasitic nematodes on established tree and vine crops with Telone II soil fumigant. Down to Earth, 37: 6-10.

Table I. Physical characteristics and pH of the soil.

Soil Type	Per Cent of Constitutents				
	Clay	Silt	Sand	Organic Matter	pH
Diam. Particles mm	(<0.002)	(0.002-0.05)	(0.05-2.0)		
Sandy-clay-loam	28.4	25.8	45.8	2.7	8.2

Table II. Concentrations of 1,3 D (ppm) in the soil profile, 7 days after Telone II was applied with a hand injector.

Rates of Application (1/ha)	60	90	120	180	240	Averages
Depth (cm)						
0 - 30	0.079	0.086	0.102	0.159	0.116	0.108
31 - 60	0.057	0.079	0.105	0.139	0.137	0.103
Averages	0.068	0.083	0.103	0.149	0.126	0.105

Significant Effects	L S D	
	$P = 0.05$	$P = 0.01$
Rates of Application (R)	0.025	0.036
Depth (D)	---	---
R x D	---	---

Table III. Concentrations of 1,3 D (ppm) in the soil profile 7 days after an emulsifiable Telone II was applied through irrigation water.

Volumes of Irrigation (l/m²)	10				20				30				Averages			
Rates of Application (l/ha)	60	120	240	X̄	60	120	240	X̄	60	120	240	X̄	60	120	240	X̄
Depth (cm)																
0 – 30	0.045	0.044	0.068	0.052	0.067	0.101	0.164	0.110	0.123	0.257	0.245	0.208	0.078	0.134	0.159	0.123
30 – 60	0.023	0.037	0.053	0.037	0.041	0.081	0.125	0.082	0.079	0.126	0.312	0.172	0.047	0.081	0.163	0.097
Averages	0.034	0.040	0.060	0.044	0.054	0.091	0.144	0.096	0.101	0.191	0.278	0.190	0.063	0.108	0.160	0.110

Significant Effects	L S D	
	P = 0.05	P = 0.01
Volumes of H₂O (R')	0.031	0.044
Rates of 1,3 D (R")	0.022	0.030
Depth (D)	---	---
R' x R"	0.039	0.052
R' x D	---	---
R" x D	0.032	---
R' x R" x D	0.055	0.073

Table IV. Concentrations of 1,3 D (ppm) in the soil profile 7 days after applying Telone II with a hand injector or through irrigation water.

Depth (cm)	Hand Injector				Irrigation water (30 l/m²)				Averages			
	60	120	240	X̄	60	120	240	X̄	60	120	240	X̄
0 - 30	0.079	0.103	0.116	0.099	0.123	0.258	0.245	0.208	0.101	0.180	0.180	0.152
30 - 60	0.057	0.105	0.245	0.135	0.079	0.126	0.312	0.172	0.068	0.115	0.278	0.153
Averages	0.068	0.104	0.180	0.117	0.101	0.192	0.278	0.190	0.084	0.147	0.229	0.153

Significant Effects

L S D

	P = 0.05	P = 0.01
Methods of Application (M)	0.029	0.040
Rates of Application (R)	0.044	0.067
Depth (D)	---	---
M x R	0.050	0.069
M x D	---	---
R x D	0.049	---
M x R x D	0.071	---

Evaluation of fungicides used in the treatment of citrus brown rot*

G.Lanza
Citrus Experimental Institute, Acireale, Italy

G.Cutuli
Plant Protection Observatory, Acireale, Italy

Summary

Two systemic fungicides, Metalaxyl and Fosetyl aluminum, were compared with copper oxychloride and Captafol in the treatment of citrus brown rot. Soil and foliar treatments were effected.
Efficacy of fungicides on soil was evaluated using untreated lemon fruit as hosts to detect Phytophthora spp. Foliar sprays, meanwhile, were evaluated using treated fruit on untreated soil (Disease Potential Index known).
Appreciable results were obtained using Metalaxyl in soil and foliar treatments. Similar results were obtained using Captafol in foliar sprays.

1.1 INTRODUCTION

Citrus brown rot is caused by Phytophthora spp.; particularly in wet years it is a major problem before the harvest and can have repercussions during the processing of fruit and during storage.
Treatment of this disease is preventive, normally using copper compounds sprayed around the lower canopy and the soil surface.
The phytotoxicity, which appears after repeated treatment by copper in certain combinations of soil and climate, can be avoided by using Captan.
The purpose of this research was to compare two new systemic products, Metalaxyl and Fosetyl aluminium, with a conventional copper compounds and with Captafol, using them to treat both the canopy and (with the exception of Fosetyl-Al) the soil.
Comparison of the products' fungicidal effect is difficult in drier years on account of the low incidence of the disease. Since the soil is the primary source of the inoculum, however, the "lemon trap" method was adopted to evaluate the efficacy of the fungicides.

---
* Research conducted as part of the Ministry of Agriculture and Forestry project "Development ans improvement of industrial fruit-growing, early fruit growing and citrus fruit growing", Publication n° 160.

1.2 MATERIALS AND METHODS

The "lemon trap" method was adopted to evaluate the efficacy of fungicides both spread on the soil and sprayed onto the trees.

Soil treatments

Application of granular Metalaxyl in doses of 0.2 and 0.4 g/m of soil was compared with copper oxychloride (0.5 g/m) in Cu, Captafol (0.5 g/m) in aqueous solution and with an untreated control. Tests were carried out on plots of citrus orchard measuring 32 m with a known DPI of 1/64. Samples were taken at intervals of 10, 20, 40, 60 and 80 days after treatment, and efficacy estimated by taking the percentage of lemon fruit previously selected as "trap" for Phytophthora and now affected by brown rot after 10 days of contamination at 20°C 2°C. For each product 60 fruit were arranged in compartment trays, each compartment containing 25 g of sieved soil and 150 cc of deionized water, and one fruit being placed in each.

Plant treatments

Tests compared the systemic fungicides Metalaxyl and Fosetyl-Al at concentrations of 2000 ppm with copper oxychloride (2500 ppm in Cu), Captafol (2000 ppm) and the untreated control. Each product, together with an anionic wetting agent at 0.05%, was sprayed onto five lemon trees, Citrus limon Burm., cv Femminello, bearing winter fruit.

The efficacy of the fungicides was evaluated by the same method as described above, using as a source of the infection a soil sample with a known DPI of 1/32, and taking for each product at intervals of 10, 30 and 60 days after treatment, sixty fruit to be used as "trap" for Phytophthora.

1.3. RESULTS AND DISCUSSION

Soil treatments

As the bar-chart in Fig. 1 shows, copper oxychloride and Captafol were so incapable of preventing soil-borne infection that the laboratory tests were abandoned after the 60th day.

Metalaxyl, on the other hand, showed a high degree of efficacy, particularly in doses of 0.4 g/m , in controlling the disease remarkably up to 80 days after the treatment. The percentage increase in the number of fruit affected by brown rot after 60 days of treatment can probably be attributed to a revival in the activity of Phytophthora resulting from environmental factors – rainfall and temperature – which raised the incidence of disease in the control to its highest level.

Plant treatments

Of the products tested, satisfactory results were obtained with Metalaxyl and Captafol (Fig. 2). In early stages the latter was notably effective in containing infection. The other products tested showed no

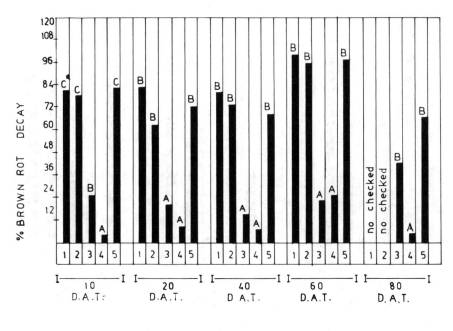

1 - Copper oxychloride   0.5 g/m^2 in Cu (soil drench)
2 - Captafol   0.5 g/m^2 ("   ")
3 - Metalaxyl   0.2 g/m^2 (granular application)
4 - Metalaxyl   0.4 g/m^2 ("   ")
5 - Untreated   -
*   Duncan's multiple range test 1% level.

Fig.1-Effect of fungicides soil treatments on brown rot
incidence of lemon fruits used as traps for Phytophthora
spp.

significant difference from the performance of the control, with the exception of copper oxychloride, which was significantly effective only for the
first test.

From this it may be concluded that new possibilities do exist for the
control of infection of citrus fruit by Phytophthora.

Particularly promising among these are the application of granular
metalaxyl at 0.4 g/m , for its lasting fungicidal effects in situ.

Finally, it does not have fatal side-effects for mycorrhizal symbiosis
(1).

In the case of foliar treatment, both products (Metalaxyl and
Captafol) reduce the infection effectively and durably.

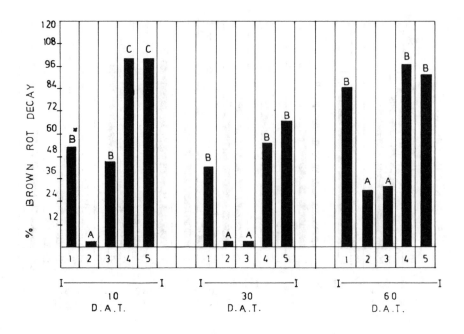

1 - Copper oxychloride 2500 p.p.m. in Cu + wetting agent 0,05%

2 - Captafol 2000 p.p.m. + " "

3 - Metalaxyl 2000 p.p.m. + " "

4 - Phosethyl Al 2000 p.p.m. + " "

5 - Unsprayed -

* Duncan's multiple range test 1% level.

Fig. 2 - Effect of fungicides foliar sprays on brown rot
incidence of lemon fruits treated and used as traps
for Phytophthora spp.

The results obtained using the two new systemic products compared,
Metalaxyl and Fosetyl-Al, must be further considered in the light of their
different fungitoxic actions. Metalaxyl has demostrated an inhibiting effect on in vitru mycelial growth and spore-formation in Oomycetes (2,3),
which is not the case for Fosetyl-Al. Neither does there seem to be a
great deal of support for the hypothesis that such compounds are activated
in the host plant, since the principal metabolite formed, phosphite, has no
fungitoxic action. On the other hand, it appears that such products
(Fosetyl-Al and its metabolite) may stimulate the host plant's defensive
reactions only in the presence of an infectious process (2). This would
explain the new systemic product's lack of efficacy in preventive treatment
by spraying.

In conclusion, it should be stated that, although the methodology adopted for evalutation of the products deliberately offered the most favourable conditions possible for development of the pathogen, in the case of certain products - Metalaxyl and Captafol - it gave valuable indications which should in all events be confirmed in full field trials in comparison with copper, whose efficacy against citrus brown rot is already the subject of considerable literature.

REFERENCES

1. BROEMBSEN, L. von (1981). The Citrus Improvement Programme Citrus Diseases and Declines. The Citrus and Subtropical Fruit Journal, October:24-26 (South Africa).
2. LEROUX, P. (1981). Les modes d'action des substances antifongiques à usages agricoles. La Défense des Végétaux, 207:59-83 (France).
3. URECH, P.A., SCHWINN, F.J. and STAUB, T. (1977). CGA 48988, a novel fungicide for the control of late blight, downy mildews and related soil-borne diseases. Proc. 1977 Br. Crop. Prot. Conf., :623-631 (England).

Phytophthora blight of ornamental citrus: Chemical control of soilborne inoculum*

G.Magnano di San Lio, A.M.Pennisi & R.Tuttobene
Plant Pathology Institute, University of Catania, Italy

Summary

Benalaxyl, captafol, captan, ethridiazole, alone and mixed with captan, fenaminosulf, furalaxyl, maneb, metalaxyl, and phosethyl-Al were evaluated as soil fungicides against Phytophthora citrophthora (Sm. & Sm.) Leon. and P. nicotianae van Breda de Haan.
Fungicides were tested in vitro and in experiments performed under screenhouse. There was no correlation between in vitro activity and performance of fungicides in eradicating inoculum of Phytophthora in soil.
Results that could be utilized in developing control stategies of Phytophthora blight, a soilborne disease of ornamental citrus, are emphasized.

1. Introduction

Blight caused by Phytophthora citrophthora (Sm. & Sm.) Leon. is the most serious disease of ornamental citrus in Sicily.

When environmental conditions are favourable to the disease up to more than 80% of plants in a nursery may be affected (6). Because infested soil is the major source of inoculum of P. citrophthora, soil treatments with fungicides have been suggested as a preventive control measure against blight. However there is a lack of information on the efficacy of fungicides in eradicating inoculum of P. citrophthora in the soil.

This paper reports results of laboratory and greenhouse tests carried out in order to evaluate the effect of fungicides on inoculum of Phytophthora spp. in soil. Preliminary results have been published (4). Data that could be utilized for developing blight control strategies are emphasized.

2. Materials and methods

2.1. Compounds tested

The following fungicides were assessed: Benalaxyl (M9834, 25% w.p.); Captafol (Orthodifolatan Blu, 80% w.p.); Captan (Orthocide 50, 50%

---

* This work has been financially supported by "Cassa per il Mezzogiorno". Research project 11/28.

w.p.); Ethridiazole (Terrazole, 25% e.c.); Ethridiazole + Captan (Ter-razole–Cap., 20 + 25% w.p.); Fenaminosulf (Bayer 5072, 70% w.p.); Furala-xyl (Fongarid, 25% w.p.); Maneb (Maneb 80, 80% w.p.); Metalaxyl (CGA 48988, 25% w.p.); Phosethyl-Al (Aliette–Ravit, 80% w.p.).

2.2. Laboratory tests

TEST FUNGI. Two species of Phytophthora, both pathogenic to citrus, were used as test fungi in laboratory assayes: P. nicotianae van Breda de Haan var. nicotianae Waterh. and P. citrophthora (Sm. & Sm.) Leon.

ACTIVITY ON MYCELIAL GROWTH. There are many reports concerning in vitro activity against P. citrophthora and P. nicotianae showed by most of the above–mentioned fungicides (3,9,10,12), but there is a lack of information on behaviour of benalaxyl, a systemic fungicide (class of acylalanines) recently developed by Farmoplant Montedison. Hence activity of benalaxyl on mycelial growth was preliminarily compared with that one of furalaxyl, another compound of this class, that in earlier works proved to be effective against Phytophthora spp. pathogenic to citrus. Activity on mycelial growth was evaluated by the standard seeded–agar bioassay technique.

Aliquotes of fungicides at varying concentrations in sterile distil-led water were added to warm (50 C°) melted potato–destrose agar (PDA) at the time of pouring plates to yeld 0.5, 1, 2, 5, 10, 50, and 100 µg a.i./ml medium. Plates (20 ml of medium/plate) were inoculated on the same day with an agar inoculum block of test fungus. Four replicates were used for each concentration. Plates were then incubated in the dark at 25° C. Diameter of colonies was recorded one week after inoculating.

ACTIVITY ON CHLAMYDOSPORE GERMINATION. It is generally assumed that chlamydospores are one of the most important form of inoculum of soil inhabitant Phytophthora spp. (8). Therefore inhibitory effect of fungici-des on chlamydospores germination was evaluated by a soil vial test (4). A sandy with pH 5.8 and 6.5% organic matter content was employed. Chlamy-dospores of P. nicotianae (11) were suspended in sterile distilled water. 0.5 ml of the water suspension were pipetted in shell vials (diameter 20 mm) previously filled with autoclaved soil. 2.5 cm of soil were then added to the vials. Each vial was drenched with 5 ml of fungicidal suspension, equivalent to about 6 liters of water/m^2. The following rates were tested: 2000, 1000, and 200 µg a.i./ml of water, equivalent to about 12, 6, and 1.2 g a.i./m^2 respectively.

After an incubation period of 24 hours at 25° C the number of viable propagules in the vials was determined by placing the soil in a BNPRAH selective medium (7).

2.3. Screenhouse trials

Efficacy of fungicides in eradicating inoculum of Phytophthora spp. in naturally infested soils was evaluated in trials perfomed under screenhouse. Pot-grown 18 months old sour orange seedlings severely affected by root and foot rot were employed. In preliminary assayes both P. citrophthora and P. nicotianae were recovered from soil of pots, by utilizing selective media. Soil of pots was a sandy with pH 5.9 and 5.9% organic matter content.

Benalaxyl, captafol, captan, ethridiazole, ethridiazole plus captan, fenaminosulf, furalaxyl, maneb, and metalaxyl were applied as soil drenches at 50 μg a.i./ml rate. Each plant was treated with a 500 ml drench per 4 liter container, equivalent to about 20 liters of water/m^2 and to 1 g a.i./m^2. Untreated control was included. Phosetyl-Al was applied as foliar spray at 2500 ug a.i./ml. Seedlings sprayed with phosethyl-Al and control seedlings were drenched with 500 ml of water at each application.

Eight applications were made at weekly intervals from May to July. 28 plants were used per treatment in four replicates of five plants each.

Inoculum density (ID) of Phytophthora spp., expressed as number of viable propagules per gram of dry soil (ppg), was determined before treatments and 28 days after the last application, by utilizing the soil dilution plate technique in conjunction with a BNPRAH selective medium (5).

In another trial all the fungicides, phosethyl-Al included, were drenched once at 15 g a.i./m^2 rate. ID of Phytophthora in the upper 10 cm soil was determined soon before and one week after the application of fungicides. Treatments were arranged in a randomized block with four replications.

3. Results and discussion

In vitro benalaxyl inhibited mycelial growth of both test fungi less than furalaxyl. Moreover P. citrophthora appeared less sensitive to either fungicides than P. nicotianae (Fig. 1 and Tab. I).

In soil vial tests the two most effective fungicides were captafol and captan which at a rate of 200 μg/ml reduced chlamydospore germination by 100 and 76% respectively.

Activity on chlamydospore germination and dosage responses to various concentrations of candidate chemicals applied as drenches on surface soil of vials are reported in Fig. 2 and Tab. II. There was no correlation between activity of fungicides in agar and in the soil vial test.

In screenhouse trials multiple applications at low rates of metalaxyl, ethridiazole, maneb, and captafol significantly (P= 0.01) reduced soil population of Phytophthora spp. as compared to untreated control but did not eradicate inoculum (Fig. 3). After the last application, mean values of ID of P. citrophthora in the upper 10 cm soil of pots drenched with the above mentioned compounds were 0.5, 2, 10, and 29 ppg (Fig. 5).

Phosethyl-Al, fenaminosulf, ethridiazole plus captan, and furalaxyl only slightly reduced ID of Phytophthora spp. in soil. Differences however were significant at 0.05 probability level (Fig. 3).

Interactions between fungicides and soil-depths and between fungicides and species of Phytophthora were not statistically significant.

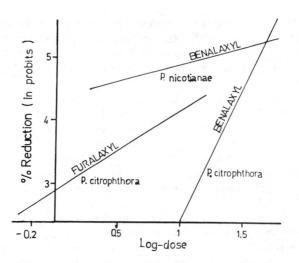

Fig. 1 – Inhibitory activity of benalaxyl and furalaxyl on mycelial
growth of <u>Phytophthora</u> spp. pathogenic to citrus. Inhibition expressed as percentage reduction (control= 100) of diameter of
colonies grown on PDA.

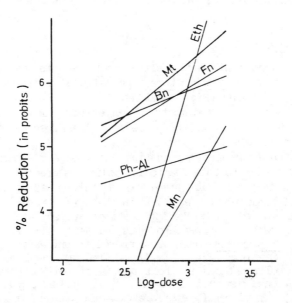

Fig. 2 – Dosage response of <u>Phytophthora nicotianae</u> to various concentrations of fungicides. Activity tested in a bioassay using chlamydospores. Bn= Benalaxyl, Eth= Ethridiazole, Fn= Fenaminosulf, Mn=
Maneb, Mt= Metalaxyl, Ph-Al= Phosethyl-Al.

474

Tab. 1 - ED$_{50}$ of benalaxyl and furalaxyl on mycelial growth of Phytophthora spp. as determined by seeded-agar test.

Test fungus	ED$_{50}$ (μg/ml)	
	Benalaxyl	Furalaxyl
P. citrophthora	40	5
P. nicotianae	16	<0.5

Tab. 2 - ED$_{50}$ of fungicides as determined by soil vial test. Test fungus: P. nicotianae.

ED$_{50}$ (μg/ml)		
<200	<1000	>1000
Benalaxyl	Maneb	Ethridiazole
Captafol		Furalaxyl
Captan		Phosethyl-Al
Fenaminosulf		
Metalaxyl		

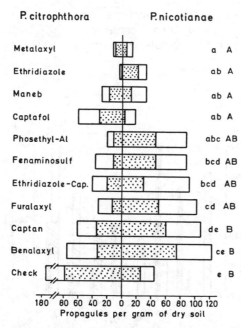

P. citrophthora   P. nicotianae

Metalaxyl — a A
Ethridiazole — ab A
Maneb — ab A
Captafol — ab A
Phosethyl-Al — abc AB
Fenaminosulf — bcd AB
Ethridiazole-Cap. — bcd AB
Furalaxyl — cd AB
Captan — de B
Benalaxyl — ce B
Check — e B

180 80 60 40 20 0 20 40 60 80 100 120
Propagules per gram of dry soil

Fig. 3 - Effect of fungicides on inoculum density (ID) of Phytophthora spp. in rhyzosphere soil of container-grown sour orange seedlings. All the fungicides were applied as soil drenches at 50 μg/ml, except phosethyl-Al, which was applied as foliar spray at 2500 μg/ml.
Legenda: ID at 10÷20 cm depth; ID at 0÷10 cm. Values followed by the some letters do not differ significantly. Duncan's multiple range test: capital letters P= 0.01; small letters P= 0.05.

475

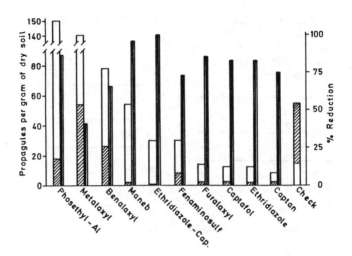

Fig. 4 – Effect of fungicides applied as soil drenches at 750 µg/ml on inoculum density (ID) of <u>Phytophthora</u> spp. in rhyzosphere soil of container–grown sour orange seedlings.
Legenda: ▢ ID before treatment; ▨ ID after treatment; ▬ % reduction (ID before treatment= 100).

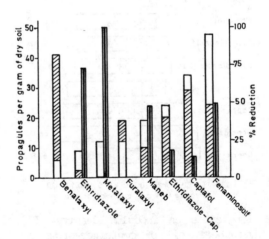

Fig. 5 – Effect of fungicides applied as soil drenches at 50 µg/ml on inoculum density (ID) of <u>P. citrophthora</u> in the upper 10 cm soil of pots.
Legenda: see Fig. 4.

In single application at high dosage all the fungicides strongly reduced population of Phytophthora in the upper 10 cm soil.

Because ID before fungicide application were not uniform in treatments it appeared difficult to compare efficacy of fungicides. Anyhow it can be concluded that the mix ethridiazole plus captan, and maneb showed the highest fungicidal activity while metalaxyl and benalaxyl were the less effective in reducing ID of Phytophthora in soil. Moreover the mix ethridiazole plus captan was more efficacious than the single components alone, thus suggesting synergy (Fig.4).

In pots drenched with benalaxyl, fenaminosulf, phosethyl-Al, and metalaxyl, ID after treatment was above 3 ppg. Therefore it could be inferred that in these pots notwithstanding treatments the likelihood of blight infections to occur kept high (6), at least supposing that fungicides should not have any residual fungistatic activity.

4. Concluding remarks

Results led to the following conclusions:

1) In vitro assayes, soil vial tests included, give only rough indications of effectiveness of fungicides in eradicating inoculum of Phytophthora from naturally infested soil.
2) Protectant fungicides performed as well as narrow spectrum systemic fungicides in eradicating inoculum of Phytophthora spp. in soil.
3) All the fungicides, with the only exception of metalaxyl, were more effective in reducing soil population of Phytophthora spp. in single application at high rate, than in repeated applications at low dosage. Metalaxyl, on the contrary, performed better in multiple application at low rates. This long-term effect could be due to antisporulant and fungistatic proprierties of this compound. On the other hand high dosages of metalaxyl could give phytotoxicity in potted citrus (2).
4) Complete eradication by chemicals of soilborne inoculum of Phytophthora is not feasible, in the practice, neither in container-grown citrus plants.
5) Previous works (1,4) provided evidence that some fungicides, as captan, metalaxyl, and phosethyl-Al, are immobilized in the upper 10 cm soil. This however does not seem an important drawback in citrus blight control because primary infections of this disease are caused mainly by inoculum harboured by surface soil.

REFERENCES

1. ABBATTISTA GENTILE, I., MONTESANO, D., and SOVERCHIA, G. (1982). Adsorbimento, migrazione verticale e persistenza del metalaxyl nel terreno. La difesa delle piante, vol. 5, 143-158.
2. DAVIS, R.M. (1982). Control of Phytophthora root and foot rot of citrus with systemic fungicides metalaxyl and phosethyl aluminum. Plant Dis., vol. 66, 218-220.

3. FARIH, A., MENGE, J.A., TSAO, P.H. and OHR, H.D. (1981). Metalaxyl and ephosyte aluminum for control of Phytophthora gummosis and root rot on citrus. Plant Dis., vol. 65, 654–657.
4. MAGNANO DI SAN LIO, G., PENNISI, A.M., and PERROTTA, G. (1984). Effects of some fungicides on population of citrus Phytophthoras in soil. Proc. Int. Soc. Citriculture, Sao Paulo, Brazil (in press).
5. MAGNANO DI SAN LIO, G., PERROTTA, G., and TUTTOBENE, R. (1983). Indagine sui funghi patogeni presenti nel terreno di semenzai di arancio amaro. Inf.tore fitopatol., vol. 33, 49–55.
6. MAGNANO DI SAN LIO, G., TUTTOBENE, R., and PENNISI, A.M. (1985). Phytophthora blight. A destructive disease of ornamental citrus. The Proceedings of the Expert's Meeting on "Integrated pest control in citrus-groves", Acireale, Italy (in press).
7. MASAGO, H., YOSHIKAWA, M., FUKADA, M., and NAKANISHI, N. (1977). Selective inhibition of Pythium spp. on a medium for direct isolation of Phytophthora spp. from soils and plants. Phytopathology, vol. 67, 425–428.
8. MITCHELL, D.J. (1978). Relationship of inoculum levels of several soilborne species of Phytophthora and Pythium to infection of several hosts. Phytopathology, vol. 68, 1754–1759.
9. SCHWINN, F.J. (1983). New developments in chemical control of Phytophthora. pp. 327–334. In: Phytophthora: its biology, taxonomy, ecology and pathology (D.C. Erwin, S. Bartnicki-Garcia, and P.H. Tsao eds.). The American Phytopathological Society, St. Paul, Minnesota, 392 pp.
10. TIMMER, L.W. (1977). Preventive and curative trunk treatments for control of Phytophthora foot rot of citrus. Phytopathology, vol. 67, 1149–1154.
11. TSAO, P.H. (1971). Chlamydospore formation in sporangium-free liquid cultures of Phytophthora parasitica. Phytopathology, vol. 61, 1412–1413.
12. WHEELER, J.E., HINE, R.B., and BOYLE, A.M. (1970). Comparative activity of dexon and terrazole against Phytophthora and Pythium. Phytopathology, vol. 60, 561–562.

Observations on the operation of a multifunctional equipment in pesticide applications

G.Baraldi
Institute of Agricultural Engineering, University of Bologna, Italy

A.Catara
Plant Pathology Institute, University of Catania, Italy

Summary

Pesticide spray tests were performed with a fixed multipurpose wind machine. Investigations included both the extent of plant canopy coverage and ground distribution, providing a diagram that demonstrates the systems performance within and outside the study area. Drop size was also evaluated. The machine was also compared to two other spray systems in current use, air-injection and a mobile pump with hand operation. The three spray methods showed different features. The effectiveness of fungicides, obtained from leaf residues, in eight (8) sections of the plant canopy showed no statistical difference between the multi-purpose wind machine and the air-injection sprayer.

1. INTRODUCTION

In Sicily pesticide applications to citrus groves are usually performed with a number of different systems. Fixed pumps with pipelines even 50-100m long are still largely used, while sprayers on tractors are utilized above all in average to large farms. This work is very often performed by third parties. This causes problems in operation, preventing the farmers from performing timely applications for the control of some particular diseases - namely the "Mal secco", which needs regular cycles of applications and, in particular, some special fungicide treatments within 48 hours from any event which can cause wounds and lacerations to the plants (hail, frost, etc.).

The need for timeliness suggests the use of mechanical means of high working potential, capable of operating even in adverse weather conditions, when entering the orchards is almost impossible.

A good solution, in this respect, is offered by A.I.D. (Agriculture Industrial Development) of Catania with the introduction of a new spraying system, mounted on a wind machine, operating from a fixed position in the farm, for the relief temperature stress in the environment.

Some observations have been made in this equipment, located in Cuticchi Farm, on 16-year-old "Tarocco" orange trees, with spacings of 6x5m (Table 1), by using one of the fixed units described above.

479

Table 1 - Plant characteristics

Canopy shape	Spherical
Age	16 years
Cultivar	Tarocco
Root stock	Sour Orange
Spacing	6 x 5
Density (trees/ha)	333
Max. canopy height (m)	3.50
Height of the first branches (m)	0.85
Min. canopy height (m)	0.35
Average diam. of the canopy (m)	3.50

---

2. DESCRIPTION OF EQUIPMENT

A description of the multipurpose equipment (machine A) is reported; an outline is also presented for the two mechanical systems used in the farm (machines B and C) previously referred to.

MACHINE A - It is an equipment known as "Polyguard H.P. 300", basically consisting of a steel tower of approx. 11 m height, having a two-blade fan of diam. 5,6 m on the top; the axis of rotation is inclined approx. 5° on the horizontal plane.

It is operated by a 104 Kw engine, through a suitable transmission (Fig. 1). The machine is equipped with special systems which make all its operating functions completely automatic.

The present version of this machine has three main functions: frost protection, cooling, pesticide applications. For this last function the machine is equipped with an octagonal structure complete with 40 nozzles operating in the air string; one auxiliary nozzle, inserted to spray at the bottom, distributes the liquid in the circular area near the machine (Fig.2).

Two separate hydraulic circuits are located on the machine: the first, at low flow rate and pressure, injects a concentrated mixture into the second, which was previously supplied only with water.

The second circuit includes a six-membrane pump connected to the engine, and is complete with a tank, a filtering unit, pressure gauges, operating and control panel, the nozzle manifold and the nozzles.

The nozzles are of the conical-flow tipe, supplied by properly positioned eccentric holes. The auto-regulating board enables the mechanical unit to

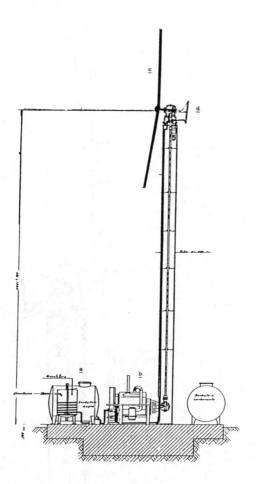

Fig. 1 - Diagram of the multipurpose wind machine
        a) tanks; b) engines and pumps; c) fan; d) spray
        octagonal structure.

automatically operate, when special conditions pre-set in the program, occur
(wind speed, temperature, humidity). The machine can be either automatically
or manually operated.

The flow-rate for each test, by operating at a pressure of 20 bar, was
1,6 l/s, equal to 4.800 l volumes supplied in one complete rotation (approx.
1.700 l/h if all the liquid is distributed within a radius of 95 m); this
rotation is completed in 50 minutes.

MACHINE B - It is a spraying machine consisting of a mobile pump unit, a
tank for the mixture, a long pipeline for the liquid with a sprayer for
manual use. The equipment was used in the tests to distribute 1.700 l/h of
liquid, at a working pressure of 30 bar.

The sprayer is complete with a nozzle with adjustable conical-flow.

481

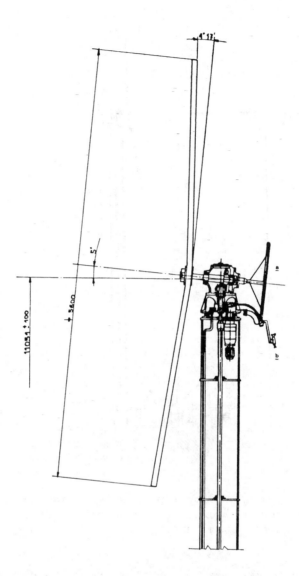

Fig. 2 - Particular of spraying structure (a) with downward
spray nozzle for distribution in the area close to
the machine (b).

MACHINE C - It is an air-injection spraying machine type "MONTANARI 1000",
complete with membrane pump, connected to a driven wheel.

The fan, diam. 620 mm, is of the eight-blade type. The machine was
used with 10 nozzles, which could be directed on two sides, and a liquid
pressure of 20 bar. The speed of displacement was 1,8 m/s. The possibility
of regulating the machines and the special way of using them allow the
distribution of approx. 1.700 l/h.

3. MATERIALS AND METHODS

A series of tests were carried out to evaluate the performance of the
equipment in pesticide applications and, particularly, to verify the
distribution diagram on the plants and the horizontal plane.

The machine was previously compared to the other two traditional
systems, taking into consideration two plants for all tests. The fixed
machine allowed wetting the plants at 60 m from the source, considering
such distance as indicating the treated area; the other two machines were
used to spray rows nearby, having characteristics similar to those
treated   with the machine A. For each plant 8 positions were taken
into consideration: high, bottom, left, right, interior and exterior front,
interior and exterior back. The compound used was SANSPOR, (a Captafol
derivative) distributed at a concentration of 400 g/hl.

12 leaves were sampled from each section and from each one a small disk
of 11 mm diam. was obtained, for the biological activity test. This was
evaluated on the basis of the zones of inhibition obtained by putting the
leaf samples in Petri plates inoculated with Penicillium sp.

In another test the distribution of the liquid on plants was evaluated
by sampling 30 leaves at the height of approx. 2 m, both on front and back
sides of the plants located in a radial direction from the spraying source,
at distances of 30, 50, 70, 90 m. The surface and, by the analysys at the
gas-chromatographer, the quantity of residues of the sprayed compounds were
evaluated for each sampling. The treatment was effected with 200 g/hl of
Tedane (a.p. Dicofol 13,5% and Tetradifon 6%).

Once verified in the laboratory the performance reliability based on
the zones of inhibition in the same way the horizontal distribution was
also evaluated in a single radial direction, at distances of 20, 40, 60,
80, 100 m from the spraying sources. Use was made of water-sensitive test-
papers fixed on the leaves of the plant at 1,5 m height, in front position
for this analysis.

The horizontal distribution, diagram within and out side the area
covered by the machine, was also evaluated when applying a treatment with
Tedane, as in the previous case, putting some filter test-papers at 0,3
and 3 m height in the treated area (0,5 x 0,5 m). The test-papers were
subsequently examinated by gas-chromatography for determining the residues.

The filters at 3 m height were located in two radial directions at
intervals of 20 m (up to 80 m from the spraying source); the filters at 0,3m
height were located on a radius with intervals of 10 m (up to 150 m).

In this last test, in the same positions, on a single radius, wind
speed at heights of 0,5 - 1,5 - 2,5 m was determined with the fan in fixed
position. In the same tests, the marks left by the drops on emulsified
films, placed on the plants at regular distances of 20 m, up to 140 m, were
also evaluated.

For each test, temperature, relative humidity and wind speed were
recorded (Table 2).

TAB. 2 - Meteorological conditions (Average Data)

Tests	Date of Test	Temp. (°C)	Relative Humidity %	Wind Speed (m/s)	Direction
- Comparison between systems with biologic tests	19/6/84	25	50	0.9-1.4	South/ South East
- Samplings on water-sensitive test-papers with biologic tests	25/7/84	19	75	0.6-1.2	West
- Samplings on leaves at 2.0 m height	13/9/84	14	90	0.9-1.2	South/East
- Samplings on filters at 3.0 m height	13/9/84	14	90	0.9-1.2	South/East
- Samplings on films at 2 m height	15/2/85	5	95	1.0-1.8	South/ South West
- Samplings on filters at 0.3 m height	25/2/85	6	95-100	0.4-0.6	North/West
- Determinations of air speed generated by the system	25/2/85	6	95-100	0.2-1.2	North/West

TAB. 3 - Compound on leaves (Dicofol + Tetradifol) sampled at heights of 2 m in East radial position.

Distance from the wind machine m	Sample position*	Distribution in mg/m^2	% (1)
30	F	10.55	100
30	B	7.32	69
50	F	10.94	100
50	B	10.70	98
70	F	11.10	100
70	B	7.73	70
90	F	9.09	100
90	B	4.02	44

* F = Front side of the tree from machine locations
  B = Back side of the tree from machine locations
(1) F value on each distance = 100

TAB. 4 - Compound sampled on the filters (Dicofol + Tetradifon) located
at 3 m height on two opposed radial positions (West and East)

| Distance | Distribution in mg/m^2 | | | |
	Direction West	Direction East	Average	% (1)
20	91.24	86.44	88.84	71.34
40	140.72	108.32	124.52	100.00
60	89.44	39.00	64.22	51.57
80	50.52	11.32	30.92	24.75

(1) Referred to highest value (at 40 m)

TAB. 5 - Compound sampled on the filters (Dicofol + Tetradifon) located
at the ground (0.3 m Height) on two opposed radial positions
(South-West and North-East)

| Distance from the wind machine m | Distribution in mg/m^2 | | | |
	Direction South-West	Direction North-East	media	% (')
10	16.72	17.20	16.96	11.16
20	52.96	72.92	62.94	41.43
30	152.10	151.72	151.91	100.00
40	166.52	74.68	120.60	79.39
50	138.12	78.96	108.54	71.45
60	39.40	43.96	41.68	27.44
70	32.80	35.80	34.30	22.58
80	25.36	14.12	19.74	12.99
90	18.48	5.72	12.10	7.96
100	9.64	4.48	7.06	4.65
110	7.16	3.56	5.36	3.53
120	4.92	2.20	3.56	2.34
130	2.88	1.20	2.04	1.34
140	1.80	0.62	1.21	0.79
150	0.65	0.52	0.58	0.38

(') Compound percentage referred to highest sampled average value (at 30m)

4. RESULTS

The data obtained were processed and reported on graphs and tables and are shown as follows in the order cited in chapter 3.

4.1 SAMPLINGS ON PLANTS

The results obtained through the comparison of the multipurpose wind machine with the other two systems demonstrated different performances.

The manual treatment showed the most even distribution on plants (Table 3).

The other two systems gave different results, depending upon their spraying sources. In particular, the air-injection spraying machine, forcing movements on two sides, showed a better efficiency on laterals, while the fixed machine covers better the front and upper positions, though there was no statistically significant difference between them.

The difference between the compound sprayed on the front and back sides of the treated plants, observed after the analysis of the leaf residues (table 4) showed a clear effect of penetration of the product into the canopy. A better coverage was obtained, however in the front part than the back. Furthermore, a regular distribution of the compound was obtained in almost all the treated area.

4.2 SAMPLINGS ON THE GROUND

The results of the tests done at the various distances with water-sensitive test-papers, and the biological evaluation of the residues, showed a good distribution, with a constant decrease after 60 m (Fig. 4).

The examination of the residues on the filters located at 3 m height shows a curve of distribution with a peak at approx. 40 m from the spraying source. This clearly shows a decrease from that point up to 100 m (Table 5).

A difference greatly due to the influence of wind was shown on the two radii considered.

The results obtained with the filters located at 0,3 m height showed a curve of distribution increasing up to approx. 30 m and then decreasing down to extremely low values (Table 6 and Fig. 6). Such event is more clearly influenced than the previous one by the presence of the canopy and by plant location in relation to the radii.

Wind speed at heights 0,5 - 1,5 - 2,5 m reached a max at approx. 50 to 60 m (respectively about 4, 6 and 12 m/s), than rapidly declining to a level close to 0 beyond 100 m (Fig. 7).

4.3 DROP CHARACTERISTICS

The size of the marks left by the drops when impacting the treated plants, at distances of 20 and 40 m, was 360 μm, with a variability coefficient of 20%.

The impossibility to read data from all pre-set positions, due to the high relative humidity, did not allow verifying this size in all the treated area.

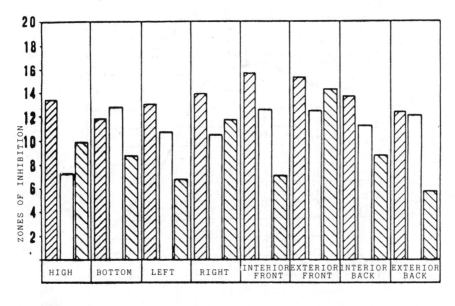

Mobile pump with hand operation

Air - injection sprayer

Multipurpose wind machine

Fig. 3 - Zones of inhibition (in ordinates) for the different
sides of the plant treated with the multipurpose wind
machine, the air-injection sprayer and the mobile
pump with hand operation.

5. CONCLUSIONS

From the observations and the results obtained in a correct use of the
multipurpose wind machine, some interesting conclusions can be drawn
concerning the spraying applications with pesticides:

- The product distribution on plants at examined distances is satisfactory,
  taking also into account the compensation effects due to the use of a
  multiple working system.
- Comparisons in product application rates between the multipurpose fixed
  machine and the two traditional mobile systems, show the characteristic
  pattern of each equipment.
- On the horizontal plane, the application rate decreases from 30–40 m from
  the spraying source, gradually at first, then rapidly, in particular
  beyond 90 m, dropping to a level close to 0, at 150 m. This effect can be
  considered as "drift" only at the edge of the treated area; it can be
  easily modified or even avoided by suitably reducing the range of action

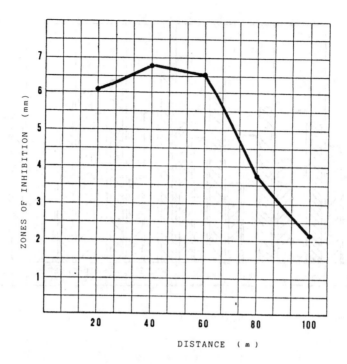

Fig. 4 - Zones of inhibition observed at the examined distances
from the spraying source.

of the machine, or sectioning the distribution area as foreseen in the
computer treatment program.
- Since the machines are installed in triangle-spacings of 190 m, the
surface covered by each machine is estimated in approx. 3 h. Each
application being completed in 50 min, the time working rate of the
machine is 3,6 h/hour.
- The distribution can be referred to as high volume spray application.
- Lastly, it must be pointed out also that the spray system of the fixed
machine and its automatic operation allow timely pesticide applications,
in all crop and soil conditions.
- The same features allow the operators to avoid any contact with the
compounds sprayed in the field.

6. REFERENCES

1. BARALDI G. (1984). Caratteristiche e prestazioni delle macchine irroratri
ci per trattamenti antiparassitari. Giornate Fitopatologiche 1984 -
Sorrento (NA).
2. CARMAN G.E. (1975). Spraying procedure for pest control on citrus. In
Citrus, 28-34. Ciba-Geigy Ltd, Basle.

3. JOWSTONE D.R. (1978). Statistical description of spray drop size for controlled drop application. Symposium on Controlled Drop Application, 35-42.
4. MATTHEWS G.A. (1979). Pesticide Application Methods. Longman, London and New York.

Cultural control methods

A.Nucifora
Institute of Agricultural Entomology, University of Catania, Italy

Summary

Various cultural pratices are reviewed (pruning, manuring, tilling, irrigation, herbicides, nontillage periods, wind-breaks) and their influence is examined on the development of insects, mites and fungi.

1. Introduction

The topic with which I have been entrusted is certainly one of the most difficult, since the reviewer must collect his anthological material here and there and make of it a unified whole enabling interested workers to take their studies further and to prepare them for future research. The matter has been entrusted to me for the sole reason that with research to a well-established cultural technique, that of the forcing of lemons for the production of summer lemons, I showed the effects of the early interruption of the dry period on Prays citri Mill., using it with certain precautions against this phytophage. Hence, starting from this fact, I found myself involved in the development of this topic, looking for other cultural practices which affect in one way or another the defence of our citrus groves.

Certainly there are things which can be studied for their consequences for acaro-entomatic population dynamics and for other biological factors of the agricultural ecosystem, since there is nothing which does not have its effect in partially modifying the ecosystem in which it acts.

Modern cultural practices differ considerably with respect to those practised in the past. One may cite the way of tilling the soil, the possible use of herbicides, the different means of irrigation, the practice of nontillage periods, the use of synthetic fertilisers instead of the manures of bygone days, the living or artificial wind-breaks, the reduced possibility of pruning and of trimming now as compared to before, interplanting or association of plots with copses which are fast disappearing, the sometimes heavy use of synthetic phytopharmaceuticals, certain of which induce changes in the cellular metabolism and in the physiological habits of plants and of the zoospecies living on them, and mulching. All of these topics represent chapters of a treatise which I would find it difficult to write both because of my limited knowledge and because of my lack of specific competence in the sectors specified above. I shall therefore treat the subject in a general way, leaving to others the making of specific scientific contributions in each case, if there exist people who are interested in this global or partial view of the problem.

It is obvious that someone like me who today wishes to treat a subject as vast and complex as cultural control methods cannot remain within his own highly specialised field of competence but must necessarily invade the fields of others; I hope

491

I do not offend my colleagues if several of the topics which I have been asked to touch on would have been better treated by them instead of me. In any case, how could I avoid other people's fields of work when for example, I speak of pruning, when this is performed annually in current agricultural practice and there are those who say that it is the only technique used against "dry rot"? How can one keep away from the field of weed control if one needs to speak of herbicides and their repercussion for usefull fauna in the ground and out of it? How can I avoid intruding on the competence of the agronomist, if I wish to discuss the influence of various types of irrigation on mites, insects and fungi? I give these examples to show the delicacy of the task which has been entrusted to me and the difficulties I experience in perfoming it.

Nevertheless I did not want to refuse the proposal, nor avoid an investigation of the question, because I believe in the actual possibility of "integrated control" in citrus fruit growing, and indeed certain of the cultural control practices have been examined and applied by me experimentally.

Furthermore, I believe in the future possibility of "integrated plant production" in the performance of which the various fields of agricultural science will partecipate, with the common goal of allowing agriculture to give the best of itself; so, we work harmonoiusly with what Nature has provided without violating the biolo gical principles which, when known and assisted, lead to the biological principles which, when known and assisted, lead to the natural fulfilment of our needs.

I have written many times of this hope of mine, and today I am offered another opportunity to make my own small contribution, presenting in a unifiels whole a col lection of cultural practices which will have their part to play in the attainment of this "integrated plant production".

2. Description

2.1. Practice of the forcing and early interruption of the dry period

I shall begin by speaking of the early interruption of the dry period in the practice of the forcing of lemon trees for the production of summer lemons, as a cultural control technique against attacks of the citrus flower's moth (Prays citri Mill.) (Nucifora et al., 1984). This insect attacks the flower openings, and although it is active in the open field from the spring, it is only on the summer lemon flowers, from summer onwards, that it is totally destructive. All attempts to control this phy tophage by chemical means have had little effect (Mineo et al., 1979) or have been inapplicable. The lack of effectiveness is because of the fact that the action and persistence of possible phytopharmaceuticals are limited in time, while the flowering period last for about a month, both in spring and in summer, when recourse is had to the practice of forcing. The inapplicability arises because no entomologist or plant health specialist worthy of the name would suggest its use during the period of the flowering of the plants, because of the damage which would be caused to the pollinators which are so numerous in this period on growths of this sort.

Since, then, one cannot count reliably on the use of insecticide treatments during the period, one should examine what else may be done. The anticipation of the forcing period in this case had shown itself to be a practical alternative, capable of resolving the problem satisfactorily. From an analysis of the facts and the results obtained by lemon-growers who apply forcing to lemon trees and from suitably performed experimental tests (Nucifora, 1984), it may be seen that where the

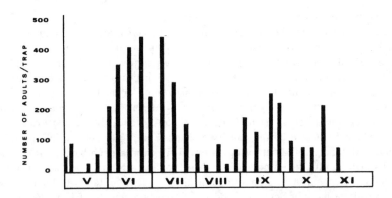

Fig. 1 – Prays citri – State of male captures from May to November 1964 in a lemon orchard by sexual pheromone traps.

breaking of the dry period was performed during the last ten days of July there resulted in August a forced flowering which partially escaped the attack of P. citri.
A study of the problem using sexual pheromone traps shows a reduced number of
flights in July and August. The histograms of fig. 1 show the progression of this
phenomenon. After this stasis the flights became numerous again, increasing progressively during September and in the first half of October. As a result, the flowering which occurs in September is considerably damaged and that which should
have occurred in October is totally destroyed, as has sometimes been seen in eastern Sicily, in the area of Messina, which is considerably behind the flowerings
of Catania and Siracusa. For those flowerings which occur in August, however,
one is able to save enough from the attack so that there are sufficient flower openings and protofruits to ensure good production levels (Nucifora, l.c.).

It follows that one only needs to work in such a way that the practice of forcing is brought forward as compared to the normal practice of most of the lemon-gro
wing areas of the Island, so that it occurs within the last twenty days of July, because one then obtains agronomically satisfactory control effects. Nowadays the
farmer finds it less difficult to bring this practice forward than it was in the past,
because although the production which results is not perfectly typical it is however
equally as easy to sell without any particular difficulty as compared with the other
more typical production which would occur with forcing at the end of July and in
August, if the Prays attacks allowed any production to be obtained at all. As is
well-known, forcing is practiced by stopping spring irrigation, so that the ground
dries out and the plant undergoes hydric stress until, more or less, the initial
dryinf out of its leaves. At this point one acts by irrigating and spreading manure
with a mainly nitrogen composition. After about five days from the first irrigation
the ground is irrigated a second time. A third irrigation is carried out after a further five days. Two weeks after the last irrigation the normal pattern of irrigations
is restarted. As a response to this hydric   stress the plant begins to
flower from about 15–20 days after the first irrigation(i.e. the breaking of the dry
period) and finally the summer lemon appear.

493

2.2. The practice of trimming and surgical pruning

Another crop-growing practice in citrus-growing management techniques and in particular of lemons is that of pruning. A thorough pruning with the removal of branches (better called "trimming" of rejuvenating) was once commonly practiced frequently; now economic factors have led to a reduction in its use. The cv Tarocco orange, which has a triennal rhythm, must however be pruned thoroughly; it is pruned lightly in full summer, between two successive trimmings. Trimming is usually carried out in spring as soon as the harvest is completed. With such a practice the plant can produce fruit of a better quality and the thinning out of the foliage is ensured, which hinders of other entomopathies. This can be seen by a comparison with what happens with cv "Valencia" which is more rustic and does not suffer from delays in the application of this practice, so that the farmer tends to let it grow. Because the foliage has not been reduced and there is a jumble of branches this suffers from consistent attacks of diaspid scales (Parlatoria pergandei Comst. and Aonidiella aurantii Mask.) and of coccids (Saissetia oleae Bern.).

A thorough pruning with a triennal rhythm is also necessary for clementines, if one wants to have a better production and to stop the development on them of S. oleae and Ceroplastes sinensis Del Guercio. The crop for which pruning has even more practical phytosanitary value is the lemon. Here we must distinguish between trimming pruning and surgical pruning used for the control of dry rot("mal secco"). The first is very useful, but not indispensable, in the sense that it can be delayed for one or more years. The second must be carried out annually if the plant under attack is not to die. In the 5 years (1979–1983) during which the CEC 0730 project was being carried out in Sicily on integrated control for lemon crops, the "trimming" technique was applied in the pilot demonstration businesses, leading to the rejuvenation of the plant, with no negative indications at the level of parasite, insect and mite attacks. The populations of these last two groups of arthropods was always permanently contained under the threshold levels by the action of phytoseids, which flourish on the green and reinvigorated foliage of the plant, and by other predators and parasites (Rodolia cardinalis (Muls.) against Icerya purchasi Mask., Anagyrus pseudococci (Girault)) against a few rare colonies of Planococcus citri Risso). The aphid population (Toxoptera aurantii Bayer d. Fosc.) did not undergo percentage increases, when compared to the non-pruned control areas. The attack of Dialeurodes citri Ashm. remained permanently at levels far below the threshold, so that no manifest infestations were recorded. In a word, no contraindications for the application of this crop practice emerged, but only considerable increases in production. On the other hand it has been known for a long time that the thinning and rejuvenation of the foliage, so that sufficient aeration is ensured, works favourably against the development of various insects. Thus Mytilococcus beckii (New.) swarms on poorly pruned and well covered plants.

Aphids develop in abundance spreading the infestation on sucker (buds or very luxuriant branches but which have no gemmae or flowers), so that it is useful to eliminate them by removing them (green pruning) (Tremblay, 1981). The trimming of foliage is also very effective against Heliothrips haemorroidalis (Bouche).

The trimming may be carried out in March – April, before flowering, or preferably during the summer period, when infection by fungi of the wounds left by cutting is more difficult.

Surgical pruning for the removal of dry rot ("mal secco")(Phoma tracheiphila (Petri)) is, in common agricultural practice, carried out in the middle of summer,

in the period of greatest heat. This is because at this time the rot is fully expressed and the high temperatures discourage the activity of the pathogen which stops
its developing (Perrotta, pers. comm.), so that the quantity of inoculant in the environment is reduced and the cutting operations are less liable to the risk of rein-
fection. The cut materials is burnt.

This type of surgical removal as a unique solution is carried out in common
practice in place of the technique which it would be more correct to carry out, i.e.
continual pruning from spring and onwards and removal of the shoots which present
symptoms, thus serving to stop the downward path of the infestation (Nucifora & Sa
lerno, 1983), which for reason of cost is rarely done.

This surgical or improvement pruning on adult plants, with the disease in the
chronic state and the infection diffused in the latent state also, often causes a strong
vegetation response and a more rapid colonization of the healthy branches by the pa
thogen, if it is not completely removed. Equally, very considerable importance
must be attributed to the extirpation of residual stocks of diseased plants, because
these easily emit many shoots on which the periodic outbreaks form preferentially
and in very large numbers (Perrotta, conference on "dry rot" ("mal secco"), 16
March, 1985, Sheraton, Catania).

It is not my job to discuss this subject exhaustively by suggesting other techni
ques (treatments, re-establishment, use of less susceptible cvs.) in the control
against this disease. We may however look at another crop practice which also has
results and repercussions, both on the epidemiology of "dry rot" and on the develop
ment of entomatic entities or of mites which interest us more closely as entomologists: i.e. the methods of fertilizing.

2.3. Fertilizing and its effects

Today plant pathologists do not recommend in the practice of forcing lemons
the use of nitrogen-based fertilizers with a quick effect (ammonium nitrate or sulphate), which in the past were used in very high doses, and suggest that they should
be replaced by balanced fertilizers which have a balanced ratio of macroelements.
Although this on the one hand slows down the vegetation growth of the plant, inducing slower growth and less flowering, on the other hand the plant is less easily attacked by the dry rot fungus which establishes itself easily, penetrating the tender
tissues of the rapidly growing shoots. The use of these complex fertilizers instead
of nitrogen-based fertilizes which would lead to higher production, may be considered one of the indirect costs which lemon growing must bear as a result of dry rot.

This same practice is however useful in another way against the attacks of insects and phytophagous mites which develop and multiply better and more rapidly in
the presence of nitrogen-rich lymph than of lymph with a normal composition.

The belaviour of the coccid populations, for example, is influenced not only
by the chemical and physical condition of the soil but also by fertilizers. Thus
Parlatoria ziziphus (Lucas) develops abundantly on plants in nitrogen-rich soil, Mytilococcus beckii (New.) is greatly encouraged by nitrogen-potash fertilization.

Aphids more than other sucking insects respond quickly and positively to increases in free aminoacids in movement in the plant. An excess of nitrogen fertilizing may lead to large aphid infestations on the buds, at the tips of which these ami
noacids concentrate (Tremblay, l.c.).

Leaf fertilizing with uric substances, zinc sulphate and manganese sulphate is
however a crop and phytomedical practice which contributes to the health of the

495

plant, without creating controindications at the level of parasite, vegetable and plant attacks. In the pilot control lemon growing plots in Sicily it has been annually applied in May – middle of June and has led to a marked regreening of the fo liage and to the disappearance of symptoms of a lack of zinc and manganese depriva tion, without however having encouraged in any obvious way increases of entomopathies or acaropathies.

2.4. Flow or droplet irrigation below and above the foliage

Irrigation is another crop-growing practice which is indispensable for citrus-fruit cultivation, but which may have some repercussions in the form of attacks by fungi and insects. The circulation of water in the soil, in fact, causes an alteration in osmotic pressure and variations in the lymph concentration, which influence the behaviour of the coccid and aphid populations. Thus plants which are usually susceptible may become resistant because of an increase in the lymphotic concentration which makes the aspiration of the liquid through the very small stylet openings difficult or impossible (Trembaly, l.c.). In these contingent circumstances the mor tality rate of scale neonates, in particular seasonally-dry conditions in addition to the dryness caused by the sun, may reach very high values or may even be total. This is the phenomenon of phenotypical and temporary phenoimmunity (Tremblay, l.c.).

For this reason the irrigations should be carried out according to individual requirements and, if suitably manipulated, can act as an indirect means of control. Thus too-frequent irrigation encourages the swarming of sucking insect populations in the same way as do nitrogen-based fertilizers.

For traditional flow irrigation no indications can be given about its direct action against insects; droplet irrigation, however, which has been used recently in many of our citrus groves, may have negative or positive repercussions. It can be carried out above and below the foliage. The former, which is widespread in Ca labria, is not as common in Sicily because of the nature of the waters, which are normally rich in chlorides which are deposited on the leaves and induce phytotoxicity phenomena. From the entomological point of view this type of water supply discourages colonization by insects and mites on the foliage, because washing-away occurs, with the mechanical removal of the mobile forms. It washes away the honeydew, when these are present, and renders the specimens which remain in the plants more open to attack by parasites and predators. This practice can thus be recommended phytoiatrically. Certainly it has disadvantages, including those due to drift or to possibilities of washing-away following treatment with phytopharmaceuticals, so that preferably it should be carried out at night, when there is less wind and evaporation effects are reduced and it should not be administered for a certain period of time after application of the pharmaceutical, which cannot always be done easily. Irrigation under the foliage, on the other hand, does not have an effect on the phytopathy of mites or insects, but may sometimes favour the development of Phytophthora spp., especially with pressure systems, because of the beating effect of the sprays on the earth and consequent splashes of earth containing spores onto the tree trunks or the fruit. It is obvious that suitable precautions and the adoption of falling systems, rather than those operating under pressure, with irrigators for single plants, would reduce the inconveniences complained of.

2.5. Tilling, non-tillage and weeding

Weeding is usually followed by greater infestations of the weeded crop because ground-dwelling polyphagous species can flow on it, or useful species may be lost which developed on these weeds at the expense of mites and insects. This is true, at least in general. The same can be said of the practice of shrub weeding. I will return to the repercussion of this when talking of windbreak hedges.

Di Martino (pers.comm.) has informed me of his observation on the basis of which in a sour lemon orchard in Mascali, there was a violent attack of Thrips tabaci Lind. in the fruit (in June 1984), which was the direct result of a chemical weeding carried out on the crop at the end of April – beginning of May.

Many studies have been carried out in the last ten years on the importance of spontaneous vegetation within and outside the crop in defending it against the phytophages which attack it (Van Emden et Williams, 1974; Tresh, 1981; Paoletti, 1984). Altieri et Letourneau (1982) discuss the general principle on the basis of which the spontaneous vegetation of the agrosystem must be controlled, but not destroyed, because it is a useful deterrant in reducing the excesses of destruction caused by phytophages in the crop. Thus, as an example, the red spider (Tetranychus urticae Koch), which develops on weeds or shrubs (e.g. castor-oil plants in Sicily) ensures a useful exchange between the predators which live on it on spontaneous flora and the phytophage mite which lives on the crop. One is often dealing with Phytoseiulus persimilis A.H., Stethorus punctillum (Weise) and Therodiplosis persicae Kieffer, to cite the most frequent and useful.

Now as to the method of performing the weeding, which should be partial and never total, today chemicals are often used. There are orange-growing firms in the Siracusa area in which weeding is done at the same time as a non-tillage system. In some cased this has been the practice for almost thirty years, without any negative consequences it seems; the weeds today are kept down with localized applications of weed killer, but there is not much need of this because the system has been applied for a long time and because there are today far fewer spontaneous weed seeds in the citrus orchard. Annually the land is irrigated normally, there are balanced applications of fertilizer in spring and the trees are trimmed and pruned. No appreciable disandvantages in the form of attacks by animal or vegetable agents have emerged, so that today the system seem to be in a stable state of equilibrium. Environments like these could certainly be studied to discover the dynamics of their equilibria. Production is definitely satisfactory.

Leaving land untilled cannot however be applied to lemon growing because of attaks of "dry rot" ("mal secco"), which would destroy the lemon orchard, greatly encouraging an explosion of "fulminating disease" or "black rot". This is due to infections of Phoma tracheiphila (Petri) from below, through the roots, which is much encouraged by the non-tillage practice; this encourages the roots, which are no longer disturbed by tilling of the soil, to spread in the layers closest to the surface, and hence to come close to the masses of inoculum which are to be found at the surface.

For this reason, even in normal cro-growing conditions, it is advised that in lemon orchards all tilling operations should be stopped between November and March of the next year, when the danger of infection through wounds is the greatest (Nucifora & Salerno, l.c.). The continuance of the weeds during this period, from the middle of the autumn to the beginning of the next spring, has no disadvantages from the zoopathy point of view and indeed may be useful for the exchange of predators with the animal occupants of the plants. The maintenance of the grass cover is useful against the "allupatura" of the fruits, because of the same species of Phytophthora which give the "elasticity of the collar"; the development of this same disease may however be encouraged by the presence of grass, which ensures a greater humidity.

2.6. Windbreaks, copses and sterile uncultivated areas

The windbreak is useful in both newly-planted citrus orchards and in mature orchards. The wind is a very important element in the passive natural abiotic dispersion of mites and insects; on the other hand, hot winds considerably discourage the development of some scale insects. Windbreaks limit the effects of this transport. It is known, however, that Aphids, Tisanoptera, Acarids, Microditterids, neanides and neonate larvae which are transported by the wind accumulate behind the obstacles (i.e. in the "wind shadow") (Tremblay, l.c.). Windbreaks, either dead or alive, form such obstacles which become greater the more they are impermeable to the wind. It is found that the damage by insects and acarids to plants protected by the windbreak decreases because of the number of exemplars which would have reached the plants and which instead have been stopped by the obstacle.

Apart from this, in a scheme of direct chemical control actions it is very important to know these phenomena; one can at suitable times apply local treatments in the lee of the barriers which can halt in time the invasion and infestations on the crops.

One should however also consider the nature of the windbreak. Apart from "dead" windbreaks using plastic netting, which are quite costly and short-lived, one should set up living windbreaks which respect the essential conditions that they do not compete with the crop and do not play host to harmful species which could damage the crop. Thus for citrus fruit the use of olive trees as windbreaks has been abandoned, because of the welcome which they offer to Saissetia oleae, which finds little threat in nature from efficient parasitoids which would succeed in containing them. Consequently they become sources of infestation by this insect for the citrus groves. Nowadays one tends to use the cypress (Cupressus arizonica) with good results and without counterindications.

Polyphytic hedges, certain of which may act as windbreaks, are not seen under the conditions of our citrus crop; coppices and sterile untilled areas are often useful for the crops.

Paoletti (l.c.) observed and quantified in one case the exchange between a thick polyphytic windbreak hedge, as tall as a man, and a field of maize, and between growths in ditches and copings and the same crop. It is known that plants with broad tomentose leaves, such as Arctium sp. and Urtica sp. in particular, give permanent shelter to fair-sized populations of phytoseids, which may then transfer to the crop. Thus the windbreak hedge was revealed as a reservoir of predators and useful parasitoids able to control swarms of red spider and of aphids. It guaranteed the availability of predators (Oligata flavicornis (Boisd. & Lacard), S.punctillum) which came onto the crop at rather low prey densities. These predators are known for their high migratory performance both active and passive.

It follows from what has already been described that paying attention to spontaneous vegetation may lead to results beneficial for agriculture on the dynamics of the populations living in its crops.

3. Conclusions

We may draw clear conclusions from the facts described above and their effects on the state of health of citrus orchards. Cro-growing practices must be considered in the running of the business as direct and indirect control methods and may be suitable inserted into integrated control programmes, of which they may form a component of primary importance.

It is useful and necessary to experiment on individual phytophages and on suitable means to halt infestation, but one should then pass to a global, not a sectorial framework of phytosanitary actions, so that all the crop may be defended in the most suitable way and according to modern viewpoints, as recent experimentation (Nucifora, l.c.) allows us to do.

REFERENCES

1. ALTIERI, M.A. and LETOURNEAU, D.K. (1982). Vegetation management and biological control in agrosystem. Crop protection, Butterworths 1, 405-30.
2. MINEO, G., SINACORI, A. and VIGGIANI, G. (1979). Contributi per la lotta integrata nel limoneto. 1. Valutazione del danno dovuto a Prays citri Mill (lep. Plutellidae). Boll.Lab.Ent.agr.Portici, XXXVI, 31-37.
3. NUCIFORA, A. (1984). Integrated Pest Control in lemon groves in Sicily: Five years of demonstrative tests and present feasibilities of transferring results. Eds Cavalloro R. & Piavaux A., C.E.C. Programme on integrated and biological control, Luxembourg, EUR 8689, 129-46.
4. NUCIFORA, A., CALABRETTA, C. and CALTABIANO, G. (1984). La tecnica del mass-trapping e la pratica della rottura anticipata della secca contro Prays citri Mill. in limonicoltura: 1° contributo. Atti Giornate fitopatologiche 1984, 325-36.
5. PAOLETTI, M.G. (1984). La vegetazione spontanea dell'agrosistema ed il controllo dei fitofagi del mais. Giornate fitopatologiche 1984, Vol.II: 445-56.
6. NUCIFORA, A. and SALERNO, M. (1984). Agrumi - IX ed. Guida ai trattamenti antiparassitari.Informatore Agrario, 1-288.
7. THRESH, J.M. (1981). Pests, pathogens and vegetation; the role of weeds and wild plants in the ecology of crop pests and diseases. Pitman pubbl. Co., 1-517.
8. TREMBLAY, E. (1981). Entomologia applicata, Vol.II, parte prima. Ed. Liguori, 1-310.
9. VAN EMDEN, H.F. and WILLIAMS, G.F. (1974). Insect stability and diversity in agro-ecosystems. Ann.Rev.Entom., 19: 445-75.

Phytopathological implications determined by cultural practices in citriculture

G.Cutuli & S.Privitera
Plant Disease Observatory, Acireale, Italy

Summary

Tillage, nontillage, soil fertilization, irrigation, pruning, and other cultural practices in citriculture are reviewed in this investigation, where recurrent diseases and injuries resulting from orchard management are considered.

1.1 Introduction

The importance of cultural practices on the occurrence and/or development of the diseases is easily understood considering that the disease is the result of the interation among pathogene, host and environment.

Including in the diseases the disorders of abiotic origin, i.e. the non fungal diseases, we notice that cultural practices can have a more specific meaning, frequently determining, or contributing to determine, the same pathogenic factor.

The present work is intented to review the more important cultural practices carried out in citriculture. Its object is to point out the phytopathological involvements or the dangerous effects of mistaken practices that sometimes can impair the production or also the life of the plant.

1.2 Description

Tillage and nontillage

The more dangerous and recurring diseases related with these cultural practices are the brown rot, the root infection of mal secco (mal nero), the Phytophthora gummosis, the Phytophthora root rot and the Armillaria root rot.

The brown rot development is aided by the soil coltivation in autumn and winter period or by nontillage, carried out by weedkillers (7).

The elimination of the weeds,in both cases, for effect of rain allows the splashing of zoospores from the soil to the skirts of canopy.

Soil coltivation in late autumn and winther can cause root infections of mal secco. Nontillage has also proved to be detrimental because it allows development of the susceptible sour orange roots near soil surface and increases the chances for root injuries and infections (6).

501

Heavy damages by root infections of mal secco occur where to the non-tillage practice follows the cultivation of the soil in periods favorable to infections (8).

Furthermore mechanical injuries on the crown can induce infections by Phytophthora gummosis or by Armillaria root rot (10).

The cultivations that accumulate soil around the base of the trunk increase Phytophthora gummosis (17).

Fertilizing

In connexion with studies of citrus in sand and solution coltures CHAPMAN and BROWN (2) have observed a number of cases of parasitic diseases (brown-rot gummosis, infections by Thielaviopsis basicola, Alternaria citri) induced by nutritional contitions.

KLOTZ et al. (11) reported an increase in fibrous root rot from high percentage of soil organic matter.

Mineral fertilization plays an importan role in affecting the predisposition of plants to mal secco. High rates of nitrogen fertilization makes the plants more susceptible and favours a more rapid development of the disease. The use of fertilizer with a low nitrogen content, alone or, still better, with the addition of phosphorus and potassium fertilizers, slows disease progress and facilitates the surgical treatments (16).

To the occurence of some disorders, as puffing (15), creasing (4,13) and splitting of fruit (12), contributes also nutritional unbalance deriving from mistaken in fertilizing.

Irrigation

Some citrus diseases are strictly linked to amount of water and with the irrigation systems.

Thus, whereas the water stresses contribute to determine endoxerosis of lemon fruits, water excesses aid root rot and Phytophthora gummosis.

The irrigation systems that allow the watering of the trunk can aid infections by Phytophthora spp., especially on the bud union point, wherever a good affinity scion-rootstock is lacking.

Overhead irrigation systems aids the infections of Botrytis cinerea (Botrytis blossom blight), especially on the summer blossom of the lemon; furthermore splinker irrigation sprays the inoculum of Botryosphaeria ribis, Phomopsis citri and Phoma tracheiphila.

Pruning

The amount of pruning may exert an evident influence on the occurrence of some fungal diseases.

No or light pruning can induce attacks of Botryosphaeria ribis particularly on older lemon trees which are low in vigor (10); unbalance among canopy and root system can weaken the roots and make them more susceptible to Phytophthora spp.

An excessive crowding of canopy can aid infections of Colletotric hum gloeosporioides Penz. to twigs and stems of clementine fruits (5).

In similar condition and in moist situations an high presence of lichens can occur also on small twigs.

Tree pruning may be influent on the incidence and severity of mal secco. A light pruning "hardens" the trees and makes them less susceptible to the disease (16). The mal secco infections may also be increased by the time of pruning for the lesions conseguent to cutting.

Topworking

Topworking practice presents considerable ptytopathological problems, particularly for the virus diseases involvements not well known by the growers. The more evident implications occur when the new top is grafted on the interstock.In this case, may occur disaffinity, physiogenetic disorders as shell bark (3), virus and fungal diseases localized in the iterstock (concave gum, blind pocket, Dothiorella gummosis).

Apart could be considered the problems linked to the height of the grafting dependent from the preceding combination.

Other practices

Other cultural practices can, finally, aid the development of some diseases: in the case of mal secco there are relations among disease and time of planting, intercrops (14), planting distance (8), trees used for windebreaks.

The development of foot rot and root rot can be increased by depth of planting (10).

REFERENCES

1.   BROADBENT P. (1981). Armillaria root rot in New South Walles,Australia. Proc. Int.Soc. Citricolture 1:351-353
2.   CHAPMAN H. D. and BROWN S. M. (1942). Some fungal infections of citrus in relation to nutrition. Soil Sci.:303-312
3.   CUTULI G. (1968). Considerazioni sulla incidenza dello "Shell bark" e suscettibilità del limone alla malattia nella Sicilia orientale. Tecnica Agricola,20:445-464
4.   CUTULI G. (1968). Il "creasing" delle arance. Risultati di prove sperimentali. Annali Ist. Sperim. Agrum. 1-2:117-139
5.   CUTULI G. (1968). Una cascola parassitaria di frutti di Clementine comune. Annali Ist. Sperim. Agrum. 1-2:141-146
6.   CUTULI G. (1972). Il "mal nero", una particolare forma di "mal secco" osservata su specie diverse di agrumi. Annali Ist. Sperim. Agrum. 5: 281-290
7.   CUTULI G. (1972). Osservazioni di campo e di laboratorio sul marciume bruno dei frutti di agrumi da Phytophthora ibernalis. Annali Ist. Sperim. Agrum. 5:99-105

8. CUTULI G., SALERNO M. (1977). Il significato di alcune pratiche colturali nella lotta contro il "mal secco" degli agrumi. Annali Ist. Sperim. Agrum. 9-10: 223-230

9. CUTULI G. (1981). Casi di marciume radicale da Armillaria mellea in agrumeti del versante orientale dell'Etna. L'Inform. Agrario 37 (51): 18539-18541

10. CUTULI G.,SALERNO M.(1984).Possibilità di lotte diverse da quelle chimiche nelle malattie degli agrumi.(in litteris)

11. KLOTZ L. J., DEWOLFE T. A., WANG P.P. (1958). Decay of fibrous roots of citrus. Phytopath. 48: 616-622

12. KOO R. C. J. (1961). Potassium nutrition and fruit splitting in Hamlin orange. Univ. Fla. Agr. Exp. Sta. Am.Reptr. 223-224

13. RACITI G., SCUDERI A. (1971). Influenza del fattore idrico sulla manifestazione del "creasing". Annali Ist. Sperim. Agrum. 3-4:263-286

14. RUGGIERI G. (1948). Fattori che condizionano o contribuiscono allo sviluppo del "mal secco" degli agrumi e metodi di lotta contro il medesimo. Annali Sperim. Agraria 2: 145-195

15. SALERNO M. (1963). Nuove acquisizioni sperimentali sulla "spigatura" delle arance in Sicilia. Tecnica Agricola 15: 196-201

16. SALERNO M., CUTULI G. (1981). The management of fungal and bacterial diseases of citrus in Italy. Proc. Int. Soc. Citriculture 1:360-362

17. TIMMER L. W., LEYDEN R. F. (1976). Effect of irrigation and soil management practices on the incidence of Phytophthora foot rot of citrus. J. Rio Grande Walley Hort. Soc. 30: 12-25

18. WHITESIDE J. O. (1971). Some factors affecting the occurrence and development of foot rot on citrus trees. Phytopath. 61: 1233-1238.

Genetics applied to the control of insect pests

R.J.Wood

Department of Zoology, University of Manchester, UK

Summary

This lecture reviews areas of work where genetical concepts and techniques are applied in the control of insect pests. One such area concerns the mass release of males of the pest species carrying particular genes or chromosome arrangements. Genetic techniques are required for separating the sexes at an early stage in development so that only males are produced. Genetic methods also aim at improving the competitive ability of the mass-bred insects so that they achieve maximum effectiveness when released. Other work aims at a better understanding of how insects become resistant to insecticides, so that improved strategies for slowing down this process can be developed. The lecture concludes by considering the impact of recombinant DNA techniques on developments in this field.

1. Introduction

The idea that genetic variation in a pest may be turned to human advantage has become increasingly acceptable in recent years. The trend towards studying the genetics of pests was first stimulated by the experience of resistance to the organic insecticides. Later genetics was applied in other aspects of pest control. This lecture sets out to explore these aspects. It then returns to the question of resistance and finally considers the possible impact of recombinant DNA techniques on pest control.

2. Genetic methods of pest control

Genetic control implies the mass rearing and release of specially bred pest insects carrying genes which reduce or eliminate the pest population. The sterile insect technique (SIT) is the big success story in this area. SIT is associated primarily with the name of E.F. Knipling and the control of the screw worm fly Cochliomya hominovorax in the USA and Mexico. Knipling conceived the idea of rearing screw worm flies in large numbers, treating them with a dose of X rays sufficient to cause lethal changes to the chromosomes of their spermatozoa and releasing the sterilised flies over infested areas from the air. He predicted that the sterilised males would seek out natural females, mate with them and thereby cause them to lay sterile eggs. His prediction has been amply confirmed by events. The pest has now been eliminated from the United States and much of Mexico, and year by year the area of control is being extended (10).

The success of the screw worm programmes encouraged the setting up of similar schemes for other pests. A case in point is the Mediterranean fruit fly (or medfly) Ceratitis capitata a worldwide pest of commercial fruits of many types and a longstanding problem in southern European countries. This pest has become a particular problem in Central America in recent years and took a temporary hold in California in 1980/81. The presence of even a small infestation of the species is a serious and potentially very expensive matter because apart from destroying a crop locally, it can hold up exports to certain countries until the quarantine authorities are satisfied that the infestation has been totally removed. Were it not for the successful control of the species in Mexico since it appeared there in 1978, hundreds of millions of dollars in exports to the USA and elsewhere would have been lost annually by that country ($800 million/annum is the most recent estimate). Potential annual revenue loss to California of an uncontrolled medfly outbreak has been put at $3000 million and to Florida as $400 million.

The medfly in Mexico has now been effectively eliminated, mainly by SIT, based on a breeding plant at Tabachula, Chiapas, producing 500-800 million sterile flies per week. The release of sterile flies over a wide area has been combined with localised applications of malathion bait sprays to deal with the few fertile flies remaining after the sterile releases. The drive against the medfly has now moved south into Guatemala (8). A smaller scale programme is in operation in two valleys in Peru. The feasibility of applying SIT against the medfly in Europe has been demonstrated on the island of Procida (4).

Two other fruit fly species have also received attention. In Taiwan, the oriental fruitfly Dacus dorsalis has been controlled in several areas by a combination of SIT and the application of insecticide mixed with the male attractant methyl eugenol. Loss of citrus fruit has been reduced by this treatment from 6% to 0.02% and loss of mangoes from 26% to 2% (11). The other species is Dacus cucurbitae the melon fly which is established on the Japanese Ryuku islands and threatens to invade the main Japanese islands to the north. SIT has been used to eliminate the species entirely from one of the Ryuku islands and a factory has been constructed to produce 100 million sterile flies per week in order to extend the programme to the other islands (9). A number of other schemes for different insect pests are in various stages of development, most of which have been reviewed recently by Curtis (5, see also 15).

The species against which SIT has been seriously applied so far occupy only two insect orders: Diptera and Lepidoptera. The former are more suitable for the technique because they can be sterilised with lower dosages of X-rays (2-15 krads) than are necessary for the Lepidoptera (20-50 krads). The latter suffer more somatic damage and consequent loss of fitness.

The result of X-ray treatment applied to dipteran species is a "dominant lethal mutation" often due to a chromosome break followed by bridge formation between chromatids, assymetrical separation and loss of the acentric fragments. Such an event cannot occur in Lepidoptera because the centromere of lepidopteran chromosomes is diffuse. However, despite the problems encountered in sterilising Lepidoptera, it has still been worthwhile to develop SIT for at least two species: the pink bollworm Pectinophora gossypiella in California (10) and the codling moth Cydia pomonella in several countries (16).

The economic advantage of SIT is often stressed. It is seen as being particularly useful for pests which are difficult to attack with insecticides. The released males search out the native females and may inseminate several of them. In some species a female can be inseminated only once which might be thought an advantage but is not essential. If females receive a mixture of fertile and sterile sperm, the effect on the

506

Fig.1 THE EXPECTED EFFECT OVER ONE GENERATION
OF A RELEASE OF STERILE MALES
(Modified after Curtis 1985)

$$W_{n+1}= \left\{ W_n f \left(\frac{W_n m + (R.F.C.)}{W_n m + (R.C)} \right) + I \right\} \cdot (D)$$

where:--

W = WILD POPULATION OF EITHER SEX
n & n+1 REFER TO GENERATIONS
f & m REFER TO FEMALE AND MALE
R = NUMBER OF RELEASED MALES
C = COMPETITIVENESS OF RELEASED MALES
F = RESIDUAL FERTILITY OF RELEASED MALES
I = NUMBER OF IMMIGRANT FERTILISED FEMALES
D = REPRODUCTIVE POTENTIAL OF THE POPULATION
INCLUDING RECOVERY DUE TO DENSITY DEPENDENT
FACTORS

next generation still depends on the proportion of sterile to fertile
males. The relationship between various factors governing the success of
SIT is shown in Fig. 1. The number of insects of either sex in the
generation following release (W_{n+1}) depends principally on the ratio of
released males (R) to wild males ($W_n m$) and the number of natural females
($W_n f$). W_{n+1} is also affected by any residual fertility of the
released males (F) and their relative competitiveness (C) compared with
wild males. The impact of immigration of fertilized females (M) must also
be considered, as well as the reproductive capacity of the species which
may increase as density is reduced (represented by (D) in Fig. 1). The
impact of M and of (D) is expected to increase as the density of the
population decreases.

SIT is applied most easily in an isolated area such as an island.
The first demonstration of SIT was in 1954 on the island of Curaçao
against the screw worm fly. Small areas of cultivation surrounded by
different crops can be another suitable location for SIT. A good example
of this is the onion fly Delia antiqua in the Netherlands where the area
protected by SIT has increased from 50 ha in 1981 to 1100 ha in 1984 (5).
Success has also been claimed for SIT control of riverine species of
tsetse flies in West Africa although SIT control of Savannah species in
East Africa has proved more difficult (5, 12).

The most remarkable demonstration of SIT is when it is used as a
barrier to the penetration of a pest from one area to another. Behind
such a barrier the remnants of the pest can be eliminated by more
conventional methods, as in the case of C. capitata in Mexico.

Based on the Mexican experience an even larger project is now being
started, to clear C. capitata from the Nile valley (12). About 5% of
Egyptian fruit is lost to this species and it has been calculated that
after 5 years of the SIT, the project will be paid for. The production of
sterile flies, expected to reach one billion per week from a breeding
factory to be built near Alexandria, will then be available for export to
other European countries.

507

A new generation of genetic methods of control (known collectively under the title of inherited sterility - IS) is being developed but requires further research to be properly established. Some of the proposed methods may be used to change a pest to a form which is less troublesome (population replacement - PR). Among the techniques which have potential for both IS and PR is meiotic drive (28). Meiotic drive is where a particular genetic character confers an advantage to the chromosome which carries it so that the latter segregates preferentially at meiosis, enters more gametes than the alternative chromosome and thus increases in frequency in the population. Population experiments have shown that a marker gene can be linked to a meiotic drive gene and thereby introduced into a pest population (25). "Useful" genes linked to a meiotic drive gene might be used to destroy a population or modify it to a less troublesome form. This is a subject for future research but neither meiotic drive nor any other IS or PR technique is yet ready to be relied upon for practical control (5).

3. Genetic sexing techniques

Whenever it is proposed to release sterile males in SIT programmes, the problem arises of what to do about the females. These are not required, they are expensive to rear, to sterilise and to transport, they may divert the attention of the sterile males and they may damage crops or transmit diseases. Hence there is an interest, more or less urgent in different species, in finding ways to separate the sexes as early as possible in development.

The search for sexing mechanisms has revealed a number of natural sexual dimorphisms which can be utilised in various species. In addition, there are autosomal sex-determining genes and heat-sensitive sex-linked recessive lethals in houseflies and meiotic drive genes in mosquitoes, all of which produce male-distorted sex ratios.

However, for many species such convenient genes are not available and it is necessary to produce a genetic sexing technique to order. The approach has been either to chemically induce heat-sensitive lethals or to use an irradiation technique to translocate a known autosomal gene onto the Y chromosome (23).

The only practical method so far available for an agricultural pest is one developed for the medfly, based on pupal colour. Several mutants affecting pupal colour have been isolated. Rössler (18, 19) showed the feasibility of separating the sexes by causing them to be of different colours (females black, males brown). This was achieved by linking the brown (wild type) allele onto the Y chromosome by an X-ray induced translocation. Following the same principle and using another of Rössler's mutants, Robinson and Riva (20) produced a genetic sexing strain in which the females were white and the males brown. In this case the sexes were more strongly contrasted in colour and therefore more easily separated. The separation is achieved automatically by use of a modified seed-sorting machine.

So far it has not been possible to separate the sexes of any pest species by colour earlier than the pupal stage which is undesirably late because of the wasted effort of rearing so many female larvae. But techniques have been developed in mosquitoes, based on Y-translocated insecticide resistance genes which can be used to separate the sexes in first instar larvae (23). The incorporation of an inversion into the system makes the separation more efficient by reducing the recombination across the translocation junction.

A big priority at the present time is to find an efficient early-

508

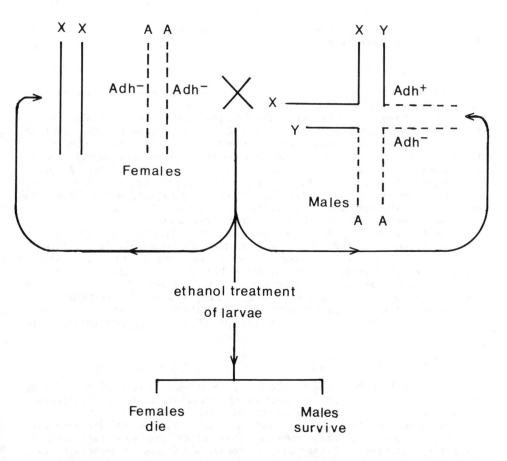

Fig. 2 Translocation of an alcohol dehyrogenase (Adh⁺) mutant into the Y chromosome to produce a sex separating system which allows Adh⁻ females to be eliminated from mass culture prior to the males being sterilised and released.

acting sexing technique for the medfly. It would be worth an estimated \$2 million annually at the Mexican medfly factory at Tabachula. Experiments to develop a sexing system based on insecticide resistance are now in progress. A resistance gene has been selected (2) but a suitable translocation has yet to be isolated (Wood and Saaid unpublished).

Ethanol is an alternative chemical recently used to separate the sexes in D. melanogaster (22). Ethanol resistance depends on the enzyme alcohol dehydrogenase, the gene for which (Adh) is translocated onto the Y chromosome. The males are thus Y-Adh⁺/Adh⁻ and because Adh⁺ gives resistance, the males survive treatment with ethanol (Fig. 2). The medfly is naturally tolerant to ethanol (Adh⁺). The problem therefore is to isolate an Adh⁻ ("Adh null") mutant. Robinson and Riva (21) have isolated several but these are unsuitable, being lethal in the homozygote.

However they have proposed a method for using one of these nulls to isolate a more suitable null.

4. Quality control

Artificially bred insects intended for release must be productive in the laboratory yet successful also when released in the wild. From the first generation of laboratory colonisation, selection will act in the direction of adapting them to the artificial environment. The need to ensure that this laboratory-adapted genotype will be competitive under field conditions is central to the success of SIT and other genetic contol programmes, and also to the release of parasites and predators. A number of examples are quoted in the literature where there appears to have been a loss of competitiveness after laboratory colonisation, which raises the question of whether it is possible to minimise the change, i.e. to exercise "quality control" of mass reared insects.

There are sound theoretical reasons for believing that the less a genotype is modified through its stay in the laboratory the better it will compete in the wild under natural conditions. Yet it is also possible that, in certain aspects, the phenotype best able to achieve control may be one in which performance surpasses the wild type. The geneticist is therefore required to consider two matters (a) ways of maintaining, as near as possible, a qualitatively and quantitatively "natural genotype" in laboratory colonies, (b) the possibility of improving upon nature where necessary.

4.1 Monitoring genetic changes

The genetic structure of laboratory colonies may be monitored (a) by observing chromosomes (b) by estimating heritabilities of continuously varying traits (c) by studying discontinuous variations.

Chromosomal changes can prove detrimental to fitness and are sensitive indicators of evolutionary change. Inversions are sometimes lost in laboratory colonies of Drosophila, a process which can be associated with an increase in the expression of recessive lethals. Useful information on chromosomes can be derived from polytene studies (where possible) and from Giemsa banding but does not appear so far to have been applied to pests (23, 24).

Heritability estimates may help to determine at a quantitative level whether genetic changes have occurred in the laboratory in relation to traits such as flight activity, mating frequency or response to pheromones. But once again, there is no evidence that pest species have been studied this way and therefore no idea yet whether such studies would have real value for quality control.

The best discontinuous phenotypes to study are those closest to the primary action of genes i.e. proteins which are monitored by electrophoresis. An extensive study of changes in electromorphs following laboratory colonisation of the olive fly Dacus oliae has been made by Zouros and Loukas et al (13,30). Their studies show that the laboratory environment tends to exert selection against the most common allele at some loci, leading to greater genetic heterogeneity. In studies on the screw worm fly, it was observed that an unusual electromorph, a variant of a flight muscle protein, had been selected in factory bred strains (3). The rate of selection was rapid and the change was claimed as an explanation for the relative failure of the screw worm programme from 1972–76 compared with the years before and since. The factory strain is now renewed regularly with wild caught flies and there has been no further breakdown of control. The full significance of factory selected genetic

510

change in relation to performance in the field is by no means clear and requires further study.

4.2 Assortative mating

Field tests of mating competitiveness have usually shown sterile males of several species to be rather less effective than wild males in fertilizing the wild females (5). In D. cucurbitae the competitiveness apparently declined from 80% to 20% during 13 years of laboratory rearing (9). A study on the mosquito Culex tarsalis (17) showed competitiveness of 29% with wild females but 353% with laboratory-bred females. The latter example gives cause for concern because it indicates a high degree of assortative mating. Recent studies on C. capitata support this finding. Investigations in field cages on the island of Procida by Robinson, Cirio, Hooper and Capparella (unpublished) showed that assortative mating occurred between released sterile males and sterile females, and that when sterile males were released alone, their ability to sterilise wild females was much improved. This experiment alone provides a powerful argument for avoiding the release of sterile females and therefore the need for an effective sexing technique. The technique used by Robinson et al was one based on pupal colour (see above).

4.3 Strain improvement

Improvement of mating competitiveness by laboratory selection is another possible direction to follow. Studies have shown that mating activity can be increased by selection (1, 7). The danger of selection producing adverse correlated effects has, of course, to be faced. Yet the experience of animal breeders leads to the conclusion that such problems can be side-stepped when selection programmes are approached in a strictly empirical way. In regard to selection for quality traits, this implies the need for comprehensive testing of the selected product under field conditions.

4.4 Resistance to the SIT

The possibility of field populations evolving resistance to released males has been discussed (24, 26). Resistance might occur by any change which would increase assortative mating between wild females or wild males. Resistance has not been observed in practice although observations in Guatemala (29) suggest that the potential may be there in medflies, based on differences between wild and laboratory males in sensitivity to light and temperature. Zapien and co-workers observed that wild males had a greater tolerance to marginal conditions (low light intensity and suboptimal temperature) than the factory flies and initiated courtship earlier.

5. Insecticide resistance

Concepts of population genetics are used in the construction of computer models of pest populations. These are used to devise pest control strategies especially those aimed at preventing or minimising the development of resistance. The simulations, considered together with the results of field studies, have revealed a number of critical factors determining the rate of evolution of resistance.

One genetic factor which enters into the resistance equation is the extent of dominance of the gene (or genes) responsible for resistance. It is well known that when such a gene is recessive, resistance will evolve more slowly than when the gene is fully dominant. The practical significance of this lies in the fact that dominance can often be manipulated by varying the dose of insecticide which the target insect picks up. The reason for this flexibility is that the heterozygote is usually intermediate in resistance between the two homozygotes. Thus a dosage of insecticide sufficiently low to allow the heterozygote to survive renders the R gene effectively dominant while a higher dosage, one which kills heterozygotes but not homozygous resistants, makes the gene effectively recessive. The conclusion to be drawn from this is the desirability of ensuring that the exposed insects receive a high enough dose to kill all heterozygotes (6).

However this strategy is effective only under certain conditions which have been demonstrated by computer simulations. Two major conditions are (a) that the R gene should be initially at a low frequency (b) that a small proportion of the population (e.g. 0.1%) should escape exposure altogether (27). Other conditions can also enhance the proposed strategy and have been studied (14). The practical reality of making a resistance gene effectively recessive under field conditions has been investigated in malarial mosquitoes in Pakistan and found to be possible.

5.1 Resistance in "useful species"

One important aspect to be considered in pest control is the presence of parasites and predators of the pest and whether these are resistant to the insecticides being used. In some cases it has been possible to manipulate the natural situation by introducing resistant predators from another location. In the case of one species of predatory mite Metaseiulus occidentalis in California, resistant strains have been produced artificially in the laboratory by selective breeding and shown to persist when released in the wild (23). Recent research in France has centred on cloning a resistance gene (taken from a mosquito) with a view to producing a strain of resistant predatory mites by "genetic engineering". The gene is an acetylcholinesterase (AChE) mutant, resistant to a wide range of carbamate and organophosphate insecticides. The probe used for identifying the gene in the mosquito genome was the wild type AChE gene already isolated from Drosophila. (Pasteur, personal communication).

5.2 New pesticides

While it is hoped that pesticides will be used less in the future because of the environmental stress they can cause, there is every reason to believe they will still be required. Therefore the search for new ones (to replace those to which pests have become resistant) must go on. One approach to devising new compounds lies in understanding the mechanisms of resistance which insects have already developed towards existing agents. The traditional method is to use genetically pure strains to screen for new effective agents. There is also the possibility of using recombinant DNA techniques to isolate and clone resistance genes so as to study their properties more intensively.

Another approach to devising new insecticides is to understand the mechanisms by which wild plants combat their own pests. Among the most interesting are the leafy ferns and especially the ubiquitous bracken. This plant is distasteful and harmful to most insects and is avoided by them. Among the chemicals identified are feeding deterrents, insect growth hormone mimics and enzyme inhibitors. The possibility exists for

mass producing such compounds by biotechnological methods in the future if
the genes responsible can be cloned.

6. Genetic engineering in pest control

Insect pathologists have identified a wide variety of micro-organisms
pathogenic to insects and other arthrpod pests.   The micro-organisms
include viruses, bacteria, fungi and protozoa.   They represent an
important natural resource and there is a constant search going on for
more virulent strains for pest control.   Recombinant DNA techniques
represent one obvious approach to bringing this about e.g. for the
production of more effective bacterial toxins or hypervirulent viral
strains.   But these methods have to be weighed for cost effectiveness
against more conventional genetic techniques.

The cloning of R genes has already been mentioned in connection with
(a) the development of a resistant parasitic mite (b) studying the
properties of R genes more intensively.   They might also be valuable in
developing probes to be used in identifying and monitoring resistance in
the field while the R gene is still rare.

Other uses for this technology will doubtless arise in the future and
it is certain that its full potential for pest control has not yet been
realised.

REFERENCES

1. BOLLER, E.F. and CALKINS, C.O. (1984).   Measuring, monitoring and
   improving the quality of mass-reared Mediterranean fruit flies,
   Ceratitis capitata (Wied) 3. Improvement of quality by selection.   Z.
   ang. Ent. 98: 1-15.
2. BUSCH-PETERSEN, E. and WOOD, R.J. (1983).   Insecticide resistance as a
   prospective candidate for the genetic sexing of the Mediterranean
   fruit fly, Ceratitis capitata (Wied).   In Fruit Flies of Economic
   Importance   Ed. R. Cavalloro) CEC/IOBC Symposium/Athens/Nov. 1982,
   Balkema, Rotterdam 1983, pp. 182-189.
3. BUSH, G.L., NECK, R.W. and KITTO, G.B. (1976).   Screw worm
   eradication: inadvertent selection for non-competitive ecotypes during
   mass rearing. Science 193: 491-493.
4. CIRIO, U. (1975).   The Procida medfly pilot experiment.   In
   Controlling Fruit Flies by the Sterile-Insect Technique.   I.A.E.A.,
   Vienna, 1975, pp. 39-49.
5. CURTIS, C.F. (1985). Genetic control of insect pests: growth industry
   or lead balloon? Biol. J. Linn. Soc. (in press).
6. CURTIS, C.F., COOK, L.M. and WOOD, R.J. (1978).   Selection for and
   against insecticide resistance, and possible methods of inhibiting the
   evolution of resistance in mosquitoes.   Ecological Entomology 3: 273-
   287.
7. HARRIS, D.J., WOOD, R.J. and BAILEY, S.E.R. (1983).   Studies on mating
   in the Mediterranean fruit fly Ceratitis capitata Wied.   In Fruit
   Flies of Economic Importance (Ed. R. Cavalloro) CEC/IOBC
   Symposium/Athens/Nov. 1982. Balkema, Rotterdam 1983, pp. 197-202.
8. HENDRICHS, J., ORTEZ, G., LIEDO, P. and SCHWARZ, A. (1983).   Six years
   of successful medfly program in Mexico and Guatemala. In Fruitflies of
   Economic Importance (Ed. R. Cavalloro) CEC/IOBC Symposium/Athens/Nov.
   1982. Balkema, Rotterdam 1983, pp. 353-365.
9. ÎTO, Y. and KOYAMA, J. (1982).   Eradication of the Melon Fly: role of

population ecology in the successful implementation of the sterile insect release method. Protection Ecology 4: 1-28.

10. KNIPLING, E.F. (1982). Present status and future trends of the SIT approach to the control of insect pests. In Sterile Insect Technique and Radiation in Insect Control. I.A.E.A., Vienna 1982, pp. 3-23.

11. KOYAMA, J. (1982). The Japan and Taiwan projects on the control and/or eradication of Fruit Flies. In Sterile Insect Technique and Radiation in Insect Control. I.A.E.A., Vienna 1982, pp. 39-51.

12. LINDQUIST, D.A. (1984). Atoms for pest control. International Atomic Energy Bulletin, 26: 22-25.

13. LOUKAS, M., ECONOMOPOULOS, A.P., ZOUROS, E. and VERGINI, Y. (1985). Genetic changes in artifically reared colonies of the olive fruit fly (Diptera: Tephritidae). Ann. Entomol. Soc. Amer. (in press).

14. MANI, G.S. and WOOD, R.J. (1984). Persistance and frequency of application of an insecticide in relation to the rate of evolution of resistance. Pestic Sci. 15: 325-336.

15. PATTERSON, R.S. and MILLER, J.A. (1982). The sterile insect technique in integrated pest management programmes for the control of stable flies and horn flies. In Sterile Insect Technique and Radiation in Insect Control. I.A.E.A., Vienna 1982, pp. 111-122.

16. PROVERBS, M.D. (1982). Sterile insect technique in codling moth control. In Sterile Insect Technique and Radiation in Insect Control. I.A.E.A., Vienna 1982, pp. 85-99.

17. REISEN, W., MILBY, M.M., ASMAN, S.M., BOCK, M.E., MEVER, R.P., MCDONALD, P.T. and REEVES, W.C. (1982). Attempted suppression of a semi-isolated population by the release of irradiated males: a second experiment using males from a recently colonised strain. Mosquito News 432: 565-575.

18. RÖSSLER, Y. (1979). Automated sexing of Ceratitis capitata: the development of strains with inherited sex-limited pupal colour dimorphism. Entomophaga 24: 411-416.

19. RÖSSLER, Y. (1980). "69-apricot". A synthetic strain of Mediterranean fruitfly Ceratitis capitata with sex-linked pupal colour and an eye colour marker. Entomophaga 25: 275-281.

20. ROBINSON, A.S. (1984). Unexpected segregation ratios from male-linked translocations in the Mediterranean fruitfly, Ceratitis capitata (Diptera: Tephritidae). Genetica 62: 209-215.

21. ROBINSON, A.S. and RIVA, M.E. (1984). A simple method for the isolation of allelic series using male-linked translocations. Theor. Appl. Genet. 67: 305-306.

22. ROBINSON, A.S. and VAN HEEMERT, C. (1980). Genetic sexing in Drosophila melanogasta using the alcohol dehydrogenase locus and a Y-linked translocation. Theor. Appl. Genet. 59: 23-24.

23. WOOD, R.J. (1983a). Genetics applied to pest control. Folia biologica 29: 188-199.

24. WOOD, R.J. (1983b). Genetic studies on fruitflies of economic importance. In Fruitflies of Economic Importance (Ed. R. Cavalloro) ECE/IOBC Symposium/Athens/Nov. 1982. Balkema, Rotterdam 1983, pp. 139-147.

25. WOOD, R.J. COOK, L.M., HAMILTON, A. and WHITELAW, A. (1977). Transporting the marker gene re (red eye) into a laboratory population of Aedes aegypti using meiotic drive at the M^D locus. J. Med. Entomol. 14: 461-464.

26. WOOD, R.J., MARCHINI, L.C., BUSCH-PETERSEN, E. and HARRIS, D.J. (1980). Genetic studies of Tephritid flies in relation to their control. In Fruit Fly problems (Ed. J. Koyama) National Institute of Agricultural Sciences of Japan, 1980 pp. 47-54.

27. WOOD, R.J. and MANI, G.S. (1981).   The effective dominance of resistance genes in relation to the evolution of resistance. Pestic. Sci. 12: 573-581.
28. WOOD, R.J. and NEWTON, M.E. (1982).   Meiotic drive and sex ratio distortion in mosquitoes in Recent Developments in the Genetics of Insect Disease Vectors (Ed. W.W.M. Steiner et al) Stipes Publ. Co. Champaign, Ill. USA, pp. 130-153.
29. ZAPIEN, G., HENDRICHS, J., LIEDO, P. and CISNEROS, A. (1983). Comparative mating behaviour of wild and mass reared medfly Ceratitis capitata (Wei`) in a field cage host tree II.   Female mate choice, In Fruitflies of Economic Importance (Ed. R. Cavalloro) CEC/IOBC Symposium/Athens/Nov. 1982.   Balkema, Rotterdam 1983, pp. 397-409.
30. ZOUROS, E., LOUKAS, M., ECONOMOPOULOS, A. and MAZOMENOS, B. (1982). Selection at the alcohol dehydrogenase locus of the olive fruit fly Dacus oliae under artifical rearing. Heredity 48: 169-185.

Session 5
Methodologies and strategies of control

Chairman: L.Brader

Address of the session chairman

L.Brader

FAO, Plant Production & Protection Division, Roma, Italy

Just before this session a detailed discussion took place within the FAO Commission Agriculture the establishing Code of Conduct on the distribution and use of pesticides. As a result of this it now seems most probable that the next session of the FAO Conference in November 1985 will adopt this Code. This will be of direct significance to the objectives of this meeting. Because the Code while setting clear principles for the future use of pesticides, it enphasizes that its objectives can best be achieved through increased support for the development and application of biological and integrated pest control.

The theme of this session, methodologies and strategies of control, brings us very closely to the practical implementation of IPM. For this it is essential that the research results are critically evaluated from a practical and economical point of view. Althought IPM has now been studied for over three decades, and a significant amount of research findings are available the extent of application in practice is still rather disappointing. This is mainly due to the fact that farmers have not yet gained enough confidence in this approach. It seems that the solutions proposed are still too complex, or otherwise said, they lack user-friendliness.

The basic principles of IPM have been tested and applied in a wide variety of crops situations. It can definitely be concluded that the approach offers the only possibility for an improved and more stable plant health situation.

But for the widespread introduction into practice further efforts to simplify the application of IPM are urgently needed. Therefore I would like to urge the speakers of this morning to test their contributions against this criterium of user friendliness.

Methodologies and strategies of pest control in citriculture

D.Rosen

Faculty of Agriculture, University of Rehovot, Israel

Summary

Citrus pests may rise through invasions into new areas, ecological changes or socio-economic changes. Modern chemical pesticides are effective in controlling citrus pests, but their use involves serious problems of cost, resistance and toxicity. Biological control has been highly successful in citriculture, but is not always feasible. Integrated pest management (IPM) provides a reasonable compromise, taking into account both the desirability of biological control and the need for some form of chemical control. Effective IPM may be achieved through a vigorous program of applied biological control, in combination with judicious use of relatively selective pesticides, only when absolutely necessary and in the least disruptive modes of application. Other selective tactics — cultural, autocidal, physical, mechanical — should be incorporated in the program whenever applicable. Effective implementation of IPM requires the establishment of reliable action thresholds and monitoring systems for all the actual and potential pests in the citrus ecosystem. Special emphasis should be placed on integrative techniques, such as the development of pesticide-resistant natural enemies. Effective extension services, backed up by supervised control systems, are essential for the success of IPM. Finally, efforts should be made to re-educate consumers and change the marketing standards for citrus fruit.

1. Introduction

Citrus is an important cash and export crop in many tropical and subtropical countries. In its native range in the Far East it is grown quite extensively for local consumption, whereas most large-scale, export- and industry-oriented citriculture is located in regions into which it has been introduced — in the Mediterranean Basin, the Americas, South Africa, etc.

Wherever citrus is grown, it is attacked by numerous pests. In the Far East, various lepidopterous and coleopterous species are the most injurious, whereas Homoptera — scale insects, aphids and whiteflies — are usually kept under effective control by natural enemies (10,17). In modern commercial citriculture, on the other hand, Homoptera usually are major pests, along with mites and fruit flies, sometimes also thrips and various other insects (2, 10).

521

2. Rise of Citrus Pests

What causes an arthropod population to exceed the economic threshold and attain pest status? The factors are many and varied, but they can be grouped into three main categories: invasions, ecological changes and socio-economic changes.

Invasions are, unfortunately, all too common. Intercontinental travel and commerce during the last several centuries, and especially the deliberate transfer of numerous citrus plants, have provided ample opportunity for pest species to migrate and become established in new habitats. Indeed, a majority of the important pests in modern citriculture usually are species of foreign origin. Even in this sophisticated day and age, invasions repeatedly occur. Thus, the Mediterranean fruit fly, Ceratitis capitata (Wiedemann), has invaded both Florida and California several times during the present century, whereas recent invasions into Israel have resulted in the establishment of such notorious citrus pests as the spirea aphid, Aphis citricola van der Goot, the citrus whitefly, Dialeurodes citri (Ashmead), the Japanese bayberry whitefly, Parabemisia myricae (Kuwana), and the citrus red mite, Panonychus citri (McGregor). Regulatory control in the form of diligent quarantines is therefore of primary importance in preventing pest problems.

Ecological changes are another major cause of pest outbreaks. The history of agriculture has been the history of constant ecological change. By creating monocultures, by selecting high-yielding plant varieties, by eliminating competitors or natural enemies through various agrotechnical practices, etc., man has often inadvertently created conditions favorable to certain species and has thus induced a manyfold increase in their populations. Recent changes in the citrus fauna of Israel may serve to illustrate this point. The California red scale, Aonidiella aurantii (Maskell), had been known in the past mainly as a pest of young citrus groves in Israel, having usually been competitively displaced from mature groves by the Florida red scale, Chrysomphalus aonidum (L.). However, when the Florida red scale was brought under complete biological control by an introduced parasite, the California red scale — which, unfortunately, is much more difficult to control — has become a major pest. Then, as large citrus acreages, planted in the 1950's, came to maturity, the chaff scales, Parlatoria pergandii Comstock and Parlatoria cinerea Doane & Hadden, habitual pests of older groves, gradually became more injurious (20).

Finally, socio-economic changes are often as important in affecting pest status as are actual changes in the physical environment of the citrus ecosystem. Economic and action thresholds are determined by such factors as the marketing value of the crop and the cost of available control measures, as well as by consumer habits and taste. Changes in public attitude, for instance, towards pest-damaged agricultural produce may drastically lower the thresholds and thus cause a hitherto insignificant organism to be considered a serious pest, although its actual population density has remained unaltered. To quote another recent example from Israel, the opening of the Japanese market, with its stringent quarantine restrictions, to Israeli citrus has brought about a general lowering of action thresholds and a general intensification of pest control operations on citrus in Israel. On the other hand, a severe freeze in Spain or Morocco, resulting in a reduced supply of citrus fruit on European markets, would enable Israeli growers to raise the culling threshold for, say, scale-insect damage on fruit exported to the European Community.

One way or another, serious pest problems are a prominent fact of life in modern citriculture and do not show any sign of letting off in the foreseeable future. Effective control measures are therefore an absolute necessity in any modern citricultural endeavor, and are made possible by an impressive array of chemical, biological, cultural, mechanical and autocidal techniques.

3. Chemical Control

Chemical pest control has been a salient feature of citriculture ever since the early days of HCN fumigation, botanical insecticides and mineral oil sprays; but the modern era of chemical control began with the advent of synthetic organic pesticides after World War II. First came the chlorinated hydrocarbons, with their broad spectrum of activity and long residual effects. These were soon followed — and eventually replaced — by the organophosphorus and carbamate pesticides, with shorter residual effects but with considerably higher toxicity to all living organisms, and more recently by the new synthetic pyrethroids. By and large, these modern pesticides have provided citrus growers with effective, and usually quite reliable, means for controlling arthropod pests. However, as is now widely recognized, this has been a mixed blessing. Not only do chemical pesticides, at best, afford only temporary relief from pest problems, but their massive overuse and frequent misuse have often resulted in grave problems. Ever-increasing cost has been just one of these problems. The high price of modern pesticides, and the large volumes required for adequate coverage of a mature citrus grove, have posed serious threats to the profitability of citriculture in many areas.

Rapid development of resistance to pesticides has been another problem, known almost since the turn of the century, when citrus scale insects were found to be resistant to HCN (16). Early hopes that the new powerful biocides would not be vulnerable to the development of resistance have proved unfounded, and we are now faced with increasing numbers of insect and mite species that have become resistant to virtually all available pesticides (11).

Last, but certainly not least, are the hazards of pesticidal toxicity. Many of the modern pesticides are highly toxic to humans, and numerous poisoning accidents have occurred among pest-control operators and others exposed to them. Continued use of such pesticides on citrus may cause serious marketing problems, as certain importing countries have recently imposed very strict residue tolerances. Toxicity to citrus itself is also quite common, certain citrus varieties, such as the Jaffa orange, being exceedingly sensitive to certain pesticidal formulations, such as oil/organophosphorus combinations. Finally, broad-spectrum pesticides have been notorious for upsetting the biological equilibrium in many citrus ecosystems, causing severe resurgences of "old" and "new" pests through decimation of their natural enemies (7,8). These and other drawbacks of chemical pest control are becoming increasingly evident in citriculture.

4. Biologcial Control

On the other hand, biologcial pest control, i.e., the utilization of natural enemies, has been highly successful on citrus (1). Ever since the first spectacular success of classical biologcial control —

achieved against the cottony-cushion scale, Icerya purchasi Maskell, on citrus in California nearly a century ago — more such efforts have been made on citrus than on any other agricultrual crop. Although the high proportion of successes on citrus may simply reflect this great amount of effort, it is also partly due to the fact, mentioned earlier, that most of the serious pests of citrus are introduced species, notably coccids and aleyrodids which, for various reasons, are perhaps more amenable to biological control by importation of natural enemies than other groups of organisms (9). Landmark projects on citrus have included the biologcial control of the Florida red scale and the California red scale, the purple scale, Lepidosaphes beckii (Newman), the Mediterranean black scale, Saissetia oleae (Olivier), the citrophilus mealybug, Pseudococcus fragilis Brain, the citriculus mealybug, Pseudococcus citriculus Green, and the citrus blackfly, Aleurocanthus woglumi Ashby, more recently also the rufous scale, Selenaspidus articulatus (Morgan), the citrus snow scale, Unaspis citri (comstock), the arrowhead scale, Unaspis yanonensis (Kuwana), the citrus whitefly, Dialeurodes citri (Ashmead), and the woolly whitefly, Aleurothrixus floccosus Quaintance. In fact, projects aimed at the control of citrus pests have played a major role in the development of biological control as a science, having contributed heavily to the establishment of some of its most important concepts, principles and methodologies. The crucial importance of sound systematics and biology, the ecological basis for importation policy, methods of augmenting natural-enemy populations, the effects of adverse factors on natural enemies and some methods for their mitigation, check methods for evaluating the effectiveness of natural enemies, etc. — these and others have been demonstrated in various projects on citrus (1,6,22,23).

When successful, biological control is an ideal form of pest management — permanent, inexpensive, non-hazardous. It is by far the most promising, and most desirable, alternative to wholesale chemical control available in our arsenal. Unfortunately, however, it cannot be expected to provide an immediate solution to all pest problems in citriculture at the present time, because no effective natural enemies are currently available for certain "key" pests, including, for instance, the Mediterranean fruit fly, Ceratitis capitata (Wiedemann), in Israel (20), the citrus thrips, Scirtothrips citri (Moulton), in California (26) and the citrus leaf miner, Phyllocnistis citrella Stainton, in India (24). Until such enemies are discovered, chemical pesticides or some other means of pest control will continue to be required.

5. Integrated Pest Management

Integrated pest management (IPM) provides a reasonable compromise, taking into account both the desirability of biological control and the need for some form of chemical control. Like diplomacy, IPM is the art of the possible. It represents a holistic approach, recognizing the unity of the ecosystem and harmonizing all available measures in an attempt to optimize pest control and crop production. Effective IPM may be achieved through development of a vigorous program of applied biologcial control, in combination with judicious use of relatively selective pesticides, only when absolutely necessary and in the least disruptive modes of application. Other selective tactics should be incorporated in the program whenever applicable (21). The following discussion will elaborate on these points.

In my opinion, biological control is not just another tactic of pest control. In view of its unique advantages and wide applicability, the utilization of natural enemies should be regarded as the backbone of any IPM program in citriculture. The value of naturally-occurring natural enemies cannot be overemphasized, and their conservation should be the first goal of IPM. However, merely introducing modifications in existing chemical control programs to conserve natural enemies does not in itself constitute IPM. Other forms of applied biological control, such as the various methods of augmentation of natural enemies, should also be included whenever feasible. In particular, any viable program of IPM should include the active importation and establishment of exotic natural enemies as one of its major components. There is an enormous, untapped reservoir of parasites and predators that have not yet been discovered, which should be sought and brought to active use in citriculture.

Microbial pest control is another aspect of biological control that should be greatly encouraged, especially against lepidopterous pests. Although similar to chemical control in application, microbial control enjoys the decided advantages of being much more selective and free of environmental hazards.

Selective chemical pesticides should also have their place in IPM. Selectivity may sometimes be physiological, some pesticides being much less toxic to certain beneficial organisms than to certain pests. All pesticide formulations suggested for use on citrus should, therefore, be routinely tested for their possible effects on natural enemies. Such tests have led, inter alia, to the replacement of sulfur with more selective acaricides on citrus in Israel (18,19).

This is not always easy. Mineral oils, for instance, are highly selective and are therefore recommended for the control of citrus scale insects. However, their effecitve use requires complete coverage of the trees and careful timing of applications, and may be problematic when several species infest the same grove. Also, mineral oils may retard the development of yellow fruit color in early citrus varieties, and may cause serious injury — including defoliation and fruit drop — to certain sensitive varieties during summer. No wonder, therefore, that citrus growers have often turned to supposedly more effective, but definitely non-selective, organophosphorus and carbamate insecticides for the control of armored and soft scale insects. Efforts should be made to reverse this undesirable trend.

Few conventional chemical pesticides exhibit true physiological selectivity. The new insect growth regulators offer a renewed hope in this respect. Some of these novel insecticides appear to be highly effective against citrus scale insects but rather harmless to their parasites (14,15). The search for such compounds should definitely be encouraged. However, even a broad-spectrum pesticide may sometimes be applied in ways that would render it ecologically selective. The use of malathion in bait sprays against the Mediterranean fruit fly may serve as a good example. Aerial strip sprays of very small amounts of poison bait, containing protein hydrolysates as a powerful attractant for female flies and malathion as a poison, have proved to be far superior in effectiveness, and much less disruptive to the citrus ecosystem, than full-coverage applications of chlorinated hydrocarbons. The exact method of application may be of crucial importance here. Thus, when the same aerial bait sprays are applied at ultra-low volume, the tiny droplets are evenly distributed on the trees, forming a "micro cover spray" that may be rather detrimental to certain natural enemies (20). Studies now in progress in Israel are threfore aimed at increasing drop size without a corresponding increase in the total volume of the bait sprays.

Using chemical pesticides "only when absolutely necessary" should not remain an empty phrase. For effective implementation, this approach requires the establishment of reliable action thresholds for each of the actual and potential pests in the citrus ecosystem. This should be regarded as one of the most important tasks of citrus entomologists (4). It is also one of our most difficult tasks, especially for an export crop, when the marketing standards are determined by foreign markets, often long after control decisions have been made in the grove. Effective monitoring systems, based on accurate, sophisticated sampling methods, should also be developed for all the species under consideration. Pest detection and monitoring may be accomplished by simple visual counts, which may be rather time-consuming, or by employing various specific attractants. Some powerful attractants have been in use in citriculture for years. Decision-making in the control of the Mediterranean fruit fly in Israel, for instance, is based on a dense network of traps containing Trimedlure, an effective synthetic attractant for male flies. A major breakthrough has been achieved with the recent identification and synthesis of numerous arthropod pheromones, which may now be utilized in the early detection and monitoring of various citrus pests, including scale insects and mealybugs (13,27).

An IPM program on citrus should encompass agricultural areas adjacent to citrus groves. Efforts should be made to prevent pesticidal drift from neighboring fields, which may have pronounced adverse effects on the biological equilibrium in the citrus ecosystem. In Israel, appropriate legislation has been enacted for this purpose, prohibiting aerial spraying of cotton and other crops with non-selective pesticides within a distance of 200m from a citrus grove (20).

Selective alternatives to chemical control may include various cultural, autocidal, physical and mechanical techniques.

Cultural control is defined as changing the methods employed in culturing the crop, in order to reduce or eliminate pest damage. I interpret this in the broadest sense, to include also the use of pest-resistant plant varieties, as well as various agrotechnical practices such as tillage, crop rotation, water and fertilizer management, sanitation, timing of planting or harvesting, etc. Some of these have been employed in citriculture for centuries. Early harvesting of citrus fruit, for instance, has helped citrus growers to escape the spring peak of Mediterranean fruit fly infestation (2). Other cultural practices should be encouraged. In particular, the use of resistant plant cultivars is perhaps the ideal means of controlling a pest. Provided, of course, that such varieties are not inferior in any other ways, their introduction is by far the safest, most effective and most economical control method, and is highly compatible with all other methods. Regrettably, although there seems to be ample evidence that certain citrus species or varieties are resistant to such serious pests as the California red scale (5) or the citrus leaf miner (24), very little has been done so far in this promising direction. Cultural controls should certainly receive much more emphasis as a major component of IPM.

Autocidal control is defined as the manipulation of a pest's population in such a way that it brings about its own control. I include in this category both the sterile-male technique and other genetic control methods, as well as the use of pheromones in pest control. The sterile-male technique has been, in my opinion, greatly over-emphasized in recent years. Although this sophisticated technique has produced some spectacular results, especially against the screw-worm fly and certain fruit flies (3), its applicability is rather limited. Ecologically safe but rather expensive, it should be recommended whenever it appears to be feasible, but should not be promoted out of proportion to its actual potential in citrus pest control.

Pheromones, on the other hand, are now coming into extensive use not only in monitoring but in actual pest control, and have already accounted for significant reductions in the use of broad-spectrum pesticides in a variety of agro-ecosystems. The recent development of a selective control program for the citrus flower moth, Prays citri Milliere, in Israel may serve as an example. From an earlier method of population monitoring by traps baited with virgin females, a rather sophisticated system has been perfected whereby sticky traps baited with a synthetic pheromone are used for mass trapping of male moths, resulting in effective control (25). This promising approach should receive high priority in IPM.

Physical and mechanical controls may range from the extraction of wood borers with a hooked wire, or the use of insect-proof constructions to prevent the access of pests to a crop or to stored products, to highly sophisticated — but usually still experimental — uses of various forms of electromagnetic energy. Sometimes a very simple method may be highly effective. A fine example in this category is the use of refrigeration to prevent the passage of the Mediterranean fruit fly with exported citrus fruit. By simply shipping the fruit under controlled temperatures in refrigerated boats, every single immature fly is exterminated during the voyage. Another simple technique, the removal of scale insects and sooty mold from citrus fruit by high-pressure rinsing in the packinghouse, appears to have the potential of revolutionizing pest control practices in citriculture. If this new technique indeed proves as effective as early trials indicate, it may eventually permit us to raise economic thresholds in the grove to much higher levels: rather than control light scale-insect infestations that cause cosmetic damage to the fruit, we would then be concerned only with infestations endangering the tree itself, and this would drastically reduce the intensity of pesticidal treatments.

All these selective techniques, and others, should be promoted within the framework of IPM. Special emphasis should be placed on integrative approaches, such as the development of pesticide-resistant strains of natural enemies. The continued use of broad-spectrum pesticides has been a major roadblock to increasing the utilization of natural enemies in citriculture. Selecting for resistance to pesticides in otherwise effective natural enemies would be an obvious approach to increasing their role and efficacy in IPM programs. Genetic improvement of natural enemies is an old idea that has received little interest until recently. A breakthrough in this direction has been achieved with the recent development and practical implementation of pesticide-resistant phytoseiid mites (12). This approach is certainly applicable to other groups of natural enemies, and should be recommended as one of the most important ways of achieving truly integrated pest management.

Success in IPM is obviously dependent on a thorough understanding of the citrus ecosystem. This includes adequate systematics of both the pests and their natural enemies, and detailed knowledge of their biology and intricate ecological interrelationships. Although certain aspects of

IPM can be successfully introduced into underdeveloped areas, a high level of education among farmers would, of course, be an invaluable asset to the implementation of this sophisticated approach. In Israel, progress in IPM has been greatly aided by the enthusiastic cooperation of our enlightened, well-educated citrus growers, who are remarkably receptive to new concepts and ideas and very quick to adopt new methods and technologies. But even with such farmers, the success or failure of IPM may largely depend on the availability of adequate extension services. Competent extension officers, who carry research developments to the farmer and instruct him in sophisticated methods of pest control, are an indispensable link in any IPM program. It is a well-trained government officer, not a pesticide salesman, who provides the Israeli citrus grower with unbiased advice on all pest control matters. These excellent extension officers, who are held in high esteem by the farmers, are thoroughly versed in the modern concepts of integrated pest management. They maintain close contacts with both farmers and research institutions, providing an invaluable mechanism for bilateral feedback that may rightfully be regarded as the cornerstone of our citrus IPM program (20).

One of the problems of modern citriculture is that the grower does not spend as much time in his grove as he should, whereas the extension officer requires accurate information about each grove in order to allocate his time effectively. An experimental system of supervised control, now being tested in Israel, may provide a solution to this problem. It consists of a network of local field scouts, who take detailed bi-weekly counts of the pest populations on representative sampling trees designated in each citrus plot. The data from an entire district are then tabulated and presented to the extension officer upon his periodic visit to the local packinghouse, so that he can devote his attention to the trouble spots, where his expert advice is most urgently needed. Similar systems should be considered for inclusion in other IPM programs in citriculture.

Finally, re-educating the public should also be a major goal of IPM. For decades, the consumer has been brainwashed, by entomologists and marketing organizations alike, to believe that citrus fruit should be selected mainly by eye appeal, and that no amount of pest damage should be tolerated, not even external, "cosmetic" blemishes. The artificially high marketing standards that have resulted do not permit the development of reasonable control programs. A world-wide campaign for more rational standards would be a great boost to the further development of integrated pest management in citriculture.

REFERENCES

1. BENNETT, F.D., ROSEN, D., COCHEREAU, P. and WOOD, B.J. (1976). Biological control of pests of tropical fruits and nuts. Chapter 15 in: HUFFAKER, C.B. and MESSENGER, P.S., Editors, Theory and practice of biological control. Academic Press, New York, pp. 359-395.
2. BODENHEIMER, F.S. (1951). Citrus entomology in the Middle East. Dr. W. Junk, Publishers, The Hague, 663 pp.
3. CAVALLORO, R., Editor (1983). Fruit flies of economic importance. Published for the Commission of European Communities by A.A. Balkema, Rotterdam, 642 pp.
4. CAVALLORO, R. and PROTA, R., Editors (1981). Standardization of biotechnical methods of integrated pest control in citrus orchards. Published for the Commission of European Communities, Direction

generale Marche de l'Information et Innovation, Luxembourg, 206 pp.

5. COMPERE, H. (1961). The red scale and its insect enemies. Hilgardia, 31: 173-278.
6. DEBACH, P. (1969). Biological control of diaspine scale insects on citrus in California. Proceedings of the First International Citrus Symposium (Riverside, 1968). University of California, Riverside, 2: 801-815.
7. DEBACH, P. (1974). Biological control by natural enemies. Cambridge University Press, London, 323 pp.
8. DEBACH, P. and BARTLETT, B. (1951). Effects of insecticides on biological control of insect pests of citrus. Journal of Economic Entomology, 44: 372-383.
9. DEBACH, P., ROSEN, D. and KENNETT, C.E. (1971). Biological control of coccids by introduced natural enemies. Chapter 7 in: HUFFAKER, C.B., Editor, Biological control. Plenum Press, New York and London, pp. 165-194.
10. EBELING, W. (1959). Subtropical fruit pests. University of California, Division of Agricultural Sciences, 436 pp.
11. GEORGHIOU, G.P. and TAYLOR, C.E. (1977). Pesticide resistance as an evolutionary phenomenon. Proceedings of XV International Congress of Entomology (Washington, D.C., 1976), pp. 759-785.
12. HOY, M.A. (1985). Recent advances in genetics and genetic improvement of the Phytoseiidae. Annual Review of Entomology, 30: 345-370.
13. MORENO, D. (1983). Efficiency of pheromone traps in citrus pest detection. Citrograph 68: 77-79.
14. PELEG, B.A. (1982). Effect of a new insect growth regulator, RO 13-5223, on scale insects. Phytoparasitica, 10: 27-31.
15. PELEG, B.A. (1983). Effect of a new insect growth regulator, RO 13-5223, on hymenopterous parasites of scale insects. Entomophaga, 28: 367-372.
16. QUAYLE, H.J. (1938). The development of resistance to hydrocyanic acid in certain scale insects. Hilgardia, 11: 183-210.
17. RAO, V.P. (1969). India as a source of natural enemies of pests of citrus. Proceedings of the First International Citrus Symposium (Riverside, 1968). University of California, Riverside, 2: 785-792.
18. ROSEN, D. (1967). Effect of commercial pesticides on the fecundity and survival of *Aphytis holoxanthus* (Hymenoptera: Aphelinidae). Israel Journal of Agricultural Research, 17: 47-52.
19. ROSEN, D. (1967). Biological and integrated control of citrus pests in Israel. Journal of Economic Entomology, 60: 1422-1427.
20. ROSEN, D. (1980). Integrated control of citrus pests in Israel. Proceedings, International Symposium of IOBC/WPRS on Integrated Control in Agriculture and Forestry (Vienna, 1979), pp. 289-292.
21. ROSEN, D. (1985). Biological control. Chapter 13 in: KERKUT, G.A. and GILBERT, L.I., Editors, Comprehensive insect physiology, biochemistry and pharmacology. Vol. 12, Insect control. Pergamon Press, Oxford, pp. 413-464.
22. ROSEN, D. and DEBACH, P. (1978). Diaspididae. In: CLAUSEN, C.P., Editor, Introduced parasites and predators of arthropod pests and weeds: a world review. Agricultural Handbook 480, Agricultural Research Service, United States Department of Agriculture, Washington, D.C., pp. 78-128.
23. ROSEN, D. and DEBACH, P. (1979). Species of *Aphytis* of the world (Hymenoptera: Aphelinidae). Israel Universities Press, Jerusalem, and Dr. W. Junk, The Hague, 801 pp.
24. SINGH, S.P. and RAO, N.S. (1978). Relative susceptibilities of

different species/varieties of citrus to leaf miner, <u>Phyllocnistis citrella</u> Stainton. Proceedings of the International Society of Citriculture (Sydney, 1978), pp. 174-177.

25. STERNLICHT, M. (1978). Control of the citrus flower moth with traps based on a synthetic pheromone. Alon Ha'Notea 32: 391-397 (in Hebrew).
26. TANIGOSHI, L.K. and NISHIO-WONG, J.Y. (1982). Citrus thrips: biology, ecology, and control. United States Department of Agriculture, Technical Bulletin 1668, 17 pp.
27. TREMBLAY, E. and ROTUNDO, G. (1981). The use of sex pheromones in the control of the citrus scale insects. <u>In</u>: CAVALLORO, R. and PROTA, R. Editors, Standardization of biotechnical methods of integrated pest control in citrus orchards. Published for the Commission of European Communities, Direction generale Marche de l'Information et Innovation, Luxembourg, pp. 59-66.

Biological and integrated control of insect pests in citrus growing initiatives of the 'Casmez'

G.Argenti
'Cassa per il Mezzogiorno', Roma, Italy

Summary

The following is a description of the activities of the "Cassa per il Mezzogiorno" which is providing financial support for a special project, already in implementation, that is studying the control of citrus tree phytophages through combined biological and integrated means. It seeks both to train specialized personnel and to create suitable environments for the breeding and study of insects. It is hoped that upon completion of this project the regional administrations will take over the finanacial support that will enable this work to continue.

In 1982 the "Cassa per il Mezzogiorno" (Fund for the South) approved a project "PS 11/24" which is still being implemented and which envisages the construction of support structures for programmes which defend citrus fruit crops against phytophages, using biological and integrated control.

These structures include the formulaation of an insect bank at the Institute of Entomology of the Agriculture Faculty of Naples, Chair of Biological and Integrated Control, for the permanent rearing of stocks of phytophage predators and parasites. It would also be equipped to carry out studies on phytophages and insects at the Naples Institute of Entomology and at the Institutes of Entomology at the Universities of Palermo and Catania for the multiplication of stocks as and when required by the local control programmes.

As a demonstration a control test was planned on 305 hectares of citrus trees in Sicily and Campania.

Simultaneously with the project described above, FORMEZ "Centro di Formazione e Studi per il Mezzogiorno" (Centre for Training and Studies for the South) was informed of the need to interest people in this method, which was also inspired by the interest shown, expressive of a desire for further professional knowledge, by the Association of Graduates in Agriculture and Forestry of the provinces of Catania, Messina and Reggio Calabria. These programmes were welcomed by FORMEZ.

In establishing these training courses the Cassa intends to bring into contact the partners who have operative responsibilities, identified in the private sector of phytochemical production, free-lancers in the agricultural sector, regional local government officials interested in safeguarding agricultural production and organizations of citrus fruit producers and the sector of research.

Three institutions made a commitment to fulfil the tasks described above - FORMEZ, CASSA and UNIVERSITIES - and whenever the actions promoted are concluded, perhaps brilliantly, they will fulfil the aim of helping to inform public opinion on the possibilities of crop protection using biological and integrated control methods and will provide the first operating instruments.

All this, naturally, is not sufficient because it achieves an affirmation of the said method throughout the Southern agricultural territory and not only in citrus fruit growing.

The Universities, in fact, will never be able to operate beyond the transfer of applicative results from research by demonstrative tests. The CASSA and FORMEZ, in their turn, must remain within their own institutional responsibilities.

Who then can put into action instruments which are suitable for the territory's problems?

The phytochemical industry, which provides a product with immediate and topical action, has found the user in the farmer, who is interested only in solving the problems of his business. Crop protection with the biological and integrated control method cannot deal with individual businesses because its natural viewpoint is control of the agroecosystem.

In trying to define quantitatively and qualitatively the problem to be dealt with and identifying the subject capable of bringing into action suitable instruments, for the reasons just mentioned above it seems natural to consider that just as the agricultural businessman is the ideal subject for running the business well, so the public institution is the natural director in the actions which affect the territory. The agro-ecosystem however has vast territorial effects, the national territory has instead administration subdivisions where regional policies and social and economic interests are involved, and which are not necessarily interested in reintegrating the habitat for a better quality of life.

One is thus dealing with making economic interests and interests which want a better quality of life coincide.

Production protection actions, generally using chemical means, have been financed throughout the southern regions. Sicily, with law N° 8 of 3.1.1985 is the first which, through its own agricultural development organization, has planned to use biological control beginning with three centres which will be financed by the Cassa per il Mezzogiorno.

There is still the doubt that the public Institution, because of all its distinguishing characteristics, including the division of territory into Regions etc., can have abilities which allow it to act quickly and suitably and with a single purpose.

Current status of integrated pest management in California citrus-groves

R.F.Luck & J.G.Morse
Department of Entomology, University of California, Riverside, USA

D.S.Moreno
Boyden Fruit & Vegetable Entomology Laboratory, USDA, ARS, University of California, Riverside, USA

Summary

   Three key citrus pests, California red scale <u>Aonidiella</u> <u>aurantii</u>
(Mask.), citrus thrips <u>Scirtothrips</u> <u>citri</u> (Moult.) and citrus red mite
<u>Panonychus</u> <u>citri</u> (McG.), occur in all three of California's major citrus
districts. However, in the coastal, intermediate and interior districts,
California red scale and many of the other potential pests are under
excellent biological control. Citrus thrips is only an occasional pest
while citrus red mite is largely an insecticide-induced pest. In the San
Joaquin Valley, biological control of California red scale is lacking and
citrus thrips is a more consistent pest. Intervention thresholds are
assessed by monitoring the flight activity of male California red scale as
an index of scale density, and assessing the proportion of fruit infested
by citrus thrips and leaves infested by adult citrus red mite as an index
of thrips and mite densities.

1.1 Introduction

   Integrated Pest Management (IPM) is an ecologically based pest control
strategy that relies heavily on plant resistance and natural mortality factors, such as natural enemies, and seeks out control tactics that disrupt
these factors as little as possible. IPM uses pesticides, but only as a
last resort and after systematic monitoring of pest populations and natural
control factors indicate a need. Ideally, an IPM program considers all
available pest control actions, including no action, and evaluates the
potential interaction among various control tactics, cultural practices,
weather, other pests and the crop to be protected (7).
   A frequent complaint about IPM is its lack of an explicitly stated
methodology. However, one does exist. The approach that has most frequently led to a successful integrated control program is one based on the
scientific method (25). "Based on ideas derived from existing information,
which may be quite scanty, a hypothesis or mental model is made about the
value or harmfulness of particular components which could be manipulated to
minimize pest incidence. The hypothesis is immediately tested by critical
field experiments. An edifice of integrated control is then constructed,
based on further hypotheses and on successive experiments planned in relation to the ultimate goal of good pest control." A major point to note is
the implication that developing an IPM program is a continuing process, one
which evolves as new research information or technology becomes available.

1.2 Citrus Pest Management in California

   The current arthropod and mollusc pest complex in California is listed
in Table I. Many minor pests, those adequately controlled by natural ene-

mies, or those which occur occasionally, are not listed. Those species that are listed are key pests in the complex or occur as pests with sufficient frequency that suppression may occasionally be necessary. Many of the occasional pests differ among the several citrus growing districts of the state. Also, the tactics that are effective in these districts differ because of differences in the climates, cultivars and natural enemy complex associated with a district (Fig. 1). For purposes of discussion, the northern Sacramento Valley and desert valleys of southern California will not be considered further and the coastal, intermediate and interior districts will be combined under citrus IPM in southern California.

Table I. Arthropod and mollusc pest complex in San Joaquin Valley and southern California citrus.

Southern California

Citrus red mite	Panonychus citri (McG.)
Citrus bud mite	Eriophyes sheldoni Ewing
Citrus rust mite	Phyllocoptruta oleivora (Ashm.)
Broad mite	Polyphagotarsonemus latus (Banks)
California red scale	Aonidiella aurantii (Mask.)
Purple scale	Lepidosaphes beckii (Newm.)
Black scale	Saissetia oleae (Bernard)
Citrus thrips	Scirtothrips citri (Moult.)
Woolly whitefly	Aleurothrixus floccosus (Mask.)
Japanese bayberry whitefly	Parabemesia myricae (Kuwana)
Citrus whitefly	Dialeurodes citri (Ashm.)
Orange tortrix	Argyrotaenia citrana (Fern.)
Fruittree leafroller	Archips argyrospilus (Walkr.)
Western tussock moth	Orgyia vetusta (Bsdv.)
Citrus cutworm	Xylomyges curialis Grote
Forktailed bush katydid	Scudderia furcata Brunn
Angularwinged katydid	Microcentrum retinerve (Burn.)
Argentine ant	Iridomyrmex humilis (Mayr)
Brown garden snail	Helix aspersa Müller
Citrus mealybug	Planococcus citri (Risso)
Longtailed mealybug	Pseudococcus longispinus (Targioni-Tozzetti)
Obscure mealybug	Pseudococcus obscurus Essig

San Joaquin Valley

Citrus red mite	Panonychus citri (McG.)
Citrus thrips	Scirtothrips citri (Moult.)
California red scale	Aonidiella aurantii (Mask.)
Citricola scale	Coccus pseudomagnoliarum (Kuwana)
Brown soft scale	Coccus hesperidum L.
Potato leafhopper	Empoasca fabae (Harris)
Citrus cutworm	Xylomyges curialis Grote
Omnivorus leafroller	Platynota stultana Wlshm.
Fruittree leafroller	Archips argyrospilus (Walkr.)
Forktailed bush katydid	Scudderia furcata Brunn
Angularwinged katydid	Microcentrum retinerve (Burn.)
Western tussock moth	Orgya vetusta (Bsdv.)

534

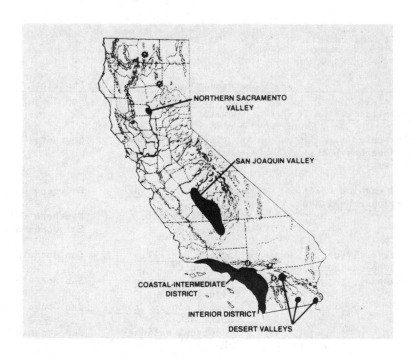

Figure 1.   The citrus districts of California.

Classical biological control has profoundly influenced citrus pest
severity in southern California. Table II lists those citrus pests that
have been substantially or completely controlled by this tactic. Fourteen
important citrus pests have been reduced to non-pest status in all or part
of their range in California. Biological control is continuing as five
biological control projects against citrus pests are active in California
(Table III).

Table II. Citrus pest insects controlled by natural enemies in one or
           more of California's citrus districts.

Pest	Natural enemy	District

Brown soft scale
    Coccus hesperidum L.
	Metaphycus luteolus
	    Timb.
		San Joaquin Valley

Citricola scale
    Coccus pseudomagnoliarum
        (Kuwana)
	Metaphycus luteolus
	    Timb.
	Metaphycus helvolus
	    (Comp.)
		Southern California
		
		Southern California

Black scale
    Saissetia oleae (Bernard)
	Metaphycus helvolus
	    (Comp.)
		Southern California

California red scale
    Aonidiella aurantii
        (Mask.)
	Aphytis melinus DeBach
	Comperiella bifasciata
	    Howard
		Southern California
		Southern California

535

Table II (continued).

Pest	Natural enemy	District
Yellow scale Aonidella citrina (CoQ.)	Comperiella bifasciata Howard	San Joaquin Valley
Purple scale Lepidosaphes beckii (Newm.)	Aphytis lepidosaphes Comp.	Southern California
Cottony cushion scale Icerya purchasi Mask.	Cryptochetum iceryae (Will.) Rodalia cardinalis (Mulst.)	Southern California Southern California San Joaquin Valley
Comstock mealybug Pseudococcus comstocki (Kuwana)	Allotropa burrelli (Muesebeck) Zarhopalus corvinus (Girault) Allotropa convexifrons (Muesebeck) Pseudophycus malinus Gahan	San Joaquin Valley San Joaquin Valley San Joaquin Valley San Joaquin Valley
Citrophilous mealybug Pseudococcus calceolariae (Mask.)	Coccophagus gurneyi Comp. Hungariella pretiosa Timb.	Southern California Southern California
Longtailed mealybug Pseudococcus longispinus (Targioni-Tozzetti)	Anarhopus sydneyensis Timb.	Southern California
Citrus mealybug Planococcus citri (Risso)	Leptomastidae abnormalis (Girault)	Southern California
Woolly whitefly Aleurothrixus floccosus (Mask.)	Amitus spiniferus (Brethes) Eretmocerus paulistus Hempel Encarsia sp. Cales noacki DeSantis	Southern California Southern California Southern California Southern California
Japanese bayberry whitefly Parabemesia myricae (Kuwana)	Eretmocerus sp. Eretmocerus sp. Encarsia n.sp.	Southern California Southern California Southern California
Brown garden snail Helix aspersa Müller	Rumina decollata (L.)	Southern California

Table III. Current biological control projects in California citrus.

Pest	Natural enemy	District
Citrus whitefly Dialeurodes citri (Ashm.)	Encarsia lahorensis Encarsia n. sp.	Southern California
Cloudy-winged whitefly Dialeurodes citrifolii (Morgan)	Just initiated	Southern California
California red scale Aonidiella aurantii (Mask.)	Aphytis sp. Pteropterix sp.	San Joaquin Valley
Citrus red mite Panonychus citri (McG.)	Euseius stipulatus Athias-Henriot	Southern California
Brown garden snail Helix aspersa Müller	Rumina decollata L.	Southern California San Joaquin Valley

1.3 Citrus Pest Management in Southern California

Probably the most important member of the pest complex is California red scale, Aonidiella aurantii (Mask.). Generally in southern California citrus districts, it is partially to completely controlled by the parasitoid, Aphytis melinus DeBach, as a result of a biological control project directed against it (20). Because of disrupted biological control, or in certain grove/citrus variety combinations, it sometimes becomes necessary to treat for the red scale. Treatments are instituted when 10% of the fruits on the tree are infested with 11 or more second instar or older scale insects. Low volume, narrow range oil sprays (NR-415 or NR-440), or organophosphates (in particular chlorpyrifos) coupled with inoculative releases of A. melinus will maintain satisfactory control of the pest. Bioassay of foliage with A. melinus adults is necessary to determine when toxic residues have abated.

A combination of surveys for scale-infested fruit (100 fruit per hectare, 20 per tree, 5 each from 0.5 to 2.5 meters off the ground in the north, east, south and west quadrants) and pheromone trapping for male California red scale is used to assess California red scale abundance. Currently, research is underway to develop a model to predict the proportion of infested fruit at harvest from the density of males caught on a pheromone trap earlier in the season. Experience is now the most frequently used guide to decisions about when to intervene.

Use of low volume, narrow range oil sprays has several advantages: it is the least disruptive of the available chemical treatments and, in addition to red scale suppression, it also satisfactorily suppresses citrus red mite, Panonychus citri (McG.), another key pest of citrus. Spring populations of this pest are normally controlled by the acarine predator, Euseius tularensis Congdon, or the coccinelid predator, Stethorus picipes Casey (16), or, with hot weather in June, by a noninclusion virus (22). Along the immediate coast E. hibisci (Chant) replaces E. tularensis as a predator of citrus red mite. Autumn populations of P. citri are the most dangerous,

537

especially when they coincide with the hot, dry, winds of autumn, because they can defoliate the tree. However, economic population densities of the mite are normally induced by improper use of pesticides. When mite densities attain two adult females per leaf, treatment is recommended. Recently, sampling schemes have been developed to estimate mite densities on oranges and lemons (11,12,26). These schemes are based on the proportion of leaves with mites. The threshold of two female mites per leaf is reached when 73% of the leaves are occupied, based on a sequential sampling design (26). Since the citrus red mite in some locations is resistant to a wide spectrum of synthetic organic insecticides and acaricides, oil sprays are one of the few effective treatment that remains (8,10). Several disadvantages of oil treatment also exist. Oil sprays tend to be more expensive than traditional, chemical treatments and soil moisture conditions in the grove need to be maintained prior to treatment: high humidity (>20%), low temperatures (<35°C) and adequate soil moisture (recent irrigation) are essential if leaf drop is to be avoided (18).

Citrus thrips, Scirtothrips citri (Moult.), is an uncertain pest problem in southern California citrus. Speculation has centered on several reasons for its presence: (1) the pest is becoming more adapted to citrus, (2) the pest is resurging due to increased pesticide use, and/or (3) the wind scars are confused for citrus thrips damage. But whatever the reason, citrus thrips varies in its occurrence with location and year. Furthermore, not much is known about its dynamics. It is a cosmetic pest of young fruit, the damage occurring in the spring to fruit less than four cm in size. When 20% of the fruit support one or more thrips, many pest control advisors recommend treatment. Two types of materials are used, dimethoate and sabadilla plus sugar bait. The latter is a botanical and less disruptive; however, it appears less able to maintain citrus thrips populations at non-economic densities when populations are generally dense throughout the grove. Citrus thrips, however, is becoming tolerant to dimethoate at some locations (Elmer, pers. comm.) so we can anticipate increasing problems in maintaining adequate suppression of this pest. E. tularensis is also an important predator of citrus thrips (23,24). This mite predator is principally a pollen feeder. Hence, to be effective it must occur at high densities. It does not respond to changes in citrus red mite or citrus thrips densities. So this species and the biological control it exerts is especially vulnerable to insecticide or acaricide disruption.

Two other pests have recently posed problems for integrated pest management in the southern California citrus districts: the woolly whitefly, Aleurothrixus floccosus (Mask.), introduced into southern California in 1966, and the Japanese bayberry whitefly, Parabemesia myricae (Kuwana), found in southern California in late 1978 (5,19). Both of these whiteflies were the subject of extensive biological control projects. The woolly whitefly has been controlled by the introduced parasitoids Amitus spiniferus (Brethes), Eretmocerus paulistus Hempel, Encarsia sp. and Cales noacki DeSantis. The whitefly is still spreading so that its ultimate distribution and the effectiveness of biological control in California is unknown.

The appearance of the Japanese bayberry whitefly led to the introduction of Eretmocerus n. sp., Eretmocerus sp., Encarsia n. sp. However, an apparently native species of Eretmocerus has been responsible for the control of this whitefly in southern California. The parthenogenic reproduction of this species coupled with its short generation time make this species particularly explosive if disrupted with improper control tactics.

IPM in southern California demands ant suppression, especially of the Argentine ant, Iridomyrmex humulis (Mayr) (2,3,4). Several species of soft scale-insects (Coccidae), mealybugs (Pseudococcidae), aphids (Aphidae) and

538

whiteflies (Aleurodidae), even though under excellent biological control in
southern California, can provide excellent sources of honeydew for ants.
Ants foraging for honeydew interfere with the natural enemies of these and
other pests (e.g. hard scales (Diaspididae) and citrus red mite). Argentine ant, where it occurs, is a key pest. Its pest status often is unrec-
ognized because it does not directly damage the tree but indirectly affects
pest control by its slow and gradual induction of increased pest species.
We strongly recommend an aggressive ant control program whenever Argentine
ant colonies are detected.

Several species of Lepidoptera larvae, collectively called orangeworms,
can cause economic injury in citrus. In the coastal district of southern
California, the principal species is the orange tortrix, *Argyrotaenia*
citrana (Fern.), while in the intermediate district this species is joined
by three others, the fruittree leafroller, *Archips argyrospilus* (Walkr.),
the western tussock moth, *Orgyia vetusta* (Bsdv.) and the citrus cutworm,
Xylomyges curiales Grote (1). Generally, these species feed on young
(recently produced) foliage and developing fruit, and on occasion, can
almost totally consume the young foliage and damage a substantial amount
(>50%) of the developing fruit. The orange tortrix is multivoltine, hence
periodic surveys for its presence in a grove must be instituted throughout
the growing season. The remaining three orangeworm species are univoltine.
Surveys for their presence must begin the first half of March. A worm survey consists of inspecting a tree for a minimum of 5 minutes using trees
in each corner and one in the center of each two hectare block of grove.
Table IV provides the economic level for each of these orangeworms. As the
survey is not random, experienced pest control advisors will be more successful in finding larvae, since they will know where to look for them.
Thus, they must adjust the intervention threshold higher than those shown
in Table IV. Since different intervention thresholds exist for each moth
species, the pest control advisor must be able to field identify the larvae
to species.

Table IV. Intervention thresholds for Lepidoptera larvae infesting
oranges, lemons and grapefruits using the time-search
method (modified from Atkins and Elmer (1)).

Species	Number of larvae/hr
Orange tortrix	30
Fruit tree leafroller	120
Western tussock moth	100
Citrus cutworm	15
Omnivorous leafroller	30

Katydids, *Scudderia furcata* Brunner and *Microcentrum retinerve* (Burn.),
occasionally scar young fruit in the spring. They too, can be evaluated in
the spring along with the Lepidoptera larvae that attack citrus, although
no intervention thresholds have been established for these species.

539

In addition to the citrus red mite, three other mite species are of concern to southern California citrus growers: the citrus bud mite, Eriophyes sheldoni Ewing, the citrus rust mite, Phyllocoptruta olievora (Ashm.) and the broad mite, Polyphagotarsus latus (Banks). The first species is a primary pest of lemons and traditionally is subjected to preventive treatments of chlorobenzilate plus oil even though an intervention threshold for this pest is lacking. The last two mite species cause fruit russetting and are pests of coastal citrus. They are only of local importance.

Several biological control projects have been instituted for pests of southern California. Citrus whitefly, Dialeurodes citri (Ashm.) remains an occasional pest. Hence additional natural enemies are being sought for its control. Another whitefly, Dialeurodes citrifolia (Morgan), has recently been introduced and is also a focus of a biological control project. It remains to be seen to what extent this whitefly species becomes a problem. Two other biological control projects have been undertaken in southern California citrus: the establishment of Euseius stipulatus (Athias-Henriot), a predacious mite from the Mediterranean region in one locality near the coast (J. McMurtry, pers. comm.) and the decollate snail, Rumina decollata (L.) against the brown garden snail, Helix aspersa Müller. E. stipulatus has displaced the native mite predator, E. tularensis and E. hibisci, but it is unclear whether it will significantly improve biological control of the mite pests (16). The decollate snail, accidentally introduced into southern California, attacks and kills younger stages of the brown garden snail. Unlike the brown garden snail, however, it will not feed on live plant material, although germinating seedlings are sometimes damaged. Introduction of the decollate snail into groves infested with the brown garden snail has resulted in displacement of the latter by the former in three to four years if adequate moisture (irrigation) is present (T. Fisher, pers. comm.). The decollate snail, however, has not been as effective in coastal citrus groves as it has in the more inland groves.

1.3.1 Citrus pest management in the San Joaquin Valley

The pest complex present on citrus in the San Joaquin Valley differs from that in southern California in several ways (Table I). With the exception of Argentine ant, the key pests in San Joaquin Valley citrus groves are the same as those in southern California: California red scale, citrus thrips, and citrus red mite. California red scale was first detected in the San Joaquin Valley in 1939 (15). Growers in the Valley responded by forming pest control districts whose purpose was to eradicate the scale. They were unsuccessful, so in 1968 the districts were disbanded (21). They did slow the spread of the scale in commercial citrus, however. California red scale is a consistent pest because it lacks an economically effective biological control agent. Both A. melinus and Comperiella bifasciata Howard are generally present in the Valley but they are unable to exert sufficient economic control of the scale-insect in commercial citrus (6). However, generally they do provide sufficient control in both dooryard citrus in the Valley and in commercial citrus in southern California.

Yellow sticky traps (75 x 125 mm) baited with a component of the synthetic pheromone are now used to monitor male flight activity and to index red scale densities in a grove (17). In the San Joaquin Valley, four periods of male flight activity have been recorded and are usually easily distinguished from one another because of their discreteness. Frequently, however, the August flight of males is suppressed, perhaps for climatic reasons. The initiation of each flight is separated by about 616 degree-days, based on a developmental threshold for the scale of 11.7°C (13). A multiple-regression model has been developed to forecast the expected proportion of infested fruit at harvest using the number of males trapped during the second flight period (late June). When 10% or more of the fruit are expected to have 11 or more second instar or older scale, treatment is suggested (6). Fruit destined for the export market has a lower intervention threshold for cosmetic reasons. The pheromone monitoring method is backed up by twig and fruit samples, as insecticide drift from adjacent groves or recent treatments can make the trap catches unreliable. With experience each grove is found to have a characteristic pattern.

Organophosphate insecticides are most frequently used to control California red scale. San Joaquin Valley growers are less inclined to use oil sprays. Navel oranges, the dominant cultivar grown in the Valley, are much less tolerant when complete coverage with oil is used, especially if the trees are moisture stressed from inadequate irrigation or excessive summer heat. Low volume, narrow range oil sprays have not been accepted either, probably because of the lack of appropriately configured equipment to deliver the material into the tree tops (18). The frequency of organophosphate treatments can often be reduced from once a year to once every second or third year by using pheromone (male) traps and a degree-day model if insecticide coverage is adequate. Failure to treat the tree tops leaves a reservoir of red scale to rain crawlers down onto the treated portion of the trees, necessitating annual scale treatments.

Citrus thrips appears to be a more frequent pest in San Joaquin Valley citrus groves, perhaps because of the dominance of the Navel orange cultivar in the district. Navel oranges are more susceptible to damage by this pest; however, little is known about the factors affecting its population dynamics. Also, evidence suggests that organophosphate insecticides increase the thrips problem by killing E. tularensis, a predatious mite that feeds on thrips. Citrus thrips is also showing increasing resistance to dimethoate, the material most widely used against it. Treatment is usually instituted when 5-10% of the fruit less than 4 cm in diameter supports one or more thrips.

The pest status of citrus red mite in the San Joaquin Valley is the subject of some disagreement, which stems from several sources. First, the status of citrus red mite as a pest in southern California is based on the interaction between its feeding damage to leaves and the presumed moisture loss this damage causes during the hot, dry winds in late spring and autumn, but principally the autumn. This interaction results in substantial leaf drop of heavily fed upon leaves. Hot, dry winds only occur in the extreme southern end of the San Joaquin Valley (Kern County). They are not a factor in other valley areas. Yet the concern with mite damage remains and the southern California intervention threshold (2 adult female mites per leaf) is used by pest control advisors in the San Joaquin Valley.

Second, rapid increases in citrus red mite density are often induced by insecticide applications. At times, as many as three post-bloom organophosphate treatments occur in San Joaquin Valley citrus: two for citrus thrips and one for California red scale. If Lepidoptera larvae are present or routinely treated for, an additional pre-bloom treatment with organophosphate can occur. Under such a treatment regime citrus red mite populations can rapidly resurge. This propensity for rapid population increase has led many pest control advisors to choose treatment levels on the order of 1/2 adult female mites per leaf. Added to the resurgence problem is that of growing acaricide resistance (8,10). Thus citrus red mite has all the indicators of a secondary (induced) pest, as originally suggested by Lewis (14) in 1948 after the initial use of DDT in San Joaquin Valley citrus.

Several additional Lepidoptera larvae are pests in the San Joaquin

541

Valley that are absent in southern California. The most important is the omnivorous leafroller, Platynota stultana Wlshm. (1). We suggest an intervention threshold of 30 larvae per hour of search using the time search method previously described. Surveys for this should be instituted throughout the growing season since the omnivorous leafroller is multivoltine.

Katydids are also present in the San Joaquin Valley, but intervention thresholds for them have not been developed. Surveys for this pest are normally begun prior to bloom in the spring.

Both California red scale and citricola scale, Coccus pseudomagnoliarum (Kuwana), are the subject of biological control projects. Should an effective natural enemy be found for California red scale in the San Joaquin Valley then citricola scale would assume key pest status. Treatments for California red scale are also effective in controlling citricola scale.

1.4 Conclusion

What has been described is the recommended integrated pest management programs for California's two major citrus growing areas. This program is described in greater detail in Publication 3303, Titled "Integrated Pest Management for Citrus," University of California, Division of Agriculture and Natural Resources, Berkeley, California, 94720, USA. Further information on how to order a copy can be obtained from this address.

With increasing numbers of pest control consultants (self-employed pest consultants hired by the grower), the area under integrated pest management is increasing, with the growers experiencing substantial savings (6,9). Depending on the district, these savings can range from $200 per hectare in the San Joaquin Valley to $600 per hectare in coastal southern California. The principal impediment to wider use of Integrated Pest Management is a supply of adequately trained pest control advisors and an effective grower education program.

REFERENCES

1. ATKINS, E. L. and ELMER, H. S. (1981). Information concerning the economic importance, life cycle, economic level and control of larvae of "orangeworms" on citrus in California. Univ. of Calif. Dept. Entomol., Riverside, Newsletter No. 74 (mimeo). 7 pp.
2. BARTLETT, B. R. (1961). The influence of ants upon parasites, predators and scale insects. Ann. Ent. Soc. Amer. 54: 643–651.
3. DeBACH, P., FLESCHNER, C. A. and DIETRICK, E. J. (1951). A biological check method for evaluating the effectiveness of entomophagus insects. J. Econ. Entomol. 44: 763–766.
4. DeBACH, P., FLESCHNER, C. A. and DIETRICK, E. J. (1951). Ants vs. biological control of citrus pests. Calif. Citrog. 36: 347–348.
5. DeBACH, P. and ROSE, M. (1976). Biological control of the woolly whitefly. Calif. Agric. (May).
6. FLAHERTY, D. L., PEHRSON, J. E. and KENNETT, C. E. (1973). Citrus pest management studies in Tulare County. Calif. Agric. (Nov.), pp. 3–7.
7. FLINT, M. L. and VAN DEN BOSCH, D. (1981). Introduction to Integrated Pest Management. Plenum Press, New York, 240 pp.
8. GEORGHIOU, G. P. and TAYLOR, C. E. (1977). Pesticide resistance as an evolutionary phenomenon. Proc. XV Int. Congr. Entomol., p. 759–785.
9. HALL, D. C., NORGARD, R. B. and TRUE, P. K. (1975). The performance of independent pest management consultants. Calif. Agric. 29(10): 12–14.
10. JEPPSON, L. R., JESSER, M. J. and COMOLIN, J. O. (1965). Response of susceptible strains of Pacific spider mite and citrus red mite to O-

isopropyl-O (2,4,5 trichlorophenyl) phosphoramidate or phosphoramdo-thioate. J. Econ. Entomol. 58: 466-467.

11. JONES, J. P. and PARRELLA, M. P. (1984). Dispersion indices and sequential sampling plans for the citrus red mite (Acari: Tetranychidae). J. Econ. Entomol. 77: 75-79.

12. JONES, J. P. and PARRELLA, M. P. (1984). Intra-tree regression sampling plans for citrus red mite (Acari: Tetranychidae) on lemons in southern California. J. Econ. Entomol. 77: 810-813.

13. KENNETT, G. E. and HOFFMAN, R. W. (1985). Seasonal development of the California red scale (Homoptera: Diaspididae) in San Joaquin Valley citrus based on day-degree accumulation. J. Econ. Entomol. 78: 73-79.

14. LEWIS, H. C. (1948). Three years' use of DDT on citrus in central California. Calif. Citrog. 33: 546-547.

15. McKENZIE, H. L. (1956). The armored scale insects of California. Bull. Calif. Insect Surv. 5. 209 pp.

16. McMURTRY, J. A., HUFFAKER, C. B. and VRIE, M. VAN DE. (1970). Ecology of tetranychid mites and their natural enemies: a review. I. Tetranychid enemies: their biological characters and the impact of spray practices. Hilgardia 40: 331-390.

17. MORENO, D. S. and KENNETT, C. E. (1985). Predictive year-end California red scale (Homoptera: Diaspididae) orange fruit infestations based on catches of males in the San Joaquin Valley. J. Econ. Entomol. 78: 1-9.

18. RIEHL, L. A., BROOKS, R. F., McCOY, C. W., FISHER, T. W. and Dean, H. A. (1980). Accomplishments toward improving pest management for citrus. In Huffaker, C. B. (ed.), New technology of pest control. John Wiley & Sons, New York, pp. 319-363.

19. ROSE, M., DeBACH, P. and WOOLLEY, J. (1981). Potential new citrus pest: Japanese bayberry whitefly. Calif. Agric. (May-April), pp. 22-24.

20. ROSEN, D. and DeBACH, P. (1979). Species of Aphytis of the World (Hymenoptera: Aphelinidae). W. Junk, Den Hague, 801 pp.

21. SCHILLING, W. E. (1968). Central California citrus pest control district terminates - Individual districts to control red scale work. Citrograph 53: 368.

22. SHAW, J. G. and BEAVERS, J. B. (1970). Introduced infections of viral disease in populations of citrus red mite. J. Econ. Entomol. 63: 850-853.

23. TANIGOSHI, L. K. and GRIFFITHS, H. J. (1982). A new look at biological control of citrus thrips. Citrograph 67: 157-158.

24. TANIGOSHI, L. K., NISHIO-WONG, J. Y. and FARGERLUND, J. (1983). Greenhouse and laboratory-rearing studies of Euseius hibisci (Chant) (Acarina: Phytoseiidae), a natural enemy of citrus thrips, Scirtothrips citri (Moulton) (Thysanoptera: Thripidae). Environ. Entomol. 12: 1298-1302.

25. WAY, M. J. (1973). Objectives, methods and scope of integrated control. In Geier, P. W., Clark, L. R., Anderson, D. J., and Nix, H. A. (eds.), Insects: studies in population management. Mem. Ecol. Soc. Austr. 1: 137-152.

26. ZALOM, F. G., KENNETT, C. E., O'CONNELL, N. V., FLAHERTY, D. L., MORSE, J. G. and WILSON, L. T. In Press. Distribution of Panonychus citri (McGregor) and Euseius tularensis Congdon on central California orange trees with implications for binomial sampling. Protect. Ecol.

Integrated pest control in citrus in Greece

L.C.Argyriou
'Benaki' Phytopathological Institute, Kiphissia, Greece

Summary

The succesful biological control of Diaspididae scales and the
decline of the population of Ceratitis capitata after using bait
sprays, in citrus groves, brought a change in the control programs
of the other citrus pests in Greece. Thus now the key pest of
citrus trees is the mealybugs. As biological control on mealybugs
failed, new factors are studied in order to determine the most
profitable pest management practices.

1.1. Introduction

Greece is one of the few countries in which native scale insects
have been parasitized by introduced parasites to such an extent that they
were capable of providing an economically successful biological control
of them. This is the case with the red scale (Aonidiella aurantii Maskel)
and the olive black scale (Saissetia oleae (Olivier)), control of which has
been obtained by the imported parasites, Aphytis melinus DeBach and
Metaphycus helvolus Compere respectively. Two other serious pests, the
purple scale Lepidosaphes beckii Newmann and Chrysomphalus dictyospermi
Morgan were controlled the first one partially and the second one almost
completely, by Aphytis lepidosaphes Compere and A. melinus respectively
(1) (2) (3). C. dictyospermi is now virtually a rare species in most
orchards, in which although it had previously occured, often as major pest.

The decline on the population of Ceratitis capitata (Wiend.), a
major pest of Citrus for years (4) encouraged us towards an integrated pest
management of citrus pests. The case of mealybugs, mainly of the Planococcus
citri Risso, is different, as very few data have been obtained concerning
its biological control. Thus the present status of integrated control for
the aforementioned pests is discussed as follows.

1.2. Present Status of Citrus pests

There are more than thirty insect pests on citrus orchards in Greece.
The current pest complex is listed in Table 1.

These pests are classified under three categories,
a) the major, Key pests : these occur as pests with sufficient frequency.
b) secondary pests : they occasionally become a problem.
c) minor pests : they are controlled by natural enemies or they occur
occasionally on citrus trees.
This classification is not stable, and it may alter due to various factors.

The most important members of the pest complex are A. aurantii, L.
beckii and P. citri. A. aurantii and L. beckii are substantially controlled
by the parasite, Aphytis melinus DeBach and Aphytis lepidosaphes Compere
respectively as a result of a biological control project (1) (2) (3), but
sometimes it becomes necessary to treat the citrus trees. When 5-10% of the

545

Table 1

a) Major pests
                   Species

Species	Family
Planococcus citri Risso	Pseudococcidae
Aonidiella aurantii Maskell	Diaspididae
Lepidosaphes beckii Newmann	Diaspididae

b) Secondary pests

Species	Family
Calocoris trivialis Costa	Miridae
Toxoptera aurantii B.d.F	Diaspididae
Parlatoria ziziphus Lucas	Diaspididae
Coccus pseudomagnoliarum Kuwana	Coccidae
Saissetia oleae (Olivier)	Coccidae
Ceroplastes rusci L.	Coccidae
Ceroplastes floridensis Comstock	Coccidae
Pulvinaria floccifera Westwood	Coccidae
Prays citri Mill.	Hyponomeutidae
Ceratitis capitata (Wind.)	Trypetidae

c) Minor pests

Species	Family
Heliothrips haemorroidalis Bouché	Thripidae
Megalurothrips kellyanus (Baynall)	Thripidae
Myzus persicae Sulzer	Aphididae
Aphis gossypii Glover	Aphididae
Aulacorthum solani Kalt.	Aphididae
Aphis craccivora Koch	Aphididae
Coccus hesperidum L.	Coccidae
Pseudococcus adonidum L.	Pseudococcidae
Icerya purchasi Maskell	Margarodidae
Carpophilus hemipterus L.	Nitidulidae
Epicometis hirtella L.	Scarabeidae
Tropinota squalida (Scop.)	Scarabeidae
Oxythyrea funesta Poda	Scarabeidae
Spodoptera littoralis (Boisd.)	Noctuidae
Archips rosanus (L.)	Tortricidae
Cryptoblabes gnidiella Mill.	Pyralidae
Ectomyelois ceratoniae Zell.	Pyralidae

fruit on the tree is infested with these or other Diaspididae scales, the
trees are treated with insecticides. Mealybugs is an increasing pest
problem in citrus groves and they are variable in their occurence from
grove to grove and year to year. Attempts to control this pest by introducing Cryptolaemus montrouziery have failed as the predator has not been
established in Greece. The native parasite Leptomastidea abnormis Gir.
(Hym. : Encyrdidae) appears in high population levels and some times keeps
the mealybugs under partial control, but the pest is resurging, due
perhaps to increasing pesticide use or other factors. Furthermore not much
is known about its dynamics, and no economic levels have been established
for these species in Greece. To press this pest, pesticide treatments are
needed. Two other pests have recently posed problems for integrated pest
management in the citrus groves of the mainland : a new invador the white-

fly D. citri (4) and the native C. trivialis. So far spread of the whitefly is restricted, to Corfu island and the Ionian coastal districts. The
introduced parasite Encarsia lahorensis (How.) (5) has given good control
results, but the whitefly is still spreading, while the parasite does not
seem to follow it in all places. So some more releases of the E. lahorensis
has to be done. Treatment with insecticides does not give good results.
The C. trivialis has a more expanded spread, all over citrus growing
areas in mainland. No effective natural enemies to it have been found.
Populations are kept below injurious density levels by means of a variety
of insecticides.
The soft scales are of variable economic importance in the difference citrus
areas in Greece. S. oleae, C. rusci, C. floridensis, C. pseudomagnoliarum
and P. floccifera have showed outbreaks here and there. In many groves S.
oleae is kept under economical level by parasites Metaphycus helvolus
Compere Metaphycus lounsburyi Howard and by its predators. The outbreak
of the rest soft scales are kept below injurious density levels by means
of insecticides.
The citrus flower moth, P. citri is very restricted to lemon growing groves
and it is an unimportant pest. The Mediterranean fruit fly, the most
destructive insect of citrus fruit, for many years in past in Greece, does
not consist a problem any more. As olive and citrus groves are interplanted
or the citrus trees occupy a belt extending on the shore line the olive
trees being behind them, the bait-sprays against the olive fly, which
are applied almost every year, have reduced the population of the Medfly.

1.3. Citrus Pest Management

Five important citrus pests have been reduced the present years to
non-pest status in Greece, by means of classical biological control or
other methods. After a long study and many other detailed investigations,
our experience suggests that classical biological control is a continuing
process which involves new research information or technology.
Now, the urgent need is to find ways of preventing occasional outbreaks
of minor and latent pests under biological control. The research for the
study of the complex interactive biocitricultural system and the Intergrated
pest management, is a good starting point. The first step is to protect
natural enemies of pests, by reducing usage and frequency of pesticide
application to the absolute minimum. Such an approach has an additional
benefit in reducing the cost of pest suppresion, the healt safety problems
in farm workers and the enviromental contamination. If some times it
becomes necessary to treat for the scale insects, the use of low volume
narrow range oil sprays will maintain satisfactory control of these pests.
Use of low volume oil sprays has the advantage that it is the least disruptive of the available chemical treatments. On the other hand oil sprays
is more exprensive than chemical treatments, and the application on
the groves needs high humidity ($<20\%$) low temperatures ($>30°$) and
adequate soil moisture with recent irrigation. All those disadvantages
lead the growers to choose the easy way with insecticides and this is the
key reason, in our citrus pests problems.

REFERENCES

1. ARGYRIOU, LOUKIA C., DEBACH P., 1968. Establishment and spread of
   Metaphycus helvolus (Compere) (Hym. Encyrtidae) in olive groves of
   Greece. Entomophaga 13(3) : 223-228

2. ARGYRIOU, L.C., MOURIKIS, P.A., 1981. Current status of citrus pests
   in Greece. Proc. Int. Citriculture. 623-627

3. DEBACH, P., ARGYRIOU, LOUKIA C., 1967. The colonization and success in Greece of some imported Aphytis spp. (Hym. : Aphelinidae) parasitic in citrus scale insects (Hom. : Diaspididae) Entomophaga 12(4) : 325-342.

4. PAPPAS, S., VIGGIANI, G., 1979. Introdotta a Corfù la Prospaltella lahorensis How. (Hym. : Aphelinidae), parassita del Dialeurodes citri (Ashm.) (Hom. : Aleyrodidae) Boll. Lab. Agr. Portici 36 : 38-41.

Feasibility of integrated pest management for citrus orchards in Spain

A.Garrido

Institute for Agrarian Research of Valencia (IVIA), Moncada, Spain

Summary

The major phytophagous present in the citrus orchards of Spain are presented with a brief study of the pest and the most significant biological characteristics and also their geographic distribution and the presence of natural enemies.

The need for conducting a control programme is set out; if not a full integrated management, at least a selective management against the citrus phytophagous, considering the influence of A. floccosus populations on the development of other arthropods, harmful to citrus.

A list of pesticides, not affecting nymphal stages of C.noacki, L. testaceipes and R. cardinalis is given and the convenience of using specific compounds —not harmful to beneficial arthropods— in controlling the phytophagous is underlined.

It is evident that selective control is possible, and, in fact, as it is being applied, is very close to integrated control, towards which all the efforts of the Government Agencies of Spain are being directed.

1.1. Introduction

Approximately 230,000 Ha are the total area of Citrus culture in Spain, the largest part of them are located in the Valencian Community, Andalucia and Murcia. There is a predominance of lemon culture in the provinces of Murcia, Alicante and Málaga and Citrus sp. of the groups of "mandarins" and "sweet oranges" are grown in the rest of districts mentioned.

The Valencian Community is in the lead for citrus culture both considering extension and production; maximum efforts are being made towards obtaining high quality products. For this reason it is necessary to have an adequate approach to management methods to control biological agents haring a detrimental incidence on production and quality: fungi, bacteria, viruses and arthropods. Such methods are to be applied similarly to the rest of citrus of Spain.

The present work aims at discussing the damages being caused by Arthropods to Citrus in Spain.

About 42 species of arthropods are cited in Spain, with varying degrees of incidence on production (21). From these, we will only present the most significant:
TETRANYCHIDAE: Panonychus citri and Tetranychus cinnabarinus (20); ALEYRODIDAE: Aleurothrixus floccosus; APHIDIDAE: Aphis citrícola, Toxoptera aurantii, Myzus persicae and Aphis gossypii (15); MARGARODIDAE: Icerya purchasi (14) COCCIDAE-LECANINAE: Ceroplastes sinensis, Coccus hesperidium, and Saissetia oleae (14); COCCIDAE-DIASPINAE: Aspidiotus

nerii, Chrysomphalus dictyospermi, Lepidosaphes beckii, L. gloverii, and Parlatoria pergandei (14); PSEUDOCOCCIDAE: Planococcus citri (14); LEPIDOPTERA: Prays citri; DIPTERA: Ceratitis capitata.

These phytophagous have traditionally been controlled by straight chemical applications. Nevertheless, the beneficial action of arthropods enabling a reduction in the populations has been always considered, together with the import and release of predators and parasites whose efficiency had been already known in other countries. With the almost exclusive application of chemical control, since the appearance of the organochlorates, followed by the organophosphorates, the little progress gained by biological control was almost forgotten and abandoned, with a few exceptions.

Eventually, like in other countries, experience gained with chemical control alone proved not to be the adequeate approach to pest control and there was an attempt to combining chemical and biological control; the latter placed on a considerable lower level than chemical control. However, the research workers were clearly determined to integrate orderly all control methods available, whether cultural, chemical or biological. Thus the first steps towards Integrated control were given, mainly since the establishment of A. floccosus in Spain.

1.2. Bio-ecology of major phytophagous and possible control methods

To achieve an effective control of the major phytophagous involved, a good understanding of their biological cycle is needed in addition to factors, whether biotic or abiotic, having an influence on their population dynamics. The following is a brief description of the arthropods:

a. P. citri and T. cinnabarinus:

In Spain, P. citri became a pest in 1981, mainly in the provinces of Alicante, Castellón and Valencia. Under our climate conditions, it is present throughout the year, in any of its stages, reaching population peaks since early or mid-September through the end of April, a period in which there is a strong decrease in population owing to natural factors, either biotic or abiotic. P. citri is found parasitising on leaves, fruits and wood, where it lays eggs; it feeds on chlorophyl, a reason why its damages appear on green leaves, shoots and fruits, showing a preference in colonizing the top, sunshine parts of the trees. The acarus parasitises all citrus varieties. On mandarins, (10) the damages are less apparent.

T. cinnabarinus is distributed all round Spain and causes the severest damages on lemon and mandarin trees, including defoliation. There are predator insects and acarus (4) such as Conwentzia psociformis, Stethorus punctillum and several species of Phytoseiidae: Euseius stipulatus, Anthoseius phialatus and Neoseiulus californicus, acting as natural enemies of the arthropod. In case of important population increases, P. citri control is based on acaricides, bearing in mind that it is necessary to wet the wood and preferably choose compounds having ovicide proporties (10).

b. A. floccosus:

This phytophagous was known in Spain long since, in the Canary Islands where it was not a threat, as there were only a few citrus plantings (16). In the Peninsula it was reported to appear in Málaga, (17) and during the summer of 1969 it was detected as a severe infestation in some citrus growing areas of the Málaga province. By the end of that year A. floccosus was found in the Alicante province and from 1974 onwards, it is present in all the citrus producing areas of Spain causing severe damages.

Adults prefer to lay their eggs on the underside of the leaves. Occasionally, if there are high populations, the eggs may be deposited on the fruits, although these are not viable. In the Valencia area, oviposition takes place the whole year round, without any dormancy on any stage of the individuals. There is only a slowing down in their evolution (8), having a coexistence of all the stages over all the seasons, decreasing the reproduction potential during the unfavorable time and attaining maximum values over July and August. Under suitable climatic conditions, the maximum reproductive potential can be maintained until September or even until part of October. Average fecundity is about 240 eggs.

A. floccosus undergoes four larval stages, and only the first one is mobile. On all the stages there is honeydew production and, as they go changing from one stage to another, there is a gradual increase of wax secretion coating that, on the maximum phase, covers the individuals completely. Five or six are the usual number of annual generations, depending on climatic conditions; if these are suitable, the biological cycle is completed within less than one month; and if these conditions are not adequate, the cycle may last more than 120 days.

White fly causes a) direct damages, consisting of sap extraction from leaves by adults and larval stages leading to weakening of sprouting; crop reduction; varying degrees of defoliation; and difficulties at picking; b) indirect damages that cause fostering of development of other biological agents affecting trees; these are sometimes more severe than direct damages, namely development of saprophytic fungi, particularly species of the Fumago, Limacinia genus; intensifying and development of other pests that find a protective and favourable harbour on down and dirt; among the arthropods, appearance of pests benefiting from this condition, such as, L. beckii, Planococcus citri, P. pergandei, Ch. dictyospermi, T. cinnabarinus, Panonychus citri and finally, the lack of pesticide effectiveness because of the down originated by the white fly, since some pest insects harbour in it and the compound cannot reach them (5). These last damages lead to consider A. floccosus as the major phytophagous attracting many others around it.

A. floccosus has a great number of natural enemies (19); In Valencia, (7) many predator species have been found feeding on eggs and larvas of white fly on their initial stages; however, control is not achieved efficiently.

In Málaga, three parasites (18) were introduced in 1970. From them, only Cales noacki was established and at present is found in all the citrus areas of Spain providing a good control of A. floccosus (7).

Wherever chemical applications are needed, butocarboxim is used, a harmless pesticide to beneficials.

c. Aphididae:

The cited aphid species are being found in all the provinces of Spain and sometimes cause damages to citrus, producing the well known characteristic deformations on leaves and limiting the growth, being particularly deleterious to young trees.

Aphididae have numerous natural enemies, being Lysiphlevus testaceipes one of the most active. In case chemical control is necessary to control aphids, specific pesticides should be used.

d. I. purchasi:

This scale can be found at small population levels in all the citrus producing areas, and this is why its damages are not important,

551

basically because the Rodalia cardinalis predator controls it. I. purchasi has three annual generations (14).

e. Lecaninae:

From the species that we are citing in this point only S. oleae is worth mentioning, since C. sinensis and C. hesperidum are not a threat because of their moderate populations which are limited by biotic or abiotic agents.

S. oleae is observed in all citrus growing areas in uneven distribution. It produces two or three annual generations (14) the first one appears in February and in March there is a complete eclosion of all the eggs. The second begins in July and the third could begin in early autumn, depending on the climatic conditions.

This phytophagous has numerous parasites and predators (13) (1) but from all of them, only Metaphycus helvolus gives a good biological control.

When sometimes the use of chemicals becomes necessary, it is advisable to apply them at egg eclosion.

f. Diaspinae:

A. nerii is basically found in those provinces predominating lemon culture; normally three yearly generations are present: (14) the first one by mid/late April; the second, in late June–early July and the third, in late September–early October. There are many parasites and predators related to A. nerii, their contribution in controlling it is not clear; Aphytis melinus is the most important parasite. If chemical treatment becomes necessary, it should be applied when the insect is in its maximum sensitive stage. Another suitable time for pesticide application, in the case of lemon, is shortly after fruit set, when the fruit begins swelling and until the calyx closes.

Ch. dictyospermi and L. gloverii are basically present in the southern part of Spain, in the provinces of Sevilla and Huelva, being found in more or less abandoned orchards (14). They produce three generations a year.

Parasites and predators have been observed on Ch. dictyospermi effectively reducing populations, therefore, no chemical treatments are required.

In 1979 our Station imported Prospaltella elongata, a parasite of L. gloverii; this beneficial was well adapted in the provinces where the releases were conducted: Valencia Alicante and Sevilla; there was no need for pesticide applications for their control.

P. pergandei is a species essentially present in adult plantings (12) and L. beckii reached heavy population levels since the establishment of A. floccosus on citrus in spain. Both scales produce two generations a year: one at the beginning of June and the other during the month of August and settle on trinks, twigs, leaves and fruits; they have large populations of parasites and predators which do not achieve the required levels of control: from our monitoring carried out over one year the mean of parasitism found was never higher than 10%, a reason why timely chemical treatments must be done in heavily infested orchards.

One common characteristic of Diaspinae is a migration of larva settled on the peduncles to the interior of calyx, beginning at petal fall; this fact was observed in Israel (12) and so we did on other species of this group in our Station. For this reason, larva having been able to harbour in these locations are very difficult to suppress through the use of insecticides, becoming a potential source of scales in future generations. It is advisable, therefore, to apply the treatment over the first generation of the insects, before the calyxes close on the fruits and peak sensitive stages are present.

g. Pseudococcidae:

P. citri is a widely spread species found all over Spain which can produce up to six annual generations (14). Its populations are being controlled by Cryptolaemus montrouzieri releases, which should be done beginning in April and, only when late infestations occur, chemical treatments should be applied, i.e.: in late September of October. An attempt to adapting Leptomastix dactilopii was done without success.

h. Lepidoptera:

P. citri preferably attacks lemon trees and is found in the provinces where this culture is predominant. Occasionally it attacks Satsumas (Citrus unshiu) (11), without serious injuries. P. citri produces three generations a year. On lemon it produces severe damages and chlorpyrifos applications are made to control it; the treatment being coincidental with the first generation of A. nerii, since these two insects inflict the most serious damages to lemons.

i. Diptera:

C. capitata is found spreading on nearly all areas of Spain (2) producing damages on a great number of fruits, particularly peach. For lack of other fruits, it attacks citrus. It can produce up to six geneations, although this is not consistent nor regular, being especially dependent on the climate. (2). On citrus, C. capitata is controlled by the use of fly-catch traps with trimedlure and phosphate and when one fly is caught per trap and day, only the area where this occurs is treated by aircraft, applications of Fenthion and hidrolyzed protein.

1.3. How to manage citrus pests?

The conditions that A. floccosus created on citrus lead to consider the need for the implementation of a rational pest management against insects and harmful parasites breeding on citrus (5), with the following conclusions:

- Treatments against citrus pests are conditioned by the presence of white fly and, in order to maintain an orchard in good health condition, we have to bear in mind this factor before making a decision on the need or convenience of a chemical application. A good assessment of the presence of harmful insects and their population levels is an important requirement as well.

- For controlling the pests we have to develop a rational management by combining the action of beneficial insects and chemical compounds with low toxicity effects on beneficial fauna, since biological control alone does not achieve adequate control on the most part of harmful insects outbreaks.

- Assessment of evolutionary stage of insect pests is a factor to be taken into consideration when deciding a treatment, since if a pesticide is to be used, its effectiveness will depend on the time of application.

From the above points we are suggesting the implementation of a selective management with the following requirements:

- Monitoring the existence of harmful insects in the orchards and assessing population levels to avoid unnecessary pesticide routine applications.

- Knowledge of the biological cycles of harmful insects, for establishing better application timings.

- Intensifying populations of beneficial insects, irrespective of population levels of the existing pest insects.

- If pesticide applications are required, those compounds less harmful to beneficial fauna should be chosen.

- Cultural practices tending to protect the presence of beneficials and to create an unfavourable environment to harmful insects should be adopted.

1.4. Pesticides and some beneficial insects

At present, there is already an important number of data concerning the effects of pesticides on beneficial fauna and, in fact, there is a work group in the O.I.L.B.: "Pesticides and Beneficial Arthropods" conducting research on this topic. We are listing here the pesticides tested in our Institute, on nymphal stages of three beneficial insects that have produced less than 60% mortality, proving to be adequate for use in programmes of selective control:

a) On C. noacki, parasite of A. floccosus (9).

INSECTICIDES: pirimicarb; butocarboxim; ethiofencarb; ethion; endosulfan; azinphos methyl; diazinon; diamethoate; oxydemeton methyl; methamidophos; dioxacarb and fenthion. ACARICIDES: cycloprate; chlorobenzilate; dicofol+tetradifon and carbophenothion. FUNGICIDES: captafol; thiram and potassium permanganate. INSECT GROWTH REGULATORS: buprofezin and fenoxycarb.

b) On L. testaceipes, parasite of aphids.

INSECTICIDES: flucythrinate; decamethrin; fenvalerate; ethiofencarb; butocarboxim; pirimicarb; carbaryl; cypermethrin; fenitrothion; oxydemeton methyl; fluvalinate; phosalone; endosulfan; phosmet; methamidophos; fenpropathrin; azinphos methyl and triazophos. ACARICIDES: cyhexatin; chlorobenzilate; dicofol+tetradifon; fenbutatin oxide; propargite and bromopropylate.

c) On nymphae of R. cardinalis, predator of I. purchasi.

INSECTICIDES: summer oil; dimethoate; fenthion; pirimicarb; ethiofencarb; etrimfos; endosulfan; tetrachlorvinphos; chlorpyrifos; ethion; fenitrothion; phenthoate; oxydemeton methyl and butocarboxim. ACARICIDES: cycloprate; dicofol+tetradifon; chlorobenzilate and carbophenothion+dicofol.

An additional explanation to the above list is that, sometimes, the coadjuvants are responsible for mortality of beneficial insects to a higher extent than active matters.

While examining the list of compounds proving less harmful to C. noacki and R. cardinalis, it was noted that Pyrethroids were not in it. Furthermore, insect growth regulator can be used to suppress insects adequately in a programme of selective control.

1.5. How should the pest control be in the future?

In this paper, only selective chemical control has been discussed, not integrated control; however, according to our criteria we are coming close to it; and if we consider the economic thresholds, we are actually doing integrated control. But although these thresholds are necessary, in our conditions they only can be taken as some guidance and cannot be applied at nation-wide level owing to the great variabilities existing for a determined phytophagous if considerable differences in abiotic conditions or cultural practices, fertilization, irrigation, pruning etc. are present. Nevertheles, since it is necessary to determine some cull level percentages from losses caused by key insect pests on citrus in 1984; the Plant Protection Agencies from Andalucia, Valencia (I.V.I.A.) and the so called Groups for Integrated Pest Management for Citrus are in the process of outlining directions on citrus economic thresholds in Spain. In Sevilla one Group for Integrated Pest Management for Citrus has been working experimentally in 1984 on 1500 ha of citrus and the results are not yet available. During the next year similiar works will be carried out in the provinces of Castellón and Valencia.

From the above, our pest management being rather bordeline integrated programmes are directed towards full Integrated control which, according to our experience, is feasible. At present our Government Agencies are applying selective control methods, this being understood that the compounds used against phytophagous be applied timely and without adverse effects to beneficials, following experience gained in our country and elsewhere.

REFERENCE

1. CARRERO, J.M. et al. (1977). Note biologique sur quelques insectes entomophages vivant sur olivier et sur agrumes en Espagne. Fruit, (32) 9: 548-551.
2. DOMINGUEZ G. TEJERO, F. (1972). Plagas y enfermedades de las plantas cultivadas, 4ª Edición. Editorial Dossat, S.A. madrid, 955 pp.
3. GARCIA-MARI, F. and DEL RIVERO, J.M. (1981). El ácaro rojo Panonychus citri (Mc. Gregor), nueva plaga de los cítricos en España. Bol. Serv. Plagas. 7: 65-77.
4. GARCIA-MARI, F. et al. (1983). El ácaro rojo Panonychus citri (Mc. Gregor): Incidencia en la problemática fitosanitarias de nuestros agrios. Bol. Serv. Plagas, 9: 191-218.
5. GARRIDO, A. (1978). La mosca blanca de los agrios obliga a actuar contra las plagas de los mismos de forma diferente a la tradicional. Levante Agricola, 202: 35-38.
6. GARRIDO, A. (1983). Moscas blancas de los cítricos en España (Bemisia citrícola Gom. Men. y Aleurothrixus floccosus Mask.) Levante Agrícola, 245: 27-34.
7. GARRIDO, A. (1983). Enemigos naturales de la mosca blanca de los cítricos (Aleurothrixus floccosus Mask) y métodos de control. Levante Agrícola, 246: 77-87.
8. GARRIDO, A. et al. (1976). Repartición y estudio poblacional de "Aleurothrixus floccosus" Mask a nivel de árbol y equilibrio con su parásito el "Cales noacki" How. An. INIA Ser. Prot. Veg. Nº 6: 89-121.
9. GARRIDO, A. et al. (1982). Incidencia de algunos plaguicidas sobre estados inmaduros de Cales noacki How., parásito de Aleurothrixus floccosus Mask. An. INIA/Ser. Agric/N. 18: 73-96.
10. GARRIDO, A. et al. (1984). Bioecología y control de Panonychus citri (Mc. Gregor) (ACARINA: Tetranychidae). Levante Agricola nºs 249-250: 26-41.
11. GARRIDO, A. et al. (1984). Evaluación de imagos de Prays citri Mill (Lep. Hyponomeutidae) con una feromona de síntesis y su correspondencia con daño. An. INIA/Ser. Agric/N 25: 147-154.
12. GERSON, U. (1977). La caspilla Parlatoria pergandei Comstock y sus enemigos naturales en Israel. Bol. Ser. Plagas, 3: 21-53.

13. LIMON, F. et al. (1976). Estudio de la distribución, a nivel de ataque, parásitos y predatores de las cochinillas Lecaninas (Saissetia oleae Bern. y Ceroplastes sinensis del Guercio) en los cítricos de la provincia de Castellón. Bol. Serv. Plagas, 2: 263-276.

14. LLORENS CLIMENT, J.M. (1984). Las Cochinillas de los Agrios. Consellería de Agricultura, Pesca y Alimentación, Valencia. 159 pp.

15. MELIA, A. (1982). Prospección de pulgones (Homoptera, Aphididea) sobre cítricos en España. Bol. Serv. Plagas, 8: 159-168.

16. MINISTERIO DE AGRICULTURA (1971). La mosca blanca de los cítricos. Dirección General de Agricultura, Servicio de Plagas del Campo, Madrid. 31 pp.

17. MINISTERIO DE AGRICULTURA (1973). Mosca blanca de los cítricos, Servicio de Defensa contra Plagas e Inspección Fitopatológica (informe), octubre 71 pp.

18. MINISTERIO DE AGRICULTURA (1975). Lucha biológica contra la mosca blanca mediante Cales noacki, Madrid, 30 pp.

19. MOUND, L.A. and HALSEY, S.A. (1978). Whitefly of the World. British Museum (Natural History) Publication Nº 787: 344 pp.

20. SERVICIO DE PROTECCION DE LOS VEGETALES, ESTACION DE AVISOS AGRICOLAS. (1983). Principales acaros rojos de los cítricos. Hoja informativa Nº 1. Consellería de Agricultura, Pesca y Alimentación. Valencia 4 pp.

21. TALHOUK, A.S. (1977). Las plagas de los cítricos en todo el mundo, tomado de Los citricos Ciba-Geigy Agroquímicos. Monografía Técnica de Ciba-Geigy Ltd. Basilea, Suiza, 21-27.

Approaches to integrated control of some citrus pests in the Azores and Algarve (Portugal)

V.Garcia

University of The Azores, Ponta Delgada, Portugal

Summary

Trichogramma mass-reared in the Azores facility are to be used in Algarve against Prays citrii. This pest is not a serious one in the Azores, but its existence in island conditions provides excellent field for research. The coccinellids mass-reared in the university facilities include a recently introduced exotic species, Harmonia axyridis, in order to control citrus aphids Toxoptera aurantii, Aphis gossypii and Aphis citricola. The med-fly, Ceratitis capitata has a great economic importance in the Azores and in Algarve. A pilot program is on the way in the Azores, where we expect to use both natural enemies (e. g. Opius concolor) and S.I.T. methods. In Madeira, Aleurothrixus floccosus is present, but it is absent from the Azores. Introduction of Cales noacki and probably Eretmocerus is expected to be of great help. The Algarve experience, where chemical control and introduction of Cales noacki has shown insufficient control, is expected to receive good improvement through the Madeira program.

1.1 Introduction

During the 19th. century, exported oranges were the bulk of the economy in the Azorean island of Sao Miguel.It's importance became minor since some pests attacked the trees and the fruits (NEWMAN 1869).Among these pests we can quote Icerya purchasi, the cottony-cushion scale,the fruit-fly Ceratitis capitata, and Coccus hesperidum. Icerya has been under control since 1915, after the introduction of the vedalia beetle,Rodolia cardinalis (CARNEIRO,1979).Ceratitis is a major source of concern,today as it was in the past.The following sentence, written in 1829 by McLEAY (in NEWMAN,op. cit.) is very significant: "Now, the decay of St.Mechael oranges,is almost universally accompanied by the presence of the larva of a small fly,which I shall show to be the cause of the evil". Almost every citrus grove in the Azores suffers from aphid outbreaks in spring and sometimes summer. This has also lead to an integrated control program, with emphasis on the introduction and releasing of mass-reared exotic species of Coccinellidae. The most commonly used pesticides for citrus pests in the Azores are trichlorfan, fozalone and pirimicarb. In Algarve, against the white-fly,butocarboxim is being recommended.

1.2 Prays citrii

Prays citrii (Lepidoptera,Yponomeutidae), the citrus moth, is not very common in the Azores (CARNEIRO 1979). In Algarve, it attacks mainly lemon trees. Considering that citrus groves represent a production of 60.000 tons and cover 6.000 hectares, this pest is one of the targets in Southern Portugal. The Azores, which have a Trichogramma mass-rearing unit with a maxi

num capacity of 1 million per day (GARCIA 1982) are supporting the development of a small mass-rearing laboratory in Faro, at the recently settled
University of Algarve. Meanwhile,we expect to rear the parasitoids in the
Azores, experiment the species under local conditions and, if the trials
are satisfactory, export them to Faro. Trichogramma to be selected against
P. citrii belong to the embryophagum group, which is known to be well adapted to trees. The releasing of Trichogramma will be synchronized with the
application of pesticides.

1.3 Citrus aphids

The most abundant species are Toxoptera aurantii,Aphis gossypii and
Aphis citricola (CARNEIRO,op.cit. ; CARVALHO 1984). Lately, the advised
pesticide has been pirimicarb, a product which we studied for a long time
on its sub-lethal effects (GARCIA 1976,79 e 80). Following the applications
of pirimicarb, mass-reared Coccinelidae of the species Harmonia axyridis
were introduced in the islands of Santa Maria and Sao Miguel. Well adapted
to predation on orchards,with a neat preference for trees (IPERTI, pers-
-comm.) this species, together with Adalia bipunctata,seems to be a good
solution for biocontrol of aphids in our citrus groves. A.bipunctata was
already studied by our team, using adults reared in the Azores, originating
from orchards of the Lisbon zone (GARCIA & FURTADO 1980).

1.4 The fruit-fly (C.capitata)

The mediterranean fruit-fly Ceratitis capitata,is widely spread in the
Azores,where it attacks mainly orange trees (Citrus sinensis) and peach
trees. It has a great economic importance (CARNEIRO,op.cit.). In the year
1965, a study of population dynamics was initiated and it went on into 1966,
based on feromone trapping.About 800.000 males of C.capitata were collected
A network of 126 traps,were put around the island of Terceira(402 Km²).Be-
sides and still at Terceira, two zones were covered,one with 56 traps (Ter
ra Chã) and another with 40 traps (Biscoitos). The results were never published (CARVÃO pers. comm.) but we intend to do so,on a broader basis,
when more and recent data are available. A new program is now on the way,
this time at Santa Maria (97 Km²). The same scheme used in Terceira to eva
luate population densities and distribution will be applied.Emphasis will
be put in bioecology. As for control,two main vectors are being followed:
a sterilized insect technique method and the releasing of mass-reared Opius
concolor. The first (S.I.T. Method) is now finishing its studies on sterilization by means of rapid neutrons,with the objective of improving quality
of the insects produced. This project is the subject of a Ph.D.thesis,to be
printed next year (GUERREIRO,pers.comm.).The biological control through
Opius is still on its first steps,for it requires the settling of a facili
ty in the Azores, which is intended to be done during 1986. The application
of pesticides will require trichlorfan,by far the most used in the islands.
In the Terceira operation (1965 and 1966) malathion was used.

1.5 The White-fly

The citrus white-fly,Aleurothrixus floccosus is a main pest of orchards
in Algarve and Madeira.It is absent from the Azores. In Madeira,A.floccosus
was first reported in 1920 (MOUND & ALSEY 1978). First reports to the Ca-
naries are dated from 1937. Aleurothrixus floccosus was first reported in
Algarve in 1977. Since then, several biocontrol operations,with the intro
duction of the aphelinid Cales noacki were carried out. Fact is that Cales
supports poorly the high temperatures of South Portugal's summers. A new
project is being studied, under our supervision,to introduce a more ther-

mal-resistant natural enemy, to be selected among the species of Amitus
and Eretmocerus. This includes the build-up of mass-rearing facilities,so
that reinforcements of the natural enemies' populations can be done when
necessary. The pesticide selected to this projects,both in Madeira and
Algarve, could easily be butacarboxim.It seems to be fairly compatible with
Cales noacki and further essays on Amitus and Eretmocerus would be of great
interest. Inclusion of these auxiliars in the testing program of the IOBC
Working Group "Pesticides and Beneficial Organisms" seems very promising.

Conclusions

1- It seems clear that insistence on four major integrated control programs
(Prays citrii, citrus aphids,Ceratitis capitata and the white-fly) is a way
to concentrate efforts and carry on a good research.
2- The use of facilities already existing for other projects (such as the
Trichogramma biofactory at the University of the Azores) gives a multi-pur
pose face to the projects, which get cheaper.
3- The islands (both the Azores and Madeira) seem a good place for field
trials on integrated control research.
4-The existence of a more or less well balanced nature and a scientific
background with an adequate tradition of biological control research is
very favourable.
5- The 4 above mentioned projects can lead to an integrated strategy of
control, which can be adapted from the islands to continental conditions.

References

CARNEIRO M. (1979). Pragas das culturas na ilha de S.Miguel,S.R.A.P.
                    Ponta Delgada (Açores)
CARVALHO P. (1984). Notas acerca de pragas das culturas dos Açores
                    Universidade dos Açores, Ponta Delgada,45 pp.
GARCIA V. (1976).   Influence de trois produits phytosanitaires sur les per
                    formances biologiques d'une Coccinelle aphidiphage.
                    Thèse de 3 ème cycle. Université de Provence,114 pp.
GARCIA V. (1979).   Efeitos de um aficida sobre as potencialidades bioló-
                    gicas de dois predadores afidífagos.Tese de Doutoramen
                    to. Instituto Universitário dos Açores, 209 pp.
GARCIA V. (1980).   Novas metodologias para avaliação dos efeitos de pesti
                    cidas específicos sobre predadores entomófagos.
                    I Congresso de Fitiatria.Instituto Superior de Agrono-
                    mia, Lisboa, p187-201.
GARCIA V. (1982).   Les Trichogrammes. Ier. Symposium International.Les col
                    loques de l'INRA, nº9. Antibes.
GARCIA V. e FURTADO M. (1980). Potencialidades biológicas dos coccinelídeos
                    afidífagos utilizados em luta integrada. "Arquipélago",
                    nº1 - Série Ciências. Universidade dos Açores,Ponta Del
                    gada, p 143-183.
MOUND,L.& ALSEY S. (1978). White-fly of the world.British Museum of Natu-
                    ral History.

Economic aspects of pest management

S.C.Misseri
Institute of Agricultural Estimate, University of Catania, Italy

Summary

With the aim of supplying a more precise scientific framework for the
economic issues involved in the control of phytophagous pests
affecting plants, animals and crops useful to man, the author has made
particular use of works published in Italian to re-examine the
thinking behind the economics of control programmes, their cost/
benefit parameters and the analytical instruments used in determining
their suitability. Attention is also paid to the operative limits of
economic models for pest control in the Sicilian citrus industry.

1. INTRODUCTION

There is, and has been, insufficient time to provide an experimental
basis for this paper on the economic aspects of the control of phytophagous
pests. In view of this quite significant limitation, our contribution is
no more and no less than a critical review of the logic, parameters and
analytical instruments used by economists, while at the same time pointing
out some differences in emphasis between the various authors writing on
this subject. Similarly, attention is drawn to some 'sins of omission' of
major economic significance when it comes to the cost of the damage
inflicted, the value of the produce saved and the cost of the measures,
together with their inter-relationships in determining the limits of
suitability – this being very difficult to establish in the case of phytophagous pest control by virtue of the wide range of variables to be taken
into account. The urgent need to reduce the number of variables in order
to develop an economic model highlights the problem of ensuring operational
validity when such a model is applied, and of assessing the significance of
the results obtained. This is of particular interest for the citrus
industry, where control schemes aim at controlling a number of pests, yet
applying specific measures against each individual pest. The situation
becomes even more complex where several methods are applied at once, as in
the case of integrated control, if this is taken to mean a multi-faceted
approach to pest control, involving the combined use of all the methods
available (biological, genetic, agronomic, chemical, etc.).

563

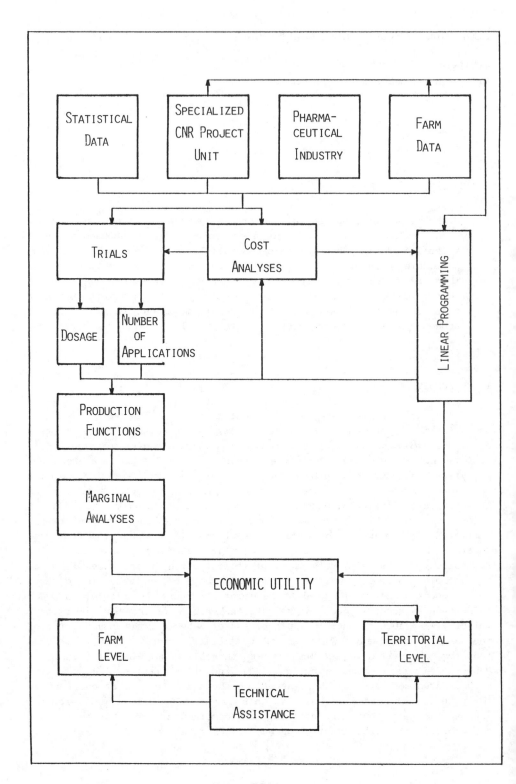

2. APPROACH, PARAMETERS, ANALYTICAL AIDS

2.1 The thinking behind the economics of phytophagous pest control.

   The argument is based on only three points.

   The first point concerns the rationality of a research programme coor-
dinating within a single vertical and horizontal sequence, the various
stages in the process in accordance with objectives that essentially amount
to an evaluation of the economic utility. Although a number of possible
procedures exist, we would like to stress the desirability of using one
common mode. Although there is a disadvantage in that it leaves less scope
for the inventiveness of research economists, this is more than offset by
the advantages of a common terminology, especially when results from different areas are being compared. A unit of the top administrative body for
scientific research in Italy - the Consiglio Nazionale delle Richerche - is
now engaged, as part of a project studying 'phytopharmaceuticals and plant
growth regulators', in a secondary study on 'prevention and cure of plant
disease in wheat, maize and fodder sorghum", intended to establish the
utility limits of pest control for a given level of technology. The plan
of action adopted is shown in the diagram on the following page in the form
of a 'flow chart for the research programme'. While of course not wishing
to urge its adoption, we would like to highlight the need/utility pointed
out by this group of researchers (1) to make use of a clearly structured
work programme. We consider it a good idea and worthy of consideration by
the body organizing this meeting, namely the Directorate-General for
Agriculture of the Commission of the European Communities.

   The second point concerns the biological range studied by economists,
which is generally restricted to the two parameters - crop and pest. This
explicit linking of the plant and the pest means that only one link in the
food chain is represented. We consider this approach inadequate, since it
excludes from scientific observation a component that we consider to be
relevant. Natural populations form a food sequence whereby a is eaten by
b, b is eaten by c, c is eaten by d, and so on. If we look at two
consecutive links and the three parameters involved in them, with the
phytophagous pest at the centre, we can focus on the food relationship
plant → phytophagous insect → parasite. In mature ecosystems, regulatory
mechanisms should ensure equilibrium and stability, particularly in terms
of population dynamics; in essence the parasite controls the phytophagous
insect. This brings us to a natural prerequisite for biological control,
but we cannot yet draw conclusions. Indeed, with due apologies for any
inaccuracies in the approach, it would seem reasonable to state that in
quantitative terms the food relationship should be plant   phytophagous
insect   parasite under conditions of equilibrium and control: reversing
this balance would lead to extinction. That being the case, the transition
from an ecosystem to the currently prevailing agrosystem is marked by a
number of deviations from a natural evolution, the specific characteristics
of which, to the extent that they are of interest here, can be summed up as
the spread of monocultures over vast areas. A degree of specificity
becomes essential in the trophic relationship plant → phytophagous insect→

565

parasite, while at the same time there is a particular divergence in the relationships between the insects and the parasites, with the former becoming quantitatively predominent to the extent that natural control is disrupted and pest control becomes essential to safeguard production. This phenomenon seems to become more acute as soon as pest control techniques, especially chemical ones, are employed. It has been noted that: 'insect-icides usually exert tremendous pressure on food chains because the further you go up the chain the greater the demand for food, and the species at the top of the chain, dependent as they are on the continued existence of preceding links, are the first to suffer from any increase in selective pressure on the community' (2). The whole topic merits more extensive research to provide a better scientific framework for biological control, in order to collect information which can then be passed on to economists for use in their calculations of the break-even point between costs and benefits that maintains an equilibrium between insects and parasites. The essential question is whether it is advisable to take pest control beyond a particular threshold at which useful parasites begin to be destroyed. Under such conditions it is nonsense to talk of biological control. Would it not be better to lower the threshold, with consequent cost-savings, and leave the residual control to the useful parasites? Under such conditions one can reasonably speak of biological control and, moreover, it clarifies the concept of an 'intervention threshold', which we would define as the pre-sence of pest populations requiring the employment of specific control techniques. This concept must be expressed in quantitative terms, and here we feel there is inadequate operational information. We would again stress the need for a more thorough review of the subject, with a view not only to making advances in agricultural entomology and plant pathology, but also to establishing the limitations and potential of biological control when combined with chemical control and to widening the economist's field of observation. In this context the only information available to us relates to attempts to provide a mathematical framework: Thompson's formulae (3) for measuring coexistence relationships from generation to generation between hosts and parasites in order to give a numerical assessment of the number of hosts with parasites in a given generation; MacArthur's statement (4) that the stability of a community is directly proportional to the logarithm of the number of links in its food chain; the Volterra/Lotka equations; the methods employed by Cavalloro and Di Cola (5). These few works are insufficient basis for any further discussion.

The third and final point is inextricably linked with economic logic. In our view the literature states unequivocally that economists are dependent on marginalistic models: the limit of utility is fixed at the point of equality between the ultimate 'quantity' of crops saved and the ultimate 'quantity' of the costs needed to achieve this. We will return shortly to the difficulty of analytical calculations of equality, but it is already evident that the economist's approach is an individual one and at the 'micro', in other words farm, level. The principle is correct provided budgetary considerations are taken into account. Can we be satisfied with this approach and achieve optimum conditions at a national level simply by totalling the equilibria on the individual farms? We intend to provide some justification for the question mark in looking at the following two options - (a) achieving larger crops by more widespread pest control, (b) a lower level of control of pest populations. These two objectives, quoted by way of example, go beyond individual and therefore micro-economic interests and indicate that there may be a limit to social utility. We would have to admit that the plausible consequences involve the location, to the right of the outer limit of economic efficiency, of a point marking the limit of technical efficiency - perhaps where 'toxicity' begins to damage the productive base of the industry. Within this range, there are quantities which are potentially producible but are not in fact produced because the costs would outweigh their value. They can only be harvested if the relative price system is modified. Marginal utility would be re-established either by an increase in product prices or by a decrease in the price of pesticide. Where the approach chosen is the control of pests in order to increase production by controlling their population, we need do no more than take 'political' decisions on the costs of the pesticides. Other alternatives, such as government subsidies for part or even all of these costs, are just variations on this theme. Another course would be to achieve a better level of technical efficiency in pest control. This would require investment in research, whether this be in biology or in work to find active ingredients. Such investment is more socially acceptable and would yield economic benefits when the results were put into practice.

2.2 The economic parameters of pest control.

Economic arguments hinge entirely on a correlation between price and cost or, in other words, between the value of the finished good after deducting the increased production cost, whereby the limits are to be found when these two marginal parameters are equal - something which will be discussed later on the basis of empirical values.

It is commonly stated in the literature that "worldwide, damage by pests is estimated to be about 35%". What interpretation should we put on this percentage? The question is approached from both a qualitative and a quantitative point of view. Qualitatively, damage would seem to be correlated with the quantity of a crop produced. If this is so, it would appear to be incorrect, since there is damage related to the crop itself, as well as non-plant-related forms of damage affecting agricultural production, such as damage to livestock. One example is the recent case of British cattle suffering from acute infestation with Hypoderma bovis and Hypoderma lineatum, which forced the government to spend almost £14,000,000 in control measures. Having taken a qualitative look at the situation, we will proceed to the quantitative aspect.

This 35% figure can in fact be interpreted in the following two ways:

- as the difference (in percentage terms) between the potential crop if pest control had been used and the crop harvested without such control;
- as the difference (in percentage terms) between the maximum potential crop and the actual harvest, with pest control being employed in both cases.

567

All the evidence points to the second interpretation being the more correct one, and this is backed up by indications in the literature, although there is insufficient information about the methodology of the calculations used to arrive at this high estimate. In view of the "political" significance (i.e. concerning political decisions) that the order of magnitude of the damage may have, this is a topic worth studying. Here three discursive corollaries can be derived and we will study them briefly.

If one accepts the truth of the above statement that, world wide, there is a 35% rate of damage, this could be defined as the efficiency limit for pest control with the current status of technology and its diffusion. Is there any way that this total loss for mankind can be obviated? It is possible to give an affirmative answer based both on improvements in pest control techniques and their wider dissemination. Indeed, geographically, damage shows an uneven distribution; its impact on possible production increases as pesticides are less widely used. With the aim of establishing a correlation, percentage damage has been compared to pesticide use (kg/ha) over large areas. The maximum damage occurs in Africa (41.6%) where pesticide consumption was 1.04 kg/ha, while the lowest (25%) was in western Europe, where pesticide consumption was 11.28 kg/ha. In Latin America and North America, values recorded were 28.7% - 2.35 kg/ha and 33% - 7.24 kg/ha respectively. Despite this disparity between the two Americas, it seems reasonable to state that a large range exists in potential technical productivity of pest control in the less advanced countries where diffusion is the on-going process, whereas by contrast it is improvements in the technology that one finds in the other countries. The third and final discursive corollary concerns the current productivity of pest control. In the absence of convincing data, the argument will be continued using symbols. Let P be the world production using current pest control levels of technology and dissemination (P is an approximate but real figure) and P_1 be the world production that could be obtained without any pest control whatever (P_1 is a hypothetical approximation). $P - P_1 = Y$, representing the productivity of pest control under existing conditions of technology and dissemination. If K represents the cost, correlations between Y and K can be made. In this way, one reaches the frontiers of abstraction at world level, and the subject will be taken up later on at a micro level.

Before we leave this subject, two comments can be made about Y. Whether at the macro or micro level, it represents the crop saved from damage and, consequently, could be described as the increment in production due to the employment of pest control: it is in this sense that it will be used throughout the rest of this paper. Secondly, and still talking of Y, is this restricted to an increase in production or are there other major economic benefits? An affirmative answer would have economic implications on a scale impossible to ignore.

Let us turn now to an analysis of costs. By their very nature, references in the literature are monetarist and almost always at the micro level (6). They all concern uses of equipment and working methods which, multiplied by their respective costs, yield the total cost to the farm and apportion an excessive share to agriculture. On this point, we wish to

568

highlight a number of hidden aspects which, although probably known, have not been systematically studied. Full accounting of the costs of pest control would comprise at least the following elements:

"Upstream" of agriculture:
- cost of scientific research (government and private) required to develop and test the active ingredient;
- cost of scientific research (government and private) required to develop and test the carrier (equipment for applying the pesticide);
- the cost of environmental pollution caused by discharges during the pesticide production cycle;
- the cost of manufacturing the pesticide;
- the cost of manufacturing the equipment;
- advertising costs;
- distribution costs.

"Downstream" of agriculture:
- conservation costs (protection from pests) while the product is on route from the producer to the consumer;
- costs of dealing with the effects of residues in man and the environment.

General costs:
- the administrative and bureaucratic costs incurred by the authorities in running pest control programmes.

To stay within the terms of reference for this discussion, attention must be drawn to the first and last items in this cost review, since they are fully borne by agriculture and, in consequence, by the farmer. In fact the purchase price for the pesticide and equipment includes provision by the companies supplying them for the whole of their costs, whereby the term "manufacturing" also covers interest and profit. The farmer's bookkeeping comprises the following entries: the purchase price of the pesticide is entered under "expenditure", while under the heading of depreciation there is the cost of the equipment and of modifications to the way the business is run. Then there are the specific labour costs of applying the pesticide. Agriculture also has to bear at least a part of the cost of research and publicity, with this occurring through the taxes levied in Italy on cadastral and farming income. It may be that this situation is a fair and just one and that it would be a true representation of the actual situation, provided that pest control did indeed yield benefits only in the form of increased production. If, however, there are other "social" benefits (control of pest populations, etc.) would it not be reasonable to ask "society" to contribute to the cost? Furthermore, there is the question of pollution, the costs of which tend to be charged to agriculture if the companies manufacturing pesticides are required to install purification plants. Looking specifically at pollution caused by residues in the human environment, we will have to leave the question unanswered, since insufficient work has been done on the health consequences. Nevertheless, this definitely impinges on the social sphere and incurs its own social costs. These points are worth further study.

2.3 Analytical economic instruments.

Having established that the aim is to identify the utility limit for pest control and disregarding all the additional or complementary observations hitherto made, so that we get back to the essential economics, one can state that the marginal approach is the only analytical instrument available in economics. In the specialized literature other terms are used: computers, linear planning, programmed budget, regressions; but in our view these are merely ways of calculating and drawing up analyses that are more representative of the results and not real analytical instruments. Having said that, let us come back to marginal theory, which presupposes the following:

- ceteris paribus, i.e. of all the means of production, only one is subject to variation - precisely the one whose utility limit you wish to determine;
- proportional divisibility of the pesticide under study;
- proportional divisibility of the product;
- stable prices for the product and pesticide.

Provided the above conditions are met, the following series can be obtained:

Q_1; Q_2; Q_3; Q_4; Q_5; Q_6 .... representing the quantities produced in correlation with

F_1; F_2; F_3; F_4; F_5; F_6 .... representing the quantities employed.

The utility limit is found at:

$$\frac{(Q_n - Q_{n-1})\, p_Q}{(F_n - F_{n-1})\, p_F} = 1$$

At a micro level, the above analytical instrument is beyond reproach in terms of its logical coherence, but it has the drawback of featuring many abstractions which hinder its operational use, as has been noted (7). Nevertheless, its use in agriculture has not been, and cannot be, excluded (8). In the same form as the above presentation, the marginal analytic approach has been used in studies conducted by the Osservatoria di Economia Agraria di Pisa in the now distant year of 1956 (9) and more recently, with a better arsenal of mathematical techniques, by the Unità Operativa del Consiglio Nazionale delle Richerche, whom we mentioned at the beginning (1). For the sake of completeness, we should say here that the earlier works cited concern the economics of combating a range of pests in fruit-bearing plants (olive, pear, melon) and herbaceous plants (beetroot, maize), while the second group are instead studies of weed control in maize. As an experimental check on the analytical instrument, the results can be considered satisfactory granted the assumed limitations, including that the pathogen is isolated (being the sole cause of the damage) and that there is proportional divisibility of both the product and of the pesticide, which under the real conditions of pest control takes the form of "treatment". The cost of this varies because it is made up of fixed and variable overheads; on a unit basis, the former will diminish and the latter increase as more treatments are applied. Things become considerably more difficult where damage during the production cycle is not due to a single cause (more than one pathogen) and where "treatments" are not uniform - a very common situation in the citrus industry.

3. ECONOMICS OF PEST CONTROL IN THE CITRUS INDUSTRY

In the case of citrus growing, an impressive quantity of information is available on the typology and biology of phytophagous insects and on the technical methods for dealing with them. Economic data is extremely rare, and this lack of interest would seem to be due to the fact that pest control accounts for only a small percentage of production costs. This is an inadequate reason. Let us look at some information and data just in the field of pesticides, on the situation in Italy and in the citrus industry worldwide. The Italian citrus industry is currently said to absorb 51% of the insecticdes used, 18.4% of the herbicides, 12.7% of the acaricides, 8.9% of the fungicides, 2.2% of the nematocides and 0.4% of other pesticides. This is a total expenditure of some 15,488 million lire, representing approximately 1.3% of the gross saleable production and approximately 92,000 lire per hectare under cultivation. These data are very relative in view of our ignorance of their influence on the value of the generated increment and bearing in mind that they represent not the entire cost of pest control, but merely that part associated with pesticides. Excellent studies have been made of the citrus industry at both the industry level and at the level of individual growers. The latter studies do provide data on the cost of control, but lump this in with the entire cost of production; they give no analytical information nor any about effects on production. Indeed, the thrust of the studies referred to here is not to determine the economic conditions under which pest control is carried out and pesticides employed; this requires ad hoc investigations which, in view of the limitations set - namely a return to pure economic data without looking at logic-based observations, parameters and analytical instruments - bring us back to the marginal approach with the objective still being the definition of the utility limit. As we see it, the topic can conveniently be split up into the following two situations: where the pest affects production and indirectly, but less significantly, the plant, and the opposite case in which the plant is affected and indirectly, but less severely, production. We will look at the first case, which is better suited to the use of the marginal approach, with attention being concentrated first on the experimental basis. We do not believe it advisable to adopt the "individual plot" as the standard; the mobility of pests has a disruptive influence, even if only a relative one, and it is better to choose the farm scale, especially since this also allows observation of all

571

the inter-relationships between farm structures. Other characteristic features are the homogeneity of the farms and the distance between reference farms with and without pest control. This proviso should make the ceteris paribus aspect acceptable. The proportional divisibility of the production increment is not worrying at a quantitative level, but one should be aware of the fact that pest control aims not so much at preserving quantity as at preserving quality, and it may be that in the Sicilian citrus industry, where the product is consumed as fresh fruit, this is economically more significant - which then raises the additional complication of the price difference between healthy and damaged fruit. The proportional divisibility of the pesticide is the thorniest aspect of all, since it does not occur in the form postulated by the theory of this analytical approach, whereby it should be uniform and applied in successive doses. In actual fact, dose should be understood to mean "treatment" (the cost of treatment) which is very often polymorphous and, in very many cases, there is a differentiation between one "dose" and another during the agricultural year. There is a similar source of a causal plurality with a corresponding plurality of targets: essentially, during a production cycle in the citrus industry, we have to be prepared to take action against almost 70 pests, of which at least 20 can cause significant damage (10).

The variety of these pests, which moreover appear at different times of the year, is reflected in the corresponding range of specific pesticides, with the subsequent complication that the "treatment" doses are not uniform "quantities", in terms of either their cost or their possible effects. Concentrating - for scientific reasons - on a correlation between one single pest and the specific treatment against it can be done, but does not seem a reasonable alternative, if only because of the much-reduced significance of the results obtained. For example, setting up a research project to study only mites and their specific treatment, on the assumption that all other pests and their respective pesticides remain the same, is a rather pointless abstraction. It is better to accept the entire "package" with experiments using a planned and rational treatment programme (11), as required by the actual pest situation, in order to determine the best technical conditions for application (the degree to which the active agents can be mixed to obtain multi-purpose treatment, etc.). The rest is nothing more than an experimental analysis and elaboration of technico-economic data. Up to now we have discussed only the pathogen/treatment (chemical) correlation. If we consider other components (biological control, guided control, integrated control) other conditions come into the theory. In each case, the response is at present more technical than economic: it is up to the entomologist and pathologist to determine the break-even points for the doses and other factors involved in the efficiency of the various pest control methods.

In conclusion, we would make a comment that is more political than techno-economic. A great deal of energy should be put into the biological control sector, which would then be entirely government run, using widespread "artificial" breeding of useful parasites; in the long term, this should lead to phytophagous organisms being less dense and less widespread, so that increased production will be possible with lower pest control costs. Associated with this conclusion is a more 'philosophical' one stressing the urgent need for man to co-exist with phytophagous organisms and to change fundamentally his approach to pest control. Instead of being an attack on pests, it should develop into the defense of crops and, more generally, of the environment (12).

All that now remains is to examine the reverse situation, mentioned above, where pests attack plants and, indirectly but less seriously, damage the crop. Although this still concerns the management of these myriad phytophagous pests, little credence can be given to the use here of the analytical instruments so far employed. The trees on a citrus farm - the plantations - are an investment in real estate; protecting them from pests is the same as maintaining their technical and productive efficiency, and the corresponding costs would probably be better reclassified under the economic heading of depreciation of assets. As a maxim, this may be wrong and open to criticism, but no discussion can disregard it.

In the absence of an experimental basis - as we mentioned at the beginning - it has not been able to make a dramatic contribution, but in providing a resumé of knowledge, the definitions we have dwelt on here may result in an improved conceptual framework, providing a more suitable approach to the economics of pest control.

REFERENCES

1. POLELLI, M. and SEGALE, A.(1980). La convenienza economica nell'impiego dei fitofarmaci in agricoltura. Transactions of the Società Agraria di Lombardia. Issue No. 2.
   POLELLI, M. and SEGALE, A. (1981). Metodologia per la valutazione delle perdite causate da patogeni e ottimizzazione dell'uso dei fitofarmaci. Transactions of the Convegno del Consiglio Nazionale delle Recherche.
   SEGALE, A.(1981). Rapporti ottimali tra costi dei trattementi e livelli produttivi. Transactions of the Convegno del Consiglio Nazionale delle Richerche.
2. CELLI, G. (1980). I limiti e i pericoli dell'impiego degli insetticidi in agricoltura. Consiglio Nazionale delle Ricerche.
3. THOMPSON, W.R. (1939). Biological control and the theories of the interactions of populations. Parasitology. 31; 299-388.
   THOMPSON, W.R. (1956). The fundamental theory of natural and biological control. Ann.Rev.Ent. 1; 379-402.
4. Taken from (2).
5. CAVALLORO, R. and DI COLA, G. (1980). Alcuni metodi matematici nel controllo di una popolazione di Ceratitis capitata Wied. Standardization of biotechnical methods of integrated pest control in citrus orchards. Commission of the European Communities.
6. PATUELLI, V. (1955). Costi di trattamenti antiparassitari (Based on studies of Bolognese orchards). Frutticoltura. November/December.
7. MEDICI, G. (1954). L'Economia dell'impiego degli antiparassitari. Rivista di Economia Agraria. Issue No. 1.

8. NACAMULI, S. (1955). Alcune considerazioni sull'economia dell'impiego degli antiparassitari in Agricoltura.Rivista di Economia Agraria. X(1).

9. PANATTONI, A. (1956). L'economía della difesa antiparassitaria in agricoltura. Osservatorio di Economia Agraria. Pisa.

10. BRUN, P.(1985).Lutte intégrée en vergers d'agrumes: problèmes posés par la lutte chimique.I.N.R.A.-Station de Recherches Agronomiques de Corse.

11. by MARTINO, E. and CUTULI, G. Guida dei trattamenti agli agrumi. Terra e Vita. N.1.

    BARBAGALLO, S. and PERROTTA, G. (1977). Orientamenti di lotta contro i parassiti animali e vegetali degli agrumi. Tecnica Agricola, N.3.

    LONGO, S. (1984). La difesa degli agrumi. Terra e Sole. N.508.

    DE SENA, E. and RAPISARDA, C. (1983). La defesa degli agrumi. Terra e Vita, N.33.

12. BREGOLI, A. (1984). Costo della difesa fitosanitaria: incide meno ora che in passato. L'informatore Agrario. XL(9).

Closing session

Chairman: R.Cavalloro

Address of the session chairman

R.Cavalloro
CEC, Joint Research Centre, Ispra, Italy

This meeting, with the valuable and numerous papers which have
been presented and with the wide exchange of ideas and information, has
certainly allowed us to discover the state of the art of phytosanitary
protection of citrus fruit crops.

The widening of basic knowledge on citrus pests has involved a
very vast sector which ranges from the Insects <u>Homoptera</u> and <u>Thysanop-</u>
<u>tera</u> to the <u>Lepidoptera</u>, <u>Diptera</u>, <u>Coleoptera</u>, from the Mites to the
Nematodes, from fungal, bacterial, viral diseases to weeds.

Moreover, the exhaustive view of the methods and means of control
has clearly indicated the pre-eminent value of biological control and
the great possibilities offered by the use of biotechnical, genetical
and cultural means, and the real possibilities of using integrated control in extensive applications of pest management.

The economic aspects of proper phytosanitary protection in citrus
groves have also been examined, and here a lack of information has been
found.

There is certainly still much to be done and above all an operational infrastructure, which is lacking at present, is needed, but I
believe that it has become evident to everyone that it is feasible to
protect citrus groves with efficient and selective means, as part of a
rational integrated control. The direction taken by the activities
promoted in the European Countries by the Commission of the European
Communities in protecting citrus groves, as I described in the opening
session of this meeting, has thus been confirmed.

We are now entering the last phase of our four days of intensive
work: the concluding reports or each session which will be presented by

577

the session chairmen, to whom I should like to express my thanks, on on behalf of all the participating experts, for having carried out their task with skill and wisdom.

Before handing over to them, I should like to thank all those who have cooperated to make this meeting important and scientifically valid through their reports, communications and contributions - in a word through their experience and ideas.

A special thanks goes to the local Authorities, who have welcomed us so cordially in this important citrus experimental Institut of Acireale, not only for the impeccable organization but also for the human warmth which thay have so kindly shown.

Insect pests

P.Jourdheuil

INRA, Zoological & Biological Control Station, Antibes, France

After the general reports and the papers presented at the session on harmful insects, it now seems possible to present the following conclusions and recommendations:

1) The progress noted for the faunistic survey confirms the great richness and extreme specific diversity of entomocenosa associated with citrus trees in various parts of the world. With the intensification and acceleration of tourism and trade, the potential or actual risk of accidently introducing pests into the Mediterranean area is becoming higher and higher and would certainly justify greater vigilance in the countries concerned and better information on the dangers they represent and ways of limiting them. Fortunately, the biocenotic richness emphasized earlier extends to the secondary consumers, in particular to entomophaga. This offers effective solutions of control by acclimatization or regular release of parasites and predators; nevertheless, it implies that recording and assessing the potentiality of these enemies can be carried out on an international scale, hence the importance which we must continue to give to systematics in its classical or modern forms.

2) Substantial progress has also been made in the bioecological and ethological characterization of the various phytophageous species, in particular on alimentary specificity, the development cycle and the various components of the biotic potential. Nevertheless, if the relationships with the climate and the complex of indigenous or introduced natural enemies begin to be analysed properly, there will still be a lot to cover in the area of insect-plant relations: analysis of the factors of varietal response, impact of the nature of vegetation and its development conditions (fertilization, feeding in water, size, etc.) on the biotic potential of pests, and importance and conditions of appearance of direct or indirect damage (in particular vectors disease). At a time when everyone is convinced of the necessity of developing and optimizing systems of integrated control for the various types of citrus growing, it would seem advisable to develop research into quantitative ecology which would give us a better understanding of entomocenosa, leading to mathematical or other models for forecasts or decisions, and finally provide a more rational basis for management and phytosanitary protection.

Other pests

E.Di Martino
Citrus Experimental Institute, Acireale, Italy

This session would not have existed a few decades ago: research on mites was disdained by some entomologists of yesterday and it seemed that it could never reach the dignity which was attributed to entomology.

Immediately after the war there was even some doubt abroad about some announcements made in Italy, but today the mites have become part of the group of primary phytophages and hence their placing in this forum is not incorrect.

As far as nematodes are concerned, their study began about thirty years ago: today it is included among the fundamental studies which have direct influence on the relationships between plants and soil and, in this case, between citrus and nematodes.

When dealing with weeds, a short time ago nothing was done beyond a mere botanical classification of the species, but the associated problems which have been noted recently in the studies on their implication in the citrus agroecosystem were not considered for citrus, apart from the competition for the removal of mineral elements and water.

As far as mites are concerned, the Vacante report gave fundamental news on the number of species existing in our citrus orchards with new indications on both phytophages, predators and indifferents, at least at the moment.

Later communications have provided views and aspects of the problem in Greece, with Souliotis and Delrio in Sardinia giving considerable information on the ethology of the Panonichus and its relationships with Phytoseids, also in connection with control situations.

Finally, one should note the enquiry of Nucifora and Vacante in the approach of ethology, especially of Tarsonemus waitei.

One should stress the weight assumed, studied and reported during the meeting by the phytoseid acarids in Italy, Sicily, Calabria and Sardinia.

What future is there for the subject? In various countries the enquiries on the knowledge of mites supply the knowledge which must be available in the least detail if one is to implement an efficient integrated defence of the citrus groves.

When dealing with nematodes on citrus, to stress the increased interest one need only remember Lo Giudice's findings: 8 species in 1949, 39 species in 1968.

The depressing effects of nematodes in citrus, especially for Thylenchulus semipenetrans, are often masked and cannot be readily identified. In fact, the plants which are genetically vigorous or well-fed

581

in favourable environmental conditions may give economically satisfactory production which is in reality less than that which could be possible. Acting againt nematodes in some cases, on plants which seemed to be healthy and highly productive, has increased their production by 20% to 25%.

Present-day research tends to show that suitable management of plants infested by nematodes may help to improve the development and the productivity of citrus plants. One should not forget, however, that each "management" must be programmed depending on the type of terrain, the source of infestation and the crop-growing practices adopted. The aim of this type of programming in controlling nematodes must mainly tend to the avoidance of all types of stress on the citrus fruit plants.

Hence it seems indispensable, before adopting any type of control, to have a clear vision of the biology, the ecology and the relationships of the nematodes both with the plants and with other organisms, and it is thus necessary that research on this subject continue.

Maugeri has illustrated the flowering and vegetative characteristics of weeds, their biology, and the influence of infesting weeds on the productivity of citrus fruits.

The choice between constantly maintaining a natural grass lawn, or seeding a summer or winter green manure, or eliminating weeds periodically by working or definitely by weedkillers does not appear to be an easy one given the innumerable situations which may occur in the various citrus growing areas of the world.

The choices, therefore, depend on the chemical and physical characteristics of the terrain, the annual rainfall, the possibilities for irrigation, the lie of the land, the type of flora and on many other local factors. Several enquiries are being carried out on the real role of weeds in citrus orchards, both concerning their effect on the terrain and on the productivity of the plants, and in relation to the various biotic and abiotic factors.

It is felt that in respecting the ecological balance it would be useful to keep the weeds and learn about their biology to obtain whatever they can give which could be positive and to eliminate them, with the most suitable means for the individual situations, when they are shown to be useless. Finally, one should evaluate the benefits from them in relationship to the negative effects, such as reduced plant development and production, in a long term view which takes account of the vitality of the land.

We propose that the close connections which exist between the ecosystems of the herbaceous plants, which are the guests of the citrus orchards, and the citrus trees themselves should be investigated more closely to discover the relationships which exist between the microfauna of the former and the latter to avoid imbalances which may aggravate the situation of the citrus phytophages.

In this field, we should also continue the studies on their relationships with the phytophages for the close connections which occur between the so-called weeds and the citrus fruit crops.

Diseases

C.C.Thanassoulopoulos
Regional Direction of Agriculture, Patras, Greece

Session three was devoted to diseases of citrus. Ths session was divided into two parts. In the first part there were reviewed, in brief, the main fungal and bacterial diseases of citrus and presented six papers of which two were dealing with <u>Phytophthora</u> and four were devoted to Mal secco disease. In the second part there were discussed several aspects of virological problems and presented a very good general review concerning insect-transmitted viruses. Finally, in one paper there were presented some problems of physiological disorders.

From the first part presentation and the discussion which followed, it can easily be concluded that root and trunk gomoses, fruit rots and, particularly, Mal secco are the main problems which give serious trouble to growers and all those involved in citriculture. It is evident that diseases which destroy the tree itself are extremely dangerous as the losses per capita are tremendous. Mal secco seems to be yet "the" disease problem of citriculture and particularly of lemon groves. From papers presented it was clear that this disease has still many unexplored sites, which unfortunately make quick progress difficult. The results in three of the papers indicated the great significance of lesions from any reason, such as hail, frost, wind and other agents, on the epidemiology and progress of the disease. There were also clear evidences that Mal secco is greatly influenced by weather conditions, which could increase the severity of the disease. The influence also of wounds on fruit in rot development has been discussed. The necessity of sprays to protect fruits has been shown.

The great need of citrus protection from these diseases creates a lot of new ideas in a control program leading to their prevention, as it is clear that in many cases control measures are coinciding. Thus the research directed towards the following apsects is considered necessary.

1. Studies on the possibility of Mal secco control with biological agents, such as bacteria or mild strains of the fungus or others.
2. The same studies should be made for Phytophthoras fruit rots.
3. Further studies of epidemiological factors in nature, favouring disease progess.
4. Intensive surveys mainly in areas where very old lemon trees exist, such as in Poros, Greece, to find probable sources of resistance, in combination, of course, with the good quality of fruit.
5. Research toward finding frost resistant lemon cultivars which would probably eliminate disease incidence.

6. Studies on the contamporates disease control on citrus groves to avoid excess spraying and curtail treatment expenses.

In the second part of the session the main subject was virus diseases. The most significant thoughts on this subject were the evidences that new citrus threats are "ante portas" in the Mediterranean basin. Insect-transmitted viruses or other diseases are the most dangerous ones among disease agents, as insect vectors are able to destroy the phytosanitary control measures taken for citrus protection. Diseases, as Greening, a bacterial disease not yet known in Europe, are found in areas around the Mediterranean countries, as well as the psyllid vectors Diaporthe citri and Trioza erytreae. The increasing communication among the countries has greatly increased the danger of transferring unknown disease in new areas, which probably will be still more virulent enemies in crops than in their native environment, as there were the cases of Downy mildew on grape vines or Hemileia vastatrix on Brazilian coffee. The increasing measures in phytoquarantine is a practice which should be more emphasized in the near future, as well as the selection of resistance sources. The research efforts to virological problems or other similar to them, such as insect-transmitted bacteria or mycoplasmas, appear to responds to the following:

1. Studies for the economic evaluation of eradication of infected trees.
2. Studies on the protection methods of virus vectors to invade the European territory. Eradication or decrease of population dynamic in vector's native habitat is probably more economic, in long-life term, than control after the vector's invasion of the European continent.
3. Research in order to find resistant graft-cultivars to virus diseases.
4. Studies on biological control of vectors leading to their possible eradication.

It is well known that phytophathological control action, has not only been the practice of spray programs. The control methods include a lot of other treatments, such as cultural practices, resistant varieties, changing environmental conditions, cross-protection and several other methods, without using the term "integrated control". This term, quite new as a term for plant pathology, is very old as a practice, probably before plant pathology was established as a separate science. However, considering new approaches of this science and other sister sciences, we are now under new ideas about our work. Such a rather new point of view for plant pathology is the biological control method which in the last decade has gained much attention among plant pathologists. Most of the papers presented in Session 3 imply in their conclusions the biological control idea, and the workers expect a new future for plant disease control.

Concerning the research of our science which is now under development, it is easy to observe that plant pathologists are in front of new difficulties arising from the rapid and ever-changing agriculture. The mechanization of agriculture a few decades ago and the present electronic automation, the rapid changes of traditional cultural practices, the over-population of cultivated lands and, consequently, the need for continuous cultivation in the same areas, and numerous others, set aside all traditional knowledge and experience of plant pathology, creating needs for new approaches and for new ideas. The traditional old plant pathologist is now a memory of the past. The new plant pathologists, if they like to make a successful career should work together with plant pathologists, geneticists, biochemists and statisticians. The plant pathology of potted plants is over. A new dynamic mass computer study in combination with the electronic laboratory equipment is the modern approach of our science. But here is the main point. Plant pathology was, is, and will continue to be necessary for agriculture. It is the master key for a successful and productive crop. But for the success of the new type of work, funds are needed - without them nothing could be done. Training of scientists in new methods, research of new types, exchange of ideas by personal contact among world scientists of plant pathology are the new practices without which plant pathology will remain in the dusty past.

Authorities of world organizations dealing with social and agricultural problems should understand that they need us, more than their fathers needed ours. If their target is the production of more food for hungry people, they have to be generous in their offerings for plant pathology developent. This is the most crucial point. The old solitary research man is done with. New considerations are in progress. Team work with brighter spirits of our science, regardless of nationality, is the focus of our times. Invention of new agricultural equipment in phytopathological work, planning of new research ways, collaborating for the control of the inter-national threatening of our food, is the work of today. All these require a lot of money. What the world will get as a reward will be several-fold more. The plant pathologists are ready to offer their knowledge and help participating in the world's famine problems. They need help, finanacial help. The authorities of the European Community should be our protectors, aiding us in our struggle against plant diseases.

We plant pathologist should be very happy if the Community would like generously to contribute to helping us to achieve more important and fruitful results towards plant disease research and control, not only in citriculture, but in other problems concerning plant pathology, too.

Means of control

R.Prota
Institute of Agricultural Entomology, University of Sassari, Italy

In this session five reports and twenty-one communications have been presented.

Of the indications suggested, from traditional chemical means to the most sophisticated genetic techniques, it should be remembered that if taken individually they do not solve the problem of crop protection, which must satisfy highly demanding economic and ecological questions.

This, however, does not mean that we should not follow up aims which are always more specifically scientific in every area, when the thrust of the research is towards integrated techniques, towards the solution required by the problems.

The results described in the session have answered what was indicated in the introduction. From all the subjects dealt with there emerges the need to move towards an integration which is correctly supported by a well-defined bio-ecological picture, without which it is difficult to do more than talk.

A) Biological means have a predominant place in future development, but at present they represent an important heritage of scientific institutions; one may note the lack of structure adapted for mass cultivation and of centres intended for the dissemination of the biological materials produced.

B) In the ambit of biotechnical means, considerable progress has been made in the study of visual stimuli, while olfactory stimuli are still to a large extent unknown. The discovery of volatile substances which govern the finding of the host and the localization of the fruit will be extremely important, as will be the discovery of the principles which determine the departure from the host.
The refining of electrophysiological techniques will help considerably in the identification of the individual components of the active substances emitted by the plants.
The combination of volatile substances with various stimuli may represent an important path to follow in developing highly efficient control systems. The combination, for example, of olfactory attractors with selective shapes and colours offers a promising approach which may be considered without forgetting the simultaneous use of deterrents to egg-laying.

The technical problems which arise from the complex use of the sterile male technique must be considered carefully and must be dealt with after thorough ecological studies of the ecosystem to be protected and the quality of the insects produced.

Overall one may therefore say that biotechnical control is at present not much applied but has highly favourable prospects.

C) As far as chemical means are concerned, it should be recognised that they are still, at least on the scale of business representative of our agriculture, the most widespread means with costs which are competitive with other types of action, apart from the sterile male technique. It is unfortunately of great importance for the economy of our agricultural businesses, at least until most of the cost of so-called social "clean" actions is assumed by the community.

D) Cultural means, although limited by agricultural and economic conditions, may prove decisive, at least for controlling certain noxious species. a Considerable results may be obtained from the creation of resistant varieties and in this field genetics may play a decisive role.

E) Genetic means will certainly become more and more important in the future.

The ever clearer understanding of evolutionary changes of insects and above all of the need to gain knowledge of the way in which harmful arthropods respond to man's actions, have allowed large strides to be taken in this sector.

Important results have been obtained in improving the quality of the insects to be used in specific operations or, for example, in the sex separation technique.

To conclude, it can be said that the means of control of insects included in citrus-fruit growing continue to be enriched by new discoveries.

We now wait for organisations and scientific laboratories to bring together the results of individual studies to give more definitive conclusions.

We are all in agreement in feeling that research, coordination, technical innovation and standards, are the corner-stones of a new policy where the economic aspect may be closely linked to the defence of the environment in the widest sense of the term.

Methodologies and strategies of control

L.Brader
FAO, Plant Production & Protection Division, Roma, Italy

The papers presented in the session on methodologies and strategies of control provided excellent examples of the current status of knowledge on biological and integrated pest management in citrus groves.

The following conclusions may be drawn.

- Under various growing conditions biological control usually offers excellent control possibilities for a large number of important pest species.

- However, for a limited number of pests biological control does not offer yet adequate opportunities for economically adequate control. This is mainly due to the lack of knowledge concerning effective parasites and predators.

- The above conclusions indicate clearly that in any IPM system in citrus groves biological control, either on the basis of making full use of locally present natural enemies or through their introduction from elsewhere, must constitute the major control approach. Consequently the introduction and application of additional control measures has to be determined by the extent to which they can be successfully combined with biological control.

- Of the additional control measures so far studied the most useful and promising seem to be:

  . selective use of pesticides

  . selective pesticides

  . pheromones, mainly for monitoring, but also for control of certain pest species

  . varietal resistance

  . modified cultural practices

. introduction of pesticide resistant natural enemies

. autocidal control for some pest species

. mechanical control measures, for example the washing off of scales
  through high pressure washing.

- In particular the extensive application of IPM in Californian citrus
  groves has shown that significant cost reductions can be achieved,
  when these systems are properly implemented.

- It is evident from the various contributions to the session that the
  application of IPM in citrus groves may in certain cases be a rather
  complex matter, requiring the assistance of skilled extension services and proper training of farmers and others. For example there is
  still a considerable need for the further development of more effective and pragmatic monitoring systems, and techniques that can be
  used in a rather simple manner. In addition, non chemical control measures need to be further developed that can be effectively integrated
  with biological control while not requiring the continuous assistance
  of skilled man-power.

Economical considerations on I.P.M. in citrus-groves

S.C.Misseri

Institute of Agricultural Estimate, University of Catania, Italy

After the presentation, the experts, Salerno (Italy), Bassino (France) and Lo Giudice (Italy), took part in a discussion that concentrated on two points.

The first (Salerno, Lo Giudice) concerned the assessment of the damage, particularly in terms of future prediction (where reductions will be required) and in the field - as is the case with some nematode infestations - when the grower is forced to change crops. The second point (Bassino), taking up observations made by Brun (France), referred to the way the growers' economic behaviour is influenced by variations in the technology of pest control.

The author noted these comments and dwelt particularly on the second, where he pointed out that, in the final analysis, the economic impact of pest control is largely dependent on the way it is applied by the grower, and for this reason it is worthwhile investing in the provision of technical assistance, bearing in mind that manpower training offers the greatest potential for investment. His final comments concerned the desirability of international coordination of pest control to ensure widespread control of pests, and underlined the imminent danger that the Mediterranean citrus industry faces from certain aphids, tree hoppers and psyllids, carriers of very serious diseases, which have been studied by Barbagallo (Italy) and Bové (France). In this way one can also talk of spatial differentiation in the economics of pest control.

List of participants

France

BASSINO J.P. - A.C.T.A., 149, rue de Bercy - 75595 PARIS

BENASSY C. - I.N.R.A., Station de Zoologie et de Lutte Biologique - Laboratoire "E. Biliotti" - 06560 VALBONNE

BRUN P. - I.N.R.A., Station de Recherches Agronomiques de Corse - 20230 SAN NICOLAO

BOVE' J.M. - I.N.R.A., Laboratoire de Biologie cellulaire et moléculaire - Université de Bordeaux - 33140 PONT DE LA MAYE

MALAUSA J.C. - I.N.R.A., Station de Zoologie et de Lutte Biologique - Laboratoire "E. Biliotti" - 06560 VALBONNE

ONILLON J.C. - Station de Zoologie - Centre National Recherche Agronomique - Route de St Cyr - 78000 VERSAILLES

PANIS A. - I.N.R.A., Station de Zoologie et de Lutte Biologique - Laboratoire "E. Biliotti" - 06560 VALBONNE

Great Britain

WOOD R.J. - Department of Zoology - Manchester University - Oxford Road - MANCHESTER M1398L

Greece

ALEXANDRAKIS V. - Institute of Subtropical Plants and Olive Trees - 73100 CHANIA (CRETE)

ARGYRIOU L. - "Benaki" Phytopathological Institute - 14561 KIPHISSIA

CHITZANIDIS A. - "Benaki" Phytopathological Institute - 14561 KIPHISSIA

SOULIOTIS P. - "Benaki" Phytopathological Institute - 14561 KIPHISSIA

THANASSOULOPOULOS A. - Regional Direction of Agriculture - 26110 PATRAS

THANASSOULOPOULOS C. - Regional Direction of Agriculture - 26110 PATRAS

TSIROPOULOS G. - "Democritos" Nuclear Research Centre -
    14341 AGHIA PARASKEVI (ATTIKIS)

Israel

ROSEN D. - Faculty of Agriculture - University of Rehovot -
    P.O. Box 12 - 76302 REHOVOT

Italia

ALBANESE G. - Istituto di Patologia Vegetale - Università degli Studi
    - Via Valdisavoia 5 - 95123 CATANIA

ARGENTI G. - Cassa per il Mezzogiorno - Ripartizione Progetti
    Promozionali - Divisione 3a - Via del Giorgione 2a - 00147 ROMA

BARALDI G. - Istituto di Meccanica Agraria - Università degli Studi -
    Via Filippo Re 8 - 40126 BOLOGNA

BARBAGALLO S. - Istituto di Entomologia Agraria - Università degli
    Studi - Via Valdisavoia 5 - 95123 CATANIA

BASILE M. - Istituto di Nematologia Agraria - Università degli Studi -
    Via Amendola 165/A - 70126 BARI

BENFATTO D. - Istituto Sperimentale Agrumicultura - Corso Savoia 190 -
    95024 ACIREALE

BERTOLINI P. - Dipartimento per la Protezione e lo Sviluppo -
    Università degli Studi - Via Filippo Re 8 - 40126 BOLOGNA

BUTERA N. - Agriculture Industrial Development SpA - Zona Industriale
    Blocco Palma I - 95100 CATANIA

CALABRETTA C. - Istituto di Entomologia Agraria - Università degli
    Studi - Via Valdisavoia 5 - 95123 CATANIA

CARBONE G. - Osservatorio Fitopatologico - 88100 CATANZARO

CARUSO A. - Istituto Sperimentale Agrumicultura - Corso Savoia 190 -
    95024 ACIREALE

CATARA A. - Istituto di Patologia Vegetale - Università degli Studi -
    Via Valdisavoia 5 - 95123 CATANIA

CIANCIO A. - Istituto di Nematologia Agraria - Università degli Studi
    - Via Amendola 165/A - 70126 BARI

COCO M. - Sindaco di - 95024 ACIREALE

CONTINELLA G. - Istituto di Coltivazione Arboree - Università degli
    Studi - Via Valdisavoia 5 - 95123 CATANIA

CUPPERI L. - Agriculture Industrial Development SpA - Zona Industriale
    Blocco Palma I - 95100 CATANIA

CUTULI G. - Osservatorio Malattie delle Piante - Via Martinez 13 -
    95024 ACIREALE

DAVINO M. - Istituto di Patologia Vegetale - Università degli Studi -
Via Valdisavoia 5 - 95123 CATANIA

DE CICCO V. - Dipartimento di Patologia Vegetale - Università degli
Studi - Via Amendola 165/A - 70126 BARI

DELRIO G. - Istituto di Entomologia Agraria - Università degli Studi -
Via Enrico de Nicola - 07100 SASSARI

DI MARTINO ALEPPO E. - Istituto Sperimentale Agrumicultura -
Corso Savoia 190 - 95024 ACIREALE

DI MARTINO E. - Istituto Sperimentale Agrumicultura - Corso Savoia 190
- 95024 ACIREALE

FERLITO G. - Istituto di Entomologia Agraria - Università degli
Studi - Via Valdisavoia 5 - 95123 CATANIA

FLORIS F. - Agriculture Industrial Development SpA - Zona Industriale
Blocco Palma I - 95100 CATANIA

GRANATA G. - Istituto di Patologia Vegetale - Università degli Studi -
Via Valdisavoia 5 - 95123 CATANIA

GRASSO G. - Istituto di Patologia Vegetale - Università degli Studi -
Via Valdisavoia 5 - 95123 CATANIA

INSERRA S. - Istituto di Entomologia Agraria - Università degli
Studi - Via Valdisavoia 5 - 95123 CATANIA

INTRIGLIOLO F. - Istituto Sperimentale Agrumicultura -
Corso Savoia 190 - 95024 ACIREALE

LAMBERTI F. - Istituto di Nematologia Agraria - Università degli Studi
- Via Amendola 165/A - 70126 BARI

LA MALFA G. - Istituto Sperimentale Agrumicultura - Corso Savoia 190 -
95024 ACIREALE

LANZA G. - Istituto Sperimentale Agrumicultura - Corso Savoia 190 -
95024 ACIREALE

LA ROSA R. - Istituto di Patologia Vegetale - Università degli Studi -
Via Valdisavoia 5 - 95123 CATANIA

LAVIOLA C. - Istituto di Patologia Vegetale - Università degli Studi -
Viale delle Scienze 2 - 90128 PALERMO

LIOTTA G. - Istituto di Entomologia Agraria - Università degli Studi -
Viale delle Scienze 13 - 90128 PALERMO

LO GIUDICE V. - Istituto Sperimentale Agrumicultura - Corso Savoia 190
- 95024 ACIREALE

LONGO S. - Istituto di Entomologia Agraria - Università degli Studi -
Via Valdisavoia 5 - 95123 CATANIA

MAGNANO DI SAN LIO G. - Istituto di Patologia Vegetale - Università degli Studi - Via Valdisavoia 5 - 95123 CATANIA

MARTELLI S. - Azienda Agricola Sperimentale Pantanello - S.S. 106 Km 448.200 - 75010 METAPONTO

MAUGERI G. - Istituto di Botanica - Università degli Studi - Via Antonino Longo 19 - 95125 CATANIA

MELILLO VITO A. - C.N.R. - Istituto di Nematologia - Via Amendola 165/A - 70126 BARI

MINEO G. - Istituto di Entomologia Agraria - Università degli Studi - Viale delle Scienze 13 - 90128 PALERMO

MISSERI S. - Istituto di Estimo - Università degli Studi - Via Valdisavoia 5 - 95123 PALERMO

NICOSIA C. - Osservatorio Fitopatologico - Via Tommaso Gulli - 89100 REGGIO CALABRIA

NUCIFORA A. - Istituto di Entomologia Agraria - Università degli Studi - Via Valdisavoia 5 - 95123 CATANIA

ORTU S. - Istituto di Entomologia Agraria - Università degli Studi - Via Enrico de Nicola - 07100 SASSARI

PATTI I. - Istituto di Entomologia Agraria - Università degli Studi - Via Valdisavoia 5 - 95123 CATANIA

PENNISI A.M. - Istituto di Patologia Vegetale - Università degli Studi - Via Valdisavoia 5 - 95123 CATANIA

PERROTTA G. - Istituto di Patologia Vegetale - Università degli Studi - Via Valdisavoia 5 - 95123 CATANIA

PRIVITERA S. - Osservatorio Malattie delle Piante - Via Martinez 13 - 95024 ACIREALE

PROTA R. - Istituto di Entomologia Agraria - Università degli Studi - Via Enrico de Nicola - 07100 SASSARI

QUACQUARELLI A. - Istituto Sperimentale Patologia Vegetale - Via C.G. Bertero 22 - 00156 ROMA

RAPISARDA C. - Istituto di Entomologia Agraria - Università degli Studi - Via Valdisavoia 5 - 95123 CATANIA

REFORGIATO RECUPERO G. - Istituto Sperimentale Agrumicultura - Corso Savoia 190 - 95024 ACIREALE

ROMEO M. - Istituto Sperimentale Agrumicultura - Corso Savoia 190 - 95024 ACIREALE

RUSSO A. - Istituto di Entomologia Agraria - Università degli Studi - Via Valdisavoia 5 - 95123 CATANIA

RUSSO F. - Istituto Sperimentale Agrumicultura - Corso Savoia 190 -
95024 ACIREALE

SALERNO M. Dipartimento di Patologia Vegetale - Università degli
Studi - Via Amendola 165/A - 70126 BARI

SCHILIRO' A. - Azienda Costantina - Paterno' - Viale Artale Alagona
37-39 - 95126 CATANIA

SCIPIONI L. -I.T.P.A. - Via Caio Mario 27 - 00192 ROMA

SOLLIMA A. - C.L.A.M. Sezione italiana - Viale XX Settembre 51 -
95128 CATANIA

SPINA P. - Istituto Sperimentale Agrumicultura - Corso Savoia 190 -
95024 ACIREALE

STARRANTINO A. - Istituto Sperimentale Agrumicultura -
Corso Savoia 190 - 95024 ACIREALE

TERRANOVA G. - Istituto Sperimentale Agrumicultura - Corso Savoia 190
- 95024 ACIREALE

TIRRO' A. - Istituto di Patologia Vegetale - Università degli Studi -
Via Valdisavoia 5 - 95123 CATANIA

TUTTOBENE G. - Istituto di Patologia Vegetale - Università degli Studi
- Via Valdisavoia 5 - 95123 CATANIA

VACANTE V. - Istituto di Entomologia Agraria - Università degli Studi
- Via Valdisavoia 5 - 95123 CATANIA

ZUCKER W.V. - Agriculture Industrial Development SpA - Zona
Industriale - Blocco Palma I - 95100 CATANIA

Portugal

GARCIA V. - University of the Azores - Rua da Mae de Deus - 9500
PONTA DELGADA

Spain

BELLO A - Consejo Superior de Investigaciones Cientificas - 28006
MADRID

GARRIDO A. - Instituto Nacional de Investigaciones Agrarias - Crida 07
- MONCADA

JIMENEZ A. - Instituto Nacional de Investigaciones Agrarias - Crida 06
- MADRID

MELIA A. - Servicio de Defensa contra Plagas e Ispeccion fitopatologica - Avenida Rey Don Jaime 74 - CASTELLON DE LA PLANA

MONER DUALDE J. - Servicio Proteccion Vegetales - Valencia - Apartado
65 - ALMAZORA

MUNIZ M. - Consejo Superior de Investigaciones Cientificas - 28006
  MADRID

NAVAS A. - Consejo Superior de Investigaciones Cientificas - 28006
  MADRID

Switzerland

BOLLER E. - Eidgenössische Forschungsanstalt für Obst-, Wein- und
  Gartenbau - Station for Agriculture - 8820 WÄDENSWIL

U.S.A.

LUCK R. - Division of Biological Control - University of California -
  RIVERSIDE CA 95121

International Organizations

C.E.C.

CAVALLORO R. - Commission of the European Communities - "Integrated
  Plant Protection" Programme - Joint Research Centre - 21020 ISPRA
  (ITALY)

ROTONDO' P.P. - Commission of the European Communities, Directorate
  General XIII/A2 - 2920 LUXEMBOURG

F.A.O.

BRADER L. - F.A.O. - Plant Production and Protection Division -
  Via delle Terme di Caracalla - 00153 ROMA (ITALY)

I.O.B.C.

JOURDHEUIL P. - I.N.R.A., Station de Zoologie et de Lutte Biologique -
  06600 ANTIBES (FRANCE)

List of authors